# SYNOPSIS OF CASTAÑEDA'S INTERVENTIONAL RADIOLOGY

# SYNOPSIS OF CASTAÑEDA'S INTERVENTIONAL RADIOLOGY

### Editors

**Hector Ferral, M.D.**
Assistant Professor of Clinical Radiology
Louisiana State University
School of Medicine in New Orleans
New Orleans, Louisiana

**Haraldur Bjarnason, M.D.**
Associate Professor of Radiology
University of Minnesota Medical School—Minneapolis
Minneapolis, Minnesota

**Zhong Qian, M.D.**
Assistant Professor of Radiology
Louisiana State University
School of Medicine in New Orleans
New Orleans, Louisiana

A **Wolters Kluwer** Company
Philadelphia • Baltimore • New York • London
Buenos Aires • Hong Kong • Sydney • Tokyo

*Acquisitions Editor*: Beth Barry
*Developmental Editor*: Michael Standen
*Supervising Editor*: Mary Ann McLaughlin
*Production Editor*: Allison Spearman, Silverchair Science + Communications
*Manufacturing Manager*: Benjamin Rivera
*Cover Designer*: QT Design
*Compositor*: Silverchair Science + Communications
*Printer*: Maple Press

**© 2001 by LIPPINCOTT WILLIAMS & WILKINS**
**530 Walnut Street**
**Philadelphia, PA 19106 USA**
**LWW.com**

All rights reserved. This book is protected by copyright. No part of this book may be reproduced in any form or by any means, including photocopying, or utilized by any information storage and retrieval system without written permission from the copyright owner, except for brief quotations embodied in critical articles and reviews. Materials appearing in this book prepared by individuals as part of their official duties as U.S. government employees are not covered by the above-mentioned copyright.

Printed in the USA

---

**Library of Congress Cataloging-in-Publication Data**
Synopsis of Castañeda's interventional radiology / editors, Hector Ferral,
Haraldur Bjarnason, Zhong Qian.
    p. ; cm.
    Includes bibliographical references and index.
    ISBN 0-683-30094-6
    1. Radiology, Interventional--Handbooks, manuals, etc. 2. Surgery,
Operative--Handbooks, manuals, etc. I. Ferral, Hector. II. Bjarnason, Haraldur.
III. Qian, Zhong. IV. Interventional radiology.
    [DNLM: 1. Radiography, Interventional. 2. Surgical Procedures, Operative. WO 500
S993 2000]
RD33.55 .I58 1997 Suppl.
617'.05--dc21

00-032742

---

Care has been taken to confirm the accuracy of the information presented and to describe generally accepted practices. However, the authors, editors, and publisher are not responsible for errors or omissions or for any consequences from application of the information in this book and make no warranty, expressed or implied, with respect to the currency, completeness, or accuracy of the contents of the publication. Application of this information in a particular situation remains the professional responsibility of the practitioner.

The authors, editors, and publisher have exerted every effort to ensure that drug selection and dosage set forth in this text are in accordance with current recommendations and practice at the time of publication. However, in view of ongoing research, changes in government regulations, and the constant flow of information relating to drug therapy and drug reactions, the reader is urged to check the package insert for each drug for any change in indications and dosage and for added warnings and precautions. This is particularly important when the recommended agent is a new or infrequently employed drug.

Some drugs and medical devices presented in this publication have Food and Drug Administration (FDA) clearance for limited use in restricted research settings. It is the responsibility of health care providers to ascertain the FDA status of each drug or device planned for use in their clinical practice.

10 9 8 7 6 5 4 3 2 1

# Contents

Contributing Authors .................................................... xi
Foreword ............................................................... xiii
    *Wilfrido R. Castañeda-Zúñiga*
Preface ................................................................. xv

1. Patient Care ........................................................ 1
   *Hector Ferral*
   - A. Patient Preparation and Informed Consent ..................... 1
   - B. Laboratory Parameters ........................................ 1
   - C. Intravenous Sedation ......................................... 2
   - D. Antibiotics .................................................. 4
   - E. Anticoagulation .............................................. 6
   - F. Management of Associated Medical Problems .................... 8
   - G. Contrast Media ............................................... 9
   - H. Postprocedure Care .......................................... 10

2. Radiation Control in Interventional Radiology ..................... 11
   *Richard A. Geise*

3. Access Techniques and Hemostasis .................................. 19
   *Leonard R. Bok, Hector Ferral, and Haraldur Bjarnason*
   - A. Access Sites ................................................ 19
   - B. Access Techniques and Devices ............................... 19
   - C. Diagnostic Catheters and Wires .............................. 23
   - D. Puncture Hemostasis ......................................... 24
   - E. Complications: Diagnosis and Management ..................... 25

4. Vascular Embolotherapy ............................................ 27
   *Haraldur Bjarnason*

5. Percutaneous Treatment of Liver Tumors ............................ 47
   *Zhong Qian and Yuchen Jia*

6. Gastrointestinal Bleeding ......................................... 55
   *Osarugue A. Aideyan and Haraldur Bjarnason*
   - A. Diagnostic Approach ......................................... 55
   - B. Embolization and Infusion Therapy ........................... 57

| | | |
|---|---|---|
| 7. | Portal Hypertension: Evaluation and Treatment .................... | 67 |
| | *Hector Ferral* | |
| 8. | Percutaneous Transluminal Angioplasty ........................... | 77 |
| | *Hector Ferral* | |
| | A. Basic Concepts................................................ | 77 |
| | B. Mechanism of Angioplasty .................................. | 78 |
| | C. Devices and Techniques..................................... | 80 |
| | D. Indications and Outcome.................................... | 82 |
| |     1. Angioplasty of Supraaortic Vessels.......................... | 82 |
| |         *Hector Ferral and S. Murthy Tadavarthy* | |
| |     2. Celiac and Superior Mesenteric Artery Angioplasty............. | 89 |
| |         *Hector Ferral, S. Murthy Tadavarthy, Joseph W. Yedlicka, Jr.,* | |
| |         *and Wilfrido R. Castañeda-Zúñiga* | |
| |     3. Percutaneous Transluminal Angioplasty of the Renal Arteries ....... | 93 |
| |         *Hector Ferral* | |
| |     4. Aorto-Iliac and Peripheral Arterial Angioplasty .................. | 101 |
| |         *Hector Ferral and Wilfrido R. Castañeda-Zúñiga* | |
| 9. | Problems and Management of Hemodialysis Access .................... | 117 |
| | *Hector Ferral* | |
| 10. | Percutaneous Atherectomy ....................................... | 129 |
| | *Jae-Kyu Kim, Hector Ferral, and Zhong Qian* | |
| 11. | Intravascular Stents ............................................ | 137 |
| | *Jorge E. Lopera, Zhong Qian, and Hector Ferral* | |
| | A. Overview ................................................... | 137 |
| | B. Arterial Stenting............................................. | 143 |
| | C. Venous Stents............................................... | 152 |
| 12. | Stent-Graft Techniques .......................................... | 157 |
| | *G. Michael Werdick, Michael S. Rosenberg, and Haraldur Bjarnason* | |
| | A. Use of Stent-Grafts for Aorto-Iliac Aneurysmal Disease ............ | 157 |
| |     *G. Michael Werdick* | |
| | B. Stent-Grafts for Peripheral Occlusive Disease..................... | 160 |
| |     *Michael S. Rosenberg* | |
| | C. Use of Covered Metal Stents for Transjugular-Intrahepatic Portocaval Shunts............................................. | 163 |
| |     *Haraldur Bjarnason* | |

| | | |
|---|---|---|
| 13. | Diagnostic Techniques................................................ | 167 |
| | A. Noninvasive Vascular Evaluation of Peripheral Vascular Disease ........ <br> *Mark Wofford and Thomas R. Beidle* | 167 |
| | B. Evaluation of Deep Venous Thrombosis ........................... <br> *Mark Wofford and Thomas R. Beidle* | 170 |
| | C. Other Doppler Applications .................................... <br> *Mark Wofford and Thomas R. Beidle* | 173 |
| | D. Carbon Dioxide Digital Subtraction Angiography.................... <br> *James G. Caridi, Irvin F. Hawkins, and Bret N. Wiechmann* | 175 |
| | E. Intravascular Ultrasound........................................ <br> *Wilfrido R. Castañeda-Zúñiga* | 179 |
| 14. | Fibrinolytic Therapy ................................................ <br> *Haraldur Bjarnason* | 187 |
| | A. Introduction to the Hemostatic Cascade........................... | 187 |
| | B. Anticoagulation Agents ........................................ | 190 |
| | C. Thrombolytic Agents .......................................... | 191 |
| | D. Hypercoagulable State ......................................... | 192 |
| | E. Thrombolysis ................................................ | 193 |
| | F. Laboratory Monitoring......................................... | 200 |
| | G. Complications................................................ | 201 |
| 15. | Mechanical Thrombectomy........................................... <br> *Zhong Qian, Hector Ferral, and Xiaoping Gu* | 203 |
| 16. | Inferior Vena Cava Filters ........................................... <br> *Steven F. Millward and Randall V. Olsen* | 213 |
| 17. | Central Venous Access............................................... <br> *Maria Rodrigues Gomes and Gwen K. Nazarian* | 225 |
| 18. | Foreign Body Retrieval .............................................. <br> *Zhong Qian and Hector Ferral* | 237 |
| 19. | Vascular Interventional Procedures in the Pediatric Age Group ............ <br> *Hector Ferral* | 251 |
| 20. | Genitourinary Interventions........................................... <br> *Hector Ferral* | 263 |
| | A. Percutaneous Urologic Techniques ............................... | 263 |
| | B. Special Techniques............................................ | 270 |

|     |     |     |
| --- | --- | --- |
|     | C. Percutaneous Nephrostomy Access for Stone Removal | 277 |
|     | D. Prostatic Intervention | 280 |
| 21. | Biliary Tract Intervention | 283 |

*Pablo A. Gamboa and Haraldur Bjarnason*
(Original manuscript by Joseph W. Yedlicka, Jr., S. Murthy Tadavarthy, Janis Gissel Letourneau, Flavio Castañeda, and Wilfrido R. Castañeda-Zúñiga)

|     |     |
| --- | --- |
| A. Interventional Techniques in the Hepatobiliary System | 283 |
| B. Gallbladder Intervention | 299 |

22. Gastrointestinal Tract Intervention ............................................ 305
    *Hector Ferral and Haraldur Bjarnason*
    A. Percutaneous Gastrostomy and Jejunostomy......................... 305
       *Haraldur Bjarnason*
       (Original manuscript by José Llerena, Elias Górriz, Manuel Maynar, Ricardo Reyes, Juan M. Pulido-Duque, Francisco Mallorquin, and Wilfrido R. Castañeda-Zúñiga)
    B. Management of Benign Esophageal Strictures ...................... 314
       *Ho-Young Song*
    C. Management of Malignant Esophageal Strictures.................. 318
       *Ho-Young Song*
    D. Interventional Treatment of Enteric Strictures....................... 322
       *Haraldur Bjarnason*
       (Original manuscript written by Gordon K. McLean, Dana R. Burke, and Steven G. Meranze)
    E. Colonic Stent Placement ........................................................ 325
       *Hector Ferral and Michael Wholey*
    F. Percutaneous Cecostomy for Ogilvie's Syndrome: Laboratory Observations and Clinical Experience..................................... 329
       *Haraldur Bjarnason*
    G. Volvulus of the Sigmoid Colon: Treatment by Transrectal Fluoroscopic Catheterization and Stenting ............................. 332
       *Haraldur Bjarnason*
       (Original manuscript by Xose C. Prego, Esperanza A. Monton, Xesus E. Cimadevila, Pedro C. Prieto, and Ana C. Noya)

23. Interventional Therapy of Infertility ........................................... 339
    *Zhong Qian, Amy S. Thurmond, and Haraldur Bjarnason*
    A. Fallopian Tube Catheterization for Treatment of Female Infertility ...... 339
    B. Transcatheter Embolization of the Internal Spermatic Vein for Treatment of Varicocele ................................................... 341

24. Lacrimal Duct Intervention .................................. 347
    *Ho-Young Song*

25. Interventional Procedures in the Thorax ...................... 359
    *Haraldur Bjarnason*
        (Original chapter written by Lisa Diethelm, Jeffrey S. Klein, and Haibo Xu)
    A. Biopsies of Thoracic Structures ............................ 359
    B. Percutaneous Drainage of Pleural Collections and Pneumothorax ........ 367
    *Timothy V. Myers*

26. Percutaneous Biopsy of Abdominal Masses ..................... 377
    *Joseph L. Higgins, Jr., and Janis Gissel Letourneau*

27. Drainage of Abdominal Abscesses and Fluid Collections ................. 385
    *Gregory B. Snyder, George C. Scott, and Haraldur Bjarnason*
        (Original chapter written by George C. Scott, Janis Gissel Letourneau, Joel M. Berman, and Thomas R. Beidle)

Appendixes:

1. Angiography: Infusion Rates ................................. 401
   *Haraldur Bjarnason*

2. Normal Laboratory Values ................................... 403
   *Darren W. Postoak*

Subject Index ..................................................... 405

# Contributing Authors

**Osarugue A. Aideyan, M.D.**
*Medical Fellow Specialist*
*Department of Radiology*
*University of Minnesota Academic Health Center*
*Minneapolis, Minnesota*

**Thomas R. Beidle, M.D., R.V.T.**
*Associate Professor of Radiology*
*Louisiana State University*
*School of Medicine in New Orleans*
*New Orleans, Louisiana*

**Haraldur Bjarnason, M.D.**
*Associate Professor of Radiology*
*University of Minnesota Medical School—*
  *Minneapolis*
*Minneapolis, Minnesota*

**Leonard R. Bok, M.D., M.B.A.**
*Head, Department of Radiology*
*Louisiana State University Medical Center*
*Lafayette, Louisiana*

**James G. Caridi, M.D.**
*Assistant Professor of Cardiovascular and*
  *Interventional Radiology*
*University of Florida College of Medicine*
*Gainesville, Florida*

**Wilfrido R. Castañeda-Zúñiga, M.D., M.Sc.**
*Professor and Chairman of Radiology*
*Louisiana State University*
*School of Medicine in New Orleans*
*New Orleans, Louisiana*

**Hector Ferral, M.D.**
*Assistant Professor of Clinical Radiology*
*Louisiana State University*
*School of Medicine in New Orleans*
*New Orleans, Louisiana*

**Pablo A. Gamboa, M.D.**
*Assistant Professor of Radiology*
*Department of Vascular and Interventional Radiology*
*Ohio State University College of Medicine*
*Columbus, Ohio*

**Richard A. Geise, Ph.D.**
*Associate Professor of Radiology*
*University of Minnesota Medical School—*
  *Minneapolis*
*Minneapolis, Minnesota*

**Maria Rodrigues Gomes, M.D.**
*Assistant Professor of Radiology*
*University of Minnesota Medical School—*
  *Minneapolis*
*Minneapolis, Minnesota*

**Xiaoping Gu, M.D.**
*Associate Professor of Radiology*
*University of Minnesota Medical School—*
  *Minneapolis*
*Minneapolis, Minnesota*

**Irvin F. Hawkins, M.D.**
*Professor of Radiology*
*University of Florida College of Medicine*
*Chief of Cardiovascular and Interventional*
  *Radiology*
*Shands Hospital*
*Gainesville, Florida*

**Joseph L. Higgins, Jr., M.D., Ph.D.**
*Associate Professor of Clinical Radiology*
*Louisiana State University*
*School of Medicine in New Orleans*
*Vice-Chairman of Clinical Affairs*
*Section Head, Abdominal Imaging*
*Director of Computed Tomography*
*Department of Radiology*
*Louisiana State University Medical Center*
*New Orleans, Louisiana*

**Yuchen Jia, M.D.**
*Department of Radiology*
*Changhai Hospital, the Second Military Medical*
  *University*
*Shanghai, China*

**Jae-Kyu Kim, M.D.**
*Associate Professor of Radiology*
*Chonnam University Hospital*
*Dong-Ku, Kwang Ju, Korea*

## CONTRIBUTING AUTHORS

**Janis Gissel Letourneau, M.D., R.V.T.**
*Professor of Surgery and Radiology*
*Associate Dean of Faculty and Institutional Affairs*
*Louisiana State University*
*School of Medicine in New Orleans*
*New Orleans, Louisiana*

**Jorge E. Lopera, M.D.**
*Clinical Fellow, Interventional Radiology*
*Louisiana State University Medical Center*
*New Orleans, Louisiana*

**Steven F. Millward, M.B.Ch.B., F.R.C.R., F.R.C.P.C.**
*Associate Professor of Radiology*
*University of Western Ontario Faculty of Medicine*
*London, Ontario*

**Timothy V. Myers, M.D.**
*Department of Radiology*
*St. John's Hospital*
*St. Paul Radiology, P.A.*
*St. Paul, Minnesota*

**Gwen K. Nazarian, M.D.**
*Interventional Radiologist*
*Methodist Hospital*
*St. Louis Park, Minnesota*

**Randall V. Olsen, M.D.**
*Fellow, Vascular and Interventional Radiology Section*
*Mallinckrodt Institute of Radiology*
*Washington University School of Medicine*
*St. Louis, Missouri*

**Darren W. Postoak, M.D.**
*Assistant Professor of Clinical Radiology*
*Louisiana State University*
*School of Medicine in New Orleans*
*New Orleans, Louisiana*

**Zhong Qian, M.D.**
*Assistant Professor of Radiology*
*Louisiana State University*
*School of Medicine in New Orleans*
*New Orleans, Louisiana*

**Michael S. Rosenberg, M.D.**
*Assistant Professor of Radiology*
*University of Minnesota Medical School—Minneapolis*
*Minneapolis, Minnesota*

**George C. Scott, M.D.**
*Assistant Professor of Radiology*
*Louisiana State University*
*School of Medicine in New Orleans*
*New Orleans, Louisiana*

**Gregory B. Snyder, M.D.**
*Assistant Clinical Professor of Interventional Radiology*
*University of Minnesota Medical School—Minneapolis*
*Minneapolis, Minnesota*

**Ho-Young Song, M.D., Ph.D.**
*Professor of Radiology*
*Asan Medical Center*
*University of Ulsan College of Medicine*
*Seoul, Korea*

**S. Murthy Tadavarthy, M.D.**
*Clinical Associate Professor of Radiology*
*University of Minnesota Medical School—Minneapolis*
*Cardiovascular Radiologist*
*Minneapolis Heart Institute*
*Co-Director, Cardiovascular and Interventional Radiology*
*Abbott-Northwestern Hospital*
*Minneapolis, Minnesota*

**Amy S. Thurmond, M.D.**
*Associate Professor of Obstetrics and Gynecology*
*Oregon Health Sciences University School of Medicine*
*Radiologist*
*Tualatin Imaging*
*Portland, Oregon*

**G. Michael Werdick, M.D.**
*Interventional Radiologist*
*Methodist Hospital*
*St. Louis Park, Minnesota*

**Michael Wholey, M.D.**
*Assistant Professor of Radiology*
*University of Texas Health Science Center at San Antonio*
*San Antonio, Texas*

**Bret N. Wiechmann, M.D.**
*Clinical Assistant Professor of Radiology*
*University of Florida College of Medicine*
*Vascular/Interventional Radiologist*
*North Florida Radiology, P.A.*
*Gainesville, Florida*

**Mark Wofford, M.D.**
*House Officer*
*Department of Radiology*
*Louisiana State University Medical Center*
*New Orleans, Louisiana*

**Joseph W. Yedlicka, Jr., M.D.**
*Director of Vascular and Interventional Radiology*
*Community Hospitals of Indianapolis*
*Indianapolis, Indiana*

# Foreword

More than a synthesis of the Interventional Radiology textbook, *Synopsis of Castañeda's Interventional Radiology* is a comprehensive, practical treatise on the entire spectrum of interventional imaging-guided procedures. The authors have used their extensive experience in the field to produce a practical, user-friendly manual focused on problem solving that is a must for residents, fellows in training, and practitioners in the field of interventional radiology.

*Wilfrido R. Castañeda-Zúñiga, M.D., M.Sc.*

# Preface

Writing a summarized version of Wilfrido R. Castañeda-Zúñiga's *Interventional Radiology* has not been an easy task. The material is based on the chapters in the Third Edition of Castañeda-Zúñiga's *Interventional Radiology*, but several chapters understandably have required updates. At the same time, some sections that we did not consider practical for the purpose of this book, but that are absolutely pertinent in a large reference book such as Castañeda-Zúñiga's, have been deleted. We have joined our efforts with those of several contributors to bring forth a concise text that covers the most important practical concepts in interventional radiology. Our intent is to publish a guide for interventional radiology fellows and residents rotating through the interventional radiology service as well as for the practicing interventional radiologist.

We wish to emphasize the effort and commitment of the original authors of the chapters upon which our book is based and give them the recognition they deserve, as we have summarized and modified their manuscripts to create this synopsis.

Last, but not least, we thank Dr. Wilfrido R. Castañeda-Zúñiga for allowing us to embark on this not-so-easy task of summarizing an excellent textbook in interventional radiology.

*Hector Ferral, M.D.*
*Haraldur Bjarnason, M.D.*
*Zhong Qian, M.D.*

# 1
# Patient Care

Hector Ferral

## A. PATIENT PREPARATION AND INFORMED CONSENT

Interventional radiology has developed as a predominantly clinical subspecialty. Most practicing interventional radiologists are responsible for the patient's periprocedural care, including preprocedure assessment, procedural management of the patient, potential complications, and postprocedural care and follow-up.

Every invasive procedure begins with the first interview with the patient. A problem-focused history and physical examination are extremely helpful in preparing for any invasive procedure. Medical and social histories disclose important information that may change the course or plan of the procedure to be performed. In addition, it is imperative to have a full knowledge of the patient's medical management, including medications and known allergies, at the time of the procedure. A problem-directed physical examination is necessary. Once the radiologist obtains the pertinent information, he or she must clearly explain to the patient the purpose of the procedure and provide a reasonable and straightforward summary of the expected risks, benefits, and alternative options.

The informed consent form is a legal document; therefore, a summary of the patient evaluation process and explanation of the procedure must be stated clearly in lay terms. The patient should ask questions or express his or her concerns regarding the procedure, and the physician should answer in a direct, straightforward manner. The consent form should be signed in the presence of a witness (usually a nurse) who is not involved in the performance of the procedure. Although this process seems to be very involved in some cases, we have found that patients and their families are usually appreciative. A pleasant interview with the patient enhances the patient's trust, relaxes the patient, and makes him or her understand the reasoning behind the medical decision to perform an invasive diagnostic or therapeutic procedure. By the same token, a careful patient evaluation helps decrease procedural complications related to neglected or ignored information regarding the patient's medical condition.

## B. LABORATORY PARAMETERS

A careful clinical evaluation is necessary to determine the laboratory parameters that need to be evaluated before a procedure. Pertinent laboratory parameters should be evaluated. For example, if a diagnostic arteriogram is to be performed in a 68-year-old diabetic patient, it is important to assess the blood urea nitrogen and creatinine levels; coagulation tests must be checked in a patient with chronic liver failure; and platelet counts should be assessed in patients with leukemia or lymphoma, or patients taking chemotherapeutic drugs. Once again, a problem-directed history is crucial in deciding the laboratory tests to be obtained and checked in a patient undergoing an invasive procedure.

## C. INTRAVENOUS SEDATION

Based on increasing public awareness of advances in pharmacology and monitoring, patients have developed an expectation of comfort and safety during diagnostic and therapeutic procedures. Intravenous sedation is widely used in the management of patient pain and anxiety during interventional procedures. Analgesic medications can decrease pain during painful procedures, and the use of additional specific medications can decrease oral and gastric secretions, and attenuate nausea and vomiting. The Joint Commission on Accreditation of Healthcare Organization has required that guidelines be applied for patients who receive sedation by nonanesthesiologists during diagnostic and therapeutic procedures (Table 1.1). Sedation and analgesia are not free of adverse effects, and nurses and physicians who administer these drugs should be aware of the potential complications and prepared to manage them. Certain factors must be considered before choosing the type of sedative and the dose to be given to the patient. These factors include the patient's age, gender, religion, race, history of alcohol or drug abuse, estimated length of the procedure, and complexity of the procedure. For example, a biliary drainage is expected to induce a much more painful experience than is a diagnostic arteriogram.

Certain aspects of sedation and analgesia are useful to keep in mind:

1. What is conscious sedation? The goal of conscious sedation is to achieve a minimally depressed level of consciousness that retains the patient's ability to maintain a patent airway independently and continuously and respond appropriately to physical stimulation and verbal commands. The drugs, dosages, and techniques are not intended to produce a loss of consciousness. Appropriate conscious sedation allows neurologic assessment at all points of the procedure.

2. How should patients be evaluated before sedation and analgesia? A baseline health history before the sedation should include questions about previous hospitalizations for medical illness or surgical procedures; current medications; diseases, disorders, and other health abnormalities; allergies and previous adverse drug reactions; a family history of diseases or disorders; and a review of systems. Height, weight, and vital signs, as well as airway, pulmonary, and cardiac systems, should be assessed (Table 1.2).

**TABLE 1.1.** *Joint Commission on Accreditation of Healthcare Organizations standards for anesthesia care: 1995*

Standard TX 2
   A preanesthesia assessment of the patient is performed before the operative or other invasive procedure.
Standard TX 2.1
   A plan for anesthesia is developed.
Standard TX 2.2
   Anesthesia options with attendant risks are discussed with the patient and/or family if appropriate.
Standard TX 2.3
   The patient's physiologic status is measured and assessed during the operative or invasive procedure.
Standard TX 2.4
   The patient's postprocedure status is assessed on admission to the postanesthesia recovery area and before discharge from the postanesthesia recovery area or the setting.
Standard TX 2.4.1
   The patient is discharged either by a qualified licensed independent practitioner or by the use of medical staff–approved criteria.
Standard TX 2.4.1.1
   When medical staff–approved criteria are used, compliance with the criteria is fully documented in the patient's medical record.

From Surgical and Anesthesia Services. *1995 comprehensive accreditation manual for hospitals.* Oakbrook Terrace, IL: Joint Commission on Accreditation of Healthcare Organizations, 1999:135.

**TABLE 1.2.** *Recommended assessment of the airway*

History
  Previous problems with anesthesia or sedation
  Stridor, snoring, or sleep apnea
  Dysmorphic facial features (e.g., Down syndrome, hemifacial microsomia, Pierre-Robin syndrome)
  Rheumatoid arthritis
  Cervical spondylosis, neck abnormalities
Physical examination
  Obesity
  Short neck with limited extension
  Decreased hyoid-mental distance (<3 cm or 2 fingerbreadths in an adult)
  Neck mass
  Disease of the cervical spine
  Neck trauma
  Tracheal deviation
  Small mouth
  Edentulous
  Significant overbite
  High, arched palate
  Macroglossia
  Tonsillar hypertrophy
  Small jaw
  Significant temporomandibular joint disease

From Castañeda-Zúñiga WR. *Interventional radiology,* 3rd ed. Baltimore: Williams & Wilkins, 1997, with permission.

**TABLE 1.3.** *Nothing by mouth policy*

For all ages:
  No food should be consumed after midnight the evening before surgery. This includes the following:
  • Solid food, candy, chewing gum
  • Milk, milk products
  • Orange juice and any juice with pulp
  • Carbonated beverages
  Clear fluids may be offered until 3 hours before the scheduled time of surgery. These include the following:
  • Water, apple juice, and clear juice drinks
  • Oral electrolyte solution and popsicles
  • Breast milk for nursing infants
For infants younger than 6 months of age:
  Infants younger than 6 months may be given formula up to 6 hours before surgery. Breast milk or clear liquids as listed above may be given up to 3 hours before surgery.

From Castañeda-Zúñiga WR. *Interventional radiology,* 3rd ed. Baltimore: Williams & Wilkins, 1997, with permission.

3. How should the patient be prepared? Consent for sedation is a separate consent. The patient should be informed about the proposed benefits and risks of, and alternatives to, sedation. The risks of sedation remain separate from the risks of the procedure itself. Recent food intake must be assessed. Nothing by mouth (NPO) policies must be clearly delineated (Table 1.3).

4. What is the appropriate level of monitoring? Frequent assessment of mental status through monitoring the patient's response to command or conversation is very helpful. Patients who are unresponsive could be on more profound levels of sedation. These changes must be identified early on. Continuous monitoring of heart rate, electrocardiography (ECG), and oxygen saturation is essential as well as assessment of blood pressure at specific intervals. We strongly recommend that a pulse oximeter for noninvasive monitoring of oxygen saturation be used continuously during sedation or analgesia.

5. How should the personnel be assigned, trained, and supervised? An assistant not operating in the procedure should be available from the time of administration of the sedative until recovery is judged to be adequate. Some institutionally based educational programs include training with members of the department of anesthesia in the safe use of sedatives and analgesics. The department of anesthesia should participate in the organization of an educational program to inform practitioners of guidelines for the use of sedating agents and monitoring techniques. We recommend that all personnel working in the angiography room should be familiar with the fastest methods to access expert resuscitation teams (e.g., beepers, speed dial).

6. What equipment should be available? We recommend the availability of a monitoring cart on which an ECG monitor, an automated blood pressure cuff, a pulse oximeter, and a portable source of oxygen are mounted. A system of routine inventory and resupply should be instituted for all sedation supplies, reusable as well as disposable. A self-inflating positive pressure oxygen delivery system capable of delivering at least 90% oxygen at a 15-L-per-minute flow rate for at least 60 minutes should be available, along with various bag and mask

sizes. A source of suction must be available as well as rigid suction tips (Yankauer) capable of rapid flow. A crash cart must be available and stocked with the necessary drugs to resuscitate an apneic patient.

7. What precautions should be taken when using multiple sedative and analgesic agents? Analgesics such as fentanyl (Duragesic Transdermal System) in combination with benzodiazepines have been found to produce significant respiratory depression. Continuous patient monitoring is essential to immediately identify deeper states of sedation.

8. What are the considerations for recovery from sedation? Release to less-monitored care can only be permitted when the patient has returned to presedation status with regard to his or her airway; breathing and circulation; level of consciousness; and ability to sit unaided, walk with assistance, and maintain an adequate state of hydration. If patients are discharged, they must be under the care of a competent adult.

## D. ANTIBIOTICS

Systemic antibiotic therapy for patients undergoing interventional procedures has become increasingly important. Not all patients undergoing an interventional procedure need antibiotics. The most important factor is to decide which patients need antibiotic therapy, which patients need antibiotic prophylaxis, when to consult the infectious disease specialist, and which drugs and doses should be used in any particular case. Specific guidelines for antibiotic usage based on prospective randomized trials are lacking; therefore, common sense and practical knowledge have dictated the use of antibiotics in interventional radiology practice. The reason to treat patients with antibiotics is to prevent or avoid infectious complications. One must not forget that the most important means of preventing infection is careful attention to sterile technique.

### DEFINITIONS

- *Antibiotic prophylaxis* is the administration of antibiotics in the periprocedure period to prevent an infectious complication.
- *Antibiotic therapy* is the treatment of a known infection.

### Prophylaxis

Antibiotic prophylaxis is recommended for certain interventional procedures to provide useful antibiotic blood levels, prevent infection of an implantable device, or prevent the onset of a severe septic reaction after manipulation of an infected cavity. In some cases, the indication for antibiotic prophylaxis may be controversial. Policies are determined locally in each hospital setting.

### *Prophylaxis Necessary: No Controversy*

Prophylactic antibiotics are recommended for patients having the following procedures:

- Biliary drainage and intervention (biliary obstruction with or without previous endoscopic manipulation, residual biliary stones, previous biliary surgery, liver transplant patient with biliary obstruction)
- Percutaneous nephrostomy or genitourinary interventions (percutaneous nephrolithotripsy in a patient with a staghorn calculus, bladder dysfunction, indwelling catheters, ureterointestinal anastomoses, impacted stones, previous urologic manipulation, kidney transplant with obstruction of the collecting system)
- Stent placement through a previously placed sheath or a groin hematoma (may be preceded by sheath exchange for a new sheath immediately before stenting)
- Liver chemoembolization
- Splenic embolization
- American Heart Association (AHA) high risk for endocarditis (prosthetic cardiac valves, previous bacterial endocarditis, complex cyanotic heart disease, surgically constructed pulmonary shunts or conduits)

### Prophylaxis Recommended: No Consensus

Most interventional radiologists (more than 70%) use prophylactic antibiotics in the following cases:

- Transjugular intrahepatic portosystemic shunting (TIPS)
- Ports
- Patient with aortobifemoral bypass undergoing angiography through a femoral approach

### Prophylaxis Controversial: No Consensus

The decision to use prophylactic antibiotics is entirely dependent on the operator. The practice is split. Fifty percent of radiologists or fewer use prophylaxis in the following cases:

- Procedures extending for more than 3 hours in which placement of a permanently implantable device is necessary (complex aorto-iliac stenting, venous recanalization followed by stent placement, venous recanalization followed by port or dialysis catheter placement)
- Noncomplicated placement of dialysis catheters
- Percutaneous gastrostomy tube placement
- Percutaneous drainage procedures

### Prophylaxis Unnecessary

The use of prophylactic antibiotics is not necessary for these patients. Most interventionalists (more than 70%) do not use antibiotics for the following procedures:

- Diagnostic angiography or venography in patients without high risk of AHA endocarditis
- Angioplasty
- Routine stent placement
- TIPS revision
- Inferior vena cava filter placement

## Recommended Antibiotics

### Biliary Drainage

The main concern is enterobacteria (*Escherichia coli, Klebsiella, Enterococcus, Pseudomonas, Clostridia*). Most patients with biliary obstruction have infected bile, especially those who have undergone previous endoscopic procedures in which drainage attempts were unsuccessful. Septic complications after biliary manipulation may be severe; consultation with an infectious disease specialist is recommended. Recommended antibiotics include the following: (a) ampicillin (Omnipen) + gentamicin (Garamycin); (b) piperacillin (Pipracil); or (c) second-generation cephalosporin. Antibiotics should be administered before the procedure and continued after the procedure until the infection is controlled.

### Genitourinary System

The main concern is enterobacteria (*E. coli, Klebsiella, Proteus, Enterobacter, Pseudomonas*). Recommended antibiotics include the following: (a) ceftriaxone (Rocephin), 1 g before the procedure and continuing until urine culture results are available; (b) ampicillin (Omnipen) + gentamicin (Garamycin); or (c) second-generation cephalosporin. In patients with a high risk of developing renal failure (patients with obstructive problems), the doses of aminoglycosides, such as gentamicin (Garamycin) and amikacin (Amikin), should be adjusted. Also, remember that these drugs are potentially nephrotoxic. Vancomycin (Vancocin HCL) also has nephrotoxic potential. Cautious use of these drugs, as well as consultation with an infectious disease specialist, is recommended for patients with renal failure.

### Tunneled Port

The concern is mainly skin flora (*Staphylococcus epidermidis, Staphylococcus aureus, Corynebacterium*). The recommended antibiotic is cefazolin (Ancef), 1 g single intravenous dose before the procedure.

### Transjugular Intrahepatic Portosystemic Shunting

The concern is infection by skin flora (*S. epidermidis, S. aureus, Corynebacterium*) and potentially biliary pathogens (*E. coli, Klebsiella, Enterococcus*). Recommended antibiotic: cefoxitin (Mefoxin), 1 g intravenously every 6 hours for 48 hours.

Complex stent case or prolonged venous recanalization case: The concern is skin flora. Recommended antibiotics include cefazolin (Ancef), 1 g intravenously before stent placement. If the patient is allergic to penicillin, give Vancomycin (Vancocin HCL), 1 g intravenously.

# E. ANTICOAGULATION

Anticoagulation is used routinely in interventional procedures involving angioplasty, stent placement, and other invasive endovascular techniques, especially procedures involving thrombolysis. The most common drugs used include heparin, low-molecular-weight (LMW) heparin, and warfarin (Coumadin). Other drugs that affect the coagulation system include the antiplatelet drugs aspirin, dipyridamole (Persantine), ticlopidine (Ticlid), and abciximab (ReoPro); and thrombolytic drugs. A full discussion of the coagulation mechanism is beyond the scope of this book. A brief overview of the drugs and their applications is presented.

## HEPARIN

Chemically, heparin is a glycosaminoglycan. It is found in the secretory granules of mast cells. For medical use, heparin is extracted from porcine intestinal mucosa or bovine lung. Heparin has an antithrombin activity. The anticoagulant effect of heparin is mediated by an endogenous component of plasma (antithrombin III), which is produced in the liver. Antithrombin III inhibits activated coagulation factors of the intrinsic and common pathways. Heparin binds to antithrombin and causes a conformational change in antithrombin, which accelerates the interaction of antithrombin with thrombin and activated factor X (factor Xa).

Heparin is not absorbed via the intestinal mucosa and must be given parenterally. Heparin has immediate onset of action when given intravenously. The half-life is variable and depends on the administered dose. Heparin doses during interventional procedures range between 5,000 and 10,000 units, corresponding roughly to a dose of 100 units per kg. With the usual doses given in interventional procedures, the half-life of heparin is approximately 1 hour. With higher doses (800 units per kg), the half-life may extend up to 5 hours. Heparin effect can be monitored using an activated clotting time (ACT) machine. An ACT of more than 300 seconds indicates optimal anticoagulation levels. Full-dose heparinization is usually given via an intravenous continuous infusion. Usually, the patient is given a bolus of 5,000 units, and intravenous infusion is started with 1,000 to 1,500 units per hour. Anticoagulant effect is monitored using activated partial thromboplastin time (aPTT) values every 6 hours. The heparin dose is titrated according to aPTT levels until anticoagulation is achieved. A clotting time of 1.5 to 2.5 times the normal (usually 50 to 80 seconds) is a satisfactory therapeutic level.

**Important:** Heparin does not cross the placenta and has not been associated with fetal malformations. Heparin is the drug used for anticoagulation during pregnancy.

### Problems with Heparin Therapy

- Heparin resistance
- Bleeding
- Heparin-induced thrombocytopenia (may result in thrombotic complications)
- Abnormal liver function tests
- Allergic reactions (rare)
- Overdose

### Reversal of Heparin Therapy

- Some angiographers favor the reversal of the anticoagulant effect of heparin after endovascular procedures. Protamine sulfate is used for this purpose. A dose of 1 mg of protamine is used for every 100 units of active heparin remaining in the patient. Protamine is given intravenously and should be infused at a slow rate (over 10 to 15 minutes).
- The ACT is monitored, and sheaths can be pulled when the ACT is 120 to 180 seconds.

**Important:**

- Protamine interacts with platelets and fibrinogen and may have an anticoagulant effect on

its own. This effect is especially associated with high doses of the drug (over 50 mg).
- Anaphylactic reactions may occur. Patients at higher risk are diabetic patients who have received protamine-containing insulin.

### Low-Molecular-Weight Heparins

LMW heparins are fragments of unfractionated heparin produced by chemical processes, yielding an LMW component. The main difference between heparin and LMW heparin is its relative activity against thrombin and factor Xa. LMW heparins have a greater activity against factor Xa. LMW heparins produce a more predictable anticoagulant response than unfractionated heparin, have better bioavailability, and a longer half-life (2 to 4 hours after intravenous injection and 3 to 6 hours after subcutaneous injection). Treatment with LMW heparin does not require laboratory monitoring. In addition, the use of these components is associated with fewer bleeding complications. The incidence of heparin-induced thrombocytopenia is also lower. LMW heparins do not cross the placenta; therefore, they can be administered safely to pregnant women. The use of LMW heparins has not been popularized among interventional radiologists in the United States. In Europe, several investigators have used LMW heparins as part of their anticoagulation regimens after endovascular procedures, especially femoral artery angioplasty and stent placement and renal artery stent placement.

LMW heparins include the following:

- Dalteparin (Fragmin)
- Enoxaparin (Lovenox)
- Nadroparin (Fraxiparine)
- Reviparin (Clivarine)

### Oral Anticoagulants

Oral anticoagulants are antagonists of vitamin K and therefore inhibit the production of K-dependent coagulation factors by the liver (II, VII, IX, and X). The full anticoagulant effect of warfarin after initiation of therapy is not achieved for several days. Patients who need anticoagulant therapy are typically anticoagulated with intravenous heparin first and then with oral warfarin. The usual adult dose is 5 to 10 mg per day for 2 to 4 days; after this period, an adult receives a dose ranging between 2 and 10 mg per day. Anticoagulant effect is monitored with measurement of prothrombin time (PT) and international normalized ratio.

**Important:**

- Oral anticoagulants should not be used during pregnancy. Administration of Coumadin during pregnancy is a cause of birth defects (e.g., nasal hypoplasia, stippled epiphyseal calcifications, central nervous system abnormalities) and abortion.
- Coumadin-induced skin necrosis. This complication is related to widespread thrombosis of the microvasculature, especially in the extremities. Because warfarin inhibits vitamin K–dependent factors produced by the liver, it also inhibits the production of proteins C and S, natural anticoagulants in the body. This reaction is seen more frequently in patients who are heterozygous for protein C or S deficiencies.
- Blue toe syndrome. This complication may occur 3 to 8 weeks after initiation of oral anticoagulation. It has been proposed to be related to release of cholesterol emboli from atheromatous plaques.

### Antiplatelet Drugs

#### *Aspirin*

Thromboxane $A_2$ is an inducer of platelet aggregation and a potent vasoconstrictor. Aspirin blocks thromboxane $A_2$ production by blocking cyclooxygenase. The action of aspirin on platelet cyclooxygenase is permanent, lasting for the life of the platelet, which is 7 to 10 days. Complete inactivation of platelet cyclooxygenase is reached when 160 mg of aspirin is given daily. In theory, maximum antiplatelet action of aspirin is obtained when the drug is given at low doses (160 mg to 320 mg per day).

#### *Dipyridamole (Persantine)*

Dipyridamole inhibits platelet function by increasing the cellular concentration of cyclic adenosine monophosphate (AMP). The use of this drug in combination with aspirin has not proven to be of additional benefit. The only current recommended use for this drug is primary prophylaxis of thromboemboli in patients with heart valves. The drug is given in combination with warfarin.

### *Ticlopidine (Ticlid)*

Ticlopidine interacts with platelet glycoprotein IIb/IIIa and inhibits the binding of fibrinogen to activated platelets. Ticlopidine inhibits platelet aggregation and clot retraction. The maximal effect of ticlopidine is seen after several days of therapy; abnormal platelet function persists for several days after treatment is discontinued. The combination of ticlopidine and aspirin has resulted in decreased incidence of early ischemic complications after coronary angioplasty. In addition, aspirin and ticlopidine used in combination were more effective than aspirin and coumadin in preventing thrombosis after coronary stent placement. The most important side effect of ticlopidine is neutropenia, which occurs in approximately 2% of patients. Other problems include nausea, vomiting, and diarrhea.

### *Abciximab (ReoPro)*

ReoPro is a monoclonal antibody fragment against the glycoprotein IIb/IIIa receptor of human platelets. It is a very powerful platelet inhibitor and acts by preventing the binding of fibrinogen, von Willebrand factor, and other adhesive molecules to the glycoprotein IIb/IIIa receptor sites. This drug should be administered with extreme care; the package insert should be carefully reviewed before its administration, especially to determine the contraindications for use of the drug. Most of the experience gained with this drug has been in coronary angioplasty and stent placement procedures.

### *Thrombolytics*

Chapter 14, Fibrinolytic Therapy, addresses the uses, indications, and contraindications of thrombolytic drugs.

## F. MANAGEMENT OF ASSOCIATED MEDICAL PROBLEMS

### COAGULOPATHY

Coagulation testing before an invasive procedure is only necessary for patients with clinically apparent bleeding disorders, especially patients with renal failure, liver failure, malabsorption syndromes, and patients taking anticoagulation therapy. The coagulation profile is evaluated to determine whether the patient is at a high risk of having a bleeding complication during or after the planned procedure. If an abnormality is found, the appropriate treatment should be instituted. An elevated PT may be corrected with the administration of vitamin K or the transfusion of fresh-frozen plasma (FFP) or cryoprecipitate. Vitamin K may be given intramuscularly or intravenously, and the dose ranges from 0.5 to 10 mg. Vitamin K should be given at a rate not exceeding 1 mg per minute. The PT should then be checked at 4- to 6-hour intervals to evaluate for correction of PT and to determine whether a repeat dose is necessary. Transfusion of FFP is recommended if immediate homeostasis is required to control active bleeding or if immediate correction of the coagulation abnormality is required for an emergent procedure. Two units of FFP are usually sufficient to correct an elevated PT. The corrective effect of a unit of FFP lasts approximately 6 hours because of the half-life of coagulation factor VII. Transfusion of plasma products should be coordinated so that the maximal effect is obtained during the critical steps of the procedure.

The administration of heparin increases the aPTT. Correction of aPTT is mainly dependent on reversal of heparin activity. The reversal of heparin activity can be achieved with the use of protamine sulfate (See Section E, Anticoagulation).

Correction of thrombocytopenia can be achieved with transfusion of platelet units. One unit of platelets usually raises the platelet count by $5–10 \times 10^9/l$. Platelets are now obtained from plateletpheresis donations. One unit of these provides a higher total number of platelets in the same volume, approximately equal to 6 units of whole blood–derived platelets. In some cases, the half-life of platelets after transfusion can be limited and, once again, transfusion of these cells must be coordinated so that maximal counts are achieved at the expected critical steps of the procedure. In general, it is agreed that a patient with platelet counts lower than $30–40 \times 10^9/l$ requires transfusion of these cells.

## DIABETIC PATIENT

Diabetic patients should be scheduled for early-morning procedures. If an early-morning procedure is planned, type I diabetics should take their usual morning dose of neutral protamine Hagedorn insulin or Lente insulin. If the procedure is planned for late morning or afternoon, half of the dose should be given. It is useful to keep these patients on an intravenous solution containing 5% dextrose. When treating diabetic patients, one should always be alert to hypoglycemic reactions, which are potentially fatal.

## HYPERTENSION

Hypertension not controlled by medication (diastolic pressure greater than 100 mm Hg) is a contraindication to outpatient angiography. If the procedure is an emergency, certain precautions may prove useful (e.g., single-wall puncture, smaller catheters). If the procedure is elective, the patient can reschedule to have the procedure performed when better pressure control has been achieved. Before deciding to lower the patient's blood pressure with intravenous medications, the interventional radiologist must consider the risk of inducing cardiac or cerebral ischemia if the patient develops a hypotensive reaction in response to the medication. The interventional radiologist must make sure the patient has taken his or her medications before the procedure, be aware of the medications and doses used by the patient, and know when the last dose of the medication was taken. It is important to remember that adequate sedation is sometimes enough to control the patient's blood pressure.

### Management of Hypertension

- The use of sublingual nifedipine (Adalat) is not indicated in the acute management of arterial hypertension. Certain safety issues have been raised regarding the use of this drug and, at this point, it is wiser to avoid it.
- Labetalol (Normodyne) is an alpha and beta blocker that usually is given in incremental doses of 5 mg to 10 mg by slow intravenous injection up to an initial dose of 20 mg. This drug is usually well tolerated.
- Transdermal nitroglycerin paste (Transderm-Nitro) may be useful. This drug is usually used in neurointerventional procedures to prevent vasospasm.

If blood pressure cannot be controlled with these basic measures, contact with the attending physician is recommended to decide on further blood pressure management.

## G. CONTRAST MEDIA

The use of iodinated contrast agents is associated with certain risks. One of the risks most important to consider is that of development of renal failure after contrast administration. The incidence of this problem varies from 1% to as high as 22%. The most important risk factor is preexisting impaired renal function before contrast administration. Other risk factors include male gender, diabetes, large contrast volume, age older than 60 years, dehydration, solitary kidney, and multiple myeloma. The interventional radiologist must identify these risk factors before the procedure to determine whether the patient is at high risk for developing contrast-induced renal failure and to decide what action to take.

**Remember:** Informed consent before the administration of iodinated contrast agents is highly recommended. An explanation of the potential risks involved during contrast administration should be part of the general consent for an interventional procedure.

- If diagnostic arteriography or venography is to be performed in a patient at high risk for renal failure or allergic reaction, a reasonable alternative is to use carbon dioxide as a contrast agent.
- Glucophage: If the patient is treated with metformin (Glucophage), the drug can be stopped after administration of contrast. Serum creatinine must be monitored 48 hours after the procedure. If no change in serum creatinine is identified, the drug may be restarted without any major risks.

## H. POSTPROCEDURE CARE

Patient evaluation and follow-up after interventional procedures are some of the most important tasks of the interventional radiologist. If the interventionalist invests time in careful postprocedural management, he or she improves relationships with peers and patients. We strongly recommend that practicing interventional radiologists emphasize the importance of postprocedure care and adopt it as part of their daily practice.

## SUGGESTED READINGS

Berlin L. Ionic versus nonionic contrast media. *AJR Am J Roentgenol* 1996; 167:1095–1097.

Dravid VS, Gupta A, Zegel HG, et al. Investigation of antibiotic prophylaxis usage for vascular and nonvascular interventional procedures. *J Vasc Interv Radiol* 1998;9:401–406.

Payne CS. A primer on patient management problems in interventional radiology. *AJR Am J Roentgenol* 1998;170: 1169–1176.

Shapiro MJ. Management of the coagulopathic patient. *J Vasc Interv Radiol* 1998;10:326–329.

Spies JB. Antibiotic prophylaxis in interventional radiology. *J Vasc Interv Radiol* 1998;10:321–326.

Weitz JI. Low-molecular weight heparins. *N Engl J Med* 1997;337:688–698.

# 2
# Radiation Control in Interventional Radiology

Richard A. Geise

Interventional radiology procedures can involve substantial amounts of ionizing radiation, requiring particularly close attention to radiation control. This chapter reviews radiation units, regulations, and the fundamental principles of radiation control and examines the procedures and devices designed to reduce patient and staff radiation dose in interventional radiology.

## RADIATION UNITS

The fundamental interactions of x-rays with matter produce ion pairs by photoelectric absorption and Compton scattering. Although sometimes used generically, the word *exposure* has a formal meaning. It can indicate the charge per unit mass of air produced by radiation expressed in *coulombs per kilogram* (C per kg). Previously, the *roentgen* (0.254 mC per kg) was used for this purpose.

The number of ion pairs produced in air does not directly measure the amount of energy deposited in another medium because of the differences in x-ray absorption by different materials. The *gray* (Gy) is used as a measure of the absorbed dose (energy deposited per unit mass). A gray is equal to 1 joule per kilogram. The older unit of the rad is equal to 0.01 Gy. These units are of fundamental importance in patient dosimetry.

Ionizing radiations other than x- and gamma rays, such as heavy particles or neutrons, may induce a greater biological effect for a given absorbed dose. To quantify this observation, the *sievert* (Sv) is used to measure the equivalent dose. The sievert is equal to the number of grays multiplied by a factor that expresses the biological effectiveness for a particular type of ionizing radiation relative to that of x-rays. This factor is 1 for x- and gamma radiation. The older unit, the *rem* (roentgen equivalents man/mammal), is equal to 0.01 Sv. This unit is often used in monitoring radiation of personnel. The term *dose* is sometimes used generically for absorbed or equivalent dose.

Because dose is an expression of energy per unit mass, one must express the volume of tissue over which the energy is being averaged when quoting equivalent dose. For example, the dose to an internal organ in the x-ray beam is significantly less than that to the skin at the entrance of the beam. The dose averaged over the entire body is much less than either. The effective dose, also expressed in sieverts, describes an average equivalent dose to the entire body, weighted according to the relative risk of harm to the various organs. To compare the risk of cancer and genetic changes from fluoroscopy with that from natural background radiation or that received by survivors of nuclear bombing, one must use effective doses.

These radiation units are summarized in Table 2.1. Recognizing the equivalence of *dose* and *equivalent dose* for x-rays, the sievert is used whenever possible in this chapter.

## RADIATION BIOEFFECTS

The biological effects of radiation can be separated into two categories: *stochastic* and *deterministic*. Stochastic effects, such as cancer, have only been verified directly above approximately 0.1 Sv; however, it is generally assumed that there is a finite probability of the existence of these effects at any dose above zero. The risk that a radiation worker will die of cancer from an effective dose of 10 mSv received during a working lifetime has been estimated at 0.04%,

**TABLE 2.1.** Radiation units

| Radiation quantity | Traditional unit | Système Internationale unit | Conversion |
|---|---|---|---|
| Exposure | roentgen (R) | coulomb (C)/kg | 0.258 mC/kg = 1 R |
| Absorbed dose | rad | gray (Gy) | 10 mGy = 1 rad |
| Equivalent dose | rem | sievert (Sv) | 10 mSv = 1 rem |
| Effective dose | rem | sievert (Sv) | — |

assuming a simple proportional relationship between dose and risk. The actual relationship is unknown and the subject of much debate.

Deterministic effects, such as cataracts, erythema, epilation, and sterility, appear to have a threshold dose below which they are not detectable (roughly 2 Sv from acute exposure and 6 Sv from chronic exposure at low dose rates). The severity of these effects increases with increasing dose. For example, in the skin, erythema is short term at an acute equivalent dose below approximately 6 Sv, and causes desquamation above approximately 10 Sv and dermal necrosis above approximately 18 Sv. To produce cataracts in humans, approximately 6 Sv from x-rays used in imaging received over several weeks is necessary. It may be that absorbed doses of approximately 15 Sv are necessary to induce cataracts in workers in the diagnostic radiology setting.

## RADIATION PROTECTION REGULATIONS

The Center for Devices and Radiological Health within the Food and Drug Administration (FDA) oversees regulations concerning equipment design and manufacture. Individual states' departments of health and human services place additional regulations on users of x-ray equipment. There is some variation from state to state, but most states pattern their regulations after the recommendations of the National Council on Radiation Protection and Measurements. The National Council on Radiation Protection and Measurements has developed recommendations that have become de facto standards for the safe use of ionizing radiation and include occupational dose limits (Table 2.2). These quarterly limits are intended to account for sporadic or variable, rather than continual, exposure. Additionally, doses should always be kept as low as reasonably achievable. Although the operator's hands and forearms may be exposed during interventional radiology procedures, doses to the eyes and thyroid are generally of greater concern and closely monitored. Other sources give further details of the general philosophy of radiation protection, as well as specific recommendations for particular situations.

It is prudent to monitor the staff and use all reasonable means to decrease head and neck exposures in interventional radiology. These concerns often prompt the use of helmetlike protective devices, lead-acrylic eyeglasses, ceiling-mounted lead plastic shields, and thyroid shields.

## STAFF RADIATION DOSE MONITORING

Monitoring devices should be worn if a person might receive 25% of the regulatory limits in the discharge of his or her duties. Anyone routinely in the fluoroscopy suite during fluoroscopic procedures should wear the devices as well.

**TABLE 2.2.** Maximum permissible dose equivalents (mSv)

| Area | 13 wk | Yearly | Cumulative lifetime |
|---|---|---|---|
| Effective dose | 12.5 | 50 | Age × 10 |
| Lens of eye | 37.5 | 150 | — |
| Other organs | 125 | 500 | — |
| Fetus | 0.5 during gestation, 0.05 per mo | — | — |

The most common method of monitoring is the *film badge,* because it is practical and economic. Each person must wear at least one film badge, preferably at the collar level, outside the lead apron. With the badge outside the apron, calculation of effective dose is more accurate and determination of thyroid or eye dose more direct. There is a trend toward wearing two film badges if interventional procedures are performed. Ring badges containing thermoluminescent dosimeters may also be worn to monitor hand radiation dose. In some states, badge requirements are defined by regulation.

Doses from film badges and rings are often not known until 1 or 2 months later, when the monitor is analyzed. Using a personal (pocket) dosimeter overcomes this problem. These devices are often expensive, fragile, and generally limited to monitoring of pregnant personnel.

## BASIC RADIATION PROTECTION

To provide perspective on operator exposure in interventional radiology, consider the following: A fluoroscopist who performs several procedures that average 30 minutes of fluoroscopic time, each in a typical radiation field of 3 mSv per hour, without any protective equipment, would reach the quarterly effective dose limit of 12.5 mSv after only eight procedures. Clearly, this level must be reduced using the basic methods of time, distance, and shielding.

### Time

Radiation dose is directly proportional to exposure time. When exposure time is cut in half, patient and staff doses are also cut in half. Personnel who do not need to be in the fluoroscopy suite during all or part of a procedure can reduce their exposure time by leaving the area. For other individuals, the time is controlled by the fluoroscopist. To reduce exposure time, the fluoroscopist should never depress the foot switch unless observing the fluoroscopic image. If the fluoroscopic equipment provides last image hold capability, radiation time can further be reduced to only that needed to view motion.

Many x-ray equipment manufacturers offer pulsed fluoroscopy (PF) as an option. PF appears to offer an improved view of rapidly moving contrast in vessels by stopping the action. Using PF, the number of frames recorded per second can be the same as in continuous fluoroscopy or can be reduced, usually by a factor of 2 or 4. This is sometimes a dose reduction method, but beware. It is possible to increase the output of the x-ray generator in the PF mode so that patients and personnel receive more, not less, dose. The fluoroscopist ne eds to be familiar with the radiation output of any piece of equipment used in all its operating modes.

### Distance

Because an x-ray beam diverges as it passes through space, radiation intensity (I) decreases as the inverse square of the distance (d) from the radiation source.

$$I_2 / I_1 = d_1{}^2 / d_2{}^2$$

If the distance from a radiation source is doubled, the radiation intensity decreases to one-fourth its original value (Fig. 2.1). Although only approximately true, this rule is useful in reducing radiation dose to personnel exposed primarily to scattered radiation from that part of the patient being exposed to the primary x-ray beam. Personnel who do not need to be close to the patient to perform their work should always stay as far away as is reasonable from the region of the patient being imaged.

The radiation exposure of the fluoroscopist also depends on imaging geometry. Figure 2.1 shows typical isodose lines for two imaging configurations. Note the increase in operator exposure near the x-ray tube compared to near the intensifier in a lateral configuration. This increase occurs for two reasons: The overall intensity of the scattered radiation is approximately 98% greater at the radiation entrance site on the skin than at the exit site, and there is less attenuating material (e.g., image intensifier) between the patient and the operator. As a rule of thumb, the maximum operator exposure at a given distance occurs when there is an unobstructed path between an operator and the loca-

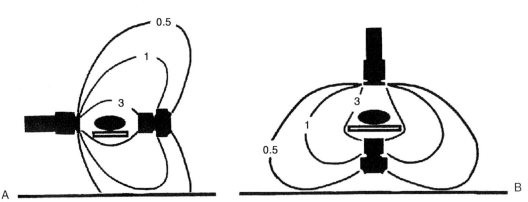

**FIG. 2.1.** Typical scatter radiation from C-arms in mSv per hour. **A:** horizontal, beam right to left. **B:** vertical, beam upward.

tion at which the x-ray beam enters the patient. For example, maximum eye exposure occurs when a person can see the area in which the beam enters the patient.

Measurements indicate that eye dose reductions of 84% are possible by changing the operator's position from table side to 30 cm from the table. Different radiation protection considerations may also be necessary, depending on a staff member's height. Such reductions depend on the system and should be verified for a particular fluoroscopy suite.

### Shielding

Because it is not always possible to change one's position relative to the beam, many devices have been suggested to reduce staff doses during interventional radiology procedures. Effective devices are often somewhat awkward, owing to time and space limitations in interventional radiology.

The reduction of the intensity of an x-ray beam as it passes through matter is roughly exponential. It can be described by the following equation:

$$I = I_o (1/10)^n$$

In this equation, $I_o$ and $I$ are the initial and transmitted radiation intensities, respectively, and $n$ is the number of tenth value layers (TVL) of the material. The *TVL* is the thickness of the material required to reduce the radiation to one-tenth of its original amount. It depends on the atomic number, the density of the material, and the energy of the radiation; therefore, small amounts of attenuating (shielding) material can greatly reduce the intensity of an x-ray beam. A 0.5-mm lead apron is from 1 to 2 TVLs for x-rays used in medical imaging, depending somewhat on how conservative the apron manufacturer is in specifying the thickness. Lead aprons should always be worn by anyone involved in any fluoroscopic procedure. Not all aprons are created the same. Some manufacturers try to lighten the weight of the apron by cutting out larger arm holes or by removing lead from the back and below the knees. Such aprons often leave some organs partially unshielded and are not appropriate for the special procedures room. Walls of shielded x-ray rooms usually contain approximately 1.5 mm of lead (approximately 4 TVLs). Other materials also provide shielding. Double-layer drywall in an unshielded room is a little less than 1 TVL. An average human body is 1 or 2 TVLs.

Surface shielding placed directly on the patient can help shield the operator's line of sight at the patient's surface rather than the operator's level. Shields can be fabricated in strips or solid pieces from lead aprons and can be sterilized for reuse. Typical radiation dose reductions with a 0.77-mm surface shield can range from 33% to 75%. The use of such devices can help to minimize staff radiation doses and comply with regulations.

Besides time, distance, and shielding, another radiation protection parameter is *x-ray beam size*. The amount of scattered radiation exposure is directly related to beam size. In addition, by reducing field size, the patient's effective dose and image quality improve. With a reduced beam size, the fluoroscopist can decrease personnel and patient doses and improve image quality. However, this effect is partially countered in some systems by the automatic brightness system that calls for more radiation to reduce image noise caused by the smaller field.

To minimize personnel doses during fluoroscopic procedures, the lowest acceptable dose rate and smallest acceptable field size should be used with the most efficacious equipment configuration. Distance from the patient should be maximized, and shielding material should be placed between the patient and personnel when possible.

## MANAGEMENT OF PATIENT RADIATION DOSE

Interventional procedures exceed all other x-ray imaging procedures in radiation dose. On average, patient skin doses are approximately 10 times more during interventional procedures than during angiography, and doses from angiographic procedures are approximately 10 times higher than from conventional gastrointestinal fluoroscopy or computed tomography scanning. Skin entrance doses have been reported in the range of 6 to 18 Gy from certain interventional procedures. Because of reports of a small number of severely injured patients, the FDA issued a warning in 1994 to all users of fluoroscopic equipment to beware of doses that can lead to severe skin damage. The FDA recommends a program of recording exposure information during certain interventional procedures, estimating of the doses and potential for injury, and considering increased patient education and modification of procedure or equipment, when possible, to avoid injuries.

Reducing doses to patients usually also reduces doses to personnel. This chapter has already provided several suggestions that reduce both. In addition, the dose to the patient's skin can be reduced by using the shortest possible distance between the image intensifier and the patient. This causes the automatic brightness system of the x-ray generator to reduce the radiation output. During lengthy procedures, it may be possible to alter the orientation of the x-ray beam so that a different area of skin is exposed for part of the procedure. Many dose-reducing equipment modifications have been suggested. Some of these have some effect on image quality and should be implemented only after deliberation.

Some fluoroscopic equipment allows for use of a high-level control (HLC) system. According to FDA regulations for equipment manufactured after May 1995, the maximum patient dose rate can be twice that of normal fluoroscopy when HLC is active. This is allowed as long as a special audible alarm is active. HLC generally is not necessary unless the patient is extremely large. Its use should be kept to a minimum.

During most interventional procedures, it becomes necessary to record images for various reasons. This may be done by cinefluorography (cine) or by digital angiography.

Exposures during digital angiography vary, depending on whether image subtraction is performed. Digital subtraction angiography (DSA) requires higher doses than conventional digital radiographs because the subtraction process increases noise in the images. DSA can easily be the primary factor in patient and personnel exposure during cerebral angiography. A series of 30 frontal DSA images of the head yields a patient skin dose in the range of 120 mSv. Typical fluoroscopy dose rates for the same view are 40 mSv per minute, so a DSA series of only a few seconds is equivalent to 3 minutes of fluoroscopy.

During cine recording, radiation doses are significantly higher for the patient and staff. Patient skin entrance dose rates typically range from 200 to 900 mSv per minute, depending on the system and image acquisition parameters. This is roughly 10 to 20 times higher than the typical 20 to 40 mSv per minute skin entrance dose rates in fluoroscopy. Conversely, a 6-second cine run produces as much radiation as 1 to 2 minutes of fluoroscopy. If fluoroscopy is performed while recording, federal law does not limit the maximum patient dose rate, nor does it require an audible warning (as in HLC). It is possible to fluoroscope in record mode on some equipment by simply inserting a videocassette in the video recorder. Because there is no convention in labeling either record or HLC modes, the operator has to be thoroughly familiar with the dose rates of all dose adjustment settings on each fluoroscopic unit used to avoid extreme patient and personnel exposures.

Radiation protection techniques should receive increased attention when recording images during fluoroscopic procedures. Image recording should only be done when it is necessary to have a record or when analyzing the recorded image can reduce fluoroscopy times by factors from 10 to 20. Image recording should be limited to the shortest possible exposure times, during which personnel move away from the site being imaged.

## SUMMARY

Interventional radiology demands an increased awareness of the fundamental radiation protection principles of time, distance, and shielding. Staff exposures can be reduced by proper use of the imaging system and ancillary shielding materials. These considerations should be strongly reinforced when image recording is used. Typical patient entrance dose rates are approximately 30 mSv per minute during fluoroscopy and 500 mSv per minute during cine. Staff eye dose rates can range from approximately 0.1 to 1 mSv per hour during fluoroscopy and from approximately 1 to 25 mSv per hour during cine.

## SUGGESTED READINGS

Allsion JD, Teeslink CR. Special procedures screen. *Radiology* 1980;136:233.

American Institute of Physics. *Evaluation of radiation exposure levels in cine cardiac catheterization laboratories.* AAPM Report No. 12. New York: American Institute of Physics, 1984.

Bush WH, Jones D, Brannen GE. Radiation dose to personnel during percutaneous renal calculus removal. *AJR* 1985;145:1261.

Bushong SC. *Radiologic science for technologists: physics, biology, and protection.* St. Louis: CV Mosby, 1984.

Code of Federal Regulations, Title 10, part 20, chapter 1, section 20.1201. Washington, DC: US Government Printing Office, 1991.

Code of Federal Regulations, Title 21, parts 1000–1050. Washington, DC: US Government Printing Office, 1985. Revision of the Radiation Control for Health and Safety Act of 1968.

Code of Federal Regulations, Title 29, part 16, chapter 17, section 1910.96. Washington, DC: US Government Printing Office, 1971.

Curry TS, Dowdy JE, Murry RC. *Christensen's introduction to the physics of diagnostic radiology.* Philadelphia: Lea & Febiger, 1984.

Gertz EW, Wisneski JA, Gould RG, et al. Improved radiation protection for physicians performing cardiac catheterization. *Am J Cardiol* 1982;50:1283.

Gilula LA, Barbier J, Totty WG, Eichling J. Radiation shielding device for fluoroscopy. *Radiology* 1985;147:882.

Gray JE, Winkler NT, Stears J, Frank ED. *Quality control in diagnostic imaging.* Baltimore: University Park Press, 1983.

Houda W, Peters KR. Radiation-induced temporary epilation after a neuroradiologically guided embolization procedure. *Radiology* 1994;193:642.

International Commission on Radiological Protection (ICRP). *Radiation protection.* ICRP Publication 26. Oxford: Pergamon Press, 1977.

Jacobson A, Conley JG. Estimation of fatal doses to patients undergoing diagnostic x-ray procedures. *Radiology* 1976;120:683.

Jeans SP, Faulkner K, Love HG, Bardsley RA. An investigation of the radiation dose to staff during cardiac radiological studies. *Br J Radiol* 1985;58:419.

Linos DA, Gray JE, McIlrath DC. Radiation hazard to operating room personnel during operative cholangiography. *Arch Surg* 1980;115:1431.

Miller DL, Vucich JJ, Cope C. A flexible shield to protect personnel during interventional procedures. *Radiology* 1985;155:825.

Miotto D, Feltrin G, Calamosca M. Radiation protection device for use during percutaneous transhepatic examinations. *Radiology* 1984;151:799.

National Council on Radiation Protection and Measurements. *Limitation of exposure to ionizing radiation.* NCRP Report No. 116. Washington, DC:1993.

National Council on Radiation Protection and Measurements. *Medical x-ray, electron beam and gamma ray protection for energies up to 50 Mev.* NCRP Report No. 102. Washington, DC:1989.

National Council on Radiation Protection and Measurements. *Recommendations on limits for exposure to ionizing radiation.* NCRP Report No. 91. Washington, DC:1993.

National Council on Radiation Protection and Measurements. *Structural shielding design and evaluation for medical use of x-rays and gamma rays of energies up to 10 MeV.* NCRP Publication No. 49. Washington, DC:1976.

Pizzarello DJ, Witcosfski RC. *Medical radiation biology.* Philadelphia: Lea & Febiger, 1982.

Rusnak B, Castañeda-Zúñiga WR, Kotula F, et al. Radiolucent handle for percutaneous puncture under continuous fluoroscopic monitoring. *Radiology* 1981;141:538.

Thomson KR, Brammall J, Wilson BC. "Flagpole" lead-glass screen for radiographic procedures. *Radiology* 1982;143:557.

Wagner LK, Eifel PJ, Geise RA. Potential biological effects following high x-ray dose interventional procedures. *J Vasc Interv Radiol* 1994;5:71.

Webster EW. Quality assurance in cineradiographic systems. In: Waggener RG, Wilson CR, eds. *Quality assurance in diagnostic radiology: medical physics monograph No. 4.* New York: American Institute of Physics, 1980.

Young AT, Morin RL, Hunter DW, et al. Surface shield: device to reduce personnel radiation exposure. *Radiology* 1986;159:801.

# 3
# Access Techniques and Hemostasis

Leonard R. Bok, Hector Ferral, and Haraldur Bjarnason

## A. ACCESS SITES

The most common site for percutaneous arterial access is the common femoral artery, using the Seldinger technique. The common femoral artery demonstrates numerous qualities that favor its use as the angiographer's vessel of choice for vascular access. The common femoral artery is of fairly large caliber, superficial in location, and is quite well fixed in position as it passes through the inguinal ligament; importantly, the femoral head lies posterior to the artery, facilitating hemostasis via direct compression after catheter removal.

Since the introduction of the Seldinger technique, the common femoral artery has been and remains the artery of choice for percutaneous vascular access; however, on occasion, the common femoral artery cannot be used, owing to obstruction of the aorta or iliac arteries. The traditional second site of choice has been the axillary artery (the high brachial artery). This approach is technically more demanding than the common femoral approach and carries a higher complication rate. Complications occur because the axillary artery is quite mobile, making it more difficult to control and compress, and the branches from the brachial plexus are in a vascular sheath with the vessel. Despite these concerns, the common femoral artery approach and the axillary artery approach remain the percutaneous approaches of choice for arterial intervention or angiography. The axillary artery can usually not take longer than 6- to 7-Fr introducers. The common femoral artery can take much larger introducers.

With the increasing use of 4 or 5 Fr–diameter high flow vascular catheters, the angiographer has more options available for vascular access. With these smaller catheters, brachial artery puncture either at the midhumerus or at the antecubital fossa has become an excellent option for vascular access. The advantage of the midarm brachial approach is larger vessel size. The advantage of the low (antecubital) brachial approach is the more superficial vessel location as well as less mobility and easier compressibility. It must be recognized that catheterization of the brachial artery places an arm, rather than a leg, at risk, and that the brachial artery is prone to spasm, particularly in younger patients. This is also true for access into the left radial artery at the wrist, which has been particularly popular among cardiologists. Using the excellent but more challenging brachial artery approach, meticulous attention to technique is necessary to achieve acceptable results. Smaller catheters also allow popliteal artery puncture, which has become useful for recanalization of superficial femoral artery occlusions. The translumbar approach to the aorta was once very popular but is used only rarely today.

## B. ACCESS TECHNIQUES AND DEVICES

In May, 1953, Sven Ivar Seldinger described a new technique for vascular access and for percutaneous intraarterial catheters. This technique and modifications of the technique have become generally known as the *Seldinger technique*. The technique involves three key steps: (a) percutaneous puncture of the artery; (b) placement of a guidewire through the needle into the artery

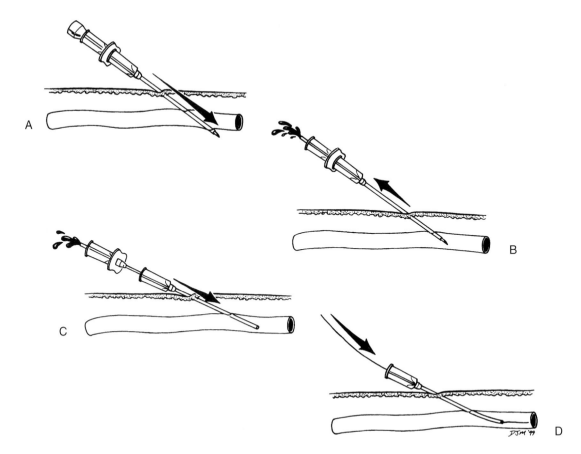

**FIG. 3.1.** Double-wall puncture. **A:** The needle has an outer plastic catheter and a stylet. It is advanced quickly though both walls of the vessel. **B:** The stylet is removed and the needle slowly pulled back until there is blood return. **C:** The plastic catheter is pushed forward of the needle into the vessel and the needle is removed. **D:** Finally, a guidewire is fed through the plastic catheter into the vessel and access has been secured.

with subsequent removal of the needle over the guidewire; and (c) placement of a catheter over the guidewire.

## PUNCTURE OF THE ARTERY

Typically a 18- or 19-gauge needle is used to puncture the artery. The puncture can be performed with either the so-called double-wall puncture (Fig. 3.1) that uses a blunt needle with a removable beveled stylet, or the single-wall puncture that uses an open beveled needle with no stylet (Fig. 3.2). Finally, vascular access can be accomplished using the more recent micropuncture set (Cook, Inc., Bloomington, IN). This third option is particularly important in percutaneous, low-brachial artery puncture or puncture of other small vessels.

## GUIDEWIRE PLACEMENT

It is important to carefully place the guidewire through the needle (access sheath) into the artery. During the subsequent removal of the needle (or access sheath), control the puncture site with the third and fourth digits of one hand while holding the guidewire between the thumb and index finger of the same hand.

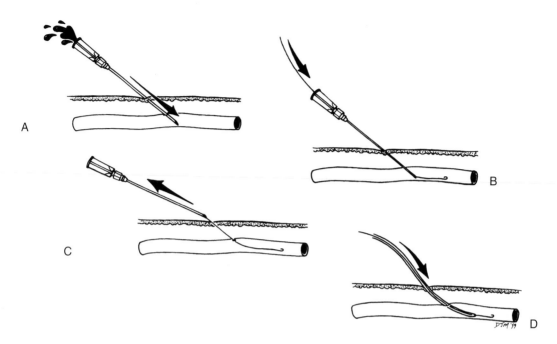

**FIG. 3.2.** Single-wall puncture. **A:** The needle is slowly advanced toward the artery. The pulsation should be felt on the needle and the needle can be slowly advanced until blood return is noted. **B:** The needle is flattened out slightly and a guidewire is placed into the artery. **C:** The needle is removed and pressure is applied on the puncture site by the third and fourth fingers, with the wire secured between the first and second fingers. **D:** An introducer or a catheter can be advanced over the wire into the vessel.

**Hint:** Note the single hand exchange technique for removing a needle or catheter over the guidewire. The hub of the needle is held between the long finger and the ring finger, while the guidewire is grasped and released between the thumb and the index finger. The needle or catheter is withdrawn progressively over the guidewire, while the operator's other hand remains free to control the puncture site and freeze the guidewire in place. The wire should then be wiped with gauze wet with normal saline.

## CATHETER PLACEMENT

Placement of the catheter (or introducer sheath) into the artery over the guidewire can now be done. Serial, coaxial predilatation of the tract may be needed if the catheter diameter is significantly greater than the needle size. For example, 4-, 5-, and 6-Fr catheters can frequently follow 18- or 19-gauge needles (or 4- or 5-Fr micropuncture sets) over a 0.035-inch-diameter guidewire without predilatation. Seven- or 8-Fr, or larger-size catheters, typically require predilatation with fascial dilators that present a more gentle taper appropriate and safe for a catheter tip.

The goal of common femoral artery puncture is to access the common femoral artery below the inguinal ligament, thereby avoiding the risk of retroperitoneal or intraperitoneal bleeding associated with puncture through the inguinal ligament. The puncture into the artery should be done at the level of the lower one-third of the femoral head. This secures puncture into the common femoral artery above its bifurcation and below the inguinal ligament in most cases. Fluoroscopy should be carried out before the puncture in all cases to localize the proper puncture site. The site

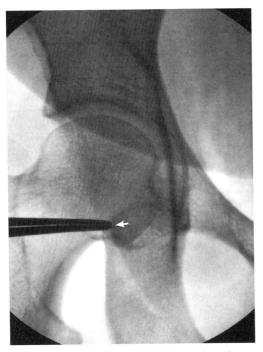

**FIG. 3.3.** A clamp (*arrow*) has been placed over the lower one-third of the femoral head using fluoroscopy. This is where the puncture into the artery should be, so the skin puncture should be slightly lower.

of the lower one-third of the femoral head is then marked with a metal clamp or other opaque object (Fig. 3.3). The skin puncture is usually made lower, because the needle points at approximately 45 degrees toward the head, making the access into the vessel higher than the skin puncture. In overweight patients in particular, the groin skin crest moves inferiorly and is not a reliable reference for common femoral artery puncture (Fig. 3.4).

Typically, the groin is shaved and prepped with povidone iodine (Betadine) or any other sterile solution [e.g., chlorhexidine gluconate (Hibiclens)] and a drape is placed. The external iliac artery runs at an angle of approximately 20 degrees from vertical, lying over the medial one-third of the femoral head in 79% of patients and over the medial one-half in 97% of patients. The deep circumflex iliac artery (lateral origin) and the inferior epigastric artery (medial origin) arise from the distal external iliac artery at the level of the inguinal ligament. The puncture should be below them (Fig. 3.4).

**Tip:** When the common femoral artery is calcified, the calcification can be used as a fluoroscopic landmark. The puncture can be made on the calcification under fluoroscopic guidance.

**Tip:** If the common femoral vein is punctured during an attempted common femoral artery puncture, the angiographer can place a guidewire or small-caliber dilator temporarily in the vessel as a marker. With the marker in place, the common femoral artery is easily punctured under fluoroscopy because it runs lateral to the vein.

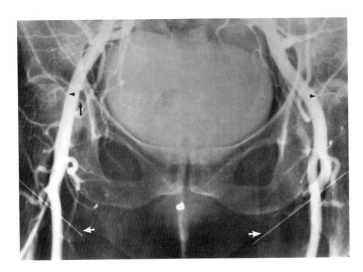

**FIG. 3.4.** On this angiogram, in an overweight patient, the inguinal crest (skin fold) has been marked with a metal marker (*white arrows*) on each side. If the skin fold had been used as a marker for puncture site, the puncture into the artery would have been much too low. Note the deep iliac circumflex artery (*arrowheads*) and the inferior epigastric artery (*black arrow*).

**Tip:** It is easy to remember the following mnemonic: *Hip navel* (*h*ip is lateral, then *n*erve, *a*rtery, *v*ein).

Real-time ultrasound can be used for common femoral artery puncture in difficult cases. This has become quite popular. The Smart Needle (CardioVascular Dynamics, Inc., Irvine, CA), a specially-designed, Doppler-based device, can be used to facilitate the puncture of difficult or non-palpable arteries.

## C. DIAGNOSTIC CATHETERS AND WIRES

Angiography catheters can be constructed from various materials, including Teflon, nylon, polyethylene, and polyurethane. Some catheters have braided, stainless steel wire mesh incorporated into their walls to improve rotational (torque) control. Other catheters now have a hydrophilic coating that makes the catheter slippery and easy to manipulate, dramatically decreasing their coefficient of friction against the tract into the vessel as well as in tortuous vessels. Catheters must be of low thrombogenicity, high tensile strength, and have low coefficients of friction to minimize the risk of stretching and resultant fracture of the catheter inside the body.

Catheter materials must be strong enough to provide high burst pressure, although they must provide the flexibility to pass through tortuous vessels and still maintain their shape at the tip. This sometimes paradoxic combination of qualities has been maximized by using different materials in the shaft and tip and fusing them together. Catheters can be produced with varying degrees of radioopacity. Radioopaque catheters are typically easier to see and manipulate on the fluoroscopic screen.

High flow injection through an end-hole catheter can cause dissection of the vessel and extravasation into the vessel wall. To achieve high flow rates (faster than 10 cc per second) without injuring the vessel, catheters with multiple side holes are required. These multiple-side-hole catheters can be obtained in varying shapes including straight, pigtail, and tennis racket. Selective injections at low flow rates (slower than 10 cc per second) can safely be performed with end-hole-only catheters. Thin-wall, high-flow-rate catheters with good torque characteristics have been important in popularizing outpatient angiography and alternative access sites.

Today, most catheters come with preformed tips of varying shapes. A number of shapes have been introduced to selectively catheterize virtually any vessel. The best shape for any given situation depends on target vessel, body habitus, vessel diameter, and tortuosity, as well as the experience and preference of the angiographer. Alternatively, flow-directed catheters can be carried to their target by the flow of blood inflating a balloon at their tip and transporting them to their destination. A good example of this is the Berman Catheter (Arrow International, Reading, PA), which is commonly used to perform pulmonary arteriograms.

Guidewires are now available in countless combinations of stiffness and flexibility. Wires produced with hydrophilic coating drastically decrease the coefficient of friction, making it easier to cross tight stenoses or negotiate tortuous vessels. Guidewires are also available in a wide range of diameters and with varying degrees of torqueability and radioopacity.

The classic two-part guidewire is constructed with a thin central stainless steel core wire and a Teflon-coated thin stainless steel wire tightly coiled around the core wire. The two wires are welded only at the proximal end, with the core wire ending a few centimeters proximal to the end the outer coiled wire. This tip can be straight or formed in a J-shaped configuration. The flexible tip (or J shape) minimizes the risk of damage to the vessel or subintimal passage of the guidewire. The movability of the spring over the core provides tremendous flexibility, whereas the core simultaneously provides considerable, firm support.

The Bentson guidewire (Cook, Inc.) is an excellent choice for initial vascular access immediately after puncture because its highly flexible tip minimizes the risk of subintimal passage of the guidewire. Traditional J-tip guidewires are often used for the same reason.

The Wholey guidewire (Advanced Cardiovascular Systems, Temecula, CA) is a high-torque, steerable guidewire with a highly radioopaque tip.

The Teflon-coated coiled-spring guidewire remains an excellent style of wire and has been produced in many variations. Wires can be produced with a movable core that allows the operator to adjust the configuration of the tip by withdrawing or advancing the core from the proximal end. Another modification is the so-called tip-deflecting wire. The wire is affixed to a proximal handle. Traction on the handle causes the tip to turn into a J shape with enough force to deflect the tip of the catheter as well as the wire, if desired.

Hydrophilic guidewires, including the Glidewire (Medi-Tech, Watertown, MA), are useful for traversing tortuous vessels, stenosed vessels, and even recanalizing occluded vessels. Hydrophilic guidewires have a markedly decreased coefficient of friction when wet.

**Tip:** The use of a hydrophilic guidewire through a needle is not recommended because the sharp edges of the needle can strip the coating from the wire, potentially depositing foreign material into the vessel.

Various stiff wires, including the Amplatz Super Stiff and Extra Stiff guidewires (Cook, Inc.), are available and can be useful to provide extra support in a number of situations. These wires aid in the placement of catheters or interventional devices across regions of tortuosity or stenosis when other wires would typically back out or buckle. Because of its stiffness, one must be careful not to perforate.

## D. PUNCTURE HEMOSTASIS

Hemostasis is typically achieved using local pressure at the site of puncture after catheter removal. Some investigators advocate the use of mechanical compression devices to apply pressure. We discourage this practice because it lacks personal observation and control in a critical phase of any percutaneous vascular procedure. Some newer devices (FemoStop, Bard, Billerica, MA) provide more tailored control and may offer promise, in particular for achieving hemostasis in situations in which compression may be prolonged and predispose users to operator fatigue. Catheter diameter, body habitus, coagulation status (platelets in particular), blood pressure level, vessel punctured, and condition of the vessel wall all affect the degree and duration of pressure required to achieve satisfactory hemostasis. These same factors should be taken into consideration when determining the duration of postoperative bedrest, and the necessity of a postoperative pressure dressing.

The goal of immediate postprocedure local pressure is to apply sufficient compression to prevent bleeding while avoiding or minimizing vascular occlusion time. This is best accomplished manually. The recommended duration of pressure is typically 5 to 10 minutes. Of course, one must recognize compression should continue until the bleeding is entirely stopped. If there is still oozing, compression must be continued rather than application of a pressure dressing. Hold pressure for 5 minutes after bleeding or oozing of blood is completely stopped, and decrease pressure progressively. Do not withdraw compression abruptly.

The early standard for postprocedure bedrest was 24 hours. Since the late 1970s and early 1980s, outpatient angiography has been common for uncomplicated cases, and bedrest of 4 to 6 hours has become standard for these uncomplicated procedures using 4- to 7-Fr catheters. Four to 6 hours appears to be a reasonable target for postprocedure bedrest.

Finally, placement of a pressure dressing locally over the puncture site has been customary in many practices; however, some have questioned its use, even arguing that it interferes with adequate observation of the puncture site for bleeding or swelling. Clearly, this is an area for personal judgment and adjustment for individual patient or clinical circumstances.

A number of puncture hemostatic devices have been introduced. These include the Perclose (Perclose, Inc., Menlo Park, CA), Angio-Seal Device (Quinton Instrument Company, Bothell, WA), VasoSeal (DataScope, Inc., Montvale, NJ), Percutaneous Hemostasis Device (MEDITECH/Boston Scientific Corp., Natick, MA), and Arterial Sealing Device (Scimed/Boston Scientific Corp., Maple Grove, MN). Although these devices offer the promise of decreased labor costs, more rapid ambulation, and earlier discharge with decreased complication rates, recent publications have demonstrated some risk of device-related complications and lack of cost effectiveness at current price levels.

## E. COMPLICATIONS: DIAGNOSIS AND MANAGEMENT

The most common local complication of angiography is local bleeding with hematoma formation. Bleeding with hematoma formation is detected visually and by palpation. Carefully applied local pressure of sufficient duration and bedrest of 4 to 6 hours should minimize this risk. If prolonged or delayed bleeding is detected, manual pressure should be reapplied for as long as necessary to establish complete control of the bleeding. In the most unusual circumstance that bleeding cannot be controlled manually (more than 2 hours of pressure), surgical repair may be necessary. Should bleeding into the thigh be severe enough to cause a fall in hematocrit, transfusion may be appropriate.

In some cases, a false aneurysm forms after a procedure. False aneurysms can be diagnosed with physical examination, but ultrasound and angiography are the definitive diagnostic tools. Treatment options include obliteration using ultrasonic visualization and directed local pressure, obliteration using transcatheter placement of occlusion coils, and open surgical repair.

Occasionally, arteriovenous fistula forms. This is more common when there has been both arterial and venous puncture at the same groin at the same time on the same day, as with right and left heart catheterization, portal hypertension work-up, or after difficult puncture that requires multiple needle puncture attempts. In general, this is best treated surgically; however, if the fistula is of sufficient length for safe coil embolization, percutaneous treatment may be a viable option.

The preceding local complications are more common with low punctures, in which the needle and catheter enter the superficial femoral artery or the deep femoral artery. Meticulous puncture technique and meticulous postprocedure compression minimize these complications. These hemorrhagic complications are also more common after interventional procedures that use larger-caliber catheters and may involve temporary or long-term anticoagulation as well as significantly prolonged catheterization durations.

In contrast to the local complications caused by low puncture, high puncture can result in potentially fatal retroperitoneal or intraperitoneal bleeding. Postoperative observation must include vigilance for the signs of decreased intravascular volume, such as decreased blood pressure and elevated pulse. The nursing staff and angiographer must be alert to the combination of increased pulse rate and stable or decreased blood pressure. If signs of shock are detected, hematocrit and hemoglobin levels must be checked and computed tomography obtained to evaluate for occult bleeding in the peritoneum or retroperitoneum. A resuscitation effort with fluid replacement must start immediately. Transfusion may be necessary; however, emergent surgical repair is often needed for both intraperitoneal and retroperitoneal hematoma. The symptoms are typically those of back pain that is different from what the patient would expect from being on the procedure table or in bed for extended periods of time.

Occlusive complications include catheter-related thrombosis, embolization, and subintimal catheter passage or subintimal contrast injection. Catheter-related thrombosis and embolization cause the immediate development of ischemic symptoms in the region selectively catheterized or distal to the site of catheterization in the extremity in which the catheter is placed. Treatment options include immediate transcatheter thrombolysis, intermediate- or long-term anticoagulation, or surgical thrombectomy. If the problem is at the puncture site, the possibility of an intimal flap must be considered. In this event, if conservative treatment is unsatisfactory, surgical repair may be necessary.

The rare occurrence of cholesterol embolization may involve compromise of gastrointestinal, renal, or pancreatic function, as well as possible skin changes in the lower extremities, developing 24 to 48 hours after angiography. The syndrome is caused by embolization of cholesterol secondary to catheter manipulation in a severely atherosclerotic aorta. Treatment is supportive.

Subintimal injection is typically a self-limited phenomenon when it is owing to a jet of contrast infusing into the extravascular tissues. This should be distinguished from subintimal passage of the guidewire or catheter, typically a more serious situation. Two distinct types of subintimal passage must be distinguished. Before further discussion, we must emphasize that the catheter and guidewire should always be advanced under fluoroscopic guidance with careful attention to tactile sensation. Catheters and guidewires should never be forced. If resistance is encountered, the operator should stop at once, withdraw, reposition, or check position with a gentle contrast injection. Every effort should be made to minimize the risk of subintimal injection. Nevertheless, during retrograde catheterization of the aorta, there is a small but unknown number of episodes of subintimal passage of the guidewire or catheter. These small retrograde subintimal passages appear to be self-limited, probably because blood flow tends to reappose these flaps against the vessel wall quite naturally; however, when a vessel has been selectively catheterized or a catheter is being advanced via an antegrade puncture and subintimal passage occurs, the antegrade blood flow may create a windsock effect. Although very small flaps of this type may remain self-limited, treatment may sometimes be judged necessary. An unrepaired intimal flap may be asymptomatic, lead to extended dissection, and lead to complete occlusion and thrombosis of the vessel. Treatments include balloon angioplasty, balloon angioplasty with stent placement, and surgical repair.

Probably the most common complication of angiography, particularly of those procedures that involve large volumes of iodinated contrast, is contrast-induced decrease in renal function. Treatment is primarily supportive. The patient should be kept well hydrated. Other measures, including preprocedure diuretics, vasodilators, and low-osmolality contrast material, remain controversial.

## SUGGESTED READINGS

Darcy MD, Kanterman RY, Kleinhoffer MA, et al. Evaluation of coagulation tests as predictors of angiographic bleeding complications. *Radiology* 1996;198:741–744.

Doby T. A tribute to Sven Ivar Seldinger. *Am J Roentgenol* 1984;142:1–3.

Dotter CT, Rosch J, Robinson M. Fluoroscopic guidance in femoral artery puncture. *Radiology* 1978;127:266–267.

Egglin TKP, O'Moore PV, Feinstein AR, Waltman AC. Complications of peripheral arteriography: a new system to identify patients at increased risk. *J Vasc Surg* 1995;22:787–794.

Hessel SJ, Adams DF, Abrams HL. Complications of angiography. *Radiology* 1981;138:273–281.

Rupp SB, Vogelzang RL, Nemcek AA Jr, Yungbluth MM. Relationship of the inguinal ligament to the pelvic radiographic landmarks: Anatomic correlation and its role in femoral arteriography. *J Vasc Interv Radiol* 1993;4:409–413.

Seldinger SI. Catheter replacement of the needle in percutaneous arteriography [Reprint]. *Am J Roentgenol* 1984;142:5–7.

# 4
# Vascular Embolotherapy

Haraldur Bjarnason

## DEFINITION

*Embolization* is defined as intravascular deposition of particulated, liquid, or mechanical agents or blood clots to cause occlusion of a vessel. This may mean mechanical obstruction of the blood flow by the injected material or chemical, inducing thrombosis in the vessel and sclerosis of the vessel wall with concomitant fibrosis and vessel shrinkage.

## HISTORICAL ASPECTS

The first embolization treatment was described in 1904, when melted paraffin petroleum was injected into an external carotid artery before surgery of a head and neck tumor. In 1930, the use of autologous muscle attached to a silver clip was described for occlusion of a carotid-to-cavernous fistula. Development in catheter techniques led to the first catheter-directed embolization of gastric bleeding using autologous blood clots in 1972. The use of detachable balloon catheters was reported in 1974. Tissue adhesives were introduced in 1972 and Gelfoam, as an embolic agent, in 1974. In 1971, polyvinyl alcohol (PVA) was introduced. Then, in 1975, the metal coil (*Gianturco coils*) with wool attached was introduced (Table 4.1).

Ethanol was the first so-called sclerosing agent and was described in 1981, followed by Dr. Amplatz's use of boiling contrast in 1982. Sodium tetradecyl sulfate (Sotradecol) was then introduced. Since then, there has been a tremendous evolution in catheter development, especially in microcatheters and coaxial systems. The development of microembolic agents, such as microcoils and various embolization and sclerosing agents, followed the evolution in catheter development.

## PATIENT CARE

### Preoperative Management

Before any embolization procedures, the operator must thoroughly review the patient's history. The most important factors depend on the nature of the embolization but usually include the following:

1. The operator should be aware of previous surgeries in the vicinity of the embolization area. This is especially important in the gastrointestinal tract where surgical procedures may have excluded normal collateral flow. This can then increase the risk of ischemia, which without previous surgery would be unlikely to occur.

2. Renal function is important because often, large volumes of contrast are used and substances are released from the embolized tissue that can be potentially harmful to the kidneys. Generous hydration should be given before, during, and after the procedure for protection of renal function.

3. Antibiotics are seldom used for embolization of gastrointestinal bleeding but are indicated for most other embolization procedures, such as tumor embolization and organ ablations. The postembolization ischemic organ is highly susceptible to infection. This is especially true for the spleen but applies for any organ ablation or major embolization. Cephalosporins are a usual choice.

4. Sedation and intravenous access are required for all patients undergoing embolization procedures. Intravenous sedation and pain medication should be given, preferably from the very beginning of the procedure, and a nurse should always be present. The most commonly used sedatives and pain medications are fentanyl

**TABLE 4.1.** *Embolization agents*

| Agent | Indications | Actions/mechanism | Actions/length |
|---|---|---|---|
| Polyvinyl alcohol | Arteriovenous (AV) malformations | Obstruction, inflammation | Lifelong |
| Coils | Bleeding, occlusion of larger vessels and AV shunts | Occlusion, inflammation | Lifelong; collaterals |
| Gelfoam particles | Bleeding where short-term occlusion is needed | Occlusion | 2 to 3 wks |
| Alcohol | Arterial malformations | Necrosis and inflammation | Lifelong |
| Sotradecol | Arterial malformations | Necrosis and inflammation | Lifelong |
| Cyanoacryl | Bleeding, AV malformations | Glue, obstruction, inflammation | Lifelong |
| Silk sutures | AV malformations | Obstruction, inflammation | Lifelong |
| Balloons | Obstruction of larger arteries | Obstruction | Lifelong |

citrate and midazolam (Versed). Meperidine (Demerol) is also used by many interventionists. The operator needs to be well informed about the drugs used for the conscious sedation. Many interventionists recommend general anesthesia for larger embolization procedures. The procedures tend to be lengthy and hard on the patient. If alcohol is used for the embolization, general anesthesia should be seriously considered because of possible systemic complications. Cardiopulmonary collapse has been described with the use of ethanol for embolization.

5. A carefully written consent, with an explanation of all aspects and possible complications of the procedure, should be given to the patient and signed before any embolization procedure. Other preparations for the procedure should be the same as for regular angiography.

### Postoperative Management

Postoperatively, the patient has to get continuous, good hydration. The patient usually still requires pain medication, and adequate analgesia has to be ensured. If antibiotics are used, they should be continued for up to 1 week, especially after splenic and tumor embolization. Renal function should be monitored.

### MATERIALS FOR EMBOLIZATION

Selection of embolization agents is critical for the success of embolization procedures. The agents can be classified in the following manner:

1. Absorbable materials
   a) autologous blood clot
   b) Thrombin
   c) Gelfoam
   d) Oxycel
2. Nonabsorbable materials
   a) particulate agents
   b) injectable fluid embolic agents
   c) sclerosing agents
   d) nonparticulate agents
3. Electrocoagulation

### Absorbable Materials

Absorbable embolization agents cause only temporary occlusion of the embolized vessel. The agents reabsorb or recanalize within a few days to weeks. Temporary occlusion is especially useful in gastrointestinal tract bleeding in which short-term pressure reduction is needed until an ulcer or sore has healed. Absorbable embolization agents are also useful for preoperative embolization of bone tumors in which surgery is planned immediately after the procedure. These agents are generally the least dangerous embolization agents, but long-term occlusion is not acquired. Embolization is rarely at the capillary level but rather is in the larger branches, allowing for collateral flow.

#### *Autologous Blood Clot*

The blood clot for embolization is made from the patient's blood. Blood is drawn from the patient and allowed to clot in a sterile container. The clotting process can be accelerated by adding thrombin or aminocaproic acid, which can be especially helpful in patients with concomitant coagulation abnormalities. The clot is then aspirated into a 1- or 3-mL syringe and injected into the catheter.

Embolization with autologous blood clots is effective. The embolization is short-lived, and recanalization occurs early. Animal studies have demonstrated complete occlusion for up to 48 hours, but at 2 weeks, approximately 50% of the vessels recanalize.

**Indications, Pros, and Cons:** Autologous clots are inexpensive and nontoxic. They can be the embolic material of choice for bleeding in areas such as the kidneys or gastrointestinal tract, in which rapid dissolution is important. Autologous clot recanalization may, however, be too rapid, causing interventionists to prefer Gelfoam because of its more permanent embolization effect.

### Thrombin

Thrombin is a protein that catalyzes the conversion of fibrinogen to fibrin. It comes from the manufacturer in 5-, 10-, and 20-mg vials containing 5,000, 10,000, and 20,000 units of thrombin. It is mixed with saline and can be injected directly into the vessel being treated. Thrombin can be used alone as an agent or in conjunction with other agents, such as blood clots, steel coils, and detachable balloons. Thrombin has been used for thrombosis of puncture site pseudoaneurysms. One-thousand international units in 1 mL of saline of the agent are injected into the pseudoaneurysm under ultrasound guidance. The injections can be repeated until thrombosis occurs.

As a stand-alone treatment, thrombin depends on anatomic circumstances, such as flow rate. Doses as small as 300 units and as high as 5,000 units have been recommended.

**Indications, Pros, and Cons:** Thrombin is a relatively simple agent to use. Reflux into non-targeted blood vessels should be avoided. If used cautiously, thrombin does not cause any untoward thrombotic or allergic manifestations.

### Gelfoam

Gelfoam is an absorbable agent derived from gelatin and commercially available in two forms. The first is a sponge sheet that is 3 cm × 7 cm × 4 mm in size. The sponge can be cut either with scissors or a knife into bites 2 to 4 mm in diameter. Individual particles can be placed into the tip of the syringe and injected separately through the catheter. The sponge can also be cut into multiple small cubicles (approximately 2 mm × 3 mm × 4 mm) that are mixed thoroughly with a iodinated contrast medium (50% contrast and 50% saline) (Fig. 4.1). A 10-mL syringe is used to mix the Gelfoam soup. For the embolization, a 3- or 1-cc syringe is used. A few particles should be injected at a time through a clear, plastic tubing, followed by injection of saline, and then contrast material, for evaluation of the embolic effect. Those particles cause occlusion of relatively large vessels.

The other commercially available form is Gelfoam powder. It comes in particle sizes 0 to 125 μm, 125 to 315 μm, 315 to 630 μm, and 630 to 800 μm. Gelfoam powder causes occlusion of much smaller vessels than Gelfoam sheets cut into cubicles. There is, therefore, less collateral flow, and ischemic complications are more likely to occur than when the Gelfoam sheets are used.

Gelfoam is thrombogenic and causes pan arteritis, which is characterized by infiltration of leukocytes into all layers of the vessel wall as well as by disruption of the intima and elastic tissues.

**FIG. 4.1.** Gelfoam emboli cut with scissors into small fragments.

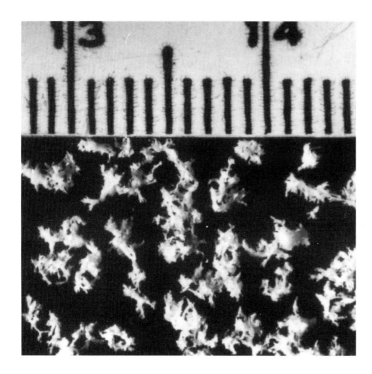

FIG. 4.2. High-power view of Ivalon particles produced by saw blade. Observe extreme variability in size, which makes separation of the particles by graduated sieves a necessity.

The arteritis usually resolves within 4 months. The embolization lasts from a few days to a few weeks. Gelfoam fragments are seldom found for more than 24 to 48 hours after embolization.

**Indications, Pros, and Cons:** Gelfoam is a good agent when temporary embolization is sought and later revascularization is desirable. The main indications are gastrointestinal bleeding—for example, bleeding from biopsy tracts in the liver and kidneys and preoperative embolization in which surgery is planned immediately after the embolization. The bleeding vessel has to be small enough to capture the Gelfoam cubicles. It is a useless agent for treatment of vascular malformation because of its quick recanalization. Gelfoam can be used in combination with other agents, such as polyvinyl alcohol and metal coils.

Gelfoam rapidly becomes contaminated with bacteria, so precautions have to be applied. Gelfoam powder should not be used for embolization of large arteriovenous communications because it can flow through and end up in the lungs. The main uses for the powder form are for tumors in which necrosis is desired.

*Oxycel*

Oxycel is a cottonlike material made from oxidized cellulose. When mixed with blood, it causes a firm clot and can be injected through the embolization catheter. Recanalization occurs within 4 months. This agent is seldom, if ever, used anymore.

### Nonabsorbable Materials

*Polyvinyl Alcohol*

PVA has been treated with formaldehyde to form a reticulated substance (Fig. 4.2) that is biologically inert and sold in small particles.

PVA comes commercially as powder in vials with typical particle sizes. The sizes include 45 to 150 μm, 150 to 250 μm, 250 to 355 μm, 355 to 500 μm, 500 to 710 μm, 710 to 1,000 μm, and 1,000 to 1,180 μm in diameter. A suspension is made of one bottle of the agent mixed with water-soluble contrast Hypaque, 60%, or Omnipaque, 300 mg per mL, equally mixed with saline, producing a total amount of 10 to 20 cc. The mixture is aspirated into a 1- or 3-mL solution and is injected directly into the catheter, either via a clear plastic tubing, to allow visualization of the particles, or directly into the catheter.

The material is then injected in small amounts at a time. The catheter should be carefully flushed with saline after each injection. A careful injection of contrast material is made for evaluation of the embolization effect. No reflux should occur.

Histologically, PVA causes the formation of fibrous connective tissue and a moderate inflammatory reaction that involves the arterial wall. Recanalization does not occur.

**Indications, Pros, and Cons:** The main indication is treatment of arteriovenous malformations, both centrally and peripherally. Care must be taken to select a particle size that will not pass through the vascular bed into the venous circulation. To determine whether particles would pass into the venous system, the PVA particles can be marked with technetium-99 and sulfur colloid. A gamma camera placed over the lungs reveals which sizes pass through the embolization area. The smallest particles that do not pass through the embolization area to the lungs are selected for the embolization.

Tumor ablation and tumor bleeding are also good indications for the use of PVA. Small particles should be used.

PVA should be used with precaution for embolization in the gastrointestinal tract or kidneys because of the risk of ischemia.

### Other Nonabsorbable Particular Materials

Multiple microparticles, such as stainless steel or ferromagnetic steel, acrylic, methacrylate, silastic, and silicones, have been and are being evaluated. Some of those agents are available and in use.

## Injectable but Nonsclerotic Embolic Agents

### Sclerosing Agents

#### Ethanol

Ethanol is a well-known sclerosing agent that induces immediate thrombosis from the capillary bed backward. This results in total tissue devitalization. Ethanol induces thrombosis by denaturing blood proteins, dehydrating vascular endothelial cells, and precipitating their protoplasm, denuding the vascular wall of endothelial cells and segmentally fracturing the vessel wall. Ethanol can induce significant pain when injected intravascularly. Proper anesthesia, such as deep intravenous sedation or general anesthesia, may be required to minimize patient discomfort. Postembolization edema always occurs with the use of ethanol. Extreme caution must be taken with its use to minimize the possibility of nontarget embolization.

**Indications, Pros, and Cons:** Ethanol has been used for high-flow lesions and for treatment of esophageal varices (e.g., varices from the left gastric vein). In the treatment of vascular malformations, ethanol has demonstrated curative potential, as opposed to the palliation seen with other embolic agents. As with Gelfoam powder, extreme caution and super-selective catheter placement are requirements when using ethanol as an endovascular occlusive agent.

#### Hot Contrast Medium

Contrast medium heated to the boiling point has been used successfully to obliterate the spermatic veins in subfertile men with varicoceles. The hot contrast causes heat damage to the vessel as well as toxic injury. Thrombosis occurs 1 to 5 days after the treatment, and the thrombus finally organizes and the vessel becomes a fibrotic cord in the majority of cases.

**Indications, Pros, and Cons:** Hot contrast medium is solely used for sclerotherapy of the spermatic vein. It causes occlusion of collateral pathways, which is important. The advantage is that the injected contrast is easy to follow fluoroscopically.

#### Sodium Tetradecyl Sulfate (Sotradecol)

Sodium tetradecyl sulfate (Sotradecol) is another sclerosing agent available in a 1% or 3% aqueous solution. Its indications are similar to ethyl alcohol, and it contains 2% benzyl alcohol. The 3% solution is used for sclerotherapy.

**Indications, Pros, and Cons:** Sodium tetradecyl sulfate has been used for sclerotherapy in similar indications as alcohol. However, the main indication is the treatment of varicose veins and venous malformations.

The same caveats apply to the use of sodium tetradecyl sulfate as to the use of ethyl alcohol and Gelfoam powder. Distal embolization and effect on neurologically sensitive areas are concerns.

### *Tissue Glue*

Tissue glue, or butyl-2-cyanoacrylates, which is similar to the so-called super glue, has been used for surgical applications since the 1960s and for radiologic embolization procedures since 1972. The glue, which is in a fluid form, is easily injected through a small catheter. Because the glue is non-radioopaque, it is mixed with different radioopaque materials for visual purposes. On contact with ionized fluids, such as blood, the glue almost instantaneously polymerizes and becomes solid. There are two types of the adhesive tissue: isobutyl-2-cyanoacrylate, which is used abroad but is not allowed in the United States because of reported carcinogenicity, and *N*-butyl-2-cyanoacrylate, which is slightly different in chemical composition but has the same adhesive properties.

**Indications, Pros, and Cons:** The indications are mainly arteriovenous malformations. The main use has been in the central nervous system for the treatment of arteriovenous malformations and intracranial aneurysms. Glue has also been used for treatment of bleeding in the kidneys, gastrointestinal tract, and other locations, but special precaution has to be applied in the intestinal tract because of ischemic risk and perforation.

The main advantage of glue is that superselective and very accurate embolization can be achieved. The glue polymerizes very quickly, and distal embolization is unlikely. One of the disadvantages is the severe local foreign body inflammatory reaction. Currently, tissue adhesives are not approved for use in humans in the United States.

### Nonparticulate Agents

### *Stainless Steel Coils*

The stainless steel coil was initially developed by Drs. Gianturco and Wallace in 1975. It is one of the most commonly used embolization agents.

The coils are short, guidewire segments formed into spirals and specified by the diameter of the spiral and the length and diameter of the coil wire. The wire diameter ranges from 0.025 to 0.038 in. To increase the thrombogenicity, wool strands were initially attached to the coils, but because of severe foreign body reaction response, the wool has been replaced with Dacron. The coils are also available without Dacron.

The inner diameter of the catheter used for delivery should dictate the diameter of the coil wire. A 0.035-in. coil should be deployed through a catheter with an inner diameter of 0.035 to 0.038 in., and it is important that the wire used to push the coil at delivery has the same diameter or is larger than the coil itself. If the push wire is smaller than the coil, it can become wedged between the coil and the catheter wall, and the coil becomes stuck within the catheter.

Smaller coils (so-called microcoils) have been available since approximately 1990. In most instances, the smaller coils are made of platinum wire. The wire diameter is 0.014 to 0.018 in. The platinum itself is highly thrombogenic as well as highly radioopaque. Platinum is biocompatible and does not cause artifacts on magnetic resonance imaging (MRI) studies. The magnetic force of MRI does not affect platinum. The same measures are used for specifications of the microcoils as for the initial coils. The microcoils come in different forms and lengths. They have been attached to silk or Dacron to increase their thrombogenicity. The microcoils are intended to be placed through microcatheters (3 Fr) like the Tracker catheter (Target Therapeutics, Boston Scientific, Natick, MA) (Fig. 4.3).

**Indications, Pros, and Cons:** The main indications for metal coils are occlusion of large vessels, mainly large arteriovenous fistulae, such as pulmonary fistulae, bleeding from large vessels after trauma, preoperative embolization of renal tumors, occlusion of esophageal varices in conjunction with the transjugular intrahepatic porto-

**FIG. 4.3.** Gianturco minicoil. **A:** Mounted over introducer wire. **B:** Partially extruded from introducer wire. **C,D:** Platinum microcoils of varied designs: straight, curved, and helical. (From Castañeda-Zúñiga WR, Zollikofer C, Barreto A, et al. A new device for the safe delivery of stainless steel coils. *Radiology* 1980;136:230–231, with permission.)

systemic shunt procedure or as a sole procedure, and for treatment of systemic-to-pulmonary shunts. The coils have also frequently been used in the treatment of spermatic vein embolization. The microcoils have been used mainly for embolization of smaller vessels, such as in gastrointestinal bleeding; smaller aneurysms, such as in the brain and gastrointestinal tract; arteriovenous fistulae in the kidneys and liver; and wherever else microsystems are needed for access and treatment.

The selection of the coil diameter is important. The diameter should be approximately 2 mm larger than the diameter of the vessel to be embolized. This is especially true for the first coil, to prevent dislodgement. With the first coils correctly sized, subsequent smaller coils will not become dislodged.

On the other hand, if the coil is too large, it cannot retain its normal form. It grows elongated and may extend out of the artery to be embolized into the feeding vessel. This may require eventual retrieval, using a snare or other modes.

Before the coil is placed, the catheter must be in a secure position in the vessel to be embolized. It is a good practice to pass a guidewire through the catheter before passing the coil to verify that the catheter does not get dislodged when the coil is passed through. A forceful injection through the catheter into the deployed coils should be avoided, because it may cause dislodgement of the coils.

If the position is thought to be inadequate, several devices allow for partial deployment of the coils and retrieval. Several mechanisms are being evaluated. One of those is the Gugliemi detachable coil (Target Therapeutics, Boston Scientific, Natick, MA). Electrical current placed through the pusher wire causes electrolytic dislodgement between the coil and the pusher on activation. The current used to dislodge the coil from the wire also potentiates thrombosis. Cook, Incorporated (Bloomington, IN) has developed a mechanism based on a screw connection between the pusher wire and the coil. Target Therapeutics has developed a system based on a wedge-interlocking system between the wire and the coil within the catheter that will dislodge as soon as the trailing end of the coil slips out of the end of the catheter.

### Detachable Balloons

Detachable balloons, made of silicone or latex, come in various sizes and shapes. The balloons are placed through a coaxial system and are floated directly to the target site from the guiding catheter. These have traditionally been used in neuroradiology but are also applicable in other areas (e.g., large arteriovenous fistulae and pulmonary arteriovenous malformations). Detachable balloons are somewhat complicated to handle. They have the disadvantage that they can deflate within several weeks and are sensitive to osmotic pressure between the fluid in the balloon and the plasma. The fluid used for inflation therefore needs to be isosmotic to plasma (Fig. 4.4).

## GENERAL TECHNICAL ASPECTS

### Catheters

An introducer sheath should be used for all embolization procedures. The sheath has several purposes. The manipulation of catheters is improved through an introducer sheath, because it causes less friction than the tract. Multiple catheters may have to be used, and an introducer sheath makes catheter exchange easier, safer, and less traumatic, and it reduces blood loss. If, for some reason, an embolization agent becomes stuck in the catheter and the catheter has to be removed without a wire through it, access to the vascular system is still secured.

Catheter selection is important. For coils, a nontapered catheter or a catheter with a uniform lumen is used. The inner lumen of the catheter should be very close to the outer diameter of the coil itself. The catheters used for coil embolization should not have side holes. Coils can get stuck in the holes, and embolization agents can come out of the side holes proximal to the intended delivery site.

Coaxial catheters or microcatheters are frequently used. These are usually 3-Fr catheters that are fed through a regular catheter (angiographic or guiding catheter) with an inner diameter of 0.035 or 0.038 in. with a rotating

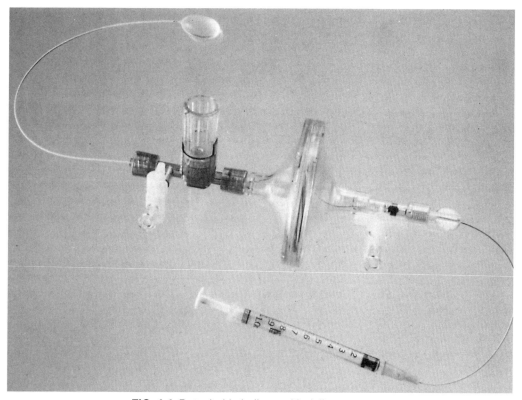

**FIG. 4.4.** Detachable balloon with delivery system.

hemostatic valve. These catheters usually accept a 0.014-in. or 0.018-in. guidewire. Target Therapeutics (Tracker), Cook, Incorporated (Microferret), and other companies have produced excellent catheters for this purpose. Through these microcatheters, coils, PVA, Gelfoam, and multiple other agents can easily be injected. Care must be taken to flush these catheters on a regular basis, and the guiding catheter should be connected to a continuous flush with high pressure to prevent development of clot inside the catheter.

### Approach

For most intravascular embolization therapies, the femoral route is selected. The brachial and axillary artery can also be used as an access. For embolization of the pulmonary artery and in the venous system, the common femoral veins, internal jugular veins, and any other accessible vein can be used for access.

## SPECIFIC EMBOLIZATION AREAS

### Bronchial Artery Embolization

The most common causes for bronchial artery bleeding in this country are cystic fibrosis in the younger population and neoplasms in the older population. Bronchiectasis related to tuberculosis used to be the most common cause and remains the common cause in less-developed areas.

Blood loss during hemoptysis is usually not large, but the main problem becomes filling of the aerated space with blood, causing asphyxia.

#### *Anatomic Considerations*

The bronchial artery anatomy is quite variable. The arteries come off the descending aorta from the anterior aspect at the level of the third through the seventh intercostal space. The most common distribution of the bronchial arteries is as follows (Fig. 4.5):

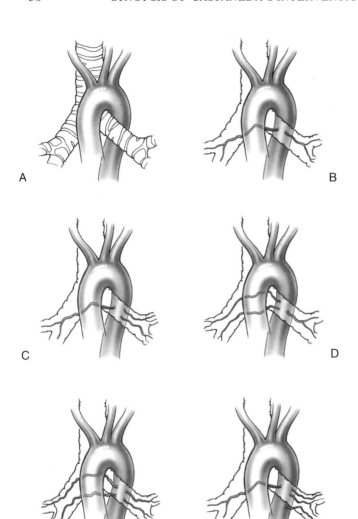

**FIG. 4.5. A:** Common anatomic variations of the bronchial arteries. Anterior view. **B:** Single vessel to each (30%). **C:** Two arteries on one side and one on the other (45%). **D:** Same as for **C**. **E:** Two vessels on each side (20%). **F:** Three on one and one on the other (4%). (Modified from Kadir S. *Atlas of normal and variant angiographic anatomy*. Philadelphia: W.B. Saunders, 1991.)

1. Bilateral single bronchial artery: Approximately 30% of individuals
2. Three bronchial arteries: Approximately 30%
3. Two left and one right: Approximately 40%
4. Two right and one left: Approximately 5%
5. Four bronchial arteries: Approximately 24%
6. Two left bronchial and two right bronchial arteries: Approximately 20%
7. Three left and one right bronchial artery: Approximately 4%
8. Five or more bronchial arteries: Approximately 1%

Approximately 60% of individuals have a single right bronchial artery, and there is a high frequency of a common origin of the right bronchial artery and the intercostal arteries on the right side. Multiple bronchial arteries are very common on the left side.

Quite commonly (in 45% of individuals), a common bronchial trunk feeds both the left and the right bronchial arteries.

Anomalous origin of the bronchial arteries may be from the internal mammary, superior intercostal, subclavian, brachiocephalic, or inferior thyroid arteries. The left bronchial artery may come from the concavity of the aortic arch in up to 15% of individuals, but a bronchial-intercostal trunk is uncommon on the left side (in fewer than 4% of individuals). The anterior and posterior spinal arteries can receive branches from extraspinal arteries, such as the intercostal and bronchial arteries.

After a surgical procedure on the chest or lung and after chronic infectious diseases that cause pleural adhesions, it is not uncommon to have transpleural collaterals from the intercostal arteries, other chest wall arteries, or mediastinal arteries. This should be considered if other sources have been ruled out.

### *Techniques*

A transfemoral approach should be used for embolization of the bronchial arteries. The most commonly used catheter is a sidewinder no. 1 catheter, an Amplatz left coronary catheter (6 Fr) that fits quite nicely into the bronchial arteries or a Debrun C catheter (Cook, Inc.).

Nonionic contrast should be used for all injections, as severe transverse myelitis has been described after bronchial arteriography and embolization. High-quality imaging is essential, and the digital subtraction technique is helpful.

Careful angiographic evaluation is necessary before any embolization is performed. Remember that bronchial to spinal and coronary artery communication has been reported. Bronchial artery embolization should not be performed when such communication is present, unless there is a strong clinical indication, a superselective position has been achieved, and reflux is avoided.

If the side of bleeding is known, a bronchial arteriogram is performed only on that side except in patients with cystic fibrosis, in whom both sides are studied and embolized. As soon as the abnormal vessels have been detected, selective (super-selective) access is gained by passing a coaxial microcatheter deep into the artery. This usually cannot be achieved with a 5- or 6-Fr angiographic catheter. Ivalon particles are then used for the embolization. The usual particle size is 250 to 500 µm. As soon as stasis is achieved, small Gelfoam cubicles can be injected (2 mm × 2 mm or 3 mm × 3 mm). Embolization of bronchial arteries with spinal artery communication has been described. One should refrain from alcohol or other liquid agents because of the risk of ischemic changes that can cause bronchial stenosis or fistulae. Metal coils should not be used either, as the coils can prevent later access to the artery if it becomes necessary.

Embolization in the treatment of cystic fibrosis can be quite successful, resulting in complete control of hemoptysis in the majority of patients, with repeated embolization in some of them. Patients with tuberculosis also have good results after embolization. The immediate success rate was, in one series, 86%, with complete remission in 50%, partial remission in 22%, and recurrent hemoptysis in 28%.

### *Complications of Bronchial Artery Embolization*

Chest pain, dysphagia, and fever may be expected in several patients. Analgesics may take care of those symptoms.

Left main bronchial stenosis and bronchoesophageal fistula have been described. Transverse myelitis, owing to inflammation secondary to the contrast dye or owing to embolization, has also been described.

## Preoperative Embolization

The main indication for preoperative embolization is to decrease perioperative bleeding during tumor resection; however, embolization may also be carried out to decrease tumor size before surgery.

### *Bone*

Preoperative embolization in bone is mainly done before tumor resection, be it primary tumor or metastasis, or before other operative manipulations, such as hip replacement or placement of internal fixation. There is no evidence that embolization agents retard healing of fracture sites. Embolization has also been used for primary treatment of benign tumors, such as aneurysmal bone cysts and giant cell tumors.

The embolization significantly decreases blood loss during surgery. It should be performed as close to the planned surgical procedure as possible and within 12 hours. Gelfoam is the most commonly used agent, but PVA can also be used, especially if the surgery will not be performed soon after the embolization. Selective injection is made into the artery, feeding the tumor after angiographic evaluation. Gelfoam slur or individual cubicles are injected until stasis has been reached in the vessel. If the selective position cannot be achieved, a metal coil can be placed into arteries feeding normal tissue and embolization performed proximal to it. Collaterals provide flow to the normal tissues, but more distal embolization is achieved in the tumor with the Gelfoam.

The most severe complications are embolization of normal vessels with possible tissue loss, skin breakdown, and ischemic neuropathy.

### *Kidneys*

Occasionally, embolization of the kidney is performed before resection. The indications are mainly to reduce the risk of bleeding from the tumor during nephrectomy and to reduce tumor size before resection; however, this is controversial. Tumors that are unresectable may become resectable after embolization. PVA, Gelfoam, and ethanol have been used for renal ablation.

## Visceral and Peripheral Arteries

Pseudoaneurysms are quite common after trauma to the extremities and neck. Other common areas for pseudoaneurysms include the arteries to the gastrointestinal tract after pancreatitis or gastrointestinal ulcers, or in the liver and kidneys after biopsies and drainage procedures. The preferred therapy is operative, but in many instances is not feasible at that time. The most commonly used agents are coils, detachable balloons, and tissue adhesives. The major advantages of endoluminal embolizations are (a) only local anesthesia is required; (b) collaterals are preserved; and (c) the operative approach can be difficult. If access cannot be gained via the endoluminal approach, the pseudoaneurysm can be punctured directly using a 22-gauge needle. Microcoils can then be injected through the needle directly into the pseudoaneurysm.

## Pulmonary Arteriovenous Fistulae

Pulmonary arteriovenous fistulae, often very large direct communications between the pulmonary artery and pulmonary veins, are most commonly (in 80% to 90% of cases) related to Osler-Weber-Rendu syndrome. Multiple lesions often are present in the lungs and are most commonly bilateral, especially when related to Osler-Weber-Rendu syndrome. This condition often causes severe hypoxia, but most important, bypasses the filter function of the lungs, causing patients to develop cerebral infarcts, abscesses, and systemic abscesses. Up to one-third of patients have evidence of a previous stroke on computed tomography (CT).

The lung lesions can be found incidentally on chest x-rays as often-lobular nodules, and the vessels feeding and draining the lesion easily can be seen. With a 98% sensitivity, a CT scan is the most sensitive way of detecting lung lesions, followed by the 60% sensitivity of pulmonary arteriography. Malformations missed by angiography are usually smaller. If smaller than 3 mm in diameter, the malformations usually are not treated.

These malformations can be grouped as simple or complex. The simple malformations have only a small, single-feeding vessel and are approximately 80% of pulmonary arteriovenous malformations, whereas the complex malformations are 20% and have multiple feeders.

The treatments of choice are surgical resection or endovascular embolization using either metal coils or detachable balloons.

A bilateral pulmonary arteriogram is initially performed, and multiple different projections may be needed. An angiographic catheter then is used to selectively catheterize the feeding arteries. The catheter is advanced beyond any branches to normal lung. Embolization is performed immediately proximal to the venous portion of the malformation. Proper positioning of the catheter in the artery, just proximal to the malformation itself, is ensured. It is important that the first coil placed has a larger diameter than the diameter of the vessel so the coil does not float through into the left side of the vascular system, where it can cause systemic embolization. After placement of the first coil, a contrast injection is performed through the catheter. Additional coils are placed until complete occlusion is acquired. Any other vessels are then treated in the same manner. Detachable balloons have also been used for treatment of this condition.

Good results are to be expected. As a rule, the arterial oxygen saturation increases and the intrapulmonary shunt decreases significantly.

Complications are not frequent. The most common complication is pleuritic chest pain that occurs in approximately 9% of patients and is relieved with nonsteroidal antiinflammatory medications in most cases. Systemic embolization of coils or balloons occurs, and several systemic embolizations have been described.

## Vasculogenic Malformations

The nomenclature for vasculogenic malformations has been unclear, but there is now general agreement on the basic classification of malformations of vascular origin. This classification system is based on endothelial cell characteristics.

Vasculogenic malformations can be divided into two major groups, *hemangiomas* and *vascular malformations*. The basic difference lies in the character of the endothelial cells and, to some extent, the mast cells. Hemangiomas are rarely present at birth, but they become apparent during the first months of life. The hemangiomas undergo rapid proliferation during the endothelial cells' rapid growth phase, which is usually during the first months of life. The hemangiomas then go through an involution phase. Most have reached near-complete resolution by 5 to 7 years of age.

Vascular malformations, on the other hand, are usually present at birth and follow the growth of the child and adult. There is no noted hyperplasia of the endothelial cells over what the growth of the individual implies.

Vascular malformations can be divided into *arterial*, *venous*, *capillary*, *arteriovenous*, and *lymphatic* malformations.

### Diagnosis

The importance of clinical history and examination cannot be overemphasized; the basic diagnosis can easily be based on them.

Plain film radiographs are usually unhelpful; Ultrasound, Doppler, and color Doppler are essential tools in diagnosing vascular malformations. They easily differentiate between high-flow and low-flow malformations. They are also important tools in following up patients after treatment of vascular malformations.

CT and MRI are extremely important modalities for evaluation and diagnosis, as well as for follow-up on vascular malformations. MRI is more helpful in most instances than CT because it easily distinguishes between high-flow and low-flow vascular malformations and makes it easy to assess the anatomic relationship to other structures, such as muscles, abdominal organs, and nerves.

On MRI, high-flow malformations typically demonstrate a signal void on most sequences. Low-flow malformations demonstrate characteristic signal intensity that is greater than skeletal muscle on T1- and T2-weighted images. The signal is less than subcutaneous fat on T1-weighted image and greater than fat on T2-weighted image.

Angiography and venography are the final diagnostic tools in the evaluation and diagnosis of vascular malformations. Arteriography is necessary for preoperative evaluation and preembolization evaluation of certain malformations. Although arteriography is necessary for high-flow lesions, venous malformations usually do not require angiographic evaluation. These malformations would most likely not be visualized on angiography. Venography can be performed by direct puncture of the malformation or by performing a regular venography. This may be important for evaluation of the deep venous system integrity of patients with syndromes such as Klippel-Trenaunay syndrome.

## Individual Vasculogenic Malformations

### Pediatric Hemangiomas

As previously stated, pediatric hemangiomas are benign lesions of childhood that affect 10%

to 12% of children before the age of 1 year. They are often referred to as birthmarks. They are nongenetically transmitted, well-circumscribed lesions that have a size of 0.5 to 5 cm in diameter. The color varies from red to deep blue, depending on how deep under the skin the lesion is. The lesions are firm on touch because of the predominance of stroma (endothelium), a trait that differentiates hemangiomas from other vasculogenic malformations, which usually are soft.

Pediatric hemangiomas usually are not present at birth, and complete regression is to be expected in 70% of the children by age 7 to 9 years; the remaining 30% experience regression by 12 years of age.

The involution starts in the center of the lesion and progresses toward the periphery. It is characterized by decreasing number of endothelial cells that become replaced with fibro-fatty tissue. The mast cell count decreases as well.

*Diagnosis*

On angiography, the lesions are densely stained, have a lobular pattern, and are usually fed by a large artery. On CT, the lesions are well defined, being densely and homogeneously enhanced with contrast.

*Treatment*

Intervention is rarely indicated unless bleeding, ulceration, or infection occurs, which usually can be treated conservatively. Steroid therapy is effective in 30% to 90% of patients.

Radiation therapy has been found to have an effect on hepatic hemangiomas, but because of cancer risk it is only used for exceptional cases. Embolization treatment is only used for those rare occasions in which the flow is so high that congestive heart failure occurs. This is most common for liver hemangiomas.

So-called Kasabach-Merritt syndrome is a self-limited condition of consumptive coagulopathy secondary to platelet trapping. The condition can be subclinical. Occasionally, severe thrombocytopenia occurs. Steroids are the first line of therapy. If the steroid therapy is not effective, embolization can be attempted. Other pharmacologic agents have also been tried, such as α-interferon, cyclophosphamide, and other agents.

### *Arteriovenous Malformations*

Arteriovenous malformations (AVMs) are congenital vascular malformations characterized by hypertrophic inflow arteries that flow into a primitive vascular nidus and drain through tortuous, dilated outflow veins. There is no capillary bed present (Fig. 4.6).

*Symptoms*

The symptoms are usually referable to the anatomic location of the lesion itself. High-output cardiac failure is more likely to occur from a centrally located lesion. Other symptoms are pain, nerve compression, disfiguration, ulceration, bleeding complication, and impaired lymph function.

*Diagnosis*

Duplex examination is the first line of diagnostic work-up, followed by MRI and occasionally CT, especially in malformations located in the pelvis and other deep and less easily accessed sites. Angiography is mainly used to plan for surgical or endoluminal therapy (embolization), but it can also be an important tool in the diagnostic process.

*Treatment*

Surgical resection used to be the main treatment option, but lately, endovascular treatment has become the treatment of choice. The ultimate goal is to destroy the nidus of the malformation. If the inflow arteries are closed without affecting the nidus itself, collateral flow develops, leaving the malformation unchanged and making later endovascular access more difficult. This is because there are now multiple collaterals feeding the malformation.

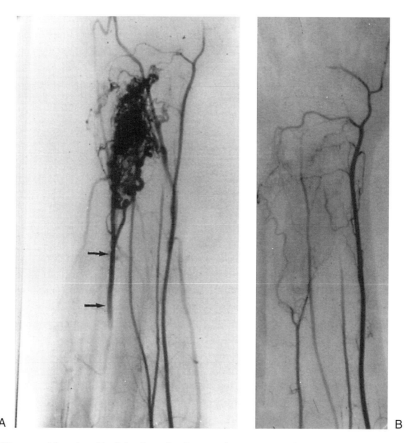

**FIG. 4.6.** 26-year-old male with right dorsal wrist arteriovenous malformation (AVM). **A:** Anteroposterior (AP) left-hand arteriogram. Note AVM nidus fed by retained primitive arterial branch arising from the brachial artery (*arrows*). **B:** AP left-hand arteriogram, 4 months posttherapy. Note obliteration of AVM. (Reproduced from Yakes WF, Parker SH, Gibson MD, et al. Alcohol embolotherapy of vascular malformations. *Semin Intervent Radiol* 1989;6:146–161, with permission.)

Coils and balloons should not be used, because they only close off the inflow arteries and leave the nidus untouched. Gelfoam reabsorbs quickly and is not used. The embolization agents of choice are alcohol, Sotradecol, PVA, and tissue adhesive. Silk also has been used. If alcohol or Sotradecol therapy is selected, superselective catheterization within the vessel immediately feeding the nidus is crucial. If it is not possible, alcohol or Sotradecol should not be used, because either could damage normal tissue. Alcohol appears to cause permanent occlusion and destruction of the nidus without neovascularization and recanalization. There is more experience with the use of alcohol than with Sotradecol, and many regard PVA as a safer agent for this type of embolization. It is nonbiodegradable, which makes it seem like an optimal agent for embolization; however, recanalization with recurrence has been observed within a relatively short time.

The tissue adhesives such as isobutyl cyanoacrylate and *N*-butyl cyanoacrylate have been used for embolizing arteriovenous malformations with good long-term outcomes.

Pelvic malformations are relatively common and difficult to treat. The feeding arteries are usually the internal iliac arteries, the middle sacral artery, and the inferior mesenteric artery or branches. The lesions become clinically apparent during early adulthood, often in conjunction with pregnancy.

These lesions should not be treated unless significant symptoms are present because of the risk of severe complications. Coils and balloons should not be used unless, in the case of preoperative embolization, they are used as preparation for surgical resection.

For preoperative embolization, the approach is similar to that for bone tumors. The feeding vessels should be occluded preferably with coils, PVA, balloons, or Gelfoam, and the procedure should be done within 12 hours before surgery. This is a safe and effective procedure.

### *Arteriovenous Fistula*

Arteriovenous fistulas (AVFs) are congenital or posttraumatic. By nature, they are high-flow lesions that can be difficult to differentiate from AVMs because, with time, the feeding artery and the draining vein often become tortuous and dilated. Multiple arteries and veins can be involved. AVFs are characterized by an artery connecting directly to a draining vein without an intervening capillary or subcapillary bed (Fig. 4.7).

The natural history for AVF is variable. Small, especially traumatic fistulas often close spontaneously, but congenital fistulas may have to be ligated or embolized. The main symptoms are high-output cardiac failure, pain, swelling, and ischemia in the affected limb secondary to steal from the normal tissue.

#### *Treatment*

To prevent collateral development, the feeding artery should be occluded close to the fistula and distal to any normal branches. Surgical ligation may be the only option, especially in fistulas that are composed of a large feeding artery and a large draining vein with a short and wide communication fistula. Long branches that lead to the fistula can easily be occluded by endovascular treatment. Stainless steel coils, detachable balloons, PVA, silk sutures, and tissue adhesives have been successfully used.

Stainless steel coils are recommended for embolization of most arteriovenous fistulas. Small traumatic fistulas, such as after biopsy or drainage in the kidney and liver, can be treated with Gelfoam plugs that will cause occlusion of the bleeding artery long enough to allow closure of the injury.

### *Venous Malformations*

Venous malformations (VMs) are congenital vascular malformations involving only the veins. The artery inflow and capillary bed are normal; therefore, the flow in the feeding arteries is normal.

Superficial lesions or lesions that are accessible to physical examination are usually soft, nonpulsatile, and pliable masses. There is no bruit, thrill, or pulse palpable. If the lesion is placed in a dependent position, or if Valsalva maneuver is performed, the malformation increases in size.

#### *Symptoms*

The symptoms are usually localized to the malformation itself. The malformation grows with the patient. Bleeding is rarely seen unless in conjunction with trauma. The VM can thrombose spontaneously and become tender.

VMs may be asymptomatic; however, they can cause cosmetic deforming and pain, induce neuropathy, ulcerate, bleed, induce changes or abnormal bone growth, cause pathologic fractures, induce thrombocytopenia, and have mixed venous lymphatic components.

#### *Diagnosis*

Duplex examination can reveal a hypoechoic cystic lesion with many connections and digits. There is slow flow or no flow within it, and the lesion is easily compressible. Occasionally, there can be areas of higher echogenicity within the spaces that represent thrombus and prevent full compression.

MRI is preferred over CT and can be very helpful for full evaluation of the malformation.

Angiography is usually of no use. The VM does not fill until during the late phase of the angiography. The feeding arteries are of normal size.

Venography is helpful before surgical or endovascular therapy in many cases and can be performed either as a direct stick or as part of conventional venography; however, the malformation may not show on conventional venography. Although direct puncture gives the best delineation of the malformation, regular venography demonstrates the normal venous anatomy.

*Treatment*

VMs that are asymptomatic or vaguely symptomatic are best left untreated. Surgical treatment has been used with good results in some cases, but the best results have been obtained by percutaneously injecting a sclerosing agent directly into the malformation. Usually an Angiocath, such as used for intravenous infusions, is placed into the VM, using ultrasound

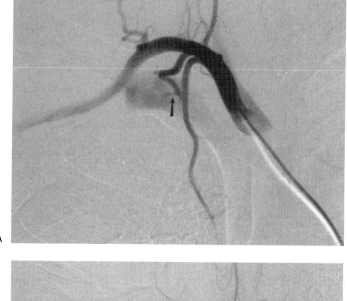

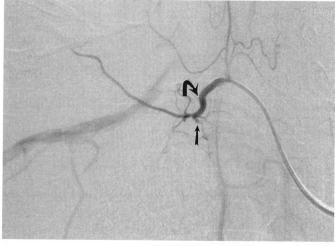

**FIG. 4.7.** 47-year-old woman who complained of right shoulder pain and had a remote history of a fall on the shoulder. Bruit was heard on physical examination, and ultrasound revealed high velocity in the area. **A:** Contrast injection into the right subclavian artery revealed a branch that drains directly into the subclavian vein (*arrow*). **B:** A 5-Fr angiographic catheter is placed selectively into the artery (*bent arrow*) and one 3 mm × 3 cm 0.035-in. coil is placed (*arrow*). The coil shuts off the flow into the vein.

guidance if necessary. Contrast is then injected, and the amount needed to fill the VM is measured. A slightly smaller amount of alcohol or Sotradecol 3%, 3 units of Sotradecol to one unit of contrast, is mixed. The same amount as previously needed to fill the malformation is injected. The fluid is then left in the VM for approximately 20 minutes. To prevent leakage into normal veins, a tourniquet can be applied where needed. This can be done during the test injections, before the injection of the sclerosing agents. After the sclerotherapy, the lesion usually becomes firm and tender. This lasts for at least 2 to 4 weeks, and the final results are usually seen at 8 weeks. Repeat therapy is commonly needed.

Compression bandages may be applied and nonsteroidal antiinflammatory agents can be given for pain. Usually, there is severe local thrombophlebitis at the site. One of the risks is extension of the thrombosis to the deep system. Care must be taken that the agents do not reach the normal venous system.

### Intramuscular Venous Malformations

Intramuscular venous malformations (IMVMs), previously named *intramuscular hemangiomas*, are an uncommonly seen subgroup of VMs. IMVMs are histologically identical to VMs and are confined to the muscles, but they may extend into nearby tissues. The extremities are most commonly affected. IMVMs differ clinically from VMs. IMVMs present at 20 to 30 years of age and may present earlier. All patients present with a growing palpable mass with or without pain.

MRI, duplex examination, and venography are the main modalities of work-up. On MRI, these lesions share the same characteristics as VMs. There is increased signal on long TR/TE. Antegrade (ascending) venography rarely defines the IMVMs. Direct puncture contrast study better outlines the various vascular compartments that are usually smaller than can be seen with VM.

On arteriography, IMVMs frequently have hypertrophied arterial inflow with a dense tissue stain but without arteriovenous shunting. Lack of rapid AV shunting, despite the hypertrophied inflow arteries and dense stain, is a typical finding.

If surgical resection is feasible from a technical and functional standpoint, preoperative transarterial PVA embolization of the feeding arteries can be performed to minimize surgical blood loss. If surgical resection cannot be performed, direct ethanol sclerotherapy can be done, as for any other VM, to shrink the abnormal venous mass.

### Lymphatic Malformations

Lymphatic malformations were previously referred to as *cystic hygroma*. They are identical histologically to VMs, except they contain lymph fluid rather than blood. The size of the vascular space can vary from large to small.

On MRI, lymphatic malformations have the same characteristics as VMs, increased signal on T2 sequences. The best sequences are T1 and T2, as well as fast spin-echo T2 fat suppression.

Percutaneous ethanol sclerotherapy gives good results, and there is rarely a concern of communication to other vascular compartments.

## SUGGESTED READINGS

Bookstein JJ, Cholasta EM, Foley D, Walters JS. Transcatheter hemostasis of gastrointestinal bleeding using modified autogenous clot. *Radiology* 1974;113:227–231.

Carey LS, Grace DM. The brisk bleed: controlled by arterial catheterization and Gelfoam plug. *J Can Assoc Radiol* 1974;25:113–115.

Cho KJ, Williams DM, Brady TM, et al. Transcatheter embolization with sodium tetradecyl sulfate: experimental and clinical results. *Radiology* 1984;153:95–99.

Cope C, Zeit R. Coagulation of aneurysms by direct percutaneous thrombin injection. *AJR* 1986;147:383–387.

Cragg AH, Rosel P, Rysavy JA, et al. Renal ablation using hot contrast medium: an experimental study. *Radiology* 1983;148:683–686.

Dawbain G, Lussenhop AJ, Spence WT. Artificial embolization of cerebral arteries: report of use in a case of arteriovenous malformation. *JAMA* 1960;172:1153–1155.

Formanek A, Probst P, Tadavarthy SM, Castañeda-Zúñiga WR, Amplatz K. Transcatheter embolization in the pediatric age group. *Ann Radiol* 1979;22:150–158.

Gianturco C, Anderson JH, Wallace S. Mechanical devices for arterial occlusion. *AJR* 1975;124:428–435.

Hayakawa K, Tanaka F, Torizuka T, et al. Bronchial artery embolization for hemoptysis: immediate and long-term results. *Cardiovasc Intervent Radiol* 1992;15:154–159.

Hunter DW, King NJ III, Aeppli DM, et al. Spermatic vein occlusion with hot contrast material: angiographic results. *J Vasc Intern Radiol* 1991;2:507–515.

Jander HP, Russinovich NAE. Transcatheter Gelfoam embolization in abdominal, retroperitoneal, and pelvic hemorrhage. *Radiology* 1980;136:337–344.

Keller FS, Rosch J, Bird CB. Percutaneous embolization of bony pelvic neoplasms with tissue adhesive. *Radiology* 1983;147:21–27.

Radanovic B, Simunic S, Stojanovic J, et al. Therapeutic embolization of aneurysmal bone cyst. *Cardiovasc Intervent Radiol* 1990;12:313–316.

Rosch C, Dotter CT, Brown MJ. Selective arterial embolization. *Radiology* 1972;102:303–306.

Salam TA, Lumsden AB, Martin LG, Smith RB III. Nonoperative management of visceral aneurysms and pseudoaneurysms. *Am J Surg* 1992;164:215–219.

Stoll JF, Bettman MA. Bronchial artery embolization to control hemoptysis: a review. *Cardiovasc Intervent Radiol* 1988;11:263–269.

Tadavarthy SM, Knight L, Ovitt TW, et al. Therapeutic transcatheter arterial embolization. *Radiology* 1974;112:13–16.

Tadavarthy SM, Moller JH, Amplatz K. Polyvinyl alcohol (Ivalon): a new embolic material. *AJR* 1975;125:609–616.

Upton J, Mulliken JB, Murray JE. Classification and rationale for management of vascular anomalies in the upper extremity. *J Hand Surg* 1985;6:970–975.

Wallace S, Gianturco C, Anderson JH, et al. Therapeutic vascular occlusion utilizing steel coil technique: clinical applications. *AJR* 1976;127:381–387.

Yakes WF, Luethke JM, Merland JJ, et al. Ethanol embolization of arteriovenous fistulas: a primary mode of therapy. *J Vasc Intern Radiol* 1990;1:89–96.

# 5

# Percutaneous Treatment of Liver Tumors

Zhong Qian and Yuchen Jia

## CHEMOEMBOLIZATION

*Hepatocellular carcinoma* (HCC) is one of the most common malignancies in the world. It is often associated with chronic liver disease, such as cirrhosis and hepatitis B or C. HCC is responsible for approximately 1,250,000 deaths each year worldwide. Although surgical resection is the only potentially curative treatment for HCC, this option is not always possible because of too far advanced tumor and severe underlying cirrhosis.

Because HCCs are primarily supplied with blood from the hepatic artery, *transcatheter arterial chemoembolization* (TACE) has been accepted as the treatment of choice for unresectable HCC. TACE is indicated for nonsurgical candidates who have a single lesion larger than 4 cm in diameter or patients who have multicentric lesions and are in Child's class A or B. The procedure can also be used as a preliminary step before liver transplantation for HCC or in combination with percutaneous ethanol injection (PEI). Patients with tumors limited to a single or a few segments, with tumors located in the subcapsular area, or with poor liver function can be treated with segmental embolization technique, in which anticancer drugs and embolic agents are injected into the tumor-bearing segment or subsegment only. Tumor thrombosis in the major portal veins used to be a contraindication to TACE.

Experience has indicated that TACE can be performed safely in patients with portal vein tumor thrombosis as long as there is adequate collateral flow around tumor thrombus and sufficient hepatic reserve. We recommend that patients with HCC invading the major portal vein be treated with the segmental technique if they are eligible for TACE. Patients who have uncontrolled coagulopathy or who cannot tolerate contrast material should be excluded from the procedure. TACE has a limited role in the treatment of borderline lesions, hypovascular tumors, and early-stage HCCs. Preoperative TACE is not recommended in patients with resectable HCCs.

The purpose of PEI is to destroy the tumor and a margin of adjacent tissue, which subsequently is replaced by a fibrous scar. PEI is performed mainly in patients with small HCC, such as a single lesion smaller than 4 cm in diameter and multiple lesions no more than 3 in number and smaller than 3 cm in diameter. PEI is particularly suitable for patients who have underlying cirrhosis or a hypovascular neoplasm in the liver. Patients with ascites or uncontrolled coagulopathy are contraindicated for the procedure. PEI may not be sufficient to eliminate all tumor tissue in some cases; therefore, the optimal outcome may be achieved by using PEI in combination with TACE. The 5-year survival rate on small HCCs treated with PEI reportedly has been similar to that with surgical resection. However, PEI proves less effective for treatment of liver metastases, whose internal consistency tends to be much firmer than HCCs. The firm metastatic lesions lessen homogenous diffusion of injected ethanol.

The beneficial effects of intraarterial I-131 iodized oil radiotherapy on HCC have been demonstrated by many authors. Indications for use of this modality include (a) an individual tumor smaller than 4 cm in diameter, (b) hypervascular tumor(s) with a dense homogeneous stain on angiogram, and (c) HCC with portal vein thrombosis.

I-131 iodized oil infusion is not suitable for patients who have arteriovenous shunts in the liver or have hypovascular neoplasm.

Local-regional treatment appears to be less effective for hepatic metastases from lung, breast, gastric, or pancreatic cancer. Favorable responses have been obtained in patients with hepatic metastasis from colorectal carcinoma treated with regional infusion of chemotherapeutic agent; however, a long-term survival rate has not been achieved. To date, chemoembolization has been effective in a relatively small number of patients with hepatic metastases.

## TRANSCATHETER CHEMOEMBOLIZATION

### Preparation

After being sedated, the patient is given local anesthesia to the groin. A 5-Fr diagnostic catheter is introduced into the celiac artery through the femoral access. A preprocedural angiography should be performed to delineate the hepatic arterial anatomy, tumor vessels, and portal flow as well as aberrant arteries or collaterals, if any. Placing the catheter with the tip in the feeding artery as close to the tumor as possible is the most important technical factor to maximize the effect on the tumor and to minimize the damage to nontumor parenchyma.

### Preparation of Anticancer Drug and Embolic Agent

Various combinations of anticancer drugs and embolic agents have been tested. Currently, the most popular formula is drug-in-iodized oil emulsion followed by use of Gelfoam particles (Upjohn, Kalamazoo, MI). Iodized oil serves as a vehicle to carry anticancer agents and may enhance the therapeutic effect by prolonged drug delivery (owing to its retention in the tumor for a substantial period of time). The most common therapeutic drugs include doxorubicin hydrochloride (Adriamycin), mitomycin (Mutamycin, Mitomycin), cisplatin (Platinol-AQ), and epirubicin hydrochloride (Table 5.1). Among them, Adriamycin is recommended as the first choice for TACE, owing to its having the best response rate.

**TABLE 5.1.** *Drugs used in chemoembolization*

| Drug | Commonly used dosage ($mg/m^2$) |
|---|---|
| Adriamycin | 20–40 |
| Mitomycin | 10–20 |
| Cisplatin | 60–100 |
| Epirubicin | 50–60 |

The emulsion is prepared by mixing iodized oil (Lipiodol, Andre Guerbet, Aulnay-sous-Bois, France) and anticancer drug in 2 to 3 mL of water-soluble contrast medium in two syringes that are connected by a three-way stopcock. The volume of iodized oil used in a patient depends on tumor size, vascularity, and liver reserve. In general, the volume of iodized oil ($D$ mL) is in proportion to the tumor diameter ($d$ cm). The best long-term results reportedly have been obtained when $D$ was equivalent to 1 to 1.5 times $d$. A larger volume of iodized oil does not necessarily yield a better long-term outcome and may impair liver function.

Injection of the emulsion should be slowly performed and closely monitored under fluoroscopy. If the small portal veins in the peripheral area of the liver are visualized during injection, emulsion administration should halt, even if the amount of administrated emulsion is less than the minimal-allowed dose level. Once injection of the emulsion is complete, the hepatic artery is embolized by slow injection of Gelfoam particles (1 $mm^3$) or powder soaked with contrast medium to eliminate washout of the chemotherapeutic agents.

### Technique of Segmental Transcatheter Arterial Chemoembolization

Super-selective segmental catheterization can be achieved by using a coaxial catheter system consisting of a 5.5- to 6.5-Fr guiding catheter and a 2.5- to 3-Fr microcatheter, such as the Tracker-18 infusion catheter (Target Therapeutics, Fremont, CA). All individual segments should be catheterized and embolized separately (Fig. 5.1). The formula for emul-

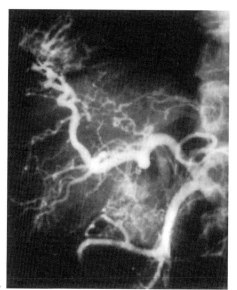 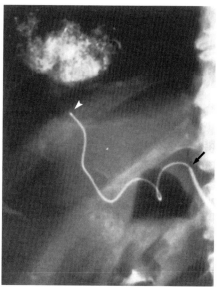

**FIG. 5.1.** Segmental TACE for recurrent HCC after segmentectomy of the lateral segment. **A:** Hepatic angiogram. Hypervascular tumor is seen in the anterior-superior segment. **B:** A Tracker-18 catheter introduced into the feeding artery. Arrow indicates the tip of the parent catheter, and arrowheads show the tip of the coaxial catheter. Iodized oil is well accumulated in the tumor. (From Uchida H, Matsuo N, Sakaguchi H, et al. Segmental embolotherapy for hepatic cancer: keys to success. *Cardiovasc Intervent Radiol* 1993;16:67–71, with permission.)

sion volume used in segmental TACE is based on the size of the tumor. When tumor diameter is larger than 5 cm, $D$ equals or is slightly smaller than $d$; when tumor diameter is smaller than 5 cm, $D$ is slightly larger than $d$. The basic rules for injection of emulsion also apply to segmental TACE. Once retention of emulsion is observed in the tumor and feeding artery, 1 to 2 mL of 2% lidocaine may be injected through the catheter. It allows an additional, small amount of emulsion to be added after dilatation of the artery. The advantages of segmental TACE technique include less damage to nontumor liver parenchyma and a favorable long-term outcome with the initial session. The major disadvantage is that detection of satellite nodules by means of iodized oil accumulation in the entire liver is not possible; however, it can be compensated for by injecting a small dose (2 to 3 mL) of iodized oil via the proper hepatic artery after treatment.

### Postprocedural Care and Follow-Up

Angiography is performed after embolization to assess the extent of intrahepatic vascular occlusion and the flow in the adjacent arteries. Patients should be monitored and hydrated for at least 24 hours after the procedure. Laboratory tests, including serum α-fetoprotein level and liver function tests, are also followed up weekly for the first month and on a biweekly basis thereafter. Outcome of treatment is evaluated by computed tomography (CT) or ultrasound on a monthly basis. TACE can be repeated every 2 to 3 months. Some authors suggest that it be repeated only when recurrence is detected after initial treatment to minimize damage to noncancerous liver tissue.

### Prevention and Management of Complications

Although major complications are uncommon after TACE, almost all patients experience some form of postembolization syndrome, which is characterized by nausea, vomiting, abdominal pain, loss of appetite, and daily intermittent fevers. These symptoms are generally mild and self-limiting. Treatment of these problems is supportive, consisting of intravenous hydration, antiemetics, and analgesia. Mild to moderate elevations of serum aminotransferase and lactate dehydrogenase are also frequently noted; however, liver function usually recovers within a few weeks.

The majority of serious complications are usually related to the use of Gelfoam powder or an overdose of iodized oil. Major complications include the following:

1. Persistent fever is usually associated with tumor necrosis. It can be managed with oral indomethacin. Unexpectedly high and prolonged fever in patients with a small tumor may indicate septic complications.

2. Acute cholecystitis or gallbladder ischemia occurs in 10% of patients having TACE. Its manifestations include right upper quadrant pain, leukocytosis, and persistent fever. Most of them do not require special treatment. The key to avoiding these complications is placing the tip of the catheter beyond the cystic artery.

3. Diffuse ulceration or bleeding of the stomach or gut is caused by nontarget embolization. Most of these cases can be managed with medical treatment; however, some severe cases may require surgical intervention. It can be prevented by avoiding reflux of emulsion and Gelfoam particles to the nontarget vessels.

4. Hepatic failure, the most devastating complication after TACE, usually occurs in cases of severe cirrhosis and has a high mortality rate. It manifests as jaundice, ascites, or encephalopathy. TACE should be cautiously performed with a super-selective technique and a reduced amount of chemoembolic agents in patients with compromised liver reserve. In patients with biliary obstruction, TACE should be performed after the biliary system has been adequately decompressed.

5. Septicemia and liver abscess have been noted in 6% to 11% of patients who underwent TACE, especially in those who had preexisting ascites, portal vein obstruction, or biliary obstruction. Prophylactic antibiotic coverage may reduce incidence of infectious complications in these high-risk patients.

6. Pulmonary embolization of iodized oil is caused by emulsion passing through normal vasculature or arteriovenous shunts in the liver. Occurrence of this complication is associated with a high dose of iodized oil used for TACE. It is recommended that no more than 20 mL of iodized oil be injected in each session.

7. Procedure-related mortality occurs in approximately 2.5% of patients, primarily owing to hepatic failure. To reduce serious complication and mortality rates, it is important to recognize predisposing factors and to strictly follow the guidelines for TACE.

## PERCUTANEOUS ETHANOL INJECTION

### Technique

Patients should fast for at least 6 hours before PEI. The procedure can be carried out on an outpatient basis. Local anesthesia (2% lidocaine) should be given at the insertion site. Under ultrasound guidance, a 21-gauge, 15-cm Chiba needle is used to puncture the tumor at one or more locations while the patient holds his or her breath.

The volume of 99.5% ethanol used to complete the whole treatment is expressed as the following equation:

$$V = 4/3\pi (r + 0.5)^3$$

$V$ (mL) is the volume of ethanol, and $r$ is the radius (cm) of the tumor. The additional 0.5 cm is provided as a safety margin, based on the concept that a certain amount of the adjacent tissue should be ablated to ensure complete tumor necrosis. Injection of the total calculated volume should be carried out in several sessions. In general, PEI is repeated two or three times per week. Three to six sessions are needed for treatment of HCC smaller than 2 cm in diameter, and seven to

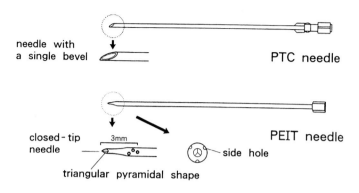

FIG. 5.2. Diagram of a PEI (PEIT) needle with multiple side holes and a conventional percutaneous transhepatic cholangiogram (PTC) needle. (From Akamatsu K, Miyauchi S, Ito Y, et al. Development and evaluation of a needle for percutaneous Ethanol injection therapy. *Radiology* 1993;186: 284–286, with permission.)

nine sessions are required when tumor diameter is larger than 3 cm. The volume and number of injections can be tailored to number, location, and size of tumors and patient's compliance.

Ethanol should be injected slowly and with great caution to avoid entering vessels or bile ducts. Injection to the deepest portion of tumor is made first, then to the central portion, and finally to the superficial portion. Ethanol diffusion can be demonstrated on real-time sonogram as a hyperechoic patch. Owing to this reason, the needle cannot be reinserted in the same session. For treatment of a large tumor, two or more needles are usually inserted at different sites. Ethanol is injected through these needles in the same session. Appearance of homogeneous hyperechogenicity, along with reduction in tumor size or complete disappearance of tumor nodule, is considered the endpoint of the procedure. To eliminate reflux of ethanol through the needle tract, the needle should be left in place for a few minutes once injection is completed.

Uneven distribution of ethanol intralesional injection is a major shortcoming of this technique. This problem may be compensated for by using a needle with multiple side holes (Hakko Electric Machine Works, Nagano-ken, Japan) (Fig. 5.2) and by using multiple needles for intra- and perilesional injections (Fig. 5.3).

### Complications

PEI is safe, simple, and effective, with a low procedure-related mortality. Minor complications occur frequently (e.g., 25% of patients) and include transient pain, fever, and mild alcohol intoxication. Management of these complications is supportive. Serious complications are rare (e.g., 1% to 3% of patients) and include tumor seeding along the needle track, partial chemical thrombosis in the portal vein branch, right pleural effusion, ascites, jaundice, pneumothorax, hemobilia, vasovagal reaction, transient hypotension, and intraperitoneal hemorrhage.

### Postprocedural Care and Follow-Up

Patients should be observed for 1 to 2 hours after the procedure. After a treatment cycle, a percutaneous biopsy can be performed to assess the efficacy of PEI. If no viable tumor cell is found, patients should be followed up monthly and then quarterly. Incomplete tumor necrosis or recurrence can be treated by repeating PEI. Because high recurrence rates (in 54% to 65% of patients) have been reported over a period of 2 to 3 years after successful PEI, regular follow-up is considered one of the most important factors influencing the prognosis of the treated patients. Common modalities used to monitor patients after PEI include ultrasound and laboratory tests (serum levels of $\alpha$-fetoprotein). MR imaging and contrast-enhanced CT have also been used to monitor the ablated lesions.

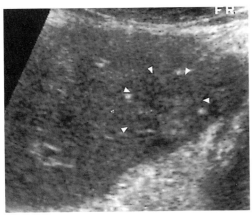 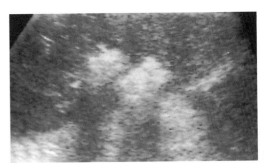

**FIG. 5.3.** PEI for small HCC in a patient with chronic hepatitis C. **A:** Approximately 2-cm mixed echogenic mass (*arrowheads*) is accidentally identified in the lateral segment by ultrasound for screening examination. **B:** Ethanol is injected into the tumor and diffused as demonstrated on real-time sonogram as hyperechoic patch. (Courtesy of Y. Nakajima, M.D.)

## INTRAARTERIAL I-131 IODIZED OIL RADIOTHERAPY

Preprocedure preparation is similar to that for TACE. Patients should be given Lugol's solution orally (2 drops 3 times a day) 1 day before the procedure and 3 days thereafter to block thyroid uptake of radioiodine.

I-131 iodized oil can be prepared using an isotopic exchange technique with a radiochemical purity of more than 95%. Its commercial product is also available as I-131 Lipiodol (Lipiocis, Cisbio International, Saclay, France) in a 2-mL vial with an activity of 15 to 40 mCi (550 to 1,480 MBq). It can be diluted with nonradioactive iodized oil to the desirable volume.

The total volume of I-131 iodized oil is determined by tumor size, vascularity, and number of tumors. In most cases, the volume used for infusion ranges from 2 to 5 mL. The therapeutic dose of I-131 is generally between 40 and 60 mCi for the initial session. The dose for repeat treatment can increase to 100 mCi (3.7 GBq). An effective half-life of I-131 in tumor is approximately 4 to 6 days. Slow injection of I-131 iodized oil through a catheter placed in the proper hepatic artery is crucial to achieve technique success. After completion of infusion, plain abdominal radiographs should be taken to confirm distribution of the infused I-131 iodized oil in the liver. Patients must be isolated in the hospital for 3 to 7 days for the purpose of radiation protection.

Neither remarkable side effects nor serious complications have been reported with this technique. Most patients experience asthenia for 1 to 3 weeks, but they do not require special medical attention.

Evaluation of biodistribution can be made with scintigraphy or a liver single-photo emission CT in 2 to 4 days after the procedure to allow for assessment of radioactivity uptake by the tumor. Like other therapeutic protocols, tumor response is monitored by CT, ultrasound, and measuring serum $\alpha$-fetoprotein level.

## SUGGESTED READINGS

Boland GW, Lee MJ, Dawson SL, et al. Percutaneous injection of ethanol in a patient with a solitary hepatocellular carcinoma. *AJR Am J Roentgenol* 1993;161:1071–1077.

Chung JW, Park JH, Han JK, et al. Hepatic tumors: predisposing factors for complications of transcatheter oily chemoembolization. *Radiology* 1996;198:33–40.

Kajiya Y, Kobayashi H, Nakajo M. Transarterial internal radiation therapy with I-131 Lipiodol for multifocal hepatocellular carcinoma: immediate and long-term results. *Cardiovasc Intervent Radiol* 1993;16:150–157.

Matsui O, Kadoya M, Yoshikawa J, et al. Small hepatocellular carcinoma: treatment with subsegmental transcatheter arterial embolization. *Radiology* 1993;188:79–83.

Pentecost MJ, Daniels JR, Teitelbaum GP, et al. Hepatic chemoembolization: safety with portal vein thrombosis. *J Vasc Interv Radiol* 1993;4:347–351.

Raoul JL, Guyader D, Bretagne JF, et al. Randomized controlled trial for hepatocellular carcinoma with portal vein thrombosis: intra-arterial iodine-131-iodized oil versus medical support. *J Nucl Med* 1994;35:1782–1787.

Takayasu K, Wakao F, Moriyama N, et al. Response of early-stage hepatocellular carcinoma and borderline lesions to therapeutic arterial embolization. *AJR* 1993;160:301–306.

Tanikawa K. Non-invasive loco-regional therapy for hepatocellular carcinoma. *Semi Surg Oncol* 1996;12:189–192.

Uchida H, Matsuo N, Sakaguchi H, et al. Segmental embolotherapy for hepatic cancer: keys to success. *Cardiovasc Intervent Radiol* 1993;16:67–71.

# 6

# Gastrointestinal Bleeding

Osarugue A. Aideyan and Haraldur Bjarnason

## A. DIAGNOSTIC APPROACH

The diagnosis of gastrointestinal (GI) bleeding is clinical in most cases. The exact location is often reached with nuclear medicine and angiographic study. The treatment in many cases can also be performed with angiographic techniques. As endoscopic techniques have developed, the need for radiologic intervention has declined; however, it still plays an important role in the evaluation and treatment of the condition. This chapter presents the diagnostic alternatives, indications, and therapeutic alternatives.

### DIAGNOSTIC TOOLS

#### Nuclear Medicine Diagnosis

Nuclear medicine imaging plays a role in the noninvasive detection and localization of GI bleeding. Two radionuclide imaging agents, technetium Tc 99m sulfur colloid (SC) and technetium Tc 99m–labeled red blood cells (RBCs), are used. Technetium Tc 99m–labeled RBCs are used more often, and their sensitivity for detection of GI bleeding is superior to that of technetium Tc 99m SC. With technetium Tc 99m–labeled RBCs, imaging can be carried out to 24 hours after injection, which is an advantage considering the intermittent nature of most GI bleeding.

Detection of lower GI bleeding with radionuclide is more sensitive than for upper GI bleeding. Endoscopy is usually performed for evaluation of upper GI bleeding; however, radionuclide imaging may be useful if endoscopy is unsuccessful for technical reasons or if a bleeding source cannot be found.

Detectable bleeding rates by radionuclide imaging in animal studies are as low as 0.04 mL per minute and for angiography as low as 0.5 mL per minute. These figures do not reflect bleeding rate sensitivities in humans; however, it is accepted that radionuclide imaging is more sensitive than angiography for the detection of GI bleeding. If radionuclide imaging is negative, angiography is unlikely to be positive.

#### Angiographic Diagnosis

Angiographic evaluation is based on the visual detection of contrast leaking out of the injured vessel into the lumen of the intestine. For most indications other than GI bleeding, diagnostic angiography requires contrast injection for 2 seconds. When looking for the bleeding site, contrast injections into either the celiac axis or the superior or inferior mesenteric artery usually must last for at least 4 seconds. In 4 seconds, enough contrast can leak out of the lesion to be visible. The injection rate is usually half of the flow rate in the vessels. For the celiac axis, the injection rate is 8 to 10 mL per second; for the superior mesenteric artery, the injection rate is 6 to 8 mL per second. The inferior mesenteric artery needs significantly smaller injections. That artery requires approximately 3 mL per second. Filming was traditionally carried out with cut-films, which are no longer used. Digital subtraction technique is now used. The filming should be rapid for the first 2 to 4 seconds, approximately three frames per second. In the following 10 to 15 seconds, one frame per second can be taken. From then on, one frame is taken every 2 to 3 seconds for a total of 25 seconds. It is imperative that the patient keeps very still during this period. Glucagon, 0.5 mg, can be given intravenously just before the injection to decrease bowel movement. Bowel movement can cause artifacts with subtraction angiography, which can imitate bleeding very closely.

Usually, it is known before angiography whether the bleeding comes from the upper or the lower GI tract. The nuclear medicine scan helps to distinguish between those sites, although it cannot always locate the bleeding site more closely.

Based on the nuclear medicine scan, the artery to be studied is selected. The celiac axis feeds the stomach and the duodenum, whereas the superior mesenteric artery feeds the entire small bowel, right colon, and transverse colon. The inferior mesenteric artery feeds the left colon, sigmoid colon, and the upper portion of the rectum. It is important to be aware of blood supply to the lower rectum by the inferior and middle hemorrhoidal arteries, which originate from the internal iliac arteries rather than the mesenteric arteries.

Different types of angiographic catheters can be used. The so-called sidewinder catheter is popular. As soon as it is engaged in the vessel, it is stable. The sidewinder is especially helpful if microcatheters are used through it; however, it is often difficult to form the catheter. It is usually formed in the aortic arch, and complications have occurred during that procedure. More recently, a smaller type of sidewinder catheter has been marketed. This catheter (SOS, Angiodynamics, Gainesville, FL) can be formed in the aorta below the arch; therefore, less severe complications are expected.

Cobra catheters are also commonly used but are not as stable in the vessel, especially if microcatheters are used.

## ENDOVASCULAR TREATMENT OPTIONS

### Pharmacotherapy

After the bleeding site is identified, an attempt to stop the bleeding, using endovascular methods, can be made. Transcatheter treatment for GI bleeding includes infusion of pharmacologic agents and deployment of embolic materials.

The main pharmacologic agent used is vasopressin. This potent vasoconstrictor causes contraction of smooth muscles in arterioles, venules, and the bowel wall. Intraarterial vasopressin infusion is complex, particularly in areas with multifocal supply, because flow from other branches may dilute or wash out the vasopressin. The infusion is given either into the target vessel or into the main artery, such as the superior mesenteric artery. The current recommended dose is 0.2 units per minute with no loading dose. Infusion is then given for 20 minutes and an angiogram is repeated. If there is no vasoconstrictive effect seen and the bleeding persists, the vasopressin infusion rate is increased to 0.4 units per minute. Higher doses are seldom required and carry a higher rate of ischemic complications. The patient is sent to the floor with continuous infusion for the following 12 hours. The next follow-up depends somewhat on the clinical status; however, if there is no clinical bleeding at 12 hours, angiography is repeated. If no bleeding is detected, the vasopressin is slowly withdrawn over a 12- to 24-hour period by halving the dose every 12 hours. The arterial catheter is left in place and heparinized saline (Heparin) is infused at 3 to 10 mL per hour at 5 units per mL for up to 12 hours. If rebleeding occurs, vasopressin can be restarted. If bleeding is arrested, the catheter can be removed. Vasopressin infusion is not as popular a treatment for GI bleeding as it used to be. Embolization is now the treatment of choice if endovascular options are discussed. If the bleeding site cannot be clearly identified, but it is known to be in a certain area, such as the superior mesenteric artery distribution, vasopressin infusion can be given into the vessel on a trial basis. Systemic infusion of vasopressin is not as effective as local infusion into the target vascular bed.

# B. EMBOLIZATION AND INFUSION THERAPY

## EMBOLOTHERAPY

GI bleeding can also be controlled by embolization. Multiple different agents are available for embolization, but only a few of them are used for treatment of GI bleeding. Those can be categorized roughly as either *temporary* or *permanent embolization agents.*

Bleeding from transient lesions, such as ulcers, erosions, diverticular bleeding, and traumatic tears, should be treated with temporary materials, which permit healing but allow for later recanalization of the embolized vessel. In this situation, decrease the blood pressure (pressure head) in the bleeding vessel temporally until the lesion has healed.

Bleeding from tumors, arteriovenous malformations, and large areas of angiodysplasia that do not heal require permanent occlusion agents. Those lesions are also usually more properly treated with surgical excision.

The temporary embolic agent most commonly used is Gelfoam. Autologous blood clot was popular for a period of time but has fallen in disfavor because of rapid recanalization. Permanent embolic agents include polyvinyl alcohol (Ivalon), coils, balloons, polymers, glues, alcohol, and hot contrast medium. The first two are characteristically used in GI bleeding; the last four are not, because the far distal occlusion they produce may induce wall ischemia and necrosis. Small Ivalon particles (particle size of 100 to 500 m in diameter) provide good end-organ embolization. Particles 1 to 2 mm in diameter can be used for occlusion of small feeding branches and in bleeding resulting from hypertrophied feeding arteries (e.g., from tumors). For large arteriovenous malformation, particles (1 to 1.5 mm) can also be used. The catheter is placed securely into the bleeding vessel, and the embolic agent is deployed until stasis is achieved. Patients with coagulopathy are more likely to have unsuccessful embolotherapy. A postembolization angiogram is always obtained.

## Anatomic Considerations for Embolization of Gastrointestinal Bleeding

GI bleeding is categorized as either upper or lower bleeding. The differentiation point is at the junction of the duodenum and the jejunum.

Organs with multiple blood supply, such as the stomach, duodenum, liver, and pancreas, are less likely to infarct after embolization. Subsequently, it is more difficult to control bleeding in those organs because of the rich collateral flow. In these organs, particulate embolization is more effective at controlling bleeding than pharmacologic treatment. It has the added advantage over infusion of a rapid single session intervention to stop bleeding.

Lower GI bleeding, namely from small bowel and colon, is managed differently because of less collateral flow; however, this may be changing with super-selective catheterization. Vasopressin infusion was the preferred treatment over particulate embolization in lower GI bleeding because of fewer complications. This has changed in favor of embolization.

## Preprocedure Care

Before performing any embolization procedure, it is imperative to be very familiar with the patient's previous medical history. Previous surgical history is important because any resection or other surgical intervention on the organ decreases the collaterals, thereby increasing the risk of ischemic complications.

1. Collect all available data, including barium and radionuclide studies.
2. Check coagulation profile. Correct as needed, except if hemorrhage is life threatening.
3. Obtain patient consent for diagnostic and possible therapeutic embolization.
4. Avoid sedative medications that may lower blood pressure, as lowered blood pressure can transiently stop or slow bleeding.

## INDICATIONS FOR INTERVENTIONS

Listed here are the major reasons for GI bleeding and endovascular intervention. For logistical reasons, the upper and the lower GI tracts are grouped separately. There is a short discussion of the individual indication and the proper treatment of each.

### Upper Gastrointestinal Bleeding

In the case of upper GI bleeding, the causes are generalized as gastritis and gastroduodenal ulcers in the majority of cases.

#### *Gastritis*

*Gastritis* is caused by drug or alcohol irritation; sometimes, however, no cause can be indicated (idiopathic causes). In gastritis, multiple sites of extravasation may be seen in the lumen of the stomach (Fig. 6.1). Evaluation should include a selective angiography of the left gastric artery to supplement a celiac artery angiography. If endoscopy shows bleeding from the distal stomach, selective gastroduodenal artery angiography may be necessary. After extravasation is identified, intraarterial vasopressin can be given to control diffuse bleeding. The infusion rates and doses outlined under Pharmacotherapy are sufficient in most cases. Extending vasopressin infusion time 24 to 36 hours after a response has been achieved before starting to decrease the dose makes rebleeding less likely. Embolization with Gelfoam particles is safe. Embolization may now be the preferred treatment for this condition. The left gastric artery can be embolized. If Gelfoam or larger polyvinyl alcohol particles are used, the same is true for the gastroepiploic artery without the risk of necrosis. The stomach has an extensive collateral supply and tolerates embolization of those vessels well. Previous surgery of the organ can make it more susceptible to ischemia.

#### *Duodenal Peptic Ulcer Disease*

Bleeding from the pylorus or duodenum is best managed with embolization treatment. The success of vasopressin infusion in this location is low at 35% to 40%, whereas embolotherapy has higher success rate at 65% to 70% (after unsuccessful vasopressin infusion). Embolization is traditionally performed with the so-called sandwich method. A microcoil is placed distal to the bleeding, and Gelfoam is placed more centrally. Finally, a microcoil is placed central to keep the Gelfoam in place. By embolizing on both sides, flow from the gastroepiploic artery is also eliminated. If the bleeding is not from the gastroduodenal artery itself, but from a small branch, embolization of that branch can be performed selectively, using Gelfoam.

#### *Gastric Ulcers*

Bleeding is usually caused by erosion into a small artery in the wall of the stomach. In addi-

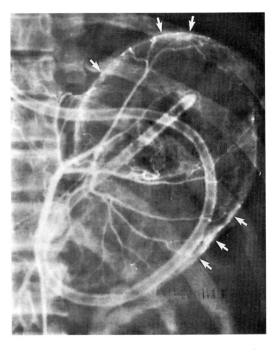

**FIG. 6.1.** Selective left gastric artery angiography in patient with diffuse mucosal bleeding in erosive gastritis demonstrated by endoscopy. Note diffuse mucosal stain (*arrows*) suggestive of mucosal extravasation of contrast medium. (From Castañeda-Zúñiga WR. *Interventional radiology*, 3rd ed. Baltimore: Williams & Wilkins, 1997, with permission.)

tion to celiac axis angiography, the angiographic work-up should include selective angiography of the left gastric, gastroduodenal, and superior pancreaticoduodenal arteries. In the case of a bleeding gastric ulcer, gastric vascularity is normal and bleeding is usually at one or two locations. Vasopressin infusion usually controls bleeding from a small branch vessel. In cases of erosion into a larger vessel (e.g., gastroduodenal or left gastric artery), vasopressin usually fails. Again, embolization is the preferred method and particulate material (e.g., Gelfoam) is used. Vasopressin infusion and particularly embolization of the left gastric, gastroduodenal, or pancreaticoduodenal arteries carry minimal morbidity because of the rich collateral flow in the stomach. However, embolization should be approached cautiously in patients with previous gastric surgery because of the risk of gastric infarction.

### *Mallory-Weiss Esophageal Tears*

Arterial supply to the esophagus is complex, with considerable variability. Primary supply of the cervical portion is from terminal branches of the inferior thyroid artery, which gives rise to multiple ascending and descending esophageal branches. Accessory cervical esophageal arteries may originate from the subclavian, carotid, and vertebral arteries, as well as from the ascending pharyngeal and caudal cervical trunk. The thoracic segment receives arterial blood from the bronchial arteries, the aorta, and the right intercostal arteries.

There are two direct unpaired aortic branches to the esophagus. The superior of these two branches is 3 to 4 cm long and usually arises from the aorta at T-6 or T-7. The more inferior branch is longer, at 6 to 7 cm, and arises at T-7 to T-8 level.

Branches of the left gastric, short gastric, and left inferior phrenic arteries supply the abdominal esophagus, which is the portion usually involved in a Mallory-Weiss bleeding.

Although complicated and technically difficult, this complex and segmental arterial supply to the esophagus makes selective arterial embolization for refractory bleeding from Mallory-Weiss tears possible. The feeding artery should be embolized as far distally as possible with Gelfoam plugs. Successful use of intraarterial vasopressin in this setting has not been reported and should not be attempted for control of the condition.

### *Bleeding Esophageal Varices*

Lower esophageal varices develop in response to high resistance to blood flow in the liver parenchyma, causing portal vein hypertension with subsequent flow reversal through the mesenteric, splenic, and coronary (left gastric) veins. This results in uphill flow in the esophageal venous plexus.

Upper esophageal or downhill varices develop secondary to superior venal caval obstruction. The flow in the esophageal blood plexus reverses to flow caudad, whereas portal vein flow is in the normal direction. Endoscopic sclerotherapy or banding is used for initial control of esophageal variceal bleeding. Embolotherapy is not used in esophageal variceal bleeding. Radiologic treatment includes transjugular intrahepatic portosystemic shunting, described in Chapter 7.

### *Postsurgical Anastomotic Bleeding*

Gastroenterostomies are studied primarily by superior mesenteric arteriography. It is frequently necessary to perform arteriography of second- and third-order branches of the superior mesenteric artery not only to identify the bleeding site without superimposition of other vessels, but also to attempt vasopressin infusion. Ileocolostomies and colocolonic anastomosis require study of the appropriate superior mesenteric arterial branches, branches of the inferior mesenteric artery, or both.

Control of the bleeding can be attempted with vasopressin infusion, which can provide good short-term results and allows time for healing. Embolization should be avoided in the case of postoperative bleeding because of decreased collateral pathways and higher risk of bowel infarction. The treatment is usually operative.

### *Vascular Tumors*

Vasopressin infusion has not been helpful for control of tumor bleeding. This outcome is not surprising, because tumor vessels are abnormal and do not have a media muscular layer. Tumors in the stomach and duodenum can be embolized with permanent agents, particularly polyvinyl alcohol, if super-selective catheterization is achieved. This not only stops the bleeding but may also cause tumor shrinkage, which can be helpful before resection. When the tumor is large and more diffuse, and super-selective embolization is not possible, subselective embolization with a temporary agent, such as Gelfoam, can be used to control acute bleeding and allow time for more elective surgery.

## Hemobilia

Fifty percent of hemobilia is thought to originate from the liver parenchyma, whereas 45% originates from the bile ducts and gallbladder. Less than 5% is estimated to come from the pancreas.

### Etiology

Trauma is responsible for 50% of all cases of hemobilia, of which two-thirds is iatrogenic. An important feature is the time lag that often exists between the initial injury and evidence of subtle bleeding. Recurrence of hemorrhage can be seen for weeks, months, or years. The major causes for hemobilia are (a) gallstones (associated with hemobilia in 25% to 37% of cases, respectively), (b) inflammatory lesions, such as cholecystitis and cholangitis, (c) ascariasis, (d) drug-induced hepatitis, (e) primary or metastatic lesions of liver, gallbladder, or bile ducts, (f) vascular abnormalities (the cause for 7% of hemobilia), including arteriovenous malformations as an example, and (g) portal hypertension.

### Clinical Presentation and Diagnosis

Hemobilia is usually self-limiting. Diagnosis is difficult, unless the condition is suspected. The combination of GI bleeding and biliary colic should raise the suspicion of hemobilia as the cause of GI bleeding. A classic triad of GI bleeding, biliary colic, and jaundice is seen in approximately one-third of patients. Other symptoms include anemia (in 70% of patients), fever, palpable mass in the right upper quadrant, and dull pain.

Percutaneous transhepatic cholangiography and endoscopic retrograde cholangiopancreatography may demonstrate filling defects, but these are not pathognomonic. Ultrasound may show hypoechoic hematoma in the liver or echogenic intraluminal mass without acoustic shadowing in the gallbladder or biliary ducts. Computed tomography (CT) may show blood clots in the gallbladder. On CT, the cause of hemobilia (e.g., cholelithiasis, tumors, trauma) may also identify. Definitive diagnosis is by direct observation of blood leaking out through the ampulla of Vater; however, endoscopic diagnosis is only made in 50% of cases.

Diagnostic angiography often includes both the celiac and superior mesenteric artery. The angiographic feature most frequently associated with hemobilia is a pseudoaneurysm. Hepatic artery to portal vein fistulae are also commonly found, but hepatic artery to hepatic vein fistulae are rare. Extravasation of contrast medium in the biliary tree is seen in only 25% of cases.

### Treatment

The only treatment option used to be surgical, but now endovascular intervention plays a more pivotal role in the treatment of hemobilia. Endovascular intervention is the first choice for treatment of relatively stable patients. Gelfoam particles are sufficient in most cases, although for larger hepatoportal or hepatobiliary fistulae, coils or detachable balloons are used. The primary goal of embolic therapy is to reduce the blood pressure at the bleeding site rather than to devascularize the liver parenchyma. Peripheral embolization with absorbable particulate material seems most effective. Embolization of the main hepatic artery or one of the main branches is only indicated if a peripheral branch cannot be catheterized, or if the patient is unstable and prompt intervention is necessary. Coils or detachable balloons can be used for these occlusions as well. Liver necrosis or infarction usually does not occur with central embolization because of the good collateral circulation of the liver and the supply by the portal vein.

Low complication rate after embolization in the liver is best explained by the dual blood supply by the hepatic artery and portal vein. The portal vein contributes approximately 75% to 80% of the total blood flow to the liver, and the hepatic artery provides the rest. The portal vein should be evaluated with Doppler before any planned embolization. If the portal vein is thrombosed, hepatic artery embolization may be risky. Transient elevation of liver enzymes can be seen in approximately 20% of patients and is indicative of minor liver damage. Major complications are rare and are associated with obstructive jaundice. In these latter patients, it is believed that decreased perfusion is associated with bile duct obstruction from the bleeding itself.

Cholecystectomy is the procedure of choice when the gallbladder is the source of bleeding.

## Bleeding Resulting from Pancreatic Disease

### Etiology

The main causes of pancreatic bleeding are (a) pancreatitis with or without a pseudocyst; (b) trauma; (c) tumors, usually carcinomas but occasionally islet cell tumors and cystadenomas; (d) iatrogenic splenic vein thrombosis after splenectomy; (e) distal splenorenal shunt; and (f) umbilical vein catheterization.

Both pancreatitis and trauma, which cause localized pancreatitis, cause weakness of the vessel wall and subsequent pseudoaneurysm formation.

Pseudocysts often form as a consequence of pancreatitis and bleeding can occur into those. The pseudoaneurysm has a tendency to enlarge and ultimately rupture into the upper GI tract, colon, abdominal cavity, or, rarely, pancreatic duct system. The hemorrhage may be confined to the pseudocyst cavity. Vessels most commonly involved appear in descending order of frequency: splenic, gastroduodenal, and pancreaticoduodenal arteries; however, any vessel affected by the pancreatitis process may be involved. Hemorrhage can be severe.

### Clinical Presentation and Diagnosis

- GI bleeding: 50%
- Recurrent abdominal pain: 25%
- Splenomegaly: 30%
- Ultrasound and CT may demonstrate blood collections or pseudoaneurysm.

Angiographic findings vary considerably. Most commonly, there is a single arterial abnormality (90%) that is either related to the rupture of the vessel into a pseudocyst (60%) or into a pseudoaneurysm formation (PSA) (48%). The source and site of bleeding are usually diagnosed by identification of the erosive arterial changes (vessel irregularity) or by demonstrating a pseudoaneurysm (Fig. 6.2).

### Treatment

Bleeding complications of pancreatic disease require prompt diagnosis. Surgery is often required; however, preoperative angiography is very important in patient management and planning of treatment. In some situations, endovascular management is appropriate. For example, in unstable patients with severe bleeding, temporary hemostasis can be obtained with embolization, allowing for a more elective surgery. Balloon occlusion catheters, coils, and detachable balloons are ideal for occlusion of large arteries. Bleeding from a small pseudoaneurysm involving intrapancreatic branches may be controlled by embolization, making surgery unnecessary. Angiographic management may also be indicated in patients with bleeding in or around the pancreas with findings on CT or ultrasound but unimpressive angiographic findings. This is mostly encountered in cases of small vessel encasement or pseudoaneurysm. Although bleeding may be seen, the source is not found. In such cases, limited embolization of branches of the splenic and left gastric arteries with Gelfoam particles immediately stops the bleeding.

## Lower Gastrointestinal Bleeding

The major causes of bleeding from the lower GI tract include the following:

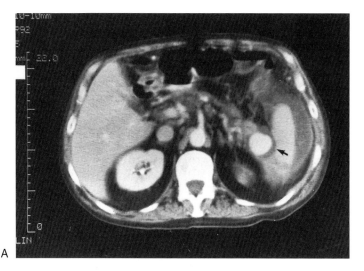

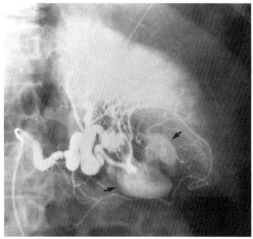 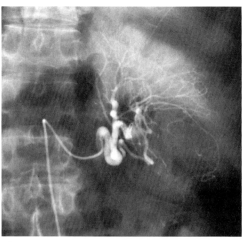

**FIG. 6.2. A:** This patient has known pancreatitis. On CT scan, a contrast-enhanced mass, presumed to be a pseudoaneurysm of the splenic artery, was detected (*arrow*). Angiography was recommended. **B:** Selective angiography of the splenic artery demonstrated a large pseudoaneurysm of the splenic artery (*arrows*). **C:** The angiographic catheter was advanced further into the splenic artery. A microcatheter was then passed through it in a coaxial fashion to the origin of the pseudoaneurysm. Microcoils were placed on each side of the defect in the vessel to prevent flow into the aneurysm from either side.

- Diverticular disease
- Angiodysplasia
- Inflammatory disease of the bowel
- Cecal ulceration (immunocompromised patients)
- Tumors of the colon
- Bleeding in Meckel's diverticulum

Combined, diverticular disease and angiodysplasia cause approximately 50% of lower GI bleeding. The small bowel accounts for 25% to 30% of lower GI bleeding.

### Diverticular Disease of the Small Bowel and Colon

Diverticulitis is the most common cause of bleeding in the large bowel. It originates more commonly on the right side of the colon. The

# 6. GASTROINTESTINAL BLEEDING

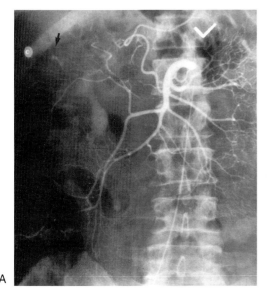
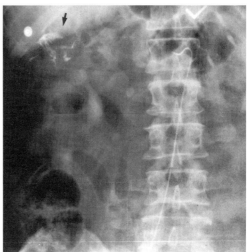
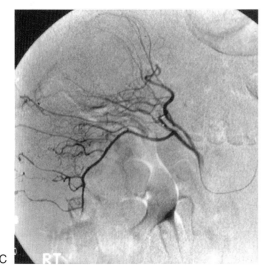

**FIG. 6.3. A:** Patient with lower GI bleeding. Angiography of the superior mesenteric artery was performed. This film is taken approximately 3 seconds after initiation of contrast injection. The bleeding site can barely be seen in the hepatic flexure (*arrow*). **B:** This is from the same study as 2A but approximately 20 seconds after initiation of contrast injection. A considerable amount of contrast has spilled into the lumen of the colon (*arrow*). **C:** A microcatheter has been passed coaxially through the angiographic catheter into the bleeding vessel. Embolization has been carried out, using two or three Gelfoam particles. There is no bleeding present. Note a missing vessel.

angiographic findings include extravasation into the diverticular site and the lumen of the colon (Fig. 6.3). Intraarterial vasopressin infusion can control diverticular bleeding; however, embolization with Gelfoam is quicker and probably more successful. The embolization should be carried out in the distal vessel as close to the bleeding site as possible (vasa recta).

## *Arteriovenous Malformations*

Arteriovenous malformations in the colon are congenital abnormalities, anatomically different from acquired angiodysplastic lesions. Angiographic features of both include early filling of draining veins, slow emptying of tortuous intramural venous channels, and dilatation of the feeding arteries. Less commonly seen are local stain of the colonic wall, tufts of abnormal vessels, and prolonged venous opacification. Congenital arteriovenous malformation is more common in the rectum and sigmoid colon. Bleeding tends to be intermittent in both types of lesions and diagnosis is difficult. These lesions are unresponsive to intraarterial vasopressin but can be managed with Gelfoam for temporary embolization or with polyvinyl alcohol particles for permanent embolization. Colonic embolization carries a significant risk of bowel infarction, especially if small particles are used for embolization. The definitive treatment is surgical excision.

One can assist the surgeon by leaving a microcatheter in the feeding artery while the patient is transferred to the operating room. Methylene blue dye is then injected into the microcatheter when the abdomen has been opened. The surgeon resects the segment of bowel that is colored by the dye.

### *Inflammatory Bowel Disease*

Bleeding is usually diffuse in inflammatory small bowel disease and similar to bleeding seen with gastritis. GI bleeding can be controlled with vasopressin, but the definitive therapy is surgical. Vasopressin may be contraindicated, particularly in Crohn's disease, because of the risk of infarcting a bowel when its blood supply is already compromised by granulomatous infiltration of the wall.

### *Cecal Ulceration in Immunocompromised Patients*

Severe GI bleeding can occur secondary to either steroid administration or to severe cytomegalovirus infection. There is no apparent lasting benefit from vasopressin infusion. Stopping immunosuppression and surgery is therapeutic. Embolization using Gelfoam can be performed successfully.

The cecum is regarded by many as the area of the GI tract most prone to ischemic complications after embolization.

### *Tumors*

Embolization can be used as a temporizing measure for bleeding from the small intestine or the colon. Surgery remains the definitive treatment.

### *Bleeding in Meckel's Diverticulum*

GI bleeding in a young patient with a negative endoscopy and positive radionuclide scan raises suspicion of a Meckel's diverticulum. Superior mesenteric arteriography can be performed to identify the bleeding site. In the case of Meckel's diverticulum, a gastric mucosal specific nuclear medicine scan can be performed. On angiography, there is a so-called Vitellin artery that supplies the diverticulum. Superselective infusion of vasopressin in the presence of bleeding can be attempted. The definitive treatment is surgical excision.

## Complications from Endovascular Treatment of Gastrointestinal Bleeding

### *Catheter-Related Complications*

Catheter-related complications are relatively uncommon. Thrombosis at the arterial puncture site is most frequent (approximately 0.1% in the United States), followed by distal embolization from indwelling catheters in long embolization procedures or with vasopressin infusion.

### *Vasopressin-Related Complications*

Abdominal pain is common. It is attributable to the effect of the hormone on the smooth muscle of the bowel, causing increased peristalsis. This pain needs to be differentiated from severe localized pain, which may indicate bowel ischemia.

The antidiuretic effect appears 6 to 8 hours after initiation of treatment. Urinary output and electrolyte status must be monitored. Diuretic therapy or electrolyte replacement may be necessary if urinary output falls or electrolyte imbalance develops.

Severe peripheral vasoconstriction is thought to be an idiosyncratic reaction to vasopressin. The extremities can become mottled and painful and may necessitate cessation of the vasopressin infusion.

A patient with known coronary ischemic disease should not be treated with vasopressin.

Significant precaution should also be taken with patients with cerebrovascular ischemia.

### *Embolization of Unintended Location*

In most cases, deposition of embolic material into vascular beds other than the target area can be prevented by careful planning and frequent contrast injections during the embolizations process. Balloon occlusion techniques can minimize the risk of embolic reflux.

### *Postembolic Syndrome*

Postembolic syndrome is a cluster of symptoms that appear soon after embolization. It is characterized by fever as high as 102°F, which is more common after extensive embolizations, local pain, and elevated leukocyte count. The syndrome can last for up to 5 days. Frank, full-thickness infarction usually presents 48 to 72 hours after the embolization and may necessitate surgery if infection causes accompanying perforation or drainage.

### *Late Complications*

Late complications include stricture formation at the site of embolization, owing to ischemia with subsequent fibrotic changes.

## SUGGESTED READINGS

Baum S, Nusbaum M, Blakemore WS, et al. The preoperative radiographic demonstration of intraabdominal bleeding from undetermined sites by percutaneous selective celiac and superior mesenteric arteriography. *Surgery* 1965;58:797.

Boudghène F, L'Herminé C, Bigot JM. Arterial complications of pancreatitis: diagnostic and therapeutic aspects in 104 cases. *J Vasc Interv Radiol* 1993;4:551–558.

Castañeda-Zúñiga WR. *Interventional radiology.* Vol 1. Baltimore: Williams & Wilkins, 1992.

Hoevees J, Nilsson U. Intrahepatic vascular lesions following nonsurgical percutaneous transhepatic bile duct intubation. *Gastrointest Radiol* 1980;5:127.

Thorne DA, Datz FL, Remley K, et al. Bleeding rates necessary for detecting acute gastrointestinal bleeding with Technetium-99m-labeled red blood cells in an experimental model. *J Nucl Med* 1987;28:514–520.

Winzelberg GG, McKusick KA, Froelich JW, et al. Detection of gastrointestinal bleeding with 99m Tc-labeled red blood cells. *Semin Nucl Med* 1982;12:139–146.

# 7

# Portal Hypertension: Evaluation and Treatment

## Hector Ferral

## TRANSJUGULAR INTRAHEPATIC PORTOSYSTEMIC SHUNT

### Pathophysiology of Portal Hypertension

Portal hypertension is characterized by an abnormal increase in portal venous pressure and the formation of portosystemic collaterals that bypass the liver to the systemic circulation. Portal hypertension is classified according to the site of increased resistance to flow (Fig. 7.1):

1. Prehepatic
2. Hepatic
   a. Presinusoidal
   b. Sinusoidal
   c. Postsinusoidal
3. Posthepatic

### *Evaluation of Portal Venous Pressure*

There are several methods to evaluate the portal pressure. Direct methods include the catheterization of umbilico-portal vessels, transhepatic portal vein catheterization, and transjugular portal vein catheterization. Indirect methods include hepatic vein catheterization and splenic pulp pressure measurement.

The hepatic vein catheterization is currently the most extensively used method for the evaluation of portal vein pressure. Hepatic vein catheterization is performed using a transjugular or transfemoral approach. An angiographic catheter is then introduced and pressures are measured in the low inferior vena cava (below the renal veins), high inferior vena cava, and right atrium. Pressures in the hepatic vein in the wedged (WHVP) and free positions (FHVP) are also obtained. The WHVP measures the intrahepatic sinusoidal pressure and is nearly equal to the portal vein pressure. The hepatic vein pressure gradient (HVPG) obtains the most important information and represents the gradient between the hepatic sinusoids and the free hepatic vein.

$$HVPG = WHVP - FHVP$$

The portal pressure can be measured in millimeters of mercury (mm Hg) or in centimeters of water (cm $H_2O$). One mm Hg is equivalent to 1.36 cm $H_2O$.

In normal individuals, the HVPG should be 3 to 5 mm Hg. Portal hypertension is defined as an increase in HVPG above 12 mm Hg; however, no linear relationship between the severity of hypertension and risk of variceal bleeding has been identified.

Patients with portal hypertension are in a state of hyperdynamic circulation. They experience increased cardiac output and decreased peripheral vascular resistance, along with sodium and water retention. The increased cardiac output is also associated with increased splanchnic circulation and portal blood flow. The total splanchnic blood flow (the blood flow to spleen, stomach, and intestines) increases by approximately 50%, presumably as a result of a decrease in splanchnic vascular resistance.

Portal hypertension is considered a multiorgan disorder associated with changes in blood flow of the systemic and splanchnic vascular beds. Portal hypertension is the product of increased blood flow as well as increased resistance to that flow.

Treatment options for patients with hemorrhagic portal hypertension include medical therapy, endoscopic sclerotherapy or variceal banding, surgical shunts or devascularization, and percutaneously created transjugular intrahepatic portosystemic shunts (TIPS).

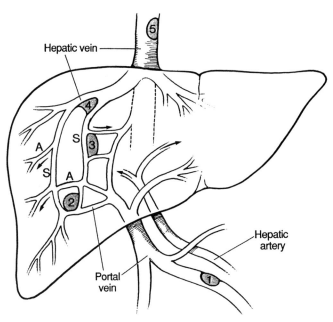

**FIG. 7.1.** Sites of increased resistance to blood flow. *1*, prehepatic; *2*, presinusoidal; *3*, sinusoidal; *4*, postsinusoidal; *5*, posthepatic. (A, anastomosis; S, sinusoid.) (From Kaplowitz N, ed. *Liver & biliary diseases*. Baltimore: Williams & Wilkins, 1996:553, with permission.)

## INDICATIONS AND CONTRAINDICATIONS FOR PERFORMING TRANSJUGULAR INTRAHEPATIC PORTOSYSTEMIC SHUNTS

### Accepted Indications for Transjugular Intrahepatic Portosystemic Shunts

1. Acute variceal bleeding that cannot be successfully controlled with medical treatment (including sclerotherapy)
2. Bleeding from inaccessible intestinal or gastric varices or resulting from severe portal hypertensive gastropathy
3. Recurrent variceal bleeding in patients who are refractory or intolerant to conventional medical management (including sclerotherapy)

### Unproven but Promising Uses

1. Therapy for refractory ascites
2. Budd-Chiari syndrome

### Unproven Uses

1. Initial therapy of acute variceal hemorrhage
2. Initial therapy to prevent recurrent variceal hemorrhage
3. Therapy for prevention of initial variceal hemorrhage
4. Reduction of intraoperative mortality during liver transplantation

In addition, many agree that the following is an indication for TIPS:

5. Acute variceal bleed control for patients waiting for liver transplants

### Absolute Contraindications

1. Right-sided heart failure with elevated central venous pressure.
2. Polycystic liver disease.
3. Severe hepatic failure.
4. Large hepatocellular carcinoma infiltrating the porta hepatis, thereby precluding the placement of the stent. Small peripheral

hepatocellular carcinoma is not a contraindication for TIPS.
5. Hepatic artery thrombosis.

### Relative Contraindications

1. Active intrahepatic or systemic infection
2. Severe hepatic encephalopathy poorly controlled by medical therapy
3. Portal vein thrombosis

### TECHNIQUE

TIPS is a technically demanding procedure that should not be performed by inexperienced interventional radiologists. The most effective way to learn the procedure is to observe cases at training centers and be assisted by an interventional radiologist with TIPS experience in the first case(s).

### Patient Preparation

1. The indication for TIPS must be discussed with the referring physician. Ideally, a transplant surgeon must be involved to determine whether the patient is a suitable candidate for transplantation in the near future.
2. A careful history and physical examination must occur to document the etiology of liver disease.
3. The Child-Pugh classification (Table 7.1) and APACHE II score are useful to determine the patient's status and prognosis.
4. Informed consent must be obtained from the patient or the family. Risks and benefits expected from the procedure must be discussed.
5. Doppler ultrasound must be taken of the liver vasculature to assess patency of the portal vein, splenic vein, hepatic artery, inferior vena cava, and hepatic veins, and to rule out the presence of cysts or large masses that might preclude the procedure. It is especially important to rule out splenic vein thrombosis, because these patients have a segmental, prehepatic portal hypertension and do not receive any benefit from a TIPS procedure.
6. It is important to improve the clinical status of the patient by administering intravenous fluids, blood, or blood products.
7. The Foley catheter must be placed in the bladder.
8. Crystalloids, vasopressors, and oxygen should be immediately available to manage hypotension and bleeding.
9. The use of a broad-spectrum antibiotic for gram-negative organisms [1 g of ceftizoxime (Cefizox)] is controversial.

The procedure is performed in an angiographic suite with high-resolution fluoroscopy and digital substraction capabilities. It is beneficial to have an anesthesiologist as part of the team whether general anesthesia or conscious sedation is used, so that the physician performing the TIPS can concentrate on the technical aspects of the procedure. General anesthesia is preferred if the patient is agitated or uncooperative from encephalopathy and if there is a danger of aspiration. Many physicians prefer to perform the procedure under conscious sedation to eliminate the toxic effects of general anesthetic agents on an already compromised liver. The most frequently used drugs are midazolam (Versed) and fentanyl (Durasgesic Transdermal System).

**TABLE 7.1.** *Child-Pugh Score*

| Variable | Points accorded | | |
|---|---|---|---|
| | 1 | 2 | 3 |
| Bilirubin (mg/dL) | <2 | 2–3 | >3 |
| Albumin (mg/dL) | >3.5 | 2.8–3.5 | <2.8 |
| Ascites | — | Easily controlled | Poorly controlled |
| Neurologic disorder | — | Minimal | Advanced |
| Prothrombin time (sec >control) | <4 | 4–6 | >6 |

Class A: 5 to 6 points; Class B: 7 to 9 points; Class C: 10 to 15 points.

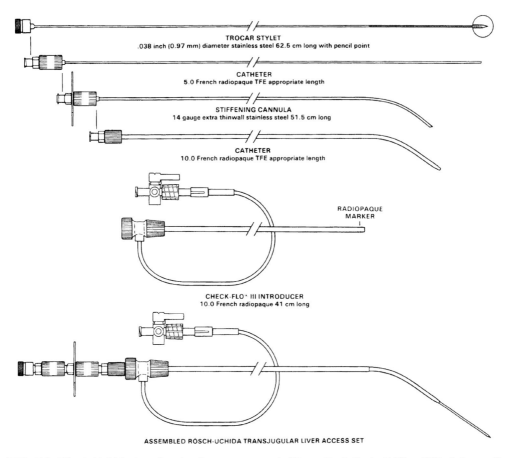

**FIG. 7.2.** Rösch-Uchida transjugular liver access set. (From Castañeda-Zúñiga WR. *Interventional radiology,* 3rd ed. Baltimore: Williams & Wilkins, 1997, with permission.)

### Required Materials

1. Micropuncture Access Set (Cook, Bloomington, IN)
2. Colapinto-Ring hepatic access set or Rösch-Uchida set (Cook) (Fig. 7.2)
3. Glidewire, angled, stiff type (MEDITECH/Boston Scientific, Natick, MA)
4. Amplatz Superstiff guidewire, 180 cm or 260 cm
5. P-308 Palmaz stent or Wallstent (usually 10 mm × 68 mm)
6. Angioplasty balloon catheters
7. Angiographic catheters

### Vascular Anatomy of the Liver

Understanding of intrahepatic vascular anatomy and its relationship to other structures within the liver is crucial to perform a successful TIPS procedure. One must understand the intrahepatic anatomy of the hepatic and portal veins in a three-dimensional format to be technically successful and obtain a safe outcome.

1. The right hepatic vein (RHV) continues to be the hepatic vein of choice; the middle hepatic vein and left hepatic vein are available in case of a hypoplastic RHV or for placement of an additional stent.
2. The biliary and arterial structures are anterosuperior to the bifurcation of the portal vein in 52% of cases, causing no major concern in the majority of patients.
3. The RHV is always posterior to the portal vein. Anatomic variants may exist. Occasion-

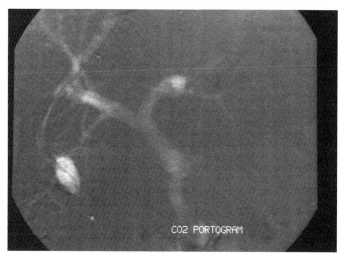

**FIG. 7.3.** Wedged hepatic venogram with $CO_2$ with excellent demonstration of the portal venous system.

ally, three RHVs exist. The most anterior RHV is anterior to the portal vein.
4. The portal vein bifurcation is intrahepatic in only 60% of the cases.
5. TIPS puncture should be performed in the posteroinferior aspect of the right portal trunk or in the posterior aspect of the main portal vein bifurcation.
6. Patients with cirrhosis have small livers; therefore, the puncture might need to be performed closer to the portal bifurcation. Arterial damage is likely to occur because of enlarged hepatic arteries.

### *Identification of the Portal Vein*

During the early development of the procedure, celiac or superior mesenteric portography with or without balloon occlusion catheters were used to demonstrate the anatomy of the portal vein. Since the mid-1990s, many centers are relying on hepatic wedge venography to visualize the portal vein, to limit the amount of contrast material used, and to avoid complications from a femoral artery puncture in patients with coagulopathies. A single-lumen angiographic catheter or a balloon occlusion catheter is wedged in the hepatic vein, and contrast material is forcefully injected into the liver parenchyma. The portal vein is visualized in at least two-thirds of cases (Fig. 7.3). Carbon dioxide wedged hepatic vein injections may be used to demonstrate and guide the puncture of the portal vein, because the gas tends to stay in the portal vein for a period of time.

### *Jugular Vein Access*

The neck is cleansed with sterile solution, covered with an adhesive dressing, and properly draped. A needle punctures the right jugular vein, blood is aspirated, and a guidewire is introduced. A 10-Fr, 35-cm sheath is advanced through the right atrium into the suprahepatic inferior vena cava. Future manipulations are done through this sheath.

A 7-Fr multipurpose catheter without side holes or a 5-Fr cobra catheter without side holes is advanced into the RHV. A hepatic venogram is obtained to assess the anatomy of the individual veins and the relationship of the hepatic vein to the branches of the portal veins. The wedge hepatic and inferior vena caval pressures are recorded to calculate the portosystemic pressure gradient before shunting. A wedge hepatic venogram can be obtained to evaluate the anatomy of the portal vein if a mesenteric or splenic arteriogram was not obtained previously.

The dilator is advanced into the hepatic vein, the metal 0.038-in. needle is introduced and locked to the 5-Fr catheter, and the 14-gauge cannula is locked to the 10-Fr catheter. The puncture site is determined by withdrawing the set to the proximal 3 cm of the RHV, cephalad to the expected portal vein bifurcation. After turning the system anteriorly, the 10-Fr catheter is wedged inferiorly against the hepatic vein wall. It is important to wedge the catheter for successful liver puncture, because if the tip is free in the hepatic vein, the needle slides along the vein wall rather than puncturing it. The puncture is made with a sharp thrust of 3 to 6 cm into the hepatic parenchyma, depending on the location and anatomy of the portal vein bifurcation. During the puncture, the needle and 5-Fr catheter are secured together and advanced simultaneously. The needle is removed, and suction is applied to the 5-Fr catheter as it is slowly withdrawn. Contrast material is injected when blood is freely aspirated to be sure that a suitable portal branch has been entered. At this time, a 0.035-in., stiff, angled guidewire is advanced into the portal and splenic or superior mesenteric veins. The 5-Fr catheter is subsequently advanced over the guidewire into the portal vein.

Next, the assembled cannula and 10-Fr Teflon catheter are advanced over the 5-Fr Teflon catheter and wire into the portal vein under fluoroscopic control. This dilates the parenchymal tract so that additional catheters may be introduced. Simultaneous advancing of the 14-gauge cannula and the 10-Fr catheter helps to overcome the resistance of the liver parenchyma and portal vein wall in hard livers. The 10-Fr sheath is moved forward until its opaque tip is against the hepatic vein wall at the puncture site. At this point, the 10-Fr catheter and metal cannula are withdrawn, while maintaining the 5-Fr Teflon catheter in the portal vein. After removal of the guidewire, pressure measurements and portal venography are performed (Fig. 7.4).

**Portal-Hepatic Vein Pressure Gradient Measurement and Tract Dilation**

It is generally accepted that a portosystemic gradient of 12 mm Hg or higher is required to increase the risk of variceal bleeding. Conversely, low portosystemic gradients with high shunt flows may significantly increase the risk of hepatic encephalopathy. Careful measurement of the portosystemic gradient is a crucial part of the procedure.

After measuring gradients, the 5-Fr Teflon catheter is exchanged over an Amplatz stiff guidewire (MEDITECH/Boston Scientific Corp.) for a low-profile 8-mm angioplasty balloon. Immediately after dilation, a portal venogram is obtained. The track is then measured and an appropriate stent is selected to adequately cover the length of the track. A self-expandable 10-mm–diameter Wallstent of adequate length is placed in the track and dilated to 8 mm.

In patients with long parenchymal tracks, or when the first stent is not placed correctly, more than one stent may be needed. The portal vein, hepatic vein, and inferior vena cava pressures are recorded; a portal venogram is obtained; and contrast material is injected into the splenic vein

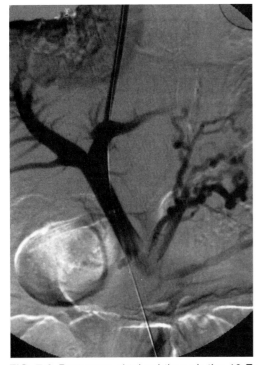

**FIG. 7.4.** Portogram obtained through the 10-Fr Rösch-Uchida sheath. The sheath has been advanced into the main portal vein. The guidewire has been advanced to the superior mesenteric vein. Variceal opacification is evident. Portal vein pressures may be obtained directly through the sheath.

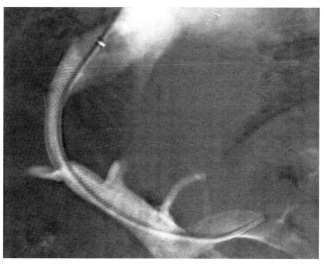

**FIG. 7.5.** Portogram immediately after stent deployment shows excellent flow through the shunt. No varices are opacified. The tip of the catheter is in the mid-splenic vein.

to assess the flow through the shunt and evaluate the persistent opacification of gastroesophageal varices (Fig. 7.5).

The issue of embolization of varices as an adjunct to a successful TIPS procedure in an acutely bleeding patient is somewhat controversial. Performing embolotherapy of large varices, along with creating a well-functioning shunt, may increase the speed of patient recovery. This is particularly true in patients with very large varices that compete with the shunt hemodynamically for portal venous flow, because flow tends to follow the path of least resistance. If the post–stent deployment portal venogram performed with the catheter tip in the splenic vein reveals equal or preferential flow through large gastroesophageal varices, it is advisable to proceed to embolize the varices to ensure adequate flow through the shunt and prevent rebleeding from the varices.

When the stent is in place, pressures across the liver are measured again, and portal venography is repeated. If the pressure gradient is sufficiently reduced and the varices no longer fill, it is not necessary to make any adjustments. If pressures remain too high, the radiologist can do the following: Balloon-dilate the stent to 10 mm to ensure maximal expansion (Fig. 7.6), or perform a second, parallel TIPS by placing a stent from another hepatic vein to the portal vein.

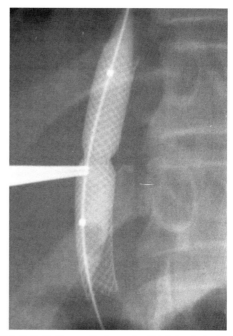

**FIG. 7.6.** Balloon dilation of a Wallstent during a transjugular intrahepatic portosystemic shunt procedure. A 10-mm balloon was used. Note the residual waist in the middle of the stent.

At this point, the procedure is finished, and bleeding should have ceased. The jugular sheath usually is removed, unless the patient may require reevaluation within 72 hours.

## RESULTS

Technical success is achieved in 93% to 100% of the patients, with the portal pressure gradient reduced by 57% immediately after shunt placement.

Early mortality (within 30 days) occurs in 3% of patients. Most deaths occur in patients who have the procedure on an emergency basis. The 1-year survival rate is 87%, which is comparable to the 70% to 90% rate reported after surgery.

Shunt stenosis occurs in 40% to 70% of the patients. Most can be treated successfully with redilation, thrombolysis, or implantation of an additional stent. Twenty-five percent of the patients develop hepatic encephalopathy. At 1-year follow-up, 82% of patients are free from variceal rebleeding. Repeat bleeding after TIPS is noted in 10% to 20% of patients and is usually associated with stenosis or occlusion. The reported survival rate is 91% at 6 months, 74% at 1 year, and 42% at 3 years' follow-up.

LaBerge et al. reported in 1993 on a series of 100 patients who had TIPS. The procedure was successful in 96% of patients, and the portal pressure was reduced from an average of 34.5 mm Hg to 24.5 mm Hg after TIPS. The 30-day mortality occurred in 13% of the patients. Of the 96 patients who had successful TIPS procedures, 26 died, 22 had orthotopic liver transplantation, and 48 are alive without liver transplantation.

## TRANSJUGULAR INTRAHEPATIC PORTOSYSTEMIC SHUNT FOLLOW-UP

Surveillance of the stents is usually done with ultrasonography or venography. The sonography examination consists of gray-scale, conventional duplex, and color Doppler sonography. The goal is to determine the patency, direction of flow, and spectral flow characteristics of stent and portal veins. Sonography plays an important role in the preoperative evaluation and in evaluating the stent after the procedure at regular intervals. Ultrasound is performed at 24 hours, 1 month, 3, 6, 9, and 12 months.

Color Doppler is optimal in detecting shunt stenoses and thrombosis. This detection is essential to the clinician in managing a patient before signs or symptoms of shunt malfunction develop.

## COMPLICATIONS

Complications associated with TIPS procedures can be classified as technical, immediate medical, and late complications. The immediate complications may be related to the technical procedure or to the underlying medical condition, primarily to the severity of liver disease. The late complications are mostly related to shunt stenosis, occlusion, or encephalopathy.

1. Overall complications occur in fewer than 10% of patients, if one excludes hepatic encephalopathy and delayed stenosis or occlusion.
2. The 30-day mortality rate ranges from 4% to 45% and is mostly dependent on the severity of underlying liver disease.
3. The incidence of encephalopathy occurs in 5% to 35% of patients.
4. Up to 75% of shunts may undergo stenosis or occlusion within a year.
5. The direct procedure mortality rate is less than 2%.

### Technical Complications

Technical complications can be summarized as follows:

1. Local puncture site complications.
2. Complications related to transhepatic needle puncture and gaining access to the portal vein. Gaining access to the portal vein is the most crucial part of the procedure. Perforation of the liver capsule with the needle can occur in 30% of patients without adverse sequelae. Several authors have recommended embolizing the transparenchymal tract with Gelfoam whenever this occurs. Again, correction of coagulation abnormalities before TIPS creates safeguards against

intraperitoneal bleeding. Puncture on the extrahepatic portal vein can be fatal. The puncture to the portal vein should be directed at least 1 cm away from the portal vein bifurcation. If this is achieved, more than 95% of the punctures occur in the intrahepatic segments of the portal vein, minimizing the risk of fatal exsanguination associated with the extrahepatic puncture.
3. Acute complications related to stent deployment.

### Immediate Medical Complications

Immediate medical complications include septicemia, intravascular hemolysis, coagulopathy, bleeding-related complications, such as aspiration pneumonia and adult respiratory distress syndrome, intraprocedural myocardial infarction, and contrast-related complications.

#### *Septicemia*

Fever without positive blood cultures has been reported in 10% of patients after TIPS procedures. Septicemia is not a common complication. The causative organisms could be gram-positive or gram-negative bacteria. The source of infection is multifactorial. Infection can be from aspiration pneumonia, infected central venous lines, septic emboli, infected biliary system, or seeding of bacteria from the mesenteric and portal circulation into the systemic bloodstream.

### Late Complications

#### *Hepatic Encephalopathy*

Hepatic encephalopathy has been reported in 5% to 35% of patients after TIPS. The factors that contribute to increased encephalopathy are increased stent diameter, patient's age greater than 62 years, advanced liver disease, and diminished or absent hepatopetal flow. In the majority of patients, diet control and oral lactulose administration can manage encephalopathy.

#### *Stent Stenosis or Occlusion*

Stent stenosis or occlusion has been reported in up to 80% of TIPS patients in 1 year. Stent or hepatic vein stenosis can be managed by balloon dilatation, atherectomy, or additional stent placement.

## SUGGESTED READINGS

Foshager MC, Ferral H, Finlay DE, et al. Color Doppler sonography of transjugular intrahepatic portosystemic shunts (TIPS). *AJR* 1994;163:105–111.

Freedman AM, Santal AJ, Tisnado J, et al. Complications of transjugular intrahepatic portosystemic shunt: a comprehensive review. *Radiographics* 1993;13:1185–1210.

Haskal ZJ, Rees CR, Ring EJ, et al. Reporting standards for transjugular intrahepatic portosystemic shunts. *J Vasc Interv Radiol* 1997;8:289–297.

LaBerge JM, Ring EJ, Gordon RL, et al. Creation of transjugular intrahepatic portosystemic shunts with the Wallstent endoprosthesis: results in 100 patients. *Radiology* 1993;187:413–420.

Mahl TC, Groszmann RJ. Pathophysiology of portal hypertension and variceal bleeding. *Surg Clin North Am* 1990;70:251–266.

Rikkers LF, Sorrell WT, Jin G. Which portosystemic shunt is best? *Gastroenterol Clin North Am* 1992;21:179–196.

Roaussle M, Haag K, Ocho A, et al. The transjugular intrahepatic portosystemic stent-shunt procedure for variceal bleeding. *N Engl J Med* 1994;330:165–171.

Sterling KM, Darcy MD. Stenosis of transjugular intrahepatic portosystemic shunts: presentation and management. *AJR* 1997;168:239–244.

Uflacker R, Reichert P, D'Albuquerque LC, et al. Liver anatomy applied to the placement of transjugular intrahepatic portosystemic shunts. *Radiology* 1994;191:705–712.

Zemel G, Katzen BT, Becher GJ, et al. Percutaneous transjugular portosystemic shunt. *JAMA* 1991;266:390–393.

# 8

# Percutaneous Transluminal Angioplasty

Hector Ferral

## A. BASIC CONCEPTS

### DISTRIBUTION OF ATHEROSCLEROTIC PLAQUES

Atherosclerotic plaques are localized or diffuse intimal growths that may impede or block blood flow to organs and tissues. Plaques develop mainly at sites of altered shear stress and flow velocity, including bifurcations, curves, and branch points. In decreasing order of frequency, plaques involve the abdominal aorta below the renal arteries, proximal coronary arteries, popliteal arteries, descending thoracic aorta, internal carotid arteries, and arteries of the circle of Willis. The pulmonary, upper extremity, mesenteric, and renal arteries are relatively free of disease, except at their ostia.

### UNCOMPLICATED ATHEROSCLEROTIC PLAQUES

The primary pathologic manifestation of atherosclerosis is the atherosclerotic plaque. Atherosclerotic plaques consist of variable quantities of (a) mesenchymal cells, (b) collagen and other connective tissue elements, (c) lipids, and (d) calcium salts. The simplest lesion is the fibrous plaque, which has a matrix of types I and III collagen, basement membrane material, elastin, and proteoglycans. Atheromatous plaques, named after the Greek word *atheroma* (gruel), contain a large central pool of lipids and cellular debris.

### COMPLICATED ATHEROSCLEROTIC PLAQUES

Most advanced, high-grade, and clinically significant atherosclerotic stenoses have undergone one or more complications. These include (a) calcification, (b) plaque surface ulceration, (c) thrombosis, (d) intraplaque hemorrhage, and (e) aneurysmal dilation.

#### Thrombosis

Acute arterial thrombosis is the most significant plaque complication. Plaque fissuring or ulceration is the precipitating cause of most thrombotic occlusions. Plaque surface ulceration presumably exposes thrombogenic substances such as collagens and lipids to the flowing blood, with resultant platelet adhesion, aggregation, release reaction, and activation of the clotting cascade.

#### Intraplaque Hemorrhage

Hemorrhage into plaques usually occurs by erosion of the plaque fibrous cap, with entry of plasma and blood cells. This may cause increased intraplaque pressure followed by plaque rupture and thrombosis.

#### Aneurysmal Dilation

Aneurysms are localized dilations of vessels, the most common cause of which is atherosclerosis. In decreasing order of frequency, atherosclerotic aneurysms involve the lower abdominal aorta, iliac arteries, ascending aorta, and descending thoracic aorta.

Aneurysms occur because the tunica media underlying plaques becomes progressively thinned and atrophic. This weakens the arterial wall, which may become stretched to form a saccular or fusiform aneurysm.

## SUMMARY

As this brief review shows, the pathology of atherosclerosis is complicated and variable. The composition of atherosclerotic plaques no doubt affects the success or failure of vascular interventions. With so many new interventional techniques becoming available, it will be possible to select therapy on the basis of plaque morphology and configuration.

## B. MECHANISM OF ANGIOPLASTY

Percutaneous transluminal angioplasty (PTA) was introduced by Dotter and Judkins in 1964. After a rather slow start, it has gained wide acceptance during the past 15 years for the treatment of both atherosclerotic and nonatherosclerotic vascular disease, even though its mechanism has not been well understood until recently.

### THEORIES OF THE MECHANISM OF TRANSLUMINAL ANGIOPLASTY

According to Dotter, the principal mechanism of angioplasty is compression and remodeling of the atheromatous plaque.

With the more frequent use of PTA, it became clear that the original theory was not tenable. Atheromatous plaques, being semiliquid or solid without true empty spaces, are virtually incompressible unless liquid is extruded.

### MORPHOLOGIC CHANGES AFTER ANGIOPLASTY

A series of experiments have been performed on human cadaver arteries and on normal and atherosclerotic dog and rabbit arteries *in vivo* to explain the mechanisms of angioplasty. Recently, studies using intravascular ultrasound (IVUS) have added important information on the mechanisms of PTA (Fig. 8.1).

The results of the first experiments on the mechanism of angioplasty confirmed the hypothesis that balloon dilation does not compress atheromatous plaque but works by intimal rupture and partial dehiscence of this layer from the media, thus producing large clefts. These longitudinal clefts corresponded well with the radiolucent linear defects seen on postangioplasty arteriograms in clinical PTA (Fig. 8.1).

IVUS has been used to analyze the changes in the walls of iliac arteries caused by angioplasty. IVUS was used to depict the full cross-sectional thickness of the arterial wall. These studies have proposed that IVUS provides a perspective similar to that of histologic examination.

Three main mechanisms participate or contribute to the increase in luminal cross-sectional area after angioplasty:

1. Plaque fractures
2. Plaque compression
3. Stretching of the arterial wall

Proportionately, plaque fracture represented the largest single factor responsible for increased luminal patency after angioplasty; 87% to 93% of the total lumen increase results from rupture of the plaque. Plaque compression was quantitatively less important, and variable amounts of plaque compression were demonstrated by IVUS. Stretching of the arterial wall was found to contribute less than plaque fracture or compression and may also participate but as a secondary mechanism.

Peripheral embolization is encountered in 3% to 5% of clinical PTA procedures and is often asymptomatic. Clinically significant embolization is the result of a completely detached atheromatous plaque or of dislocation of relatively fresh thrombi.

### RECOMMENDATIONS

Careful analysis of the angiogram is important to avoid overstretching the artery. Rupture and dehiscence of plaques should not be interpreted as complications of PTA, as long as there is no obstruction to flow or detachment of the plaque (Fig. 8.1). Angiographically and histologically, these phenomena differ from true intraluminal

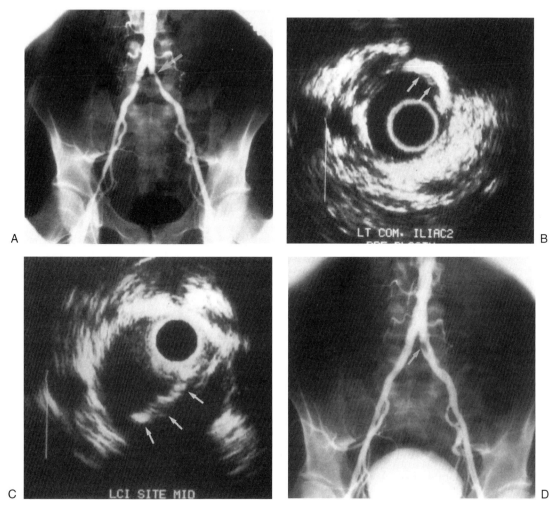

**FIG. 8.1. A:** Iliac arteriogram shows high-grade stenosis (*arrow*). **B:** Left common iliac intravascular ultrasound (IVUS) shows high-grade stenosis with extensive calcifications (*arrows*). **C:** IVUS postangioplasty shows large calcified intraluminal intimal flap (*arrows*) and the increase with luminal diameter. **D:** Pelvic arteriogram postangioplasty shows large intraluminal intimal defect at site of dilation (*arrow*). (*Continued.*)

dissections because they are confined to the dilation site and do not have a tendency to propagate.

In clinical PTA, the number of repetitions of balloon inflation and the duration for a given lesion are still arbitrary. Commonly, the balloon is inflated to an arbitrarily selected diameter for 15 seconds to 1 minute when treating renal or peripheral arteries; indeed, this recipe is often followed religiously despite the absence of scientific validation.

As a general rule, balloons with the lowest possible compliance should be used. In this way, maximum dilation force with exact definition of the balloon diameter should be achieved even with high inflation pressures. Balloon size is chosen according to the estimated diameter of the nondiseased vessel. The use of a slightly oversized balloon seems justified as long as inflation pressures do not exceed 4 to 6 atm, but when high pressures (10

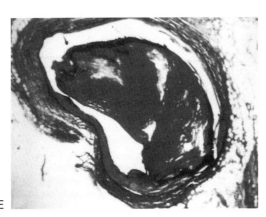 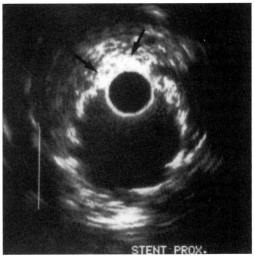

**FIG. 8.1.** *Continued.* **E:** Specimen postangioplasty shows the atherosclerotic plaque within the lumen. **F:** IVUS after Palmaz stent deployment confirms that the intimal flap has been impacted against the wall (*arrows*). (Reprinted from Castañeda-Zúñiga WR. *Interventional radiology,* 3rd ed. Baltimore: Williams & Wilkins, 1997, with permission.)

to 12 atm or more) are used, it is prudent not to use oversized balloons.

After clinical PTA, arterial spasms are not uncommon and may lead to early thrombotic reocclusion. For this reason, especially in renal and coronary angioplasty, prophylactic administration of vasodilating drugs (i.e., verapamil, nitroglycerin) is recommended.

## C. DEVICES AND TECHNIQUES

### BALLOON CATHETER TECHNOLOGY

It is important to be familiar with a series of catheter performance criteria. Some helpful concepts are briefly explained here.

### Trackability

*Trackability* refers to the characteristics of a catheter that allow it to follow over a guidewire through tortuous paths to its ultimate destination. Balloon catheters tend to have stiff segments where the balloon joins the shaft, which can make tracking difficult. Most PTA catheter shafts are 5 French (Fr) for balloon sizes from 4 to 8 mm.

### Crossability

The key subcharacteristics of crossability are the profile of the tip and deflated balloon segment, as well as frictional properties of the surfaces. *Profile* refers not only to the deflated diameter of the balloon, but also to the tip taper and abruptness of the change in diameter underneath the radioopaque marker and the beginning of the balloon taper.

### Pushability

*Pushability* refers to the ability to transfer axial force applied at the proximal end of the catheter to the tip. The most important subcharacteristic of pushability is column strength.

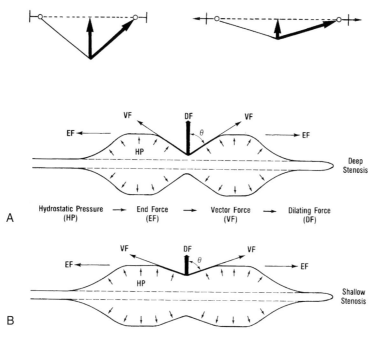

**FIG. 8.2.** Dilating force and clothesline effect. **A:** When weight hanging in the center of the clothesline is lifted by pulling on each end of the line, the force vector pushing upward decreases as the line straightens. **B:** Same principle applied to balloon dilation. If the balloon surface is indented by localized stenosis, the component force pushing outward corresponds to the dilating force. With progressive expansion of the balloon or dilation of stenosis, the dilating force decreases. (Reprinted from Castañeda-Zúñiga WR. *Interventional radiology*, 3rd ed. Baltimore: Williams & Wilkins, 1997, with permission.)

### Dilating Force

*Dilating force* refers to the radial force applied to the lesion to enlarge its inner diameter. This circumferential force is known as *hoop stress*. Larger-diameter balloons have a larger inside surface area over which pressure is applied, and because force = pressure × area, they experience more hoop stress at a given pressure than do smaller balloons. This relation is expressed by Laplace's law, HS = P × D, where HS is the hoop stress, P is the pressure in the balloon, and D is the inside diameter of the balloon.

Dilating force is the sum of the hydrostatic force plus a mechanical vector force that occurs when the circumference of the balloon, in the process of straightening out, exerts a radial component. As the depression in the circumference is straightened with increasing tension, force vectors are created, resulting in an outward, or radial, force on the lesion. This latter force vector is greatest when the balloon is more "hourglassed." The radial force vector diminishes as the line straightens (Fig. 8.2). To the person performing angioplasty, this has several important implications. First, applying more pressure to eliminate a small dent in the balloon produces little dilating force, and it is more likely to break the balloon. Second, for the same lesion at the same pressure, more force will be applied by a larger balloon because it will be "hourglassed" more, and thus the radial force vector will be greater. Third, larger balloons will require much less pressure to exert substantial dilating force than small balloons, because their large diameters generate more hoop stress. Conversely, small balloons require much more pressure to produce adequate force for dilation. Fourth, hoop stress occurs in vessels as well as in balloons. Larger vessels require less pressure to dilate—and to rupture.

## DILATING FORCE AND BALLOON COMPLIANCE

*Compliance* is a measure of how much the balloon surface stretches when force is applied to it. Because all balloon materials stretch slightly and to varying degrees, the balloon diameter increases with pressure. A low-compliance (nonstretch) balloon has the following characteristics:

- More dilating force. Balloons with the same dimensions produce considerably greater dilating force at the same pressure if the material does not stretch.
- Predictable diameter. What you see is what you get. Repeat inflations produce the same results.
- Better "feel" of the lesion. The feel at the syringe barrel is a result of changes in the lesion, not in the balloon.
- Retention of profile. A nonstretch balloon retains its sausage shape rather than expanding to dilate normal vessels on either side of the lesion or expanding in the aorta, as in the case of an ostial renal stenosis.
- Safer failure mode. The balloon does not stretch before rupture. An expanded balloon is difficult to remove without entry-site trauma.

Most manufacturers have relatively low-compliance balloons. The challenge is to produce a low-compliance balloon that is smooth, flexible, and has a low profile on deflation.

Present catheter designs have achieved many of the original objectives of the ideal balloon catheter:

- Low profile. Great progress continues to be made in this area. High-strength, ultrathin wall polymers (polyethylene, nylon, new composites, coextrusions) allow most catheters up to 8 mm in diameter to be introduced through a 5-Fr or smaller sheath.
- High pressure. High-pressure balloons (20 to 30 atm) have been available for a number of years but were relatively thick-walled. Improvements in polymers, new alloys, and new techniques in processing make it possible to produce extremely low-profile, low-compliance, high-burst-strength balloons.

## NEW DEVELOPMENTS

- Stent delivery balloons. Extra-strong, puncture-resistant materials are necessary for stent expansion. Short tapers are frequently preferred to allow stent placement close to bifurcations or angulations.
- Large balloons. Although large balloons have been available for many years, new designs are available with extra-strong, thin-walled, low-profile construction. Short tapers are particularly preferred on large sizes to reduce dilation length and avoid vessel damage.
- Cutting balloons. "Balloons with blades" have been developed to aid in dilating tough calcific lesions and to produce controlled linear dissections.

# D. INDICATIONS AND OUTCOME

## 1. ANGIOPLASTY OF SUPRAAORTIC VESSELS

Hector Ferral and S. Murthy Tadavarthy

Angioplasty of the subclavian arteries was first reported in 1980. Angioplasty of the supraaortic arteries is not commonly performed because of the fear of cerebral embolization. However, recent literature reports have shown that these procedures are being performed more often and that techniques, indications, and experience in the treatment of lesions in these vessels are rapidly evolving.

**Preliminary Investigations before Angioplasty of Supraaortic Branches**

Only interventionalists with experience in both angioplasty and cerebral catheterization should attempt supraaortic vessel dilation, because inexperience may lead to permanent neurologic deficits. Angioplasty in the brachiocephalic vessels should be limited to lesions that are clinically symptomatic. Close cooperation between interventional radiologist, neurologist, and vascular surgeon is mandatory. A uniform opinion should be sought before transluminal angioplasty, and the patient should be warned of the potential risks and complications and should be informed about other therapeutic alternatives. Preliminary evaluation should include thorough clinical examination, duplex ultrasound studies, arch aortography, and selective catheterization of the supraaortic branches with intracranial vascular studies to determine the hemodynamic significance of the vascular lesions.

**Angioplasty of the Subclavian Artery**

Stenotic or occlusive lesions of the subclavian and axillary arteries are considered to be uncommon. Patients usually complain of symptoms such as arm claudication, signs of upper extremity ischemia, and subclavian steal.

A diagnosis of subclavian steal syndrome should be made only after thoracic aortography. Stenosis of the left subclavian artery at its origin is more frequently observed than on the right side. The clinical symptoms include vertigo, ataxia, diplopia, paresis, numbness, and arm claudication. Physical examination reveals reduction of blood pressure in one arm, along with a supraclavicular bruit.

*Methodology*

Aspirin (325 mg daily) is administered orally 1 to 2 days before angioplasty. Subclavian angioplasty can be performed from the transfemoral route. In difficult cases, the transaxillary or transbrachial approach has been recommended because of tortuosity and dilation of the aorta. Complications such as axillary hematoma, brachial plexus trauma, and distal embolization of atherosclerotic plaques to the fingers are known sequelae of this approach. In cases of recanalization of proximal subclavian occlusions, the transaxillary or transbrachial access is recommended as a first choice.

Heparin bolus (3,000 to 10,000 units) is administered immediately after catheterization. The stenotic lesions are crossed with soft-tip Bentson (Cook, Inc., Bloomington, IN) or torque-controlled hydrophilic guidewires. As soon as the lesion is crossed, an exchange-length guidewire is advanced far beyond the stenotic area to be dilated. The angioplasty balloon catheters are advanced over the exchange guidewire. We prefer stronger guidewires, such as the Rosen wire, for the exchange. A tightly curved tip is better than a straight wire because it minimizes the spasm and the risk of dissection of the subclavian artery.

Occasionally, spasm of the subclavian and axillary arteries is encountered. This can be relieved with the administration of intraarterial nitroglycerin in bolus of 100 to 200 µg. The size of the balloon to be used is determined by the measurement of the normal-sized vessel both proximal and distal to the area of the stenosis. Usually, 8- to 10-mm balloons are used; 6-mm balloons are used in small-sized vessels. The length of the balloons is usually between 2 and 4 cm. Balloons are usually inflated up to 4 to 5 atm for 15 to 30 seconds. The endpoint of the dilation is indicated by the full inflation of the balloon at the site of the stenosis, which usually requires two to three inflations.

*Discussion*

Angioplasty of subclavian artery stenotic lesions is an attractive therapeutic alternative. In the reported data from more than 400 subclavian and innominate angioplasty procedures, a 92% technical success rate was achieved, with an overall 6% complication rate, including central nervous system complications.

The results of angioplasty in cases of subclavian artery occlusion have not been as encouraging. The reported technical success in cases with subclavian artery occlusion is 83%. The 6-month patency without recurring symptoms is 88% in the successfully recanalized patients. The treatment of choice in these cases is a surgical bypass. The percutaneous treatment may be attempted in patients who cannot be treated surgically or who refuse surgery.

The most feared complication of this procedure is vertebral artery embolization. After successful recanalization, the direction of blood flow within the vertebral artery does not change

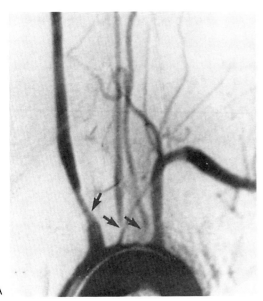 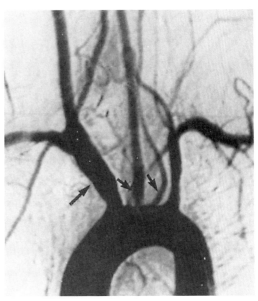

**FIG. 8.3.** Takayasu's arteritis with stenotic and occlusive lesions of aortic arch vessels. **A:** Aortic arch angiogram before angioplasty; narrow stenosis of left vertebral, left common carotid, and right common carotid arteries. Occlusion of the right subclavian artery. **B:** Aortic arch angiogram after angioplasty. Enlargement of the stenosed vessels (*arrows*). Origin of right subclavian artery is no longer occluded. (Reprinted from Théron JG. Angioplasty of supra-aortic arteries. *Semin Intervent Radiol* 1987;4:331–342, with permission.)

immediately to antegrade; it takes approximately 27 seconds to several minutes. Furthermore, it is thought that if embolization occurs, the debris would be directed across the dilated segment toward the axillary artery. Vertebral artery embolization does occur, however, and when it does it is frequently fatal.

Balloon dilation of a stenotic subclavian artery, either proximal or distal to the vertebral artery origin, does not cause occlusion of the vertebral artery, in spite of the balloon's being inflated across the origin of the vertebral artery. Vertebral artery occlusion can occur if the stenotic plaque is adjacent to its origin.

In case of proximal subclavian artery stenosis on the right side, the balloon position might compromise flow to the carotid artery. The carotid artery circulation can be protected by placing an angiographic catheter from the contralateral femoral approach.

The simplicity of subclavian angioplasty, along with the low complication rate, fewer hospitalization days, and its low cost, is appealing in comparison to the surgical alternative.

### Innominate and Carotid Artery Angioplasty

Angioplasty of the carotid arteries is controversial due to the low reported morbidity and mortality of surgical carotid endarterectomy. Furthermore, the experience with carotid angioplasty is limited, and the follow-up is short, making a good comparison of results with surgical endarterectomy almost impossible. Surgical endarterectomy has a morbidity of 1% to 4% and a mortality of 1% to 8%. Recently, there has been an increased interest in the placement of carotid artery stents.

The indications for angioplasty of the carotid arteries gathered from the few reports in the literature include the following:

- Smooth stenotic lesions at the origin of the innominate and carotid arteries, either from atherosclerotic or inflammatory processes such as Takayasu's disease (Fig. 8.3).
- Fibromuscular dysplasia of the internal carotid artery (Fig. 8.4).
- Stenotic external carotid artery in patients with an occluded internal carotid artery (Fig.

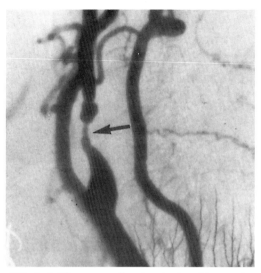 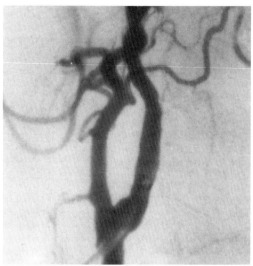

**FIG. 8.4.** Fibromuscular dysplasia of internal carotid artery. **A:** Right brachial angiogram; narrow stenosis of internal carotid artery with characteristic pattern of fibromuscular dysplasia (*arrow*). **B:** Postangioplasty angiogram of common carotid artery. Angioplasty was performed from a direct neck approach. (Reprinted from Théron JG. Angioplasty of supra-aortic arteries. *Semin Intervent Radiol* 1987;4:331–342, with permission.)

8.5). Dilation might improve the collateral circulation between the external and internal carotid artery branches.
- Weblike stenosis of the common carotid artery.
- Recurrent stenosis from myointimal hyperplasia after carotid endarterectomy (Fig. 8.6).
- Common and internal carotid artery stenosis in patients with concomitant coronary artery disease. These are high-risk patients for carotid endarterectomy and general anesthesia.

*Methodology*

Carotid artery angioplasty can be performed either from the femoral or axillary approach. Aspirin (325 mg per day) is administered orally 2 to 5 days before angioplasty. In older patients, tortuosity of the aorta and iliac vessels can hinder the advancement of catheters across the stenotic innominate or carotid artery. In these patients, an axillary approach may be beneficial, as it is easier to catheterize the acutely angled innominate and carotid arteries. The origin of the aortic branches usually requires 10-mm balloons for effective dilation. The common and external carotid arteries require 8-mm balloons. The internal carotid artery is dilated either with a 5- or 6-mm balloon. The tip of the balloon catheter should be 5 mm in length or less. This avoids complications such as spasm in the external carotid artery.

Cerebral embolic events during or after carotid angioplasty can lead to permanent neurologic deficits. This fear has stimulated various innovative ideas, which have been tried in only a handful of cases. Therefore, their real value cannot be established at this time.

A cerebral protection technique was proposed by Théron. In this technique, through a guiding catheter positioned in the common carotid artery, a thin polyethylene balloon is floated beyond the stenosis, and the internal carotid artery is occluded. If the patient can tolerate the occlusion, angioplasty is performed (Fig. 8.7). The blood that might potentially contain the debris is aspirated through the balloon catheter and flushed into the external carotid artery system.

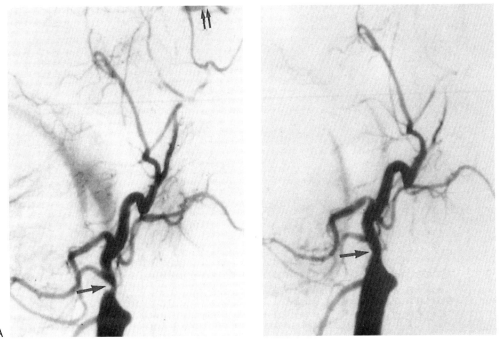

**FIG. 8.5.** External carotid angioplasty in a patient presenting with thrombosis of the internal carotid artery. **A:** Common carotid angiogram before angioplasty; narrow stenosis of external carotid (*arrow*). Carotid siphon (*double arrows*) is opacified via ophthalmic artery. **B:** Postangioplasty angiogram; enlargement of external carotid artery (*arrow*). (Reprinted from Théron JG. Angioplasty of supra-aortic arteries. *Semin Intervent Radiol* 1987;4:331–342, with permission.)

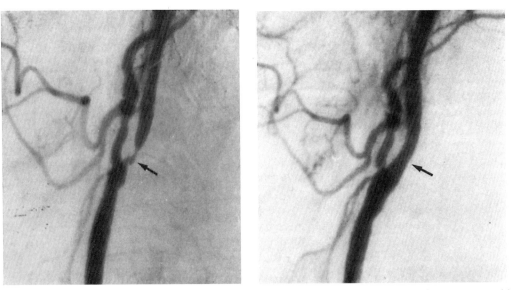

**FIG. 8.6.** Early recurrent postsurgical stenosis due to myointimal hyperplasia. **A:** Common carotid angiogram; narrow stenosis of left internal carotid artery (*arrow*). **B:** Postangioplasty angiogram shows marked improvement in diameter of internal carotid (*arrow*). (Reprinted from Théron JG. Angioplasty of supra-aortic arteries. *Semin Intervent Radiol* 1987;4:331–342, with permission.)

## Discussion

Angioplasty of the carotid arteries is controversial and at the present time is not recommended for patients who can undergo surgical endarterectomy. It is an alternative in patients with a prohibitively high surgical risk. Angioplasty has been recommended for stenotic lesions of the innominate or carotid artery origins. Surgical endarterectomy in these patients requires a thoracotomy. The surgical bypass procedures can be associated with complication rates as high as 23%, including chylothorax, wound infection, phrenic nerve palsy, and lymphatic fistulae.

Brachiocephalic trunk origin stenosis secondary to Takayasu's arteritis, fibromuscular dysplasia of the internal carotid arteries, and recurrent stenosis of the internal carotid artery from myointimal hyperplasia after surgical endarterectomy are favorable lesions for angioplasty because they are less likely to release embolic debris to the intracranial circulation. The role of intravascular stents is currently being evaluated in Europe and the United States.

Transluminal angioplasty of the carotid arteries at the present time cannot be proposed as a routine method of revascularization of the cerebral arteries. Furthermore, its precise role is uncertain because its surgical counterpart, carotid endarterectomy, is under close scrutiny in the United States and Europe.

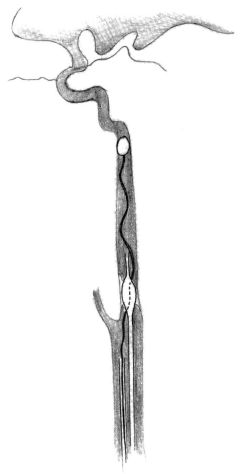

**FIG. 8.7.** Théron's cerebral protection technique. Polyethylene balloon inflated above site of dilation to occlude the internal carotid artery. Percutaneous transluminal angioplasty balloon inflated at the level of the stenosis. Debris is aspirated through percutaneous transluminal angioplasty catheter lumen. (Reprinted from Castañeda-Zúñiga WR. *Interventional radiology,* 3rd ed. Baltimore: Williams & Wilkins, 1997, with permission.)

Systemic heparinization is achieved during angioplasty by using 3,000 to 10,000 units of heparin. Successive rapid inflations and deflations of the balloon at the level of the stenotic lesions are recommended by Théron. The method seems to be as effective as leaving the balloon inflated for longer periods of time.

All patients are maintained indefinitely on aspirin (325 mg daily).

## Percutaneous Transluminal Angioplasty of the Vertebral Artery

The indications for angioplasty of the vertebral artery are poorly defined, mostly because the symptomatology is nonspecific. The symptomatology includes dizziness, diplopia, nystagmus, ataxia, cortical blindness, weakness, and memory disturbances.

Angioplasty of the vertebral artery is an attractive nonsurgical alternative, and its advantages include the following:

- Recanalization can be accomplished under local anesthesia.
- It has a low cost.
- It results in fewer in-hospital days.

- Repeat angioplasty is possible if re-stenosis occurs.
- Surgical complications are avoided.
- The endovascular approach is relatively easier in the treatment of ostial lesions.

In an uncertain situation, when the physician is confronted with symptoms of vertebral insufficiency and is not certain of its etiology in a patient with a proven stenosis of the proximal vertebral artery, angioplasty can be attempted as a therapeutic trial because of its low morbidity and mortality.

Ostial stenosis of the vertebral artery accounts for 90% of vertebral artery lesions and 40% of all vertebral basilar insufficiency cases. The natural history of vertebral basilar insufficiency indicates that 30% of patients remain asymptomatic after one or two episodes, 50% evolve, and 18% have a stroke. The ostial stenosis of the vertebral artery is usually a smooth lesion, unlike that of the carotid arteries, which is frequently ulcerated. These anatomic features are conducive to angioplasty, and the risk seems to be small.

Before angioplasty, a careful Doppler examination of the carotid-vertebral system or a four-vessel cerebral angiogram is highly recommended for the following reasons:

- To establish a diagnosis of ostial stenosis of the vertebral artery.
- To determine the dominant vertebral artery.
- To identify the abnormal flow pattern in the vertebral circulation.
- To evaluate the presence or absence of associated carotid artery stenosis.
- Manual compression of the internal carotid arteries during the Doppler study will identify the presence or absence of effective collateral circulation through the circle of Willis.

When associated hemodynamically significant carotid lesions are present, surgical correction of the carotid lesion is recommended before attempting vertebral angioplasty. Improvement of collateral circulation might alleviate the symptoms of vertebral-basilar artery disease. If the patient remains symptomatic after carotid endarterectomy, vertebral artery PTA can be performed as an alternative.

## Methodology

The angioplasty is performed under local rather than general anesthesia, using either the femoral or the axillary approach. If, during the procedure, there is an embolic event that might evolve into a stroke, the neurologic changes are easily diagnosed if the patient is awake. One of the disadvantages of using the femoral approach is the distance between the point of entry and the target site. In older people, tortuosity and ectasia might hinder the selective placement of balloon catheters from the femoral approach. When such anatomic problems exist, an axillary artery approach is recommended. The shorter and relatively straighter course between the entry point in the brachial or axillary artery and the vertebral artery ostium makes for an easier procedure. Before angioplasty, aspirin (325 mg) is administered for 2 to 5 days to inhibit the platelet activity. A bolus of 5,000 units of heparin is given as soon as the vertebral artery is catheterized. If catheterization of the artery or guidewire advancement proves to be difficult, the procedure should not be continued because potential complications may occur. When the lesion is crossed with a 5-Fr catheter, an exchange wire is advanced beyond the stenoses. The balloon size used commonly varies from 4 to 6 mm for ostial lesions. More distal lesions demand the use of smaller (3-mm) coronary angioplasty balloons. The usual balloon length is from 2.0 to 2.5 cm. The size is predetermined from the measurement of the normal vessel adjacent to the stenosis on the cut films. Rapid hand inflation and deflation for 10 to 15 seconds until the "waist" on the balloon disappears are recommended. Usually, two to three inflations are required to effectively dilate the stenosis.

In the event of complications, such as occlusion of the vertebral artery, the symptoms may not be aggravated.

Aspirin (325 mg once a day) is administered indefinitely. The postangioplasty Doppler examinations serve as a baseline for comparison with future examinations in case of recurrent stenosis.

## Results

Successful angioplasty of vertebral artery lesions is achieved in 91% of cases, when the procedures are performed by experienced neurointerventional radiologists. There is relative lack of data on long-term clinical follow-up. The role of vertebral angioplasty in the treatment of vertebral basilar insufficiency is not well established and is still evolving due to the lack of sufficient data.

## 2. CELIAC AND SUPERIOR MESENTERIC ARTERY ANGIOPLASTY

Hector Ferral, S. Murthy Tadavarthy, Joseph W. Yedlicka, Jr., and Wilfrido R. Castañeda-Zúñiga

Intestinal angina is an uncommon entity. The clinical manifestations of abdominal angina are postprandial pain and substantial weight loss. The weight loss is attributed to the fear of eating because of the anticipated abdominal pain. Nausea, vomiting, and diarrhea may be encountered but are not frequent. The disease is more common in women, in a ratio of 2:1 to 4:1. Physical findings include the presence of an abdominal bruit in 75% of patients. Patients with evidence of chronic intestinal ischemia are at increased risk to present sudden thrombosis and extensive visceral necrosis.

The majority of the patients undergo multiple diagnostic studies, including upper and lower gastrointestinal tract endoscopy, barium contrast studies, and computed tomography. The negative work-up usually prompts clinicians to obtain biplane abdominal aortography for diagnosis of visceral artery stenosis. More than two-thirds of the patients have peripheral vascular disease and aneurysmal dilation of the abdominal aorta. Coronary artery disease is a frequently associated problem, and up to 33% of the patients are symptomatic. Angioplasty of the celiac trunk and superior mesenteric artery (SMA) has been recommended for chronic abdominal angina as a nonsurgical alternative.

### Pathophysiology of Abdominal Angina

It has been said that at least two of the three visceral branches (celiac, superior mesenteric, and inferior mesenteric arteries) have to be either stenosed by 50% or occluded for manifestations of abdominal angina to occur. Most patients do not manifest intestinal angina because of the existence of significant collateral circulation.

The celiac and SMA narrowing can be secondary to atherosclerosis, fibromuscular dysplasia, or the median cruciate ligament. The median cruciate ligament crosses the origin of the celiac axis and by extrinsic compression can compromise the blood flow. Obviously, this lesion cannot be effectively managed by balloon dilation, because the compression is extrinsic (Fig. 8.8).

### Angioplasty Methodology

Aspirin (325 mg per day) is administered to minimize platelet activity 1 to 2 days before the angioplasty procedure. A lateral abdominal aortogram is obtained in inspiration and expiration. If angioplasty is indicated, the visceral branches (celiac axis, SMA, and inferior mesenteric artery) are catheterized either from the femoral or axillary artery approach (Fig. 8.9). If the origin of the vessel is acutely angled downward, it is preferable to catheterize these vessels from the axillary route because it is easier to advance the balloon catheter over the guidewire from above (Fig. 8.10). From the femoral approach the celiac and superior mesenteric arteries are easily catheterized with a 5-Fr cobra or Simmons catheter; using a soft Bentson guidewire, the stenosis is usually crossed, and the catheter is advanced over the wire. An exchange Rosen guidewire is then passed through the catheter to facilitate advancement of the balloon catheter. Alternatively, a Simmons balloon angioplasty catheter can be used to catheterize the artery. Once the ostium is engaged, a Bentson guidewire is passed across the stenosis, and the Simmons balloon is then pulled over the wire until it bridges the area of narrowing. Another alternative is the use of a coaxial system, first catheterizing the

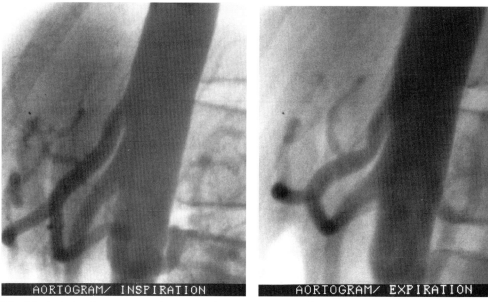

**FIG. 8.8. A:** Lateral aortogram demonstrating compression of the celiac trunk by the median cruciate ligament. Note the vascular compression evident in the expiration phase (**B**).

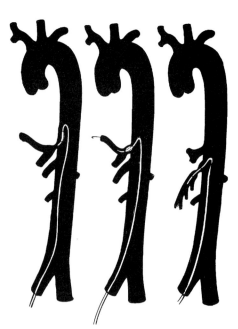

**FIG. 8.9.** Diagrammatic representation of catheterization of celiac trunk and superior mesenteric artery using a Simmons catheter from the femoral approach. (Reprinted from Castañeda-Zúñiga WR. *Interventional radiology*, 3rd ed. Baltimore: Williams & Wilkins, 1997, with permission.)

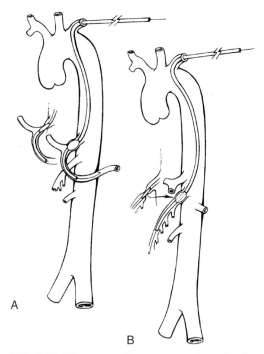

**FIG. 8.10.** Diagrammatic representation of catheterization of celiac trunk (**A**) and superior mesenteric artery (**B**) using a transaxillary approach. (Reprinted from Castañeda-Zúñiga WR. *Interventional radiology*, 3rd ed. Baltimore: Williams & Wilkins, 1997, with permission.)

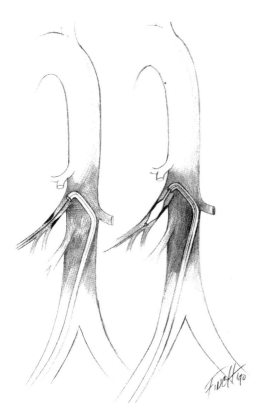

**FIG. 8.11.** Coaxial technique used to catheterize the superior mesenteric artery. (Reprinted from Castañeda-Zúñiga WR. *Interventional radiology*, 3rd ed. Baltimore: Williams & Wilkins, 1997, with permission.)

artery with the guiding catheter, crossing the obstruction with an 0.018-in. guidewire, and then advancing the balloon through the guiding catheter to the level of the obstruction over the guidewire (Fig. 8.11).

A systemic anticoagulation effect is achieved by the administration of 5,000 units of heparin either intravenously or by intraarterial injection. The size of the balloon is predetermined from the measurement of the celiac or SMA adjacent to the stenosis. Usually, it requires either a 6- or an 8-mm balloon. The stenosis is dilated in the usual fashion until the "waist" in the balloon disappears. Pressure measurements and angiographic studies are taken before and after angioplasty to evaluate the results of angioplasty and the status of the vascular bed. If necessary, a second balloon dilation can be performed with a slightly larger balloon. In patients in whom angioplasty has failed and in whom surgery is contraindicated for medical reasons, the placement of an intravascular stent should be considered (Fig. 8.12).

### Discussion

The diagnosis of abdominal angina is a challenge because it does not exclude or preclude the coexistence of nonvascular disorders that can account for the pain. Therefore, whether to revascularize is a difficult decision.

The mortality rate after surgery varies from 3% to 20%. The majority of the patients have coexistent coronary artery disease, and the immediate operative mortality is attributed to myocardial infarction. Short-term patency rates have been reported to be 70% to 90%. The mean 5-year survival rate is approximately 83%, and the 10-year survival rate is 62%.

Many interventionalists recommend angioplasty as an excellent nonsurgical alternative because of its simplicity and repeatability; initial technical success is approximately 90%, with relief of symptoms that may last from 6 to 24 months. In 50% of the patients, the symptoms recur.

Angioplasty should be the initial choice of treatment for intestinal angina in high-risk patients, as it avoids the perioperative mortality and morbidity related to the coexistent coronary artery disease. If angioplasty fails, patients can be subjected to the more invasive surgical revascularization procedures, which may involve either a thromboendarterectomy or aortovisceral bypass grafting.

### Summary

Angioplasty of the celiac axis and SMA is an interesting nonsurgical alternative treatment in the management of chronic mesenteric ischemia. Angioplasty of the celiac axis and SMA should be offered as a first therapeutic approach to patients with high surgical risk. Patients with ischemic symptoms secondary to celiac axis compression by the median cruciate ligament should undergo surgical treatment, because angioplasty in these cases will be unsuccessful.

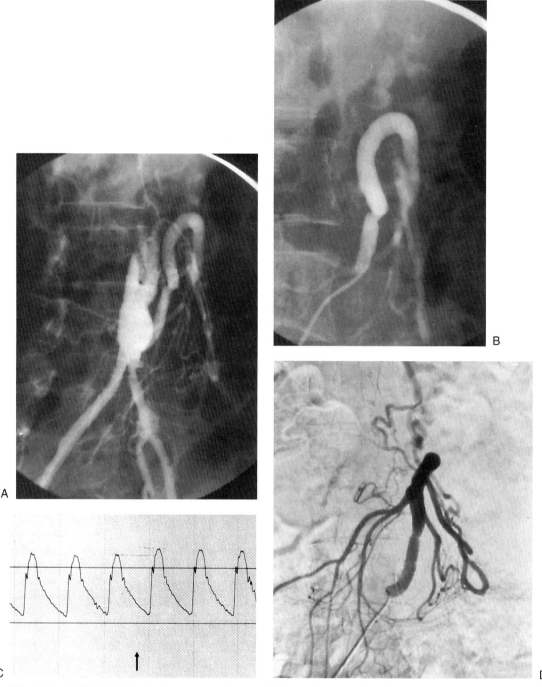

**FIG. 8.12. A:** Distal abdominal aortogram shows a high-grade stenosis of the graft-superior mesenteric artery anastomosis. **B:** Selective injection on the graft confirms the presence of a high-grade stenosis. **C:** No pressure gradient after percutaneous transluminal angioplasty/stenting with Wallstent. **D:** Selective arteriogram after stent placement shows a widely patent lumen. (Reprinted from Castañeda-Zúñiga WR. *Interventional radiology*, 3rd ed. Baltimore: Williams & Wilkins, 1997, with permission.)

## 3. PERCUTANEOUS TRANSLUMINAL ANGIOPLASTY OF THE RENAL ARTERIES

Hector Ferral

Hypertension (blood pressure more than 160/95 mm Hg) occurs in 10% to 15% of the adult population in this country. Approximately 4% of these patients have potentially correctable renovascular hypertension.

Whenever possible, the treatment of choice for renovascular hypertension is correction of the renal artery stenosis.

Grüntzig reported the first successful balloon dilation of renal artery stenosis in 1978. Renal artery angioplasty is a therapeutic alternative that has been found to be extremely useful in the management of patients with renovascular hypertension.

### RENAL ARTERY STENOSES

#### Etiology

The etiologies of renal artery stenoses include atherosclerosis (65%), fibromuscular dysplasia (30%), and miscellaneous disorders (5%).

#### Significance

*Renovascular hypertension* is defined as hypertension caused by obstruction of the main renal artery or one of its branches. True renovascular hypertension is only diagnosed retrospectively after the renal artery stenosis has been corrected and the patient's hypertension has resolved or improved.

Several tests to predict renovascular hypertension are in current use, such as renal vein renin assays, captopril renography, and aspirin injection test. However, sensitivity and specificity of these tests vary considerably from one series to another, and they are still considered to be suboptimal.

#### Indications for Renovascular Evaluation

- Patients with a documented sudden onset of hypertension
- Those without a family history of hypertension or other identifiable secondary causes of hypertension
- Young women who develop hypertension and are not taking oral contraceptives
- Patients, especially whites, who develop malignant hypertension
- Patients with long-standing hypertension who suddenly develop accelerated hypertension
- Patients who are refractory, or who become refractory, to antihypertensive drugs other than blockers of the renin-angiotensin system
- Patients with a flank bruit
- Patients who develop renal insufficiency while taking captopril

The risks and benefits of the anticipated procedure should be discussed with the patient. The aims and goals of the treatment, as well as the expected outcomes, should be clearly outlined. (Performing angioplasty in a young patient with fibromuscular dysplasia will probably have a better outcome than angioplasty in an older individual with renal failure.)

#### Indications for Renal Angioplasty

Patients with indications for renal angioplasty include the following:

- Patients at risk with a renal artery stenosis of greater than 70% demonstrated on angiography
- Patients at risk with a renal artery stenosis of 50% to 70% and a positive selective renal vein renin assay

#### *Techniques*

Renal angioplasty should be performed only by angiographers with considerable experience in dilating peripheral vessels. Inflating one balloon is simple, but crossing a tight renal stenosis with a balloon catheter requires great skill, and selection of the proper balloon size requires experience.

A high-quality preliminary midstream arteriogram is necessary to determine the approach. Unless an abdominal arteriogram has been obtained within the previous month, this study should always be obtained before catheterizing the renal arteries because profound changes may occur in a short time in the presence of a tight renal artery stenosis.

There are five percutaneous angiographic approaches to the treatment of stenoses in the renal arteries with Grüntzig-type balloons:

1. Femoral balloon catheter system via a femoral approach
2. Guided coaxial balloon catheter system
3. Axillary approach
4. Femoral balloon catheter system with the sidewinder approach
5. "Kissing balloon" technique

*Femoral Balloon Catheter System: Femoral Approach*

Access to the femoral artery is obtained using the modified Seldinger technique. We recommend the use of an introducing sheath to minimize trauma to the femoral artery. A diagnostic angiographic catheter with a cobra or Simmons-1 shape is used to achieve selective catheterization of the renal artery. The lesion is negotiated with a soft-tip Bentson or Wholey (Mallinckrodt, St. Louis, MO) guidewire. It is imperative to avoid subintimal passage of the guidewire. Once the stenosis has been crossed, the 5-Fr selective catheter should be advanced across it. A small amount of contrast medium is injected to confirm the intraluminal position of the catheter, and 2,000 to 5,000 units of heparin and a vasodilator (nitroglycerin, 100 to 200 µg, or verapamil, 2.5 mg) are injected through the catheter. A stiff wire for better support is then placed (movable-core-type J, Rosen wire, or 0.025 platinum-plus wire) through the catheter beyond the stenosis, and the diagnostic catheter is replaced with the appropriate size renal balloon catheter. The balloon catheter is chosen by measuring the renal artery proximal and distal to the stenosis (Fig. 8.13) and estimating the original size of the renal artery. If the artery is estimated to have been 5 mm in diameter, this is the balloon size that should be used. Because this method does not take radiographic magnification into account, the renal arteries are being slightly (1 mm) overdilated. The guidewire must not be moved back and forth in the branches of the renal artery when exchanging the catheters, because this may induce spasm or cause occlusion of the segmental branches.

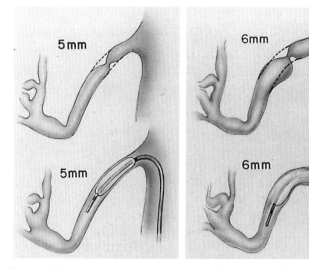

**FIG. 8.13.** A balloon catheter is selected to correspond to the original diameter of the stenotic artery. Poststenotic dilation must be taken into account. (Reprinted from Tegtmeyer CJ, Kellum CD, Ayers C. Percutaneous transluminal angioplasty of the renal arteries. *Radiology* 1984;153:77–84, with permission.)

The balloon catheter is positioned across the stenosis under fluoroscopic control, and the balloon is inflated either with an inflation device or with a syringe. If a syringe is used, a 10-mL size is probably ideal because it is capable of generating approximately 9.4 atm of pressure during inflation and sufficient negative pressure to deflate the balloon rapidly. A pressure gauge should always be used. The balloon is first inflated to 2 atm of pressure to determine its position in relation to the stenosis. If it is properly positioned, the balloon is then inflated to 4 to 6 atm and left inflated for 30 to 40 seconds. It may be necessary to repeat this three times. The progress can be monitored by watching the configuration of the balloon as it is inflated. Immediately after angioplasty, an arteriogram is performed to assess the results. Before removing the balloon catheter, it is important to completely deflate the balloon and apply suction as it is being removed.

*Angiogram Postangioplasty*

We do not recommend removing the guidewire from the renal artery before performing the control angiogram. Several alternatives can be applied:

- Placement of a pigtail catheter from a contralateral approach.
- Removal of the angioplasty catheter over a 0.018-in. wire and injection of carbon dioxide or contrast using a diagnostic catheter and a Y adaptor.
- Removal of the angioplasty catheter over a 0.018-in. guidewire and advancement of a 4-Fr pigtail via the same 7-Fr femoral sheath.

*Guided Coaxial Balloon Catheter System*

Renal angioplasty may be performed using a coaxial catheter system. The technique is as follows:

- Eight- or 9-Fr renal guiding catheter (different shapes).
- 5-Fr coaxial angioplasty balloon catheter (Fig. 8.14). The renal guiding catheter is available in three configurations (Fig. 8.15). The guiding catheter may be advanced coaxially, using a 6-

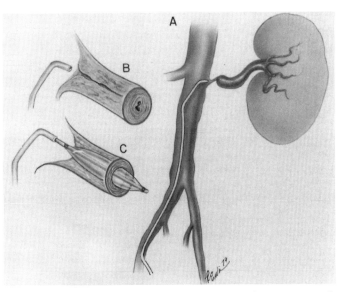

**FIG. 8.14.** Technique of renal dilation with the coaxial balloon catheter system. **A:** Stenosis in the left renal artery. **B:** Orifice of the renal artery is selected with the guiding catheter. **C:** The 4.5-Fr dilation catheter is directed through the stenosis. (Reprinted from Tegtmeyer CJ, Dyer R, Teates CD, et al. Percutaneous transluminal dilation of the renal arteries: techniques and results. *Radiology* 1980;135:589–599, with permission.)

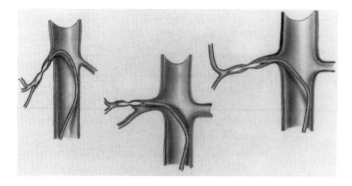

FIG. 8.15. The renal guiding catheter is available in three basic configurations. (Reprinted from Castañeda-Zúñiga WR. *Interventional radiology*, 3rd ed. Baltimore: Williams & Wilkins, 1997, with permission.)

or 7-Fr angiographic catheter as a guide and using a 0.035-in. guidewire. The orifice of the renal artery is carefully selected. The small coaxial catheter is then passed through the guiding catheter across the stenosis. The balloon catheter will accept a 0.018- or 0.035-in. guidewire, which is passed across the stenosis before advancing the coaxial catheter to facilitate traversing a tight stenosis. With the advent of the new platinum wires, which are highly visible radiologically, this is a very effective technique. In the presence of a tortuous renal vessel, a tight stenosis, or a stenosis in a branch, the fine guidewire greatly facilitates passage of the balloon catheter. The guiding catheter technique is useful because it provides support, allows contrast injection after angioplasty, and facilitates stent placement if necessary.

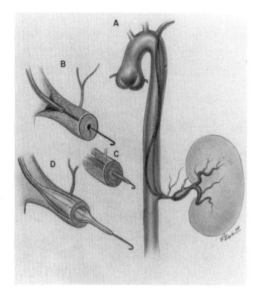

FIG. 8.16. Axillary approach. **A:** Stenotic left renal artery is selected with a diagnostic cobra catheter. **B:** The guidewire is passed through the stenosis. **C:** The selective catheter is replaced with a dilation catheter. **D:** The selective catheter is replaced with a dilation catheter, which is inflated. (Reprinted from Tegtmeyer CJ, Dyer R, Teates CD, et al. Percutaneous transluminal dilation of the renal arteries: techniques and results. *Radiology* 1980;135:589–599, with permission.)

### Axillary Approach

The axillary approach may also be used to dilate the renal arteries (Fig. 8.16). This approach greatly simplifies the procedure when the renal arteries originate from the aorta at a sharp angle, because the stenotic artery is easily selected. Once the guidewire is in place across the lesion, the dilation catheter has a natural tendency to follow its gentle downward curve. Passage of the guidewire and then the catheter through the stenosis is often facilitated by having the patient take a deep breath. The axillary approach is also useful when severe atherosclerotic disease or a bypass graft is present in the pelvic or abdominal vessels.

A left axillary approach is usually taken because it offers the straightest approach to the descending aorta and because the catheter has a tendency to buckle in the ascending aorta when the right axillary approach is attempted.

Using the axillary approach has the increased possibility of brachial plexus injury. This can be minimized by using the high brachial approach, in which the artery is entered just distal to the axillary crease. The artery is easier to control in this area because it can be compressed against the humerus.

### Sidewinder Approach

The sidewinder approach combines the advantages of the femoral and axillary approaches (Fig. 8.17). The femoral artery is punctured, but the renal artery is approached from above. The renal artery is selected with a "shepherd's crook," or "sidewinder," catheter, which is advanced across the lesion under fluoroscopic control as contrast medium is injected, with care taken to keep the tip within the lumen of the artery. Alternatively, a flexible-tip guidewire is advanced across the stenosis followed by the catheter. Once the catheter is across the lesion, heparin, 2,000 to 5,000 units, is injected. The guidewire is replaced with a movable-core J guidewire or a Rosen wire. The diagnostic catheter is then exchanged for the appropriate-size renal balloon catheter. After the sidewinder catheter has crossed the lesion, the balloon catheter usually crosses the stenosis with ease. After dilation, a midstream arteriogram is obtained to document the results.

The principal advantage of this technique is that withdrawing the sidewinder catheter advances the diagnostic catheter across a tight stenosis because the configuration of the catheter exerts considerable downward force as it is withdrawn over the guidewire. This technique is very helpful in traversing tight stenoses.

### "Kissing Balloon" Technique

If two renal arteries originate from the aorta in proximity or if the lesion involves a major bifurcation in the renal artery, dilating the lesion may occlude the adjacent vessel. In this situation, catheters are inserted through both femoral arteries, and a catheter is passed across the lesion in the involved branch. The origin of the uninvolved branch is usually protected by the diagnostic catheter. If the origin of the vessel is involved by the lesion or compromised by the procedure, the catheter is exchanged for a second balloon catheter, and the lesion is dilated.

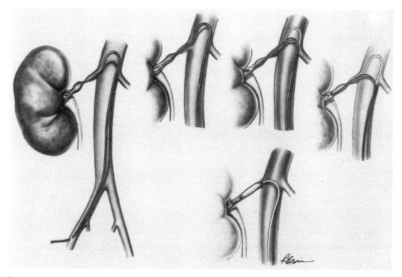

**FIG. 8.17.** Sidewinder technique. After selection of the renal artery, a flexible-tip guidewire is advanced through the lesion under fluoroscopy. The catheter is advanced by a withdrawing movement at the puncture site. Exchange for a stiffer wire can then be done and intervention performed. (Reprinted from Castañeda-Zúñiga WR. *Interventional radiology,* 3rd ed. Baltimore: Williams & Wilkins, 1997, with permission.)

## Periprocedural Management

### Preprocedure

- A team approach should be used.
- Patients should be properly selected.
- Management of blood pressure requires the close cooperation of a hypertension specialist.
- A complication may create a surgical emergency, so the procedure should be performed only when a skilled vascular surgeon is available.
- Blood pressure must be monitored especially carefully during the first 24 to 48 hours after angioplasty, because profound changes may occur.
- Consider discontinuation of antihypertensive drugs before angioplasty, thereby helping to prevent a precipitous drop in the blood pressure. If the diastolic pressure rises above 110 mm Hg, the blood pressure should be controlled.
- All patients undergoing renal angioplasty should have an intravenous line in place.

### Postprocedure

Renal angioplasty is not an outpatient procedure. Drastic blood pressure changes may occur after a successful angioplasty. The patients should be kept in observation for 48 hours.

If not contraindicated, patients receive 2,000 units of heparin subcutaneously every 6 hours, beginning 8 hours after the procedure and continuing for 2 days. The patients also receive 75 mg of dipyridamole (Persantine) orally twice a day for at least 6 months and 325 mg of aspirin once a day, beginning the day before angioplasty and continuing for life. After angioplasty, patients should be monitored by a cardiologist or a nephrologist to control high blood pressures. The patient's need for antihypertensive drugs changes after a successful procedure, so blood pressure must be monitored closely. If the blood pressure rises in the ensuing months, arteriography or intravenous digital subtraction angiography should be repeated.

## Complications

- The complication rate of renal angioplasty ranges from 5% to 10%, with most complications being minor.
- The most frequent major complication is transient renal insufficiency, which is related to the contrast medium. The frequency can be reduced by performing the diagnostic arteriogram several days before the therapeutic procedure or with the use of carbon dioxide in patients with renal insufficiency.
- Subintimal dissection with the guidewire or the diagnostic catheter may result when attempting to cross the stenosis. A small intimal flap will usually heal, but angioplasty should be postponed for 4 to 6 weeks. The use of intravascular stents is indicated in the management of intimal flaps causing significant hemodynamic obstruction.
- Thrombosis of the renal artery is an infrequent complication. This complication may be treated by infusing a thrombolytic agent into the renal artery.
- If renal artery rupture is noted immediately after deflation of the balloon, the balloon should be reinflated to occlude the proximal renal artery, and the patient should be operated on immediately.
- Distal embolization is an infrequent complication.
- The most frequent minor complication is spasm or occlusion of a renal arterial branch, usually caused by movement of the tip of the guidewire back and forth within the vessel. If possible, the guidewire should not be placed within the segmental branches. Focal spasm may also be produced in the area immediately adjacent to the angioplasty site. If spasm occurs, verapamil may be given through the arterial catheter in a dose of 2.5 to 5.0 mg. Also, because the mechanism of action of nitroglycerin differs from that of the calcium antagonists and because it has an additive effect with the calcium channel blockers, it may be useful against spasm in the renal arteries. It is injected directly into the affected renal artery (50 to 200 µg). Cal-

cium channel blockers may induce hypotension and should be used with caution in patients with known cardiac conduction defects.
- Hematomas at puncture site are common. Special care is necessary when using an axillary/brachial approach. A high brachial artery approach is recommended to avoid devastating effects, including brachial plexus injury, caused by a large hematoma in the axilla.

## *Results*

### Definitions

*Technical success* is defined as residual stenosis of less than 30% after angioplasty.

*Procedural failure* (the procedure is a failure) occurs if the procedure cannot be completed or if residual stenosis is greater than 50%.

*Clinical outcome* is described according to the World Health Organization criteria.

*Cure* is defined as diastolic pressure of 90 mm Hg or less while the patient is not receiving any antihypertensive medication.

*Improvement* requires (a) diastolic pressure of 90 mm Hg or less with the administration of equal or reduced doses of medication or (b) diastolic pressure of greater than 90 mm Hg but less than 110 mm Hg, with at least a 15 mm Hg decrease from measurements obtained before angioplasty.

### Long-Term Results

The long-term results of renal angioplasty can be assessed in three ways: the effects on vessel patency, on blood pressure, and on renal function. The effect of the procedure on vessel patency is related to the cause and characteristics of the lesion. The patients can be divided into five distinct groups:

1. Atherosclerotic renal artery stenoses or occlusions
2. Fibromuscular dysplasia
3. Renal allografts
4. Saphenous bypass grafts
5. Patients treated primarily for renal insufficiency

### *Atherosclerotic Lesions*

- Reported technical success rates vary between 78% and 83%.
- Lower technical success rates and clinical outcome are expected for ostial as compared with nonostial lesions.
- In the University of Virginia series, 94% of the 65 hypertensive patients with atherosclerotic lesions were helped by the procedure: 15 were cured, and 46 improved. Analysis of the results in these patients revealed some factors that are important to the success of renal angioplasty. First, better results are achieved with unilateral renal artery stenoses than with bilateral stenoses, because the re-stenosis rate is higher in patients with severe bilateral renal artery disease than in patients with unilateral renal artery stenosis.
- The use of metallic stents in ostial lesions has been reported, with a success of 90% and a long-term patency of 80%.

### *Fibromuscular Dysplasia*

The best results in renal angioplasty are achieved in patients with fibromuscular lesions (Fig. 8.18), which respond well to balloon dilation, usually at pressures of 4 atm or less. Thus, if the lesion can be crossed, a good result can be expected. Technical success can be expected in more than 88% of patients. Approximately 50% will be cured after angioplasty, and overall, 90% will improve. Long-term patency is very good and has been described to be as high as 85% to 87% at 10 years. Angioplasty is currently the treatment of choice for renovascular hypertension secondary to fibromuscular dysplasia.

### *Stenoses in Renal Allografts*

Hypertension in a patient with a renal transplant represents a diagnostic and therapeutic challenge. The prevalence of hypertension in patients with renal transplant can be as high as 50% to 60% 1 year after transplantation. Causes for hypertension in renal transplant patients include acute and chronic rejection, recurrence of primary disease (glomerulonephritis), renin

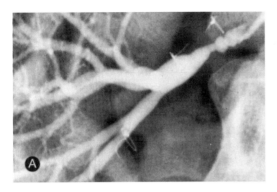
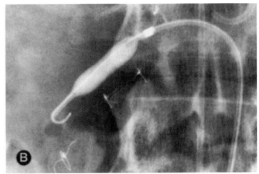
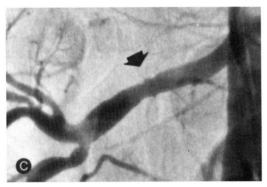

**FIG. 8.18.** Renal angioplasty for fibromuscular dysplasia. **A:** Right renal arteriogram reveals stenosis in the artery. **B:** A 5-mm balloon is used to dilate the lesion. **C:** Excellent angiographic result seen after dilation (*arrow*). (Reprinted from Castañeda-Zúñiga WR. *Interventional radiology*, 3rd ed. Baltimore: Williams & Wilkins, 1997, with permission.)

production by native kidneys, cyclosporine toxicity, and true renovascular hypertension secondary to renal artery stenosis. The problem may be caused by accelerated atherosclerosis, rejection, or fibrosis at the anastomotic site.

In the same manner, criteria for technical success and clinical outcome are difficult to describe in this group of patients. A technical success ranging from 58% to 88% has been reported.

A 6.3% cure rate at 12 months' follow-up with an 80% clinical improvement rate can be expected. Caution should be exercised when dilating lesions of renal allografts because there are no collateral vessels to the kidney. Therefore, if the vessel is occluded, operation must be undertaken immediately. A transplant surgeon should be readily available. Angioplasty is the recommended procedure in nonanastomotic renal transplant arterial stenosis because results in anastomotic lesions are poor, and complication risks are higher.

*Stenoses in Saphenous Bypass Grafts*

In the University of Virginia series, three patients had PTA of stenoses in renal saphenous bypass grafts. The procedure was successful in all cases.

*Renal Insufficiency*

Under certain circumstances, alleviation of renal artery stenosis or obstruction is important to preserve renal function. Correction of stenosis is probably worthwhile in unilateral disease if kidney size indicates potentially significant preservation of functional tissue or if bilateral renal artery stenoses are present. In the latter case, an attempt to alleviate obstruction for the larger kidney should be given priority even if the smaller kidney is secreting all the renin.

The results of angioplasty are not as dramatic in patients treated primarily for renal insufficiency as in patients treated primarily for hypertension, but they are nonetheless encouraging.

## Summary of Results

Renal angioplasty is a clinically effective means of treating renovascular hypertension. A review of the large series in the world literature shows that if the initial dilation is successful, the vessels can be expected to remain patent in 70% to 90% of the patients, and that if the vessel remains patent for at least 8 months, it is likely to remain patent for at least 5 years. The reported recurrence rate has varied from 12.9% to 22.5%. The rate is clearly higher for atherosclerotic stenoses than for other types. A significant factor is the success of the initial dilation: Lesions with a 30% residual stenosis on the immediate postdilation films are more likely to recur than lesions in which a better result has been obtained. Therefore, it is essential to obtain a good result initially. However, re-stenosis can usually be redilated, and the procedure is often easier than the initial one.

## 4. AORTO-ILIAC AND PERIPHERAL ARTERIAL ANGIOPLASTY

Hector Ferral and Wilfrido R. Castañeda-Zúñiga

The advantages of the percutaneous treatment of atherosclerotic disease of the aortic, iliac, and peripheral arteries include the following:

- In properly selected patients, the results are as good as the results achieved with traditional surgical methods.
- The procedure is safe, simple, and relatively painless.
- It can be performed with the patient in the hospital for little more than 24 hours.
- Unsuccessful angioplasty does not preclude surgical revascularization.
- Recurrent or worsening disease after an initially successful PTA can be managed with another angioplasty without the necessity of dealing with postoperative scarring.
- There is no risk of loss of sexual function as a consequence of aorto-iliac angioplasty, as there is when the surgical procedures are used.

### Patient Selection

#### Indications and Contraindications

PTA is a cooperative effort; the decision to do PTA should be made jointly by a qualified vascular surgeon, the angiographer, and, usually, an internist. It is imperative to refrain from attempting PTA when surgical cooperation is not available. Although the procedure is relatively safe, complications are inevitable in any series, and the immediate availability of a capable vascular surgeon is essential to avert catastrophes.

#### Clinical Indications

- Patients with intermittent claudication who desire symptom relief.
- Limb salvage (rest pain, infection, nonhealing ulcer). In high-risk patients, angioplasty is considered regardless of the site of disease predicted by noninvasive studies.

#### Clinical Contraindications

- The presence of acute symptoms (less than 6 to 8 weeks) or sudden worsening of lower extremity symptoms within this time is a relative contraindication to angioplasty. Fresh thrombus is likely to be present. In these cases, thrombolytic therapy should be performed before angioplasty procedures.
- Patient presenting with the "trash-foot" syndrome, indicative of peripheral embolization.
- Unstable patient.

#### Anatomic Indications and Procedure Planning

Doppler data are very useful in planning the approach for diagnostic arteriography in anticipation of angioplasty and also permit objective assessment of the immediate and long-term results of the procedure.

The single most important factor in deciding whether a patient is a candidate for PTA is the diagnostic arteriogram. The arteriogram must be of the highest quality to permit identification of ulcerations, mural thrombus, or threadlike patent channels within high-grade stenoses. Oblique views are needed for examination of such areas as the orifices of the common iliac arteries, the origins of the hypogastric arteries, and the bifurcation of the common femoral artery.

### *Anatomic Contraindications*

Certain angiographic features are said to diminish the likelihood of success at the angioplasty site. These findings may be considered relative anatomic contraindications:

- Aorta: Eccentric lesions in the proximal infrarenal region, particularly those of nonatherosclerotic etiology.
- Iliac arteries: Stenoses longer than 2 to 3 cm, eccentric stenoses, heavy calcification (particularly if eccentric), total occlusion of the common iliac artery, or stenosis at the origin of the common iliac artery.
- Superficial and deep femoral arteries: Stenoses longer than 2 to 3 cm, occlusions longer than 10 to 15 cm, or heavy calcification of the stenotic or occlusive lesion.
- Popliteal, tibial, and peroneal arteries: Long segments of disease where there is a large patent vessel at the ankle. (The latter vessel is suitable for bypass grafting, particularly in institutions in which *in situ* grafting is used.)
- Ulcerative disease with evidence of distal embolization.
- Complete occlusion of the superficial femoral artery in the presence of morbid obesity.
- Extensive disease in any segment of the aorto-iliac or femoral vasculature because long segments of intimal-medial disruption predispose to recurrent disease and distal complications.
- The demonstration of fresh mural thrombus is an absolute contraindication. Thrombolysis should be performed first.
- Poor run-off diminishes the likelihood of long-term benefit from PTA.

PTA may be attempted in the presence of unfavorable anatomy in the following circumstances:

- When the patient is at high operative risk.
- When preservation of a saphenous vein is desirable.
- When operation would be technically compromised because of an inadequate saphenous vein.
- When the patient is not expected to live much longer.
- Whenever limb salvage is the primary goal—that is, when an amputation is expected, angioplasty is attempted regardless of the anatomy if for no other reason than that it may alter the level of amputation even if it is only partially successful.
- In high-risk patients in whom a major intraabdominal operation is to be avoided, angioplasty of a flow-limiting aorto-iliac segment lesion may be used to provide inflow to the groin in preparation for a femoral-popliteal bypass graft, profundoplasty, or femoral bypass graft.

### Lesions Ideal for Angioplasty

- Focal distal aortic stenoses of atherosclerotic origin
- One- to 2-cm concentric, high-grade stenoses of the iliac arteries located some distance from a major bifurcation
- Two- to 4-cm stenoses or occlusions of the superficial and deep femoral arteries
- Relatively short, isolated stenoses or occlusions in the popliteal and tibial arteries

### Periprocedural Patient Care

### *Preangioplasty Care*

- Obtain informed consent and discuss the risks and benefits of the procedure with the patient. Discuss the probable value of an exercise program (the guidelines for which are best determined by the patient's internist or cardiologist) and the necessity of abstinence from tobacco use.
- The following laboratory data should be available: blood urea nitrogen level, serum creatinine level, platelet count, prothrombin time, and partial thromboplastin time. If the patient

has compromised renal function and a diagnostic arteriogram is obtained near the time of the planned angioplasty, the serum creatinine level should be ascertained immediately before PTA to ensure that the procedure is not begun on a patient with failing kidneys.
- Begin intravenous fluids the evening before angioplasty and continue at a rate of 50 to 100 mL per hour at least until the evening after the procedure.
- *Do not premedicate the patient heavily* if narcotic analgesics are to be used. Clinical experience has shown that a patient who feels no discomfort during inflation of the balloon in the distal aorta, common or external iliac arteries, or hypogastric arteries may have an anatomically unsatisfactory result unless a larger balloon is used. Conversely, the patient who experiences excruciating pain with minimal inflation of a balloon is at risk of excessive arterial trauma or vessel rupture and should be treated with a smaller balloon. We think it is ideal for a patient to feel moderate discomfort and to communicate this during the procedure.

### *Intraprocedural Care*

#### *Puncture Technique*

1. The optimal access for the procedure is selected based on clinical findings, noninvasive tests, and previous arteriograms.
2. Single-wall puncture technique should always be attempted first.
3. Puncture of the pulseless common femoral artery in the presence of occlusive or stenotic iliac disease can be accomplished in the following ways:
   a. Palpation of the nonpulsatile thick-walled artery (in slender patients).
   b. Fluoroscopic observation of the femoral head and puncture of the artery, which is almost always located over the medial third of the femoral head. Any calcification in the vessel wall may facilitate the puncture.
   c. Doppler localization of the vessel.
   d. Injection of contrast medium, if a catheter is present in the distal aorta from a diagnostic procedure performed via the contralateral approach, to cause collateral opacification of the common femoral artery, which can then be punctured. Road mapping is also very useful to guide the puncture.
   e. The puncture may be performed under real-time ultrasound guidance.

#### *Balloon Selection and Use*

The angiographer should be familiar with the available balloon types, balloon profile and caliber, bursting pressures, and compliance. We use a balloon equal to the measured diameter of the nearest normal caliber segment of vessel without correction for magnification. The balloon should be slightly longer than the lesion. Recurrent stenoses are most likely after incomplete dilation.

We routinely inflate the balloon to its maximum diameter and pressure and leave it in this state for 60 to 120 seconds. With this technique, a single inflation usually produces the desired result. Several inflation devices are available that can monitor inflation pressure continuously. In general, the manufacturer's recommended bursting pressure should not be exceeded to avoid balloon rupture. Diluted contrast medium (1:2 to 1:3) should be used for balloon inflation. Deflation of balloons is best performed with large-bore syringes, which can generate maximum negative pressure.

#### *Pressure Measurements*

Pressure measurement is not absolutely necessary before angioplasty in patients with lesions that are obviously hemodynamically significant; it is only in questionable lesions that pressure measurements are of value in determining whether to perform angioplasty. Under these circumstances, pressures are measured with a multi-side-hole catheter, with the patient at rest and after maximizing lower extremity flow with tolazoline (15 mg intraarterially), nitroglycerin (100 µg intraarterially), or papaverine (30 mg intraarterially). Any pressure gradient at rest is significant, and a gradient of more than 15 mm

Hg after flow augmentation warrants angioplasty of the offending lesion.

### *Pharmacologic Considerations*

During angioplasty, all patients should be systemically heparinized. Iliac artery angioplasty is adequately covered by systemic heparinization with 2,500 units given as an intravenous bolus dose. During angioplasty of the distal vasculature, higher doses (5,000 units) may be used.

Monitorization of systemic heparinization with an activated clotting time (ACT) machine is useful. A baseline ACT is obtained before any heparin is administered. Subsequent heparin doses may be regulated based on ACT results. An ACT of greater than 300 seconds is ideal for angioplasty of distal vessels.

Severe vasospasm obliterating the vascular lumen around the catheter in the external iliac artery, particularly in women, is not an uncommon phenomenon. To combat such catheter-induced vasospasm, the drug of first choice is nitroglycerin given as a 100-μg intraarterial bolus. This dose may be repeated several times unless systemic hypotension appears. The stimulus to vasospasm is the presence of the catheter and guidewire; once the spasm is broken with nitroglycerin or other drug and the stimulus is removed, it is unlikely that vasospasm will recur.

Because of the propensity of the popliteal and tibial vessels to respond with spasm to the presence of the guidewire and catheter, we routinely premedicate patients scheduled for angioplasty of these vessels with 10 to 20 mg of nifedipine orally unless there is a contraindication to its use. Sublingual nifedipine has been associated with severe clinical complications, and its use is no longer recommended.

Patients undergoing below-the-knee procedures may be given at least one 100-μg bolus of nitroglycerin as soon as the catheter is placed in the proximal popliteal artery. Nitroglycerin is used liberally whenever a spasm is noted. In the unlikely event that severe vasospasm cannot be controlled with nitroglycerin, intravascular verapamil may be given if one exercises the utmost care. Verapamil is contraindicated in the presence of congestive heart failure or bradyarrhythmias.

### *Postangioplasty Care*

The intravascular sheath can be safely removed when the ACT is less than 180 seconds. Some authors leave the sheath in until the heparin wears off and the ACT is less than 180 seconds; others choose to reverse the heparin dose with intravenous protamine sulfate (10 mg per 1,000 IU of "active" heparin). The half-life of heparin ranges between 1 and 2 hours. Some choose to remove the sheath immediately after the procedure without heparin reversal to avoid the potential reactions to the protamine. Regardless of the protocol, groin control is of great importance to avoid complications. Pressure should be held until there is no evidence of bleeding through the puncture sites.

The patient is returned to a general medical or surgical bed and maintained at bedrest with the leg extended for 6 hours. The groin is monitored for bleeding, vital signs are monitored, and pulses are checked at frequent, regular intervals. If an axillary approach was necessary, the patient is likewise maintained at bedrest, with the arm kept in a sling for 24 to 48 hours. The axilla is monitored, and the neurologic function of the arm is checked regularly for evidence of brachial plexus dysfunction secondary to an axillary hematoma, which, if not promptly cared for, could cause devastating permanent brachial plexus palsy. Systemic heparinization, if needed (below-the-knee angioplasty, low-flow states), may be restarted 2 to 3 hours after the sheath has been removed. The patient must remain on strict bedrest during this period.

### *Antiplatelet Therapy*

Antiplatelet therapy is usually started the day before the procedure. A dose of 500 mg of oral aspirin is given the day before angioplasty and on procedure day. All patients receive salicylates for at least 6 months postangioplasty; we think that salicylates should be maintained indefinitely. We administer 80 mg of aspirin once daily in combination with 50 to 75 mg of dipyridamole. The optimum dose of aspirin is uncertain, particularly when used in combination with dipyridamole, and the results of some current clinical studies are being awaited.

Recently, some investigators have used ticlopidine (Ticlid) as an antiplatelet agent. It interacts with platelet glycoprotein IIb/IIIa and inhibits the binding of fibrinogen to activated platelets; thus, it inhibits platelet aggregation and clot retraction. The usual dose is 250 mg orally twice a day for 2 weeks followed by 250 mg orally every day for 1 week. The most important adverse reaction is neutropenia, which occurs in approximately 1% of the cases.

## TRANSLUMINAL ANGIOPLASTY OF ABDOMINAL AORTA AND PROXIMAL ILIAC VESSELS

Hemodynamically significant, discrete, focal aortic stenoses are best treated with aortoplasty. The results are encouraging; the technical success rate is close to 100%, and clinical improvement is achieved in 93% of the patients. Diffuse, long stenotic lesions with calcifications and ulcerations, with or without thrombi, are best treated with surgery. Transluminal angioplasty of distal aortic stenosis with involvement of the proximal segment of the common iliac arteries can be successfully managed with the kissing balloon technique (Fig. 8.19). These patients are dilated with two balloons using bilateral retrograde femoral punctures (Fig. 8.20). The balloons must be long enough to include the segment of iliac artery that needs to be dilated as well as the segment of abdominal aorta requiring dilation.

### Technique of Aortoplasty

- The procedure is usually performed using the femoral route.

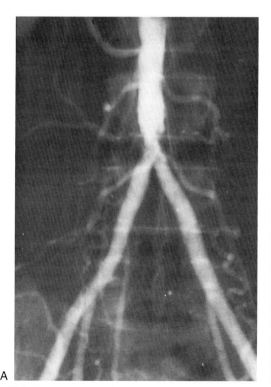
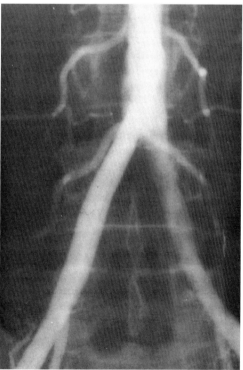

**FIG. 8.19. A:** High-grade stenosis of distal abdominal aorta extending into both common iliac arteries. **B:** Follow-up angiogram shows a patent lumen postplasty. (Reprinted from Castañeda-Zúñiga WR. *Interventional radiology*, 3rd ed. Baltimore: Williams & Wilkins, 1997, with permission.)

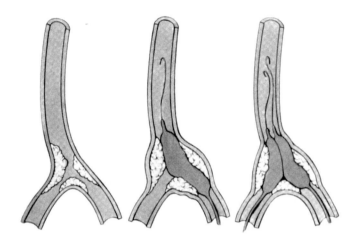

**FIG. 8.20.** Diagram illustrates the possible mechanism of plaque displacement into contralateral iliac artery by single balloon inflation. (Reprinted from Castañeda-Zúñiga WR. *Interventional radiology*, 3rd ed. Baltimore: Williams & Wilkins, 1997, with permission.)

- The diagnostic angiography is performed through the femoral artery with the stronger pulse.

Diagnostic angiography is done with a 5-Fr pigtail. The morphologic appearance is studied. Oblique views are necessary to evaluate the extent of iliac artery involvement. In some cases, it is difficult to negotiate the lesion. This can be solved by redirecting the guidewire with a torque-controlled catheter. A wide variety of guidewires can be used to cross the obstruction, including Bentson, movable core, long tapered straight, hydrophilic-coated, and occasionally tight J guidewires. Extreme caution should be used during advancement of hydrophilic-coated wires, especially the straight-tip type. Occasionally, the wire can be subintimal; this can go unnoticed because the slippery surface allows the wire to be easily advanced subintimally, up to the level of the abdominal aorta, without feeling any significant resistance. When using such wires, catheter advancement should be performed with extreme caution, and at the point of least resistance, contrast should be injected to assess intravascular position. We prefer to use a long, floppy-segment Bentson wire, which is usually atraumatic because it deflects off eccentric plaques with minimal risk of subintimal dissection.

- If the anatomy of the lesion dictates the use of a dual-balloon technique instead of a single-balloon dilation, the opposite femoral artery is catheterized.
- Once the lesion is crossed, a bolus of 5,000 units of heparin is given.
- Sizing of the balloon for effective dilation is crucial. The diameter of the dilating balloon should be equal to the diameter of the aorta above or below the level of the obstruction. Although there is an inherent magnification of 10% to 20% on cut-film angiography, the measured diameter determines the size of the balloon or balloons to be used. The resultant effect is stretching of the aortic wall slightly beyond its normal size.
- If the lesion is infrarenal and well above the bifurcation, a single-balloon catheter of larger diameter will suffice to complete dilation. The availability of 16-mm balloon catheters on 5.8-Fr shaft diameters (XXL, Boston Scientific, Natick, MA) greatly facilitates balloon dilation of the aorta.
- If the lesion involves the distal abdominal aorta and the proximal common iliac arter-

ies, the dual-balloon technique must be used.
- Before dilation, measurements of pressure above and below the stenosis are routinely obtained.
- The balloons are inflated two or three times for 30 to 45 seconds. If the kissing technique is used, balloons are inflated simultaneously.
- The endpoint of balloon inflation is the disappearance of the indentation in the balloon during inflation. Pressure measurements are routinely obtained after the aortoplasty along with follow-up arteriograms, preferably using a 5-Fr catheter.
- Patients are placed on aspirin (325 mg daily) for life and are encouraged to quit smoking and to start on a vigorous exercise program.

## Blue Digit Syndrome

The blue digit syndrome is characterized by sudden onset of pain and purple discoloration of the digits in the presence of *p*alpable *p*eripheral *p*ulses (the "three Ps" of the blue digit syndrome). The cyanotic areas can appear in the lower legs, toes, hands, or fingers and are clearly demarcated from the adjacent normally perfused skin. The pathologic alterations are attributed to atheromatous plaque or fibrinoplatelet aggregate embolization. Therefore, the blue digit syndrome should be treated as a limb salvage situation, as in other entities, such as rest pain or gangrene. Atheromatous plaque in the aorta may erode through the intima, ulcerate, and discharge cholesterol emboli into downstream distant peripheral circulation. It can cause renal failure, hypertension, gastrointestinal hemorrhage, diarrhea, transient ischemic attacks, and myocardial infarctions.

### *Etiologic Factors*

1. In the upper extremity, the source of emboli is aneurysms of the subclavian artery, bypass grafts, and radial and ulnar arteries. The mural thrombi secondary to compression from the cervical or anomalous first rib might generate the emboli (thoracic outlet syndrome). The atherosclerotic plaques from stenotic lesions of the subclavian arteries might also embolize to digital arteries.
2. Diffuse ulcerative lesions of the abdominal aorta and iliac vessels with or without aneurysms are factors, as are stenotic and occlusive lesions of the femoral-popliteal system.
3. Rarely, the thrombotic material in false aneurysms of the thoracic aorta can also shower emboli.
4. Nonatherosclerotic lesions, such as fibromuscular dysplasia of the external iliac arteries, can also generate emboli.

### *Treatment*

The traditional treatment for blue digit syndrome has been surgical eradication of the embolic source. This includes surgical endarterectomy, arterial incision with interposition of synthetic grafts, and bypass procedures with interruption of the diseased arteries proximal to the distal anastomosis, thereby preventing further embolization through an intact arterial conduit.

A lateral abdominal aortogram is extremely helpful in detecting the mural thrombi or posterior aortic wall ulcers that are responsible for showering distal emboli.

So far, the role of angioplasty has been limited and controversial. In the presence of complex ulcerated iliac disease, a therapeutic option is primary stent placement.

## COMMON AND EXTERNAL ILIAC ARTERY STENOSES

- The common and external iliac arteries are ideal for angioplasty. The best approach is the most direct approach, which, in most cases, is ipsilateral retrograde catheterization.
- The lesion is identified on angiography.
- The lesion is negotiated using angiographic catheters and guidewires. The initial attempt to negotiate the artery can be made with a standard 0.035- or 0.038-in., 15-mm J guidewire. These wires are, to some degree, steerable and may traverse the vessel with relative ease (Fig. 8.21). If this fails, a simple curved catheter with a "hockey-stick" configuration in combi-

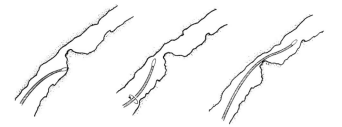

FIG. 8.21. Use of 15-mm J guidewire to negotiate through a simple iliac artery stenosis. (Reprinted from Castañeda-Zúñiga WR. *Interventional radiology*, 3rd ed. Baltimore: Williams & Wilkins, 1997, with permission.)

nation with a 15-mm J wire can be used in an attempt to steer through the diseased segment(s) of the vessel. If the vessel is severely ulcerated with multiple eccentric lesions, a Bentson wire is used in combination with the catheter because of its soft distal segment (Fig. 8.22).

- Once the lesion has been crossed with a guidewire, a two-view angiogram is obtained to characterize the lesion and the vessel is measured. An appropriate-size balloon is chosen for dilation.
- With the balloon catheter in the distal aorta, a bolus of 2,500 units of heparin is administered intraarterially. The soft-tip guidewire is then placed in the distal aorta below the renal arteries but well above the aortic bifurcation. The radioopaque markers designating the balloon position are placed within the stenosis, which, in the iliac system, is usually easy to locate without contrast medium because of the available bony landmarks. Road mapping can also be used to locate the lesion precisely.
- The balloon is inflated slowly to its maximum diameter and then to its maximum permissible pressure with diluted contrast medium and maintained in this state for one to two minutes, even if there is an initial "waist" in the balloon that "pops."
- After deflation of the balloon, a control arteriogram should be obtained. If digital capabilities are available, a high-quality study can be obtained with a simple manual injection of contrast medium through the angioplasty catheter. If digital techniques are not available, it is advisable to place a multi-hole catheter in the distal aorta and obtain a cut-film arteriogram.
- Never inject through an end-hole catheter in the vicinity of an angioplasty site. The risk of lifting the fractured intima-media is significant and may convert a successful angioplasty procedure into a catastrophe. The documentary arteriogram should always be obtained with the catheter tip remote from the angioplasty site.
- If the stenosis revealed by arteriography is of uncertain hemodynamic significance by anatomic criteria, it is appropriate to obtain pressure measurements to confirm the significance of the lesion before angioplasty. This is accomplished by placing a 0.018-in. guidewire in the distal aorta and retracting the catheter across the stenosis into the iliac artery. Pressure measurements are particularly useful after flow augmentation with 30 mg of papaverine injected intraarterially, to maximize any gradient that may be present. If a gradient cannot be

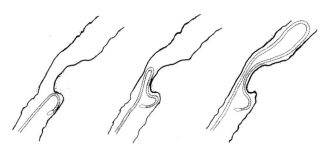

FIG. 8.22. The Bentson wire is so soft that it will deflect off eccentric, undermined lesions with little risk of intimal elevation or subintimal passage. (Reprinted from Castañeda-Zúñiga WR. *Interventional radiology*, 3rd ed. Baltimore: Williams & Wilkins, 1997, with permission.)

documented, angioplasty should not be performed.

- A residual stenosis of more than 30% in any projection or a residual gradient of more than 10 mm Hg (with flow augmentation) in a patient who experienced only mild discomfort during balloon inflation is enough to warrant exchanging the original angioplasty catheter for one with a balloon 1 mm larger and repeating the procedure. In this case, it is important to inflate the balloon slowly; if the patient experiences severe discomfort, further inflation should be approached with exceptional care.
- Lesions at the orifice of the common iliac artery may be treated by using a single-balloon catheter and the ipsilateral retrograde approach. However, many of these lesions resist angioplasty, and it appears that placement of a second, contralateral balloon catheter and simultaneous inflation of the two balloons (kissing balloon technique) in the orifices of the common iliac arteries buttress the balloon on the diseased side, helping to create the desired intimal-medial cleft and produce a successful angioplasty.
- Insertion and removal of the large-caliber balloon catheters should be done with considerable care. The catheter should be extracted while suction is applied with a large-bore syringe to ensure that the balloon hugs the catheter shaft as tightly as possible. Some manufacturers wrap the "wings" of the balloon around the catheter shaft and designate that catheter as wrapped for clockwise or counterclockwise insertion. Be aware of the design of the catheter of your choice, and if it is appropriate to rotate it as it is removed, be sure the rotation is in the proper direction.

## HYPOGASTRIC ARTERY AND COMMON ILIAC BIFURCATION

Angioplasty at any bifurcation may create a noncolinear intimal-medial cleft that spirals an intimal flap into the orifice of the untreated vessel, causing a temporary or even a permanent occlusion. In the case of the common iliac artery bifurcation, we think that it is optimal to protect the hypogastric artery, even if it does not have a significant stenosis at its origin. To do this, angioplasty is accomplished by bilateral retrograde catheterization of the common femoral arteries. On the ipsilateral side, the appropriate-size balloon catheter is placed within the stenosis using bony landmarks as guides. The contralateral femoral artery is catheterized with a catheter suitable for passage of a guidewire and catheter system around the aortic bifurcation. This crossover technique can be accomplished in at least three ways:

1. A pigtail catheter is used to engage the orifice of the contralateral common iliac artery. The guidewire is then passed down the contralateral iliac system (Fig. 8.23).

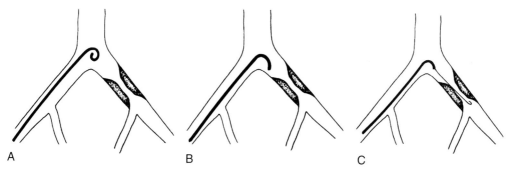

**FIG. 8.23.** Use of a pigtail catheter to engage the contralateral iliac orifice. **A:** Pigtail catheter is placed at the level of the aortic bifurcation. **B:** Pigtail catheter's loop is hooked over the bifurcation. **C:** Guidewire is advanced. (Reprinted from Castañeda-Zúñiga WR. *Interventional radiology*, 3rd ed. Baltimore: Williams & Wilkins, 1997, with permission.)

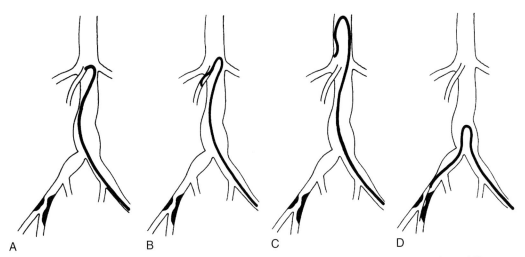

**FIG. 8.24.** Formation of the Waltman loop with a cobra catheter to engage the contralateral iliac system. **A:** Cobra catheter is engaged in the origin of the superior mesenteric artery. **B:** The tip of the cobra catheter is advanced into the renal artery. Further advancement of the catheter results in formation of a loop in the aortic lumen (**C**). **D:** The loop is manipulated across the aortic bifurcation. (Reprinted from Castañeda-Zúñiga WR. *Interventional radiology*, 3rd ed. Baltimore: Williams & Wilkins, 1997, with permission.)

2. A Simmons curved catheter is used to engage the contralateral common iliac artery origin after the catheter has been reformed in a more proximal branch of the abdominal aorta or in the descending thoracic aorta.
3. Form a loop with a cobra catheter and use this in conjunction with a Bentson wire to negotiate the orifice of the contralateral common iliac artery and ultimately the hypogastric artery (Fig. 8.24).

Once the hypogastric artery has been entered, if the aortic bifurcation is not extremely acute, a heavy-duty, 15-mm, 0.038-in. J guidewire is placed in the hypogastric artery. Angioplasty of the diseased common iliac bifurcation is performed, and a control arteriogram is obtained using digital techniques and injecting while a guidewire is left in the hypogastric artery. If there is no evidence of compromise of the orifice of the hypogastric artery, the procedure is considered complete, and the catheters are removed. If there is a stenosis of the hypogastric artery or if there is evidence of compromise of the hypogastric artery after angioplasty of the iliac bifurcation, an appropriate-size balloon catheter is placed around the aortic bifurcation over the heavy-duty guidewire and inflated simultaneously with the catheter in the common iliac artery bifurcation. Digitally acquired images are again reviewed, and if the angioplasty result is satisfactory, the catheters are removed as previously described.

When the contralateral approach is used for treatment of iliac disease, the catheter tends to buckle into the abdominal aorta rather than to follow the guidewire around the aortic bifurcation. If the guidewire can be advanced to the groin, manual compression of the common femoral artery will entrap it and reduce the likelihood of such buckling (Fig. 8.25). Access to the contralateral iliac or femoral artery may be achieved also by placing a curved "up and over" Balkin sheath (Cook, Inc.), which can be advanced to the contralateral common or external iliac arteries, providing adequate support to prevent wire or catheter buckling into the abdominal aorta and providing enough strength to advance wires and catheters through femoral occlusions or stenoses.

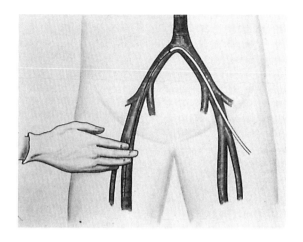

FIG. 8.25. Entrapment of guidewire at groin by manual compression provides a more rigid track for catheter to follow and decreases the likelihood of buckling in the aorta. (Reprinted from Castañeda-Zúñiga WR. *Interventional radiology*, 3rd ed. Baltimore: Williams & Wilkins, 1997, with permission.)

### Recanalization of Occluded Iliac Arteries

Recanalization of occluded iliac arteries is a controversial issue and may be attempted using thrombolysis followed by angioplasty only or angioplasty and stent placement. It can also be attempted using primary recanalization and stent placement. Regardless, these are complex cases that should be attempted only by experienced interventional radiologists.

### COMMON FEMORAL ARTERY

Lesions of the common femoral artery are frequently related to previous surgery. When atherosclerosis is present, the lesions are frequently thick, heavily calcified, and eccentric, and they may not yield optimally to balloon angioplasty. In contrast, lesions at graft anastomoses are usually caused by intimal hyperplasia and respond readily to angioplasty. Surgical management of common femoral artery disease is a well-established procedure that usually can be done at low risk. Angioplasty of the occluded common femoral artery may be attempted in certain cases. Careful evaluation of distal runoff is crucial to determine whether recanalization and angioplasty of the common femoral artery are likely to succeed. If distal runoff is extremely poor, the performance of angioplasty in this arterial segment is not indicated.

### SUPERFICIAL AND DEEP FEMORAL ARTERY STENOSES

- The optimal approach to stenoses of the superficial and deep femoral arteries is an ipsilateral, antegrade, common femoral puncture.
- Retrograde catheterization of the superficial femoral artery through a popliteal approach may facilitate the advancement of guidewires or laser probes in cases of failure of an antegrade catheterization.
- The puncture of the popliteal artery is technically easy if it can be seen fluoroscopically by contrast injections of the femoral artery. Road mapping is extremely useful in these situations. The artery can be entered if the puncture is made in the medial aspect of the fossa above the knee joint. The tibial nerve and the popliteal vein can be avoided because they cross the artery from lateral to medial at the inferior aspect of the popliteal fossa (Fig. 8.26).
- An alternative approach is to use ultrasound guidance for popliteal artery puncture. Using a 5- to 10-MHz linear transducer with sterile cover, the popliteal vessels are identified in the transverse and longitudinal views, and an optimal site for puncture is chosen, avoiding the popliteal vein. Once the vessels and their relationship are clearly identified, the puncture may be performed under real-time sonographic guidance. This technique has proven to be safe and easy to perform.

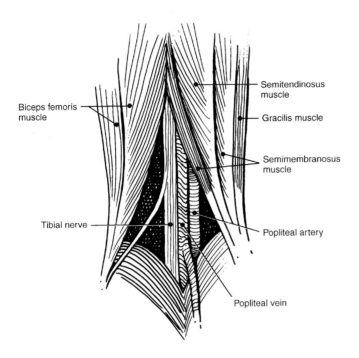

**FIG. 8.26.** Anatomic relationship at the popliteal fossa. (Reprinted from Castañeda-Zúñiga WR. *Interventional radiology*, 3rd ed. Baltimore: Williams & Wilkins, 1997, with permission.)

- Hemostasis after angioplasty may be difficult because of the deeper placement of the vessel and the loose tissue around the artery. In two of 50 angioplasties carried out using a popliteal approach, two painful hematomas were encountered. The hematomas are a potential problem and may be minimized by neutralizing the effects of heparin with protamine sulfate after a successful procedure.

## DEEP FEMORAL ARTERY

- Angioplasty of the deep femoral artery should be performed using an ipsilateral or contralateral approach. A sharp-angled aortic bifurcation and severely diseased and tortuous iliac vessels make a contralateral approach difficult or impossible. If the ipsilateral antegrade femoral approach is used, the arterial puncture must be at a sufficient distance from the PTA site so as not to dilate the arterial entry site during balloon inflation.
- Access is obtained with an antegrade puncture (preferred approach), and a 5- or 6-Fr introducer sheath is advanced.
- The lesion is located and adequately imaged, in either the superficial or the deep femoral artery.
- The lesion is negotiated using a guidewire.
- The lesion is crossed with an angiographic catheter. A small volume of contrast medium is injected to ensure intraluminal position. Heparin is administered.
- A catheter with a balloon of appropriate diameter and length is advanced over the guidewire and positioned within the lesion. Road mapping is an extremely useful tool in these cases.
- The guidewire is positioned in the distal superficial femoral artery or the distal deep femoral artery. The balloon is inflated slowly to its maximum diameter with diluted contrast medium and then to its maximum pressure, and inflation is maintained for 60 to 120 seconds.
- After angioplasty, the dilating catheter may be retracted, and a control angiogram can be performed through the sideport of the introducer sheath.
- If the result is satisfactory, the catheters can be removed and hemostasis obtained at the groin.

- If the result is not satisfactory, an angioplasty catheter with a larger balloon should be used.
- It is essential that a guidewire be left well distal to the angioplasty site until the angiographer is certain that the procedure is complete. Any attempt to recross a freshly dilated site risks intimal dissection and vessel occlusion.
- When dealing with multiple stenoses in the same vessel, all of the lesions should be negotiated with the guidewire; angioplasty should begin with the most distal lesion, working toward the most proximal lesion. More than one catheter may be necessary because of the increasing diameter of the vessel.
- Deep femoral artery angioplasty is thus presented as an alternative therapy for limb salvage in patients with severe stenosis or occlusion of the superficial femoral artery, with approximately 66% early clinical success. It is desirable to maintain patency of the deep femoral artery, as this vessel is the main collateral to the extremity in case of subsequent reocclusion or progression of occlusive disease in the superficial femoral artery.

### Technique of Angioplasty— Infrapopliteal Vessels

The technique of infrapopliteal angioplasty is relatively straightforward. Antegrade puncture of the common femoral artery is performed in the usual fashion, and a sheath is advanced into the superficial femoral artery over a J guidewire. If there are superficial femoral or popliteal artery lesions, these should be dilated before the small vessel angioplasty. The presently available low-profile balloon catheters are equipped with different-diameter balloons, ranging from 2.0 to 3.5 mm. Before the angioplasty, 200-μg bolus of nitroglycerin is administered intraarterially. Angiographers performing infrapopliteal angioplasty should be familiar with the use of vasodilator drugs:

- Nitroglycerin, 100 to 200 μg intraarterial bolus as needed
- Transdermal nitroglycerin patches
- Verapamil, 2.5 to 5.0 mg

Patients should be carefully monitored for hypotension during the administration of vasodilators. Preshaping the distal end of the guidewire to a gentle curve facilitates the catheterization of the tibioperoneal arteries. Once the balloon is placed across the area to be dilated, it is inflated to its maximum diameter and maintained at its maximum diameter/pressure for 90 to 120 seconds. After the dilation, follow-up angiography is obtained, leaving the guidewire across the stenosis and retracting the balloon catheter proximal to the site of dilation. Contrast medium is injected through the sidearm of the sheath, and digital images are obtained. If a suboptimal result is obtained, redilation can be performed by simply advancing the balloon over the guidewire or by using a larger-diameter balloon.

If vasospasm is a problem, multiple boluses of intraarterial nitroglycerin can be administered, bearing in mind that the patient should be monitored constantly for systemic hypotension. For the rare spasm that is not relieved by nitroglycerin, intravascular verapamil may be administered if there is no contraindication to its use.

On completion of the procedure, the balloon catheter and guidewire are removed. Systemic heparinization is maintained for 48 to 72 hours. Several authors recommend long-term (2 to 3 months) oral anticoagulation after below-the-knee angioplasty.

### Anastomotic and Graft Stenoses

Angioplasty is ideal for the management of stenoses at anastomoses of aorto-iliac or aortofemoral bypass grafts and at anastomoses of reverse saphenous vein bypass grafts. Stenoses at the distal anastomosis of an aortofemoral bypass graft can be approached from the contralateral groin around the graft bifurcation. This requires the use of a sheath to introduce the balloon catheter. Graft bifurcations are frequently acute, and it is usually technically difficult to reach the stenoses for angioplasty.

## COMPLICATIONS OF ANGIOPLASTY

- Groin hematomas (2% to 4% of angioplasty procedures).
- Distal embolization (2% to 5%).

- Elevation of intimal flaps (4%).
- False aneurysms (2%).
- Arterial rupture secondary to catheter or balloon trauma (approximately 3% in femoral artery angioplasty).
- Burst balloon with a circumferential tear (rare).
- Arteriovenous fistula (less than 1%).
- Pseudoaneurysm (2%).
- Occasionally, immediately after an apparently successful angioplasty, the patient's foot will appear cold, mottled, and pale, suggesting distal embolization. Before catheter removal, an angiographic check of the distal vasculature should be made. If no evidence of distal embolization is observed, a 100-μg bolus of nitroglycerin should be administered through the catheter in an attempt to relieve what is probably distal vasospasm causing the "ugly foot" syndrome.

### Results

There is concern for an accurate, objective method to report angioplasty results. The guidelines set by Rutherford et al. are straightforward and should be followed whenever percutaneous revascularization procedures are performed. Patency rates should also be reported following Rutherford's guidelines.

The 5-year patency rate in initially successful procedures involving the iliac arteries with short, focal, concentric stenoses should approximate 90%, given a 95% primary success rate. The 5-year patency rate approaches 85%. When longer segments of the iliac artery are subjected to angioplasty, the primary success rate drops to approximately 89% for stenotic lesions and as low as 33% in recanalization of occluded common iliac arteries, and the 5-year patency rate drops to approximately 55%.

Short stenoses of the superficial femoral and popliteal arteries can be dilated more than 90% of the time and will remain patent 5 years later in approximately 70% of patients. When longer or multiple stenoses are treated, a high primary success rate may be obtained, but the 5-year patency rate varies from 45% to 70%.

Angioplasty is still considered a viable therapeutic option for patients with peripheral vascular disease. It offers low morbidity and mortality, short hospital stay, good early clinical results, and very promising long-term results. Objective comparison with other therapeutic alternatives will only appear when standards for clinical status and outcome are established in prospective, randomized trials.

## SUGGESTED READINGS

### A. Basic Concepts

Davies MJ, Thomas AC. Plaque fissuring—the cause of acute myocardial infarction, sudden ischemic death, and crescendo angina. *Br Heart J* 1985;53: 363–373.

Munro JM, Cotran RS. Pathogenesis of atherosclerosis: atherogenesis and inflammation. *Lab Invest* 1988;58: 249–261.

Ross R, Glomset JA. The pathogenesis of atherosclerosis. *N Engl J Med* 1976;295:369–377, 420–425.

Ross R. The pathogenesis of atherosclerosis—an update. *N Engl J Med* 1986;314:488–500.

### B. Mechanism of Angioplasty

Castañeda-Zúñiga WR, Formanek A, Tadavarthy M, et al. The mechanism of balloon angioplasty. *Radiology* 1980;135:565.

Losordo DW, Rosenfeld K, Pieczek A, et al. How does angioplasty work? Serial analysis of human iliac arteries using intravascular ultrasound. *Circulation* 1992;86:1845–1858.

### C. Devices and Techniques

Abele J. Balloon catheters and transluminal dilatation: technical considerations. *AJR Am J Roentgenol* 1980;135: 901–906.

Barath P, Fishbein MC, Vari S, Forrester JS. Cutting balloon: a novel approach to percutaneous angioplasty. *Am J Cardiol* 1991;68:1249–1252.

Levin DC, Harrington DP, Bettmann MA, et al. Equipment choices, technical aspects and pitfalls of percutaneous transluminal angioplasty. *Cardiovasc Intervent Radiol* 1984;7:1.

Spies J, Bakal C, Burke D, et al. Guidelines for percutaneous transluminal angioplasty. *J Vasc Intervent Radiol* 1990;1:5–15.

Vorwerk D, Adam G, Muller-Leisse C, Guenther R. Hemodialysis fistulas and grafts: use of cutting balloons to dilate venous stenoses. *Radiology* 1996;201:864–867.

## D. Indications and Outcome

### Angioplasty of Supraaortic Vessels

Bockenheimer SA, Mathias K. Percutaneous transluminal angioplasty in arteriosclerotic internal carotid artery stenosis. *AJNR Am J Neuroradiol* 1983;3:791–792.

Bogey WM, Demasi RJ, Tripp MD, et al. Percutaneous transluminal angioplasty for subclavian artery stenosis. *Am Surg* 1994;60:103–106.

Düber C, Klose KJ, Kopp H, Schmiedt W. Percutaneous transluminal angioplasty for occlusion of the subclavian artery: short- and long-term results. *Cardiovasc Intervent Radiol* 1992;4:205–210.

Higashida RT, Hieshima GB, Tsai FY, et al. Transluminal angioplasty of the vertebral and basilar artery. *AJNR Am J Neuroradiol* 1987;5:745–749.

Mathias KD, Lüth I, Haarmann P. Percutaneous transluminal angioplasty of proximal subclavian artery occlusions. *Cardiovasc Intervent Radiol* 1993;4:214–218.

Selby JB Jr., Matsumoto AH, Tegtmeyer CJ, et al. Balloon angioplasty above the aortic arch: immediate and long-term results. *AJR Am J Roentgenol* 1993;16:631–635.

Smith DC, Smith LL, Hasso AN. Fibromuscular dysplasia of the internal carotid artery treated by operative transluminal balloon angioplasty. *Radiology* 1985;155:645–648.

Théron J. Angioplasty of supra-aortic arteries. *Semin Intervent Radiol* 1987;4:331–342.

### Celiac and Superior Mesenteric Artery Angioplasty

Becker GJ, Katzen BT, Dake MD. Noncoronary angioplasty. *Radiology* 1989;170:921–940.

Beebe HG, MacFarlane S, Raker EJ. Supraceliac aortomesenteric bypass for intestinal ischemia. *J Vasc Surg* 1987;5:749–754.

Bergan JJ, Yao JST. Chronic intestinal ischemia. In: Rutherford RB, ed. *Vascular surgery*, 2nd ed. Philadelphia: WB Saunders, 1984:964–972.

Furrer J, Gruentzig A, Kugelmeier J, Goebel N. Treatment of abdominal angina with percutaneous dilation of an arteriomesenteric superior stenosis. *Cardiovasc Intervent Radiol* 1980;5:367–369.

Golden DA, Ring EJ, McLean GK, Freiman DB. Percutaneous transluminal angioplasty in the treatment of abdominal angina. *AJR Am J Roentgenol* 1982;139:247–249.

McCollum CH, Graham JM, DeBakey ME. Chronic mesenteric arterial insufficiency: results of revascularization in 33 cases. *South Med J* 1976;69:1266–1268.

Odurny A, Sniderman KW, Colapinto RF. Intestinal angina: percutaneous transluminal angioplasty of the celiac and superior mesenteric arteries. *Radiology* 1988;167:59.

Reilly LM, Ammar AD, Stoney RJ, Ehrenfield WK. Late results following operative repair for celiac artery compression syndrome. *J Vasc Surg* 1985;2:79–91.

Roberts L Jr, Wertman DA Jr, Mills SR, et al. Transluminal angioplasty of the superior mesenteric artery: an alternative to surgical revascularization. *AJR Am J Roentgenol* 1983;141:1039–1042.

Stanton PE Jr, Hollier PA, Seidel TW, et al. Chronic intestinal ischemia: diagnosis and therapy. *J Vasc Surg* 1986;4:338–344.

Uflacker R, Goldany MA, Constant S. Resolution of mesenteric angina with percutaneous transluminal angioplasty of a superior mesenteric artery stenosis using a balloon catheter. *Gastrointest Radiol* 1980;5:367–369.

### Percutaneous Transluminal Angioplasty of the Renal Arteries

Losinni F, Zuccala A, Busato F, Zuchelli P. Renal artery angioplasty for renovascular hypertension and preservation of renal function: long-term angiographic and clinical follow-up. *AJR Am J Roentgenol* 1994;162:853–857.

Martin EC, Mattern RF, Baer L, et al. Renal angioplasty for hypertension: predictive factors for long-term success. *AJR Am J Roentgenol* 1981;137:921–924.

Martin LG, Cork RD, Kaufman SL. Long-term results of angioplasty in 110 patients with renal artery stenosis. *J Vasc Intervent Radiol* 1992;619–626.

Matalon TAS, Thompson MJ, Patel SK, et al. Percutaneous transluminal angioplasty for transplant renal artery stenosis. *J Vasc Intervent Radiol* 1992;3:55–58.

Rodriguez-Perez JC, Plaza C, Reyes R, et al. Treatment of renovascular hypertension with percutaneous transluminal angioplasty: experience in Spain. *J Vasc Intervent Radiol* 1994;5:101–109.

Sos TA, Saddekni S, Sniderman KW, et al. Renal artery angioplasty: techniques and early results. *Urol Radiol* 1982;3:223–231.

Sos TA, Pickering TG, Sniderman K, et al. Percutaneous transluminal renal angioplasty in renovascular hypertension due to atheroma or fibromuscular dysplasia. *N Engl J Med* 1983;309:274–279.

Tegtmeyer CJ, Dyer R, Teates CD, et al. Percutaneous transluminal dilatation of the renal arteries: techniques and results. *Radiology* 1980;135:589–599.

Tegtmeyer CJ, Elson J, Glass TA, et al. Percutaneous transluminal angioplasty: the treatment of choice for renovascular hypertension due to fibromuscular dysplasia. *Radiology* 1982;143:631–637.

Tegtmeyer CJ, Kellum CD, Ayers C. Percutaneous transluminal angioplasty of the renal arteries. *Radiology* 1984;153:77–84.

## Aorto-Iliac and Peripheral Arterial Angioplasty

Bakal WC, Sprayregen S, Scheinbaum K, et al. Percutaneous transluminal angioplasty of the infrapopliteal arteries: results in 53 patients. *AJR Am J Roentgenol* 1990;154:171–174.

Becker GJ, Katzen BT, Dake M. Noncoronary angioplasty. *Radiology* 1989;170:921–940.

Bergqvist D, Jonsson K, Weibull H. Complications after percutaneous transluminal angioplasty of peripheral and renal arteries. *Radiology* 1987;28:3–12.

Blum U, Gabelmann A, Redecker M. Percutaneous recanalization of iliac artery occlusions: results of a prospective study. *Radiology* 1993;189:536–540.

Hallisey MJ, Meranze SG, Parker BC, et al. Percutaneous transluminal angioplasty of the abdominal aorta. *J Vasc Intervent Radiol* 1994;5:679–687.

Johnston KW. Iliac arteries: reanalysis of results of balloon angioplasty. *Radiology* 1993;186:207–212.

Kumpe DA, Zwerdlinger S, Griffin DJ. Blue digit syndrome: treatment with percutaneous transluminal angioplasty. *Radiology* 1988;166:37–44.

Motarjeme A, Keifer JW, Zuska AJ. Percutaneous transluminal angioplasty of the deep femoral artery. *Radiology* 1980;135:613–617.

Rutherford RB, Flanigan D, Gupta SK, et al. Suggested standards for reports dealing with lower extremity ischemia. *J Vasc Surg* 1986;4:80–94.

Tegtmeyer CJ, Kellum CD, Kron IL, Mentzer RM. Percutaneous transluminal angioplasty in the region of the aortic bifurcation. *Radiology* 1985;157:661–665.

Tonnesen KH, Sager P, Karle A, et al. Percutaneous transluminal angioplasty of the superficial femoral artery by retrograde catheterization via the popliteal artery. *Cardiovasc Intervent Radiol* 1988;11:127–131.

# 9

# Problems and Management of Hemodialysis Access

## Hector Ferral

Approximately 129,000 patients are in a hemodialysis program in the United States. The quality of life and the survival of hemodialysis patients depend on the performance of their arteriovenous accesses. An adequately functioning hemodialysis access should have a flow rate of at least 250 to 300 mL per minute (ideally 400 to 500 mL per minute) and be able to withstand two 14-gauge needle punctures at least 150 times each year. It has been estimated that 18% of these individuals die because of a lack of adequate vascular access after all usable locations have been exhausted.

The availability of thrombolytic enzymes, transluminal angioplasty, and vascular stent placement has enabled the interventional radiologist to participate in the preservation and salvage of these accesses. Moreover, the interventional radiologist may resort to a number of extreme techniques when all conventional accesses have failed and death is otherwise inevitable.

## BASIC CONSIDERATIONS

Long-term hemodialysis accesses, such as the Brescia-Cimino fistula and bridge grafts, usually must mature for 3 weeks before being used. If a temporary hemodialysis catheter is required, the preferred access site is the right internal jugular vein.

## CHOICE OF ACCESS FOR LONG-TERM HEMODIALYSIS

The direct arteriovenous Brescia-Cimino fistula was introduced in 1966. It is constructed between the distal radial artery and the cephalic vein and is still the best access for long-term hemodialysis (Fig. 9.1). The 1- and 3-year cumulative patency rates are estimated at 85% to 90% and 60% to 85%, respectively. The development of adequate venous drainage requires a period of maturation of the fistula, usually 3 to 8 weeks, before it can be used.

Another alternative is the use of bridge grafts (Fig. 9.2). In most dialysis centers, expanded polytetrafluoroethylene (E-PTFE) has performed better than any other material, and 80% of hemodialysis patients have a PTFE graft access. The upper extremities are usually used for these grafts. Bridge grafts in the groin are associated with a high incidence of infectious complications and should be considered only for the rare patient who has no suitable vessels in the arms and in whom peritoneal dialysis is not feasible.

## PREOPERATIVE EVALUATION

Preoperative venography is recommended in patients with history of subclavian catheters. A study by Surratt et al. found a 40% prevalence of subclavian vein stenoses in patients with prior or existing temporary dialysis catheters, and no stenoses in patients without a history of dialysis catheters. An arteriovenous fistula placed on the side with a tight stenosis or thrombosis of the central veins (subclavian, innominate) will quickly fail or cause severe venous hypertension in the patient's arm.

Digital subtraction venogram is an excellent imaging method. Access to the venous system is obtained by cannulation of a superficial vein over the dorsum of the hand. A rubber tourniquet or blood pressure cuff is then placed around the upper arm to occlude venous outflow so that the

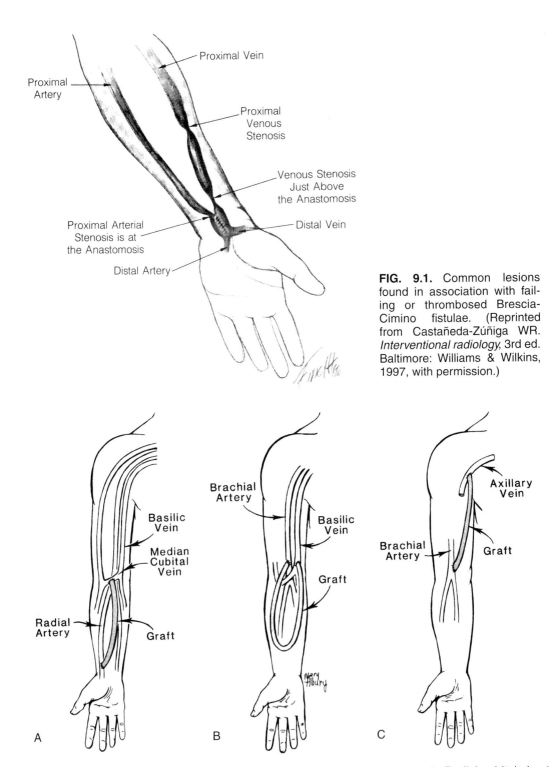

**FIG. 9.1.** Common lesions found in association with failing or thrombosed Brescia-Cimino fistulae. (Reprinted from Castañeda-Zúñiga WR. *Interventional radiology*, 3rd ed. Baltimore: Williams & Wilkins, 1997, with permission.)

**FIG. 9.2.** Common types of bridge grafts used for long-term hemodialysis. **A:** Radial-cubital shunt. **B:** Brachial-basilic shunt. **C:** Brachial-axillary shunt. (Reprinted from Castañeda-Zúñiga WR. *Interventional radiology*, 3rd ed. Baltimore: Williams & Wilkins, 1997, with permission.)

vessels can be filled with just 10 to 20 mL of contrast. Anteroposterior and lateral views of the forearm and upper arm vessels are then obtained. As the tourniquet is released, rapid sequential films should be taken of the upper arm or shoulder to rule out a more proximal lesion. Proximal obstruction may either be evident radiologically or be suggested by a delay in the clearance of the contrast medium. Venography is extremely useful in patients with history of previous catheterization for dialysis. If a stenotic lesion is found in the central veins, the affected extremity should not be used for fistula creation.

## EVALUATION AND MANAGEMENT OF ACCESS PROBLEMS

Complications related to the vascular access are the most frequent cause of hospitalization of patients on long-term hemodialysis. Approximately 26% of the initial hospitalizations in a hemodialysis population are necessitated by access problems.

There are essentially five complications: infection, bleeding, aneurysm, partial or complete obstruction, and hemodynamic problems.

The most common cause of access failure is thrombosis, which is often preceded by increasing problems during dialysis, such as difficult cannulation, withdrawal of clots, and high venous resistance or poor arterial flow. In addition, patients with arteriovenous communications may have a diminished bruit or thrill.

Infection is the second most common cause of access failure and accounts for approximately 20% of vascular access complications. The principle in the management of access-related infection is prevention; it is important to give prophylactic antibiotics and to use strict aseptic techniques in access surgery and whenever invasive radiologic procedures are used. The predominant pathogens involved in access graft infection include *Staphylococcus aureus* and, less commonly, *Staphylococcus epidermidis*.

True aneurysms are uncommon in E-PTFE grafts, although repeated traumatic punctures may create a small hole in the graft, leading to formation of a pseudoaneurysm.

### Surgically Created Arteriovenous Fistulae

The common causes of failing Brescia-Cimino fistulae are stenoses at either the proximal artery or the vein at or just above the anastomosis, or stenoses of the proximal vein at the dialysis needle puncture sites (Fig. 9.1).

### Steal Syndrome and Venous Hypertension

Although dialysis takes place only for a few hours a few days weekly, there is continuous flow through fistulas and accesses. This creates the possibility of complications caused by an imbalance of blood inflow and outflow. If there is too little inflow, the fistula or access will be ineffective. However, if there is excessive inflow, the extremity distal to the fistula or access may not get enough blood for its needs.

This siphoning off of blood is termed the *steal syndrome* and can result in digital ischemia or even gangrene. The resultant ischemia, like most other vascular complications of dialysis, is more severe in diabetic patients. The only treatment is to surgically reduce the inflow to the fistula.

Congestive heart failure secondary to a high-output fistula has occasionally been reported. It is more common in patients who have two simultaneously functioning fistulae.

Conversely, if inflow is acceptable, a stenosis or obstruction of the venous drainage may produce venous hypertension of the extremity, caused by retention in the veins of blood under elevated pressure because of the inability of that blood to empty as fast as it flows in. Venous hypertension usually is indicated by swelling, cyanosis, and ulceration of the entire hand. This condition may mandate closure of the fistula or access.

## ARTIFICIAL ACCESSES

Only 15% of dialysis patients use the Brescia-Cimino fistula. The remaining 85% must rely on artificial shunts. All have lower patency rates and a higher incidence of complications than the Brescia-Cimino fistula. Because of this frequency of complications, their widespread use, and their amenability to treatment, they are of great interest to the interventional radiologist. The current prosthetic material of choice, and by far the most widely used, is E-PTFE. One-year shunt survivals of 62% to 80% and 5-year survivals of 47% have been reported. Stenosis in these shunts commonly occurs at the level of the arterial (15% to 20%) or venous anastomosis (80%).

### Early Warning Signs of Thrombosis

The timely treatment of stenotic lesions by angioplasty can be performed on an outpatient basis and is greatly preferable to treating a thrombosed shunt or fistula. Elevated venous resistance during dialysis or recirculation rates exceeding 15% to 20% should initiate a search for a correctable outflow stenosis. In our institution, venous pressures of more than 200 mm Hg at a flow rate of 300 mL per minute trigger such a search. Angioplasty of venous lesions can easily be performed on an outpatient basis.

If thrombosis occurs, not only is immediate treatment less likely to be successful, but also long-term patency is decreased.

## EVALUATION OF DIALYSIS ACCESSES

### Duplex and Color Doppler Evaluation

Duplex and color Doppler allow identification of the most common problems of dialysis fistulae, provide physiologic information, and can also be used as tools to evaluate results after thrombectomy, surgical revision, thrombolysis, and angioplasty. A complete study can be time consuming and at times confusing, and a dedicated group of sonographers and technologists is required to complete these examinations successfully.

The patients are examined using 5- to 10-MHz linear array transducers. To simplify the examination, the flow of blood is followed. A complete examination includes the feeding artery, anastomosis, graft, and outflow vein. Analysis of the draining subclavian vein should always be included. Normal findings include monophasic flow with peak systolic velocities of 100 to 400 cm per second in the supplying arteries and throughout the graft (Fig. 9.3). The draining veins show arterial pulsations with peak velocities of 30 to 100 cm per second. With this technique, the most common and important complications presenting in these fistulae can be identified, such as thrombosis, stenotic areas, infection, aneurysms and pseudoaneurysms, and arterial steal. Diagnostic criteria of shunt stenosis include visible narrowing of the lumen and increased velocity of flow as compared with that in adjacent normal segments. Arterial steal can be clearly identified with color Doppler sonography by demonstrating retrograde flow in the distal portion of the artery forming the fistula. However, due to the rich collateral flow of the hand, only approximately 6% of these patients are symptomatic. Arterial steal is more readily demonstrated by color Doppler sonography than by angiographic evaluation.

### Angiographic Evaluation

Angiography still remains the primary modality in evaluating "failing" hemodialysis grafts and fistulae. The indications for evaluation include the following:

- Urea recirculation ratio of greater than 20%
- Loss of the thrill and bruit on compression of the distal vein below a side-to-side direct arteriovenous fistula (a sign of proximal venous obstruction)
- Fistula or graft difficult to stick
- Digital ischemia
- Aspiration of clots during dialysis
- Swelling of the extremity with the fistula

### *Brescia-Cimino Fistula*

A failing fistula is best evaluated by a venous fistulogram. Better access is obtained by cannulation of the distal vein of the fistula, which allows easy access to both the proximal artery and the vein for balloon angioplasty.

Fistulograms can be done on an outpatient basis. Consent should be obtained to perform percutaneous balloon dilation if it proves appropriate.

A blood pressure cuff is placed close to the axilla, and the extremity is prepared with antiseptic solution and draped from the hand to 6

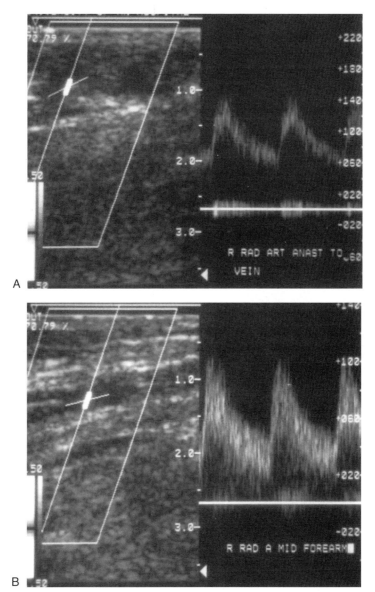

**FIG. 9.3. A:** Normal, low-resistance, monophasic, relatively high-velocity arterial inflow in a Brescia-Cimino dialysis fistula. **B:** At the arteriovenous anastomosis, the flow velocities increase slightly to 140 cm per second in peak systole. (Reprinted from Castañeda-Zúñiga WR. *Interventional radiology*, 3rd ed. Baltimore: Williams & Wilkins, 1997, with permission.)

cm above the elbow. After venous distention has been obtained by inflating the cuff to 60 mm Hg, the distal vein of the fistula is punctured with a 19-gauge butterfly infusion needle 3 to 5 cm below the fistula (Fig. 9.4D). The cuff is then deflated, and the needle is replaced with a 5-Fr angiocatheter threaded over a 0.022-in. guidewire to avoid displacement of the needle during subsequent manipulation. The catheter is advanced 2 to 3 cm beyond the puncture site so that the tip lies below the level of the proximal vessels of the fistula. Small doses of contrast

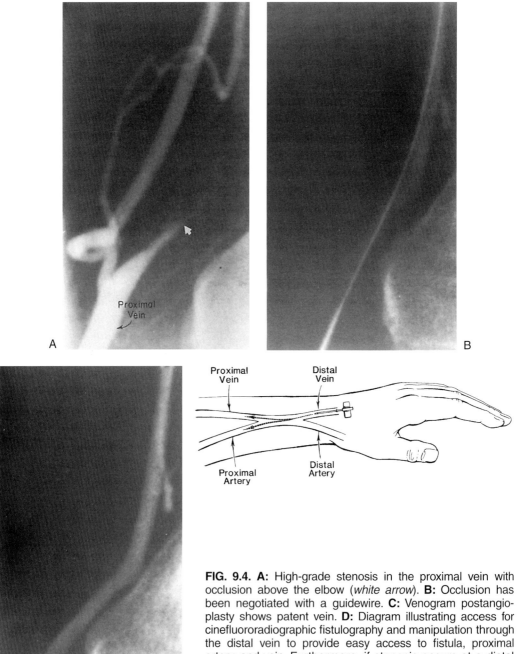

**FIG. 9.4. A:** High-grade stenosis in the proximal vein with occlusion above the elbow (*white arrow*). **B:** Occlusion has been negotiated with a guidewire. **C:** Venogram postangioplasty shows patent vein. **D:** Diagram illustrating access for cinefluororadiographic fistulography and manipulation through the distal vein to provide easy access to fistula, proximal artery, and vein. Furthermore, if stenosis occurs at a distal artery or vein, it has almost no functional significance. (Reprinted from Castañeda-Zúñiga WR. *Interventional radiology*, 3rd ed. Baltimore: Williams & Wilkins, 1997, with permission.)

medium are infused to verify the catheter position and to make a preliminary assessment of the proximal venous anatomy. A cine fistulogram is made with 15 to 30 mL of 30% contrast medium (60% diatrizoate in a 1:1 dilution) infused at the rate of 2 to 5 mL per second with the blood pressure cuff inflated 20 mm Hg above systolic pressure to obtain arterial as well as venous filling. Rapid clearing of the contrast when the cuff is deflated and the anatomic distribution distinguish the arteries from the veins. Oblique and lateral views are obtained by rotation of the arm slowly through 90 degrees to evaluate lesions that may otherwise be obscured by contrast-filled vessels. One or two injections may be necessary. The vessels should be traced to the subclavian vein level to avoid overlooking a proximal stenosis.

Occasionally, this technique will produce confusing images around the anastomotic site. In such situations, direct puncture of the feeding artery provides excellent visualization of the fistula or the arterial anastomosis and the draining veins. The brachial artery is punctured in a retrograde fashion with a 20-gauge Angiocath or a 4-Fr micropuncture set. To facilitate injection of contrast, the syringe is connected to a plastic extension tube (K-50). Hand injection is usually adequate; usually 15 to 20 mL of contrast is enough to provide adequate images of the fistula. Images may be obtained using cut-film (nondilute contrast) or digital subtraction angiography (dilute contrast).

### *Polytetrafluoroethylene Grafts*

A puncture directed toward the outflow vein is performed with an 18-gauge Angiocath or a micropuncture set. After access is obtained, multiple images of the venous outflow system, including subclavian and innominate veins, should be obtained.

To demonstrate the inflow anastomosis, a blood pressure cuff may be placed in the arm and inflated at suprasystolic pressure (20 mm Hg). Careful injection is then performed to demonstrate the inflow artery and its anastomosis with the graft. In these cases, it is important to demonstrate good outflow before assessing the inflow portion of the graft, as embolization of thrombi may occur with forceful injections of contrast within the graft if outflow occlusion or stenosis exists.

Pressure gradients are measured across the fistula to assess the functional significance of any stenoses and to reveal any lesions not appreciated on angiography. Although a gradual change in pressure may occur without indicating a lesion, an abrupt change of more than 20 mm Hg is considered significant. Pressure gradient measurement is an essential part of the evaluation and may predict the outcome of the dilation.

On completion of the contrast study and pressure measurements, the vascular access surgeon should review the findings with the radiologist and plan the therapeutic approach.

## INTERVENTIONAL RADIOLOGIC SALVAGE OF DIALYSIS ACCESSES

### Thrombolysis

The preferred drug is recombinant tissue plasminogen activator. The techniques for thrombolysis treatment of occluded dialysis grafts include the following:

- Pharmacologic thrombolysis
- Pharmacomechanical thrombolysis (pulse-spray)
- Mechanical thrombolysis/thrombectomy

The "crossed-catheter" technique for pulsed-spray infusion of thrombolytics, and early angioplasty is extensively used. Single-wall needles are used to puncture the graft. Initially, the puncture is performed very close to the arterial inflow region and directed toward the venous outflow area. An angiographic guidewire is directed to the outflow region, and the passage of the wire into the venous outflow must be confirmed before proceeding with further punctures or thrombolysis. If the venous outflow cannot be cannulated, thrombolysis is not performed, and the patient is sent for surgical revision.

Subsequently, a second puncture in the opposite direction is done, and pulsed-spray thrombolysis started. Before urokinase was withdrawn from the market, the classic technique was as follows: Initially, the contents of a 250,000-unit urokinase vial are diluted in 9 mL of sterile water and 1 mL of heparin (5,000 units per mL), yielding a concentration of 25,000 units of urokinase and 500 units of heparin per mL of solution. Injection of the solution is performed with tuberculin syringes, and increments of 0.2 mL of solution are injected as forcefully as possible. After 250,000 units of urokinase have been injected, angiographic evaluation of the graft is done, and angioplasty of the causative lesions for graft occlusion is performed. With this technique, patency was established in 98.5% of the grafts treated, and 96% were patent at 24 hours. It usually takes 30 to 40 minutes to complete thrombolysis using this method. After the procedure, the patients can be discharged or sent to hemodialysis.

Another promising trend in thrombolysis is the development of mechanical devices for delivering forceful pulsed sprays of thrombolytic agent into clot.

## ANGIOPLASTY

Dialysis access stenoses may be due to intimal hyperplasia or scarring at a surgical anastomosis or to stenotic lesions in the subclavian vein related to prior catheterization. Dilation of scar tissue requires extremely high balloon pressures and can be extremely painful. Occasionally, scarring appears to respond, but the stenosis recurs as soon as the balloon is deflated. However, some decrease in venous resistance during subsequent dialysis usually results.

Infected accesses should not receive thrombolysis or angioplasty because of the risk of inducing sepsis and the probable necessity for eventual surgical removal.

The current recommendation to provide a longer life for the graft is to use an aggressive follow-up of these patients. A close screening program with early intervention using balloon angioplasty of graft stenoses before occlusion occurs will prolong graft function.

### Metallic Stents after Angioplasty

Re-stenosis is a frequent occurrence after angioplasty. Placement of stents helps to prolong the useful life of an otherwise failed hemodialysis access. Stents have been placed most commonly in the efferent and subclavian veins. Intimal hyperplasia limits the utility of stents, although it responds adequately to angioplasty or atherectomy. In patients in whom the central veins have all occluded, secondary to multiple previous venous catheter placements, recanalization followed by angioplasty or stent placement is an alternative to maintain a venous access.

## MECHANICAL THROMBECTOMY

New devices for the treatment of occluded dialysis grafts have been developed. These new devices include the Amplatz thrombectomy device (ATD) (Microvena, Minnetonka, MN), the Castañeda Thrombolytic Brush Catheter [Microtherapeutics, Inc. (MTI), Irvine, CA], and the Arrow-Trerotola percutaneous thrombolytic device (Arrow International, Reading, PA). These devices have been designed to treat occluded dialysis grafts with minimal or no thrombolytic drugs.

### *Amplatz Thrombectomy Device*

The ATD is an 8-Fr catheter with an impeller mounted on a drive shaft inside a metal cap. This metal cap has three sideports used for recirculation of the clot particles. The impeller is activated by an air turbine that rotates at 150,000 rpm. The high speed of the impeller creates a vortex that recirculates and pulverizes the clot. The device has a sideport that allows for high-pressure infusion of saline, contrast media, or both to cool the system (Fig. 9.5).

#### *Results*

Restoration of flow in the shunt is obtained in 89% of patients after mechanical thrombectomy with the ATD. The most important factor limit-

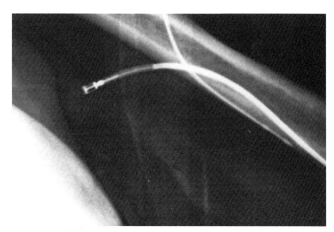

**FIG. 9.5.** The Amplatz thrombectomy device.

ing success is the presence of severe outflow lesions that will not respond to angioplasty. The 30-day primary patency is 47%, and the secondary patency is 68%.

The most important potential complication that has been described with the use of the ATD is hemolysis. However, clinical experience has demonstrated that the induced hemolysis has not been clinically significant. The use of the ATD is limited to fresh clots up to 3 weeks old.

The most important drawbacks of the current design of the ATD include the following:

- Size (8-Fr) potentially causing hemostasis problems
- Lack of steerability (Fig. 9.6)

### MTI Thrombolytic Brush Catheter

The MTI Thrombolytic Brush Catheter is a mechanical device with a 6-mm-diameter soft nylon brush that is attached to a flexible drive cable and a hand-held battery-powered motor.

The device is advanced through a 6-Fr sheath and is activated during the infusion of lytic agents into the occluded graft. This is a potentially interesting device, with the advantage of being of smaller size than the ATD. It appears to be less traumatic; however, it requires the use of thrombolytic agents, albeit at lower doses.

### Arrow-Trerotola Percutaneous Thrombolytic Device

The Arrow-Trerotola percutaneous thrombolytic device is a motor-driven fragmentation cage. The cage is attached to a stainless-steel drive cable. The fragmentation cage is housed within a 5-Fr catheter. The device is attached to a hand-held disposable rotator drive unit that rotates at 3,000 rpm. Technical success with this device approaches 95% with a relatively low (8%) complication rate. The graft patency at 3 months is approximately 40%.

## PERCUTANEOUS BALLOON-ASSISTED ASPIRATION THROMBECTOMY

Percutaneous balloon-assisted aspiration thrombectomy is a recently described percutaneous technique that involves the combination of thrombus pullback with an aspiration thrombectomy mechanism.

Two 7-Fr sheaths are introduced in a crisscross fashion. A Fogarty balloon is directed to the inflow arm of the graft (without crossing the anastomotic site) from the venous (outflow) access site. The balloon is then inflated, and a Fogarty maneuver and continuous aspiration via the 7-Fr sheath are performed simultaneously. The maneuver is then performed in the outflow arm of the graft (Fig. 9.7). The maneuver is

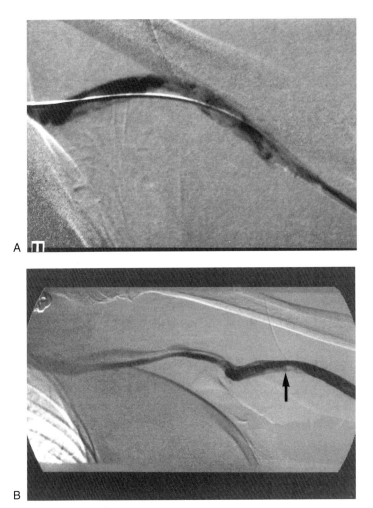

**FIG. 9.6. A:** Image shows clotted dialysis graft before treatment with Amplatz thrombectomy device. **B:** After thrombectomy with the Amplatz device, note the more than 90% clot elimination. Minimal amount of remaining clot is seen (*arrow*), mainly due to lack of steerability of the device.

repeated multiple times (two to five) until no further thrombus is recovered. If no flow is restored with these maneuvers, the arterial anastomosis is then declotted. Once flow is reestablished in the shunt, adjunctive procedures, including thrombolysis, angioplasty, or stent placement, may be performed to restore graft function.

With this technique, flow can be restored in approximately 90% of patients. Eighty-five percent of grafts are still functional 1 week after declotting.

Primary patency at 24 weeks is 33%, and secondary patency is 40%.

## COMPLICATIONS

### Thrombolysis

- Bleeding
- Embolization of the partially lysed clot
- Distal embolization
- Sepsis (if graft is infected)
- Pulmonary embolism
- Reactions to urokinase are characterized by shaking chills and restlessness [not an allergic reaction proper; usually controlled with meperidine (Demerol), 50 mg IV]

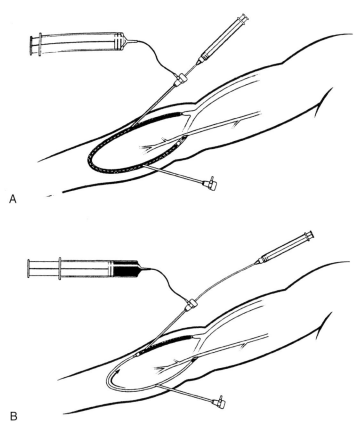

**FIG. 9.7. A,B:** Technique for percutaneous balloon-assisted aspiration thrombectomy in a clotted forearm loop graft. (Reprinted from Sharafuddin MJA, Kadir S, Joshi SJ, Parr D. Percutaneous balloon-assisted aspiration thrombectomy of clotted hemodialysis access grafts. *J Vasc Intervent Radiol* 1996;7:177–183, with permission.)

### Angioplasty and Stent Placement

- Intimal shearing
- Vein rupture
- Vein dissection
- Acute thrombosis
- Pseudoaneurysm formation
- Access site bleeding/hematoma

## DUAL-LUMEN DIALYSIS CATHETERS

Dual-lumen polyethylene or polyurethane catheters are inserted percutaneously through the internal jugular or subclavian vein.

Dialysis catheter placement using the subclavian veins should be avoided, because the subclavian catheters have been associated with a high incidence of subclavian vein stenosis or thrombosis, rendering the corresponding upper extremity unsuitable for future hemodialysis access. The dual-lumen silicone rubber catheters are particularly useful in managing adult patients in whom almost all the available vessels in the arms have been exhausted in previous access procedures and in those who are likely to suffer ischemic complications if the proximal arteries are used. These catheters may be safely placed in the angiographic suite under fluoroscopic guidance to decrease the incidence of complications such as arterial puncture and pneumothorax.

Flow inadequate for efficient dialysis may occur when (a) the catheter is too small for the patient, (b) the catheter tip is not positioned correctly in the right atrium, (c) the lumen of the catheter is compromised by a tight suture, or (d) the catheter is kinked along its subcutaneous course. A roentgenogram of the chest is essential after catheter placement to assess the position and course and to rule out technical complications such as pneumothorax or hemothorax.

## DESPERATION INTERVENTION

Translumbar venous catheterization, recanalization of occluded veins, and transhepatic hepatic vein catheterization have been used as a last resource for hemodialysis in some patients. These procedures usually provide a functional access for patients with few remaining options.

## CONCLUSION

Interventional techniques have been used successfully to maintain and also provide access for dialysis patients. Close cooperation among the nephrologist, vascular surgeon, and interventional radiologist provides the best results for these very complicated patients.

## SUGGESTED READINGS

Finlay DE, Longley DG, Foshager MC, Letourneau JG. Duplex and color Doppler sonography of hemodialysis arteriovenous fistulas and grafts. *Radiographics* 1993; 13:983–999.

Gray RJ. Percutaneous intervention for permanent hemodialysis access: a review. *J Vasc Intervent Radiol* 1997;8:313–327.

Hirschman GH, Wolfson M, Mosimann JE, et al. Complications of dialysis. *Clin Nephrol* 1981;15:66.

Lund GB, Lieberman RP, Haire WD. Translumbar inferior vena cava catheters for long-term venous access. *Radiology* 1990;174:31–35.

Matsumoto AH, Selby JB, Tegtmeyer CJ. Recent development of rigors during infusion of urokinase: Is it related to an endotoxin? *J Vasc Intervent Radiol* 1994;5:433–438.

Sharafuddin MJA, Kadir S, Joshi SJ, Parr D. Percutaneous balloon-assisted aspiration thrombectomy of clotted hemodialysis access grafts. *J Vasc Intervent Radiol* 1996;7:177–183.

Sullivan KL, Besarab A. Hemodynamic screening and early percutaneous intervention reduce hemodialysis access thrombosis and increase graft longevity. *J Vasc Intervent Radiol* 1997;8:163–170.

Surratt RS, Picus D, Hicks ME, et al. The importance of preoperative evaluation of the subclavian vein in dialysis access planning. *AJR Am J Roentgenol* 1991;156:623–625.

Trerotola SO, Johnson MS, Harris V, et al. Outcome of tunneled hemodialysis catheters placed via the right internal jugular vein by interventional radiologists. *Radiology* 1997;203:489–495.

Uflacker R, Rajagopalan PR, Vujic I, Stutley VE. Treatment of thrombosed dialysis access grafts: randomized trial of surgical thrombectomy vs mechanical thrombectomy with the Amplaz device. *J Vasc Intervent Radiol* 1996;7:185–192.

# 10
# Percutaneous Atherectomy

Jae-Kyu Kim, Hector Ferral, and Zhong Qian

Percutaneous atherectomy is a technique designed to debulk atheromatous plaque and to create a smooth luminal surface in atherosclerotic arteries using a mechanical catheter-based device. Despite its initial success, atherectomy has never become a treatment of choice in the management of the peripheral vascular diseases because the long-term efficacy remains questionable. Currently, percutaneous atherectomy is primarily performed as a complementary process to other techniques, including balloon angioplasty, stent placement, and fibrinolysis.

During the 1990s, various atherectomy devices have been used experimentally and clinically; this chapter focuses on four devices that have been approved by the U.S. Food and Drug Administration and on their clinical applications.

## AUTH ROTATIONAL ATHERECTOMY DEVICE

The Auth rotational atherectomy device (Heart Technology, Bellevue, WA), also known as the *Rotablator*, was first introduced in the peripheral applications in 1987. The device consists of an oval-shaped abrasive burr rotated at high speed over a wire (Fig. 10.1). The diamond chips (20 to 30 micrometers) are embedded on the surface of the burr and create millions of microscopic divots at high speed. The burr sizes are from 1.25 to 6 mm in diameter. The mechanism of atheromatous ablation with the device is based on the principles of differential cutting and orthogonal displacement of friction. Differential cutting preferentially removes inelastic materials such as fibrous, calcified, and fatty atheromatous tissues, while sparing healthy elastic tissues. Orthogonal displacement of friction generated by the burr permits easy advancement of the device through tortuous vessel segment. One of the most appealing features of this device is exceptional smoothness in the recanalized lumen.

### Indications

Use of the Auth device is particularly indicated in the following situations:

- Heavily calcified or eccentric atheroma, especially in diabetic patients with claudication or limb-threatening ischemia
- Long and diffuse stenotic or occlusive lesions

### Procedure

After diagnostic angiography, the Auth device is introduced through a guiding catheter, which usually has a diameter 0.004 in. larger than the burr. Once the guiding catheter is seated just proximal to the lesion, the 0.009-in. central guidewire is introduced across the lesion into the distal segment. The device is then introduced over the guidewire into the target segment. Atherectomy is accomplished by slowly advancing the device through the lesion while the burr is activated by depressing a foot pedal switch. The optimal burr spinning speed is between 160,000 and 190,000 rpm to ensure the smooth intraluminal surface. Results from early studies indicate most of the particles generated by the device are smaller than red blood cells and cleared harmlessly by the reticuloendothelial system.

### Technical Considerations

- A gradual increase in burr size is necessary for treatment of long (more than 10 mm) or

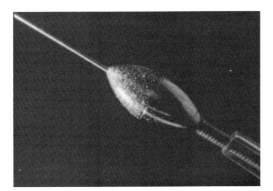

**FIG. 10.1.** Close-up enlargement of the Rotablator burr as it tracks over its 0.009-in. guidewire. Hundreds of microscopic diamond chips are embedded in the burr. When the burr rotates, each chip scoops out a microscopic amount of plaque.

diffuse lesions to avoid massive ablated particle burden.
- If the speed of the device is suddenly dropped down or resistance is encountered during atherectomy, the burr should be slightly withdrawn and readvanced.
- Adjunctive balloon angioplasty may be performed to improve luminal gains if the residual (more than 50%) stenosis remains after atherectomy.
- In case of total occlusive lesion, infusion of thrombolytics can be used in combination with guidewire manipulation to allow passage of the central guidewire across the occlusion.
- During atherectomy, excessive pushing force should be avoided to prevent vessel perforation resulting from alteration of burr slope. Especially in severe tortuous vessel, the activated device should be advanced gently but evenly in a to-and-fro fashion.
- Intermittent injection of contrast medium during atherectomy facilitates identification of the margins of the lesion and burr-working status.

### Complications and Limitations

Complications have been noted when the Auth device is used, including early thrombosis, arterial spasm, distal embolism, and hemoglobinuria resulting from hemolysis. Vessel perforation was also reported and may be associated with rapid advancement of the device. The main limitation of the device is that it is unable to bore through organized thrombus or rubbery atherosclerotic intima. Despite an encouraging initial technical (89% to 95%) and clinical success (72% to 94%) reported by several authors from multicenter investigations, a poor midterm outcome of primary patency (12.0% to 18.6%) has limited its wide applications.

## SIMPSON DIRECTIONAL ATHERECTOMY

The Simpson catheter, developed in 1986 (Devices for Vascular Intervention, Redwood City, CA), is available in two versions: AtheroCath (a fixed-wire device) and AtheroTrac (an over-the-wire catheter), with sizes ranging from 6 to 11 Fr. The catheter consists of a cylindric metal housing attached to the end of the shaft (Fig. 10.2). The housing has a 20-mm window with an opposing balloon. Atherectomy is achieved by pushing this open window against the atherosclerotic plaque using the inflated balloon. The trapped plaque is then excised and pushed by a high-speed rotating cutter into the distal collecting chamber. This mechanism permits restoration of the vessel lumen to a diameter larger than the catheter's size.

### Indications

Peripheral applications of the Simpson catheter have been focused on the iliac, superficial femoral, and popliteal stenotic lesions, which have one of these features:

- Discrete stenoses with intermittent claudication for at least a 3-month period
- Short (less than 5 cm) and eccentric atheroma
- Focal intimal hyperplasia and ulcerative plaque
- Peripheral bypass graft stenoses or anastomotic stenoses associated with fibrointimal hyperplasia

In addition, use of the Simpson can be extended to the management of hemodialysis access failure in the selected cases.

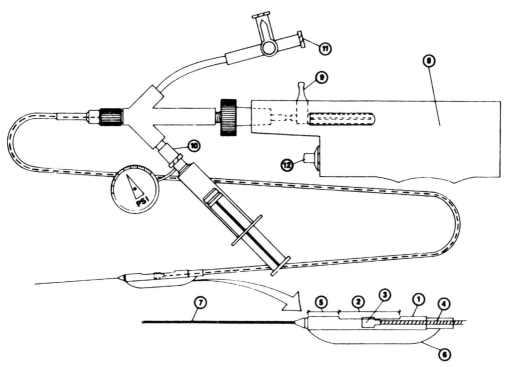

**FIG. 10.2.** Diagrammatic representation of the Simpson directional atherectomy catheter. (*1*, metallic capsule containing cutter and storage area; *2*, window; *3*, rotating cutter; *4*, rotating shaft; *5*, storage compartment; *6*, positioning balloon; *7*, floppy wire; *8*, power pack; *9*, manual cutter advance unit; *10*, sideport for balloon inflation; *11*, sideport for contrast medium injection; *12*, power unit switch.)

### Procedure

After the lesion is localized by angiography, the catheter is introduced through a check-flow sheath by an ipsilateral approach, retrograde for the iliac and renal arteries and antegrade for the femoral, popliteal, and tibioperoneal arteries. In the case of dialysis fistula, the approach may be made either on the cephalic vein or directly on the prosthetic graft. The access should be made as horizontally as possible to avoid kink of the sheath, which makes passage of the device extremely difficult. With careful manipulation of the steerable leading wire, the catheter is advanced using road mapping, to cross the lesion. Once the window is positioned against the atheromatous plaque, the balloon is inflated to 20-20 psi, forcing the plaque into the window, and the cutting blade is retracted proximal. The balloon is then further inflated to 35 psi. The cutter blade, driven by a battery-powered motor, is activated and manually pushed to shave the atheromatous plaque protruding into the window. While holding down the cutter blade, the motor is stopped, the balloon is deflated, and the catheter is rotated 10 to 20 degrees. The procedure can be repeated until the circumferential lesion is completely excised. The endpoint of the procedure should leave less than 20% residual stenosis. When the balloon is deflated or the catheter is withdrawn, the cutter blade should always be held against the distal end of the metal housing to prevent escape of resected atheromatous materials, which would cause distal embolization. Initial technical success (90% to 100%) and mid-term patency (50% to 86%) with the Simpson catheter have been promising in both iliac and femoropopliteal applications. In management of failing hemodialysis access, better outcome seems to be seen in intragraft or subclavian vein occlusion than in venous outflow stenoses.

## Technical Considerations

- It is very important to choose proper size of the device to ensure optimization of extraction of atheromatous materials. The best match is one in which the working diameter (of the metal housing plus the inflated positioning balloon) is at least equal to the diameter of the normal arterial segment adjacent to stenotic lesion.
- Excessive residual stenosis may be associated with the high risk of re-stenosis. Residual stenosis should not exceed 20%. Multiple passes of the atherectomy catheter are frequently necessary to achieve maximum removal of atheromatous plaque.
- In case of stenotic bypass graft associated with thrombosis, thrombolytic therapy should be initiated before the atherectomy procedure.
- To eliminate postprocedural hematomas at the entry site due in part to the larger vascular sheath, the effects of heparin may be reversed with protamine, and the sheath is removed when the activated clotting time is less then 160 seconds.

## Complications and Limitations

Complications associated with the Simpson catheter are relatively minor and manageable, including entry-site hematomas, distal embolization, puncture site thrombosis, and pseudoaneurysm. Its main drawback is prolonged procedure time. It is not suitable for atherectomy in the long or diffuse lesions or in the tortuous arteries.

## TRANSLUMINAL EXTRACTION CATHETER

The transluminal extraction catheter (TEC) (Interventional Technologies, San Diego, CA) is a forward-cutting atherectomy device. It consists of a semiflexible and torque-controlled catheter, an attached 125-mL vacuum bottle, and a low-speed (750-rpm) rotational cutting unit mounted on the distal tip of the catheter (Fig. 10.3). The two triangular cutter blades on the cone-shaped tip of a hollow catheter driven by a battery-powered motor shave off the surface of the atheromatous plaque while the excised materials are continuously aspirated by the vac-

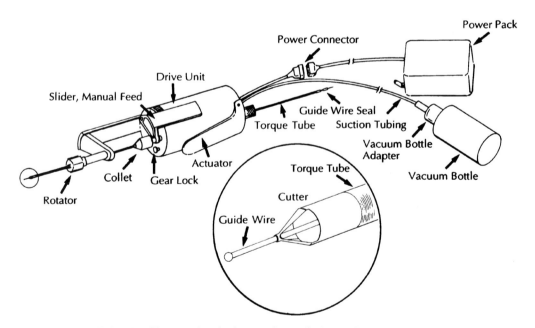

**FIG. 10.3.** The transluminal extraction catheter system.

uum effect, thereby reducing the risk of distal embolization. Unlike the Simpson catheter, a larger amount of atheromatous materials can be immediately removed from the system. However, the TEC provides no expansion ratio and therefore must be used on a one-to-one basis. Peripheral TECs are available in several sizes: 5 through 10, 12, 14, and 15 Fr, with a length of 115 cm. Insights from percutaneous angioscopy and intravascular ultrasound prove that the TEC is also able to remove intraluminal thrombus.

### Indications

- Short or asymmetric atherosclerotic stenoses
- Total occlusion or long stenotic lesions (atherectomy with the TEC as an adjunctive technique before balloon angioplasty to achieve maximum lumen gain)
- Failing hemodialysis access
- Percutaneous thrombectomy

### Procedure

The catheter is introduced over a 0.014-in. ball-tipped guidewire into the target vessel through an introducer sheath. The guidewire must be advanced to be beyond the leading part of the lesion. The cutting head then follows the wire to the segment proximal to the trailing of the lesion. The catheter is slowly advanced while activated by the motorized drive unit. The flow into the vacuum bottle should be monitored continuously. The vacuum bottle is immediately changed when it is full. The postatherectomy site is evaluated by removing the TEC device while holding the guidewire in position and injecting contrast medium through the sheath. If a large lumen is necessary, the device is replaced with one of larger diameters and the procedure is repeated. Adjunctive balloon angioplasty is usually required to achieve symptomatic relief for patients with femoropopliteal lesions. Although the technical success (92%) and immediate clinical improvement (90%) have been encouraging, long-term results are not available to determine its role in management of peripheral atherosclerotic lesions.

### Technical Considerations

- Because the catheter creates a channel with a diameter equal to that of itself, the catheter chosen should never have a diameter larger than the intraluminal diameter of the native vessel or graft immediately proximal or distal to the lesion(s).
- Although there is no absolute restriction on the number, length, or severity of the lesions, the treated vessel(s) must be crossed by the 0.014-in. leading guidewire.
- In severe stenosis, atherectomy should start with a smaller-size cutter and then upsize to a larger cutter.

### Complications and Limitations

Procedure-related complications are relatively low in experienced hands. Most of these complications manifest as puncture-site hematoma, thrombosis at the treated vessel, dissection, and intimal flaps, which can be managed by percutaneous techniques. Re-stenosis and reocclusion remain major constraints on TEC atherectomy. Heavily calcified lesions do not respond well to TEC atherectomy.

## TRAC-WRIGHT CATHETER

The Trac-Wright catheter (Dow Corning, Miami Lakes, FL), formerly known as the Kensey catheter, was developed in 1987. The device consists of a flexible polyurethane catheter and a blunt spinning cam driven by a power unit at speeds of up to 100,000 rpm (Fig. 10.4). It was designed to ablate atheromatous tissues into microparticles by a powerful fluid vortex generated by the high-speed spinning cam. Three theoretical features seem to make the device attractive: (a) The activated catheter tends to travel the path of least resistance due to fluid jets and vibration; (b) the device is able to selectively micropulverize fibrous or atheromatous material without damage to viscoelastic tissue such as vessel wall; and (c) powerful micropulverization by a combination of mechanical and hydrodynamic effects minimizes the risk of distal embolization. A

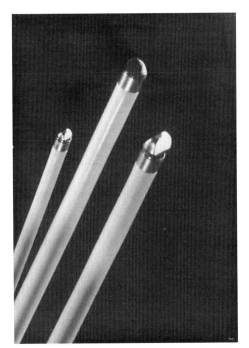

**FIG. 10.4.** Trac-Wright catheters of different sizes.

high-pressure fluid-irrigating system is used to dilute particulate density in the distal circulation, to cool and lubricate the device, to allow the addition of medications locally, and to monitor atherectomy status by injection of contrast medium. The Trac-Wright catheter is available in sizes of 5, 8, and 10 Fr.

### Procedure

After an ipsilateral femoral arterial access is established either in antegrade or in retrograde fashion depending on the level of the lesion, the Trac-Wright catheter is introduced through the sheath. Irrigating solution can be constituted by normal saline, nonionic contrast medium, and a thrombolytic agent, delivered with a flow rate of 20 to 30 mL per minute at 100 psi. The initial rotation speed of the device should be maintained at approximately 20,000 until the lesion is in direct contact with the tip. The speed is then increased to the regular operational speed (60,000 to 90,000 rpm), while the catheter is slowly advanced at 1 mm per second to permit the spinning cam sufficient time to pulverize the atherosclerotic plaque. It is recommended that the catheter pass through the proximal segment of the occluded artery several times before the leading end of the occlusion is reached. After complete penetration of the occlusion, the spinning speed should slow down to 20,000 rpm and the catheter should be slowly removed. Because the Trac-Wright catheter offers no expansion ratio, adjunctive balloon dilation is almost always needed to achieve adequate luminal gain. In case the device does not penetrate the leading end of the occlusion, recanalization may be further accomplished by a combination of steerable guidewire manipulation and balloon angioplasty.

### Technical Considerations

- For a totally occluded vessel, the infusion rate should be reduced to 10 to 20 mL per minute until the leading end of the lesion has been recanalized to reduce the risk of distal embolization by breaking off the leading end of the occlusion.
- During atherectomy, the catheter should be advanced slowly in a to-and-fro motion. The operator must watch for deflection of the tip, which may cause the catheter tip to be jammed into the arterial wall, or even vessel perforation. If this occurs, the device should be pulled back to the nonwave state, allowing the spinning tip to pulverize the deflecting lesion.
- A guidewire should not be manipulated before the device is used, because a false lumen created by the guidewire could become a path of least resistance to advancement of the catheter.
- Constant fluoroscopic observation and intermittent injection of contrast medium are most important measures to prevent vessel perforation.

### Complications and Limitations

Complications are not uncommon after application of the device. The initial conceptual advantages of the Trac-Wright catheter have not yet been corroborated by clinical observations. Overall complications range from 15% to 37% and include vessel perforation, dissections, and distal embolizations. Because of its size limitation, the Trac-Wright catheter is currently restricted for use as a mechanical adjunct to balloon angioplasty or thrombolytic therapy.

**Patient Preparation**

Patients are given diazepam and diphenhydramine orally before the procedure. A heparin bolus (10,000 to 15,000 units) is administrated intravenously during the procedure. Additional heparin is injected as needed to maintain the activated clotting time at more than 350 seconds. Intraarterial injection of vasodilators is recommended if the treated artery is below the femoral level.

**Postprocedural Care and Follow-Up**

Aspirin should continue to be given to maintain patients' hypocoagulable states for months. One day after the procedure, the treated segment of the vessel should be followed up by either angiography or intravascular ultrasound for evaluation of early outcome.

## SUGGESTED READINGS

Ahn SS, Concepcion B. Current status of atherectomy for peripheral arterial occlusive disease. *World J Surg* 1996;20:635–643.

Annex BH, Sketch MH, Stack RS, Phillips HR. Transluminal extraction coronary atherectomy. *Cardiol Clin* 1994;12:611–622.

Gray RJ, Dolmatch BL, Buick MK. Directional atherectomy treatment for hemodialysis: early results. *J Vasc Intervent Radiol* 1992;3:497–503.

Kim D, Gianturco LE, Porter DH, et al. Peripheral directional atherectomy: 4-year experience. *Radiology* 1992;183:772–778.

Maynar M, Cabrera P, Reyes R. Percutaneous atherectomy with the Simpson atherectomy device in the management of arterial stenosis. *Semin Intervent Radiol* 1988;5:247–252.

Maynar M, Reyes R, Cabrera P. Percutaneous atherectomies of iliac arteries. *Semin Intervent Radiol* 1988;5:253–255.

McLean GK. Percutaneous peripheral atherectomy. *J Vasc Intervent Radiol* 1993;4:465–480.

Mizumoto D, Watanabe Y, Kumon S, et al. The treatment of chronic hemodialysis: vascular access by directional atherectomy. *Nephron* 1996;74:45–52.

Prevosti LG, Cook JA, Unger EF. Particulate debris from rotational atherectomy: size distribution and physiologic effects [abstract]. *Circulation* 1988;78:II-83.

Reisman M, Buchbinder M. Rotational ablation. *Cardiol Clin* 1994;12:595–610.

Schwarten DE, Cutcliff WB, Krok KL. *The late results of femoral directional atherectomy.* RIPCV 2nd International Congress, Toulouse, France, 1990.

Wholey MH, Jarmolowski CR. New reperfusion devices: the Kensey catheter, the atherolytic reperfusion wire device, and the transluminal extraction catheter. *Radiology* 1989;172:947–952.

# 11
# Intravascular Stents

Jorge Lopera, Zhong Qian, and Hector Ferral

## A. OVERVIEW

Intravascular stents were developed to overcome the problems of recurrent stenosis and dissections after angioplasty. Although there are currently many different types of stents, in this chapter only the most commonly used types of stents are discussed. Intravascular stents can be classified in two different groups: balloon expandable and self-expandable (Table 11.1).

### BALLOON-EXPANDABLE STENTS

#### Palmaz Stent

The Palmaz stent (Johnson & Johnson Interventional Systems Co., Warren, NJ) (Fig. 11.1) has been extensively used in the vascular system. It is made of a special 316 L stainless steel. The large-size stents (P128, P188, P308) have a 0.005-in. wall thickness and expansion range of 8 to 12 mm and are 30 mm in length requiring a 10-Fr introducer sheath. The medium-size stents (P104, P154, P204, P294, P394) have a 0.0055-in. wall thickness and expansion range of 4 to 9 mm and are 10 to 39 mm in length requiring an 8-Fr introducer system. The medium-size stents have 30% more resistance to deformation than the larger stents. The Palmaz stent is very rigid, with high resistance forces, making it useful for highly resistant lesions. It can be deployed very precisely. Due to its rigidity, it should not be used in areas of external compression or in very tortuous vessels to avoid permanent deformation of the stent. The larger stents cannot be advanced from the contralateral approach.

### *Stent Preparation*

1. The unmounted Palmaz stent requires manual assembly on an angioplasty balloon (premounted Palmaz stents on a 5-Fr balloon catheter are available in 4- to 7-mm sizes that can be introduced through 7-Fr sheaths).

2. The Palmaz stent is mounted on the angioplasty balloon; OPTA-5 balloons (Cordis Corporation, Miami) or the 5.8-Fr Olbert balloon (Boston Scientific, Natick, MA) are recommended. The reasons are the excellent stent fixation to the balloon and minimal stent migration problems. The Olbert balloon assumes the same profile after deflation without wings, which facilitates safe removal after stent deployment, and it can be used to deploy multiple stents. This balloon uses a coaxial system that tends to pull back toward the catheter hub, displacing the stent toward the trailing end. Placing the stent 5 to 6 mm toward the leading end, above to the desired site of deployment, can prevent this problem (Fig. 11.2). Inspect the balloon to be sure that it has lowest profile and is well folded (preinflation is not recommended); negative pressure is applied with a 20-cc syringe to completely deflate the balloon.

3. The stent is mounted over the balloon and positioned exactly between the radioopaque markers of the balloon. The stent is then crimped with the fingers tightly over the balloon. A metallic crimping tool is also available to be used with the plastic crimping tube. Inspect the crimped stent to ensure adequate centering in the balloon, and perform a gentle tug to test for a secure crimp (Fig. 11.2).

4. The stent-balloon assembly is carefully placed through an introducer sheath. A metallic introducer is available to facilitate stent introduction. This introducer protects the stent when it is introduced through the hemostatic

**TABLE 11.1.** Characteristics of the stents

| Type | Alloy | Flexibility | Magnetic resonance imaging effect | Strength | Shortening | Radioopacity |
| --- | --- | --- | --- | --- | --- | --- |
| Palmaz | Stainless steel | + | ++ | ++ | 17% | ++ |
| Wallstent | Elgiloy | +++ | + | + | 20–40% | + |
| Strecker | Tantalum | ++ | – | ++ | 19% | +++ |
| Gianturco | Stainless steel | – | +++ | + | 0% | ++ |
| Cragg | Nitinol | ++ | – | ++ | 7% | ++ |

+, least; ++, medium; +++, most; –, information not available.

valve. Its use is not always necessary. Once the balloon and the stent are past the valve, the metallic introducer is pulled out of the valve and pushed to the back end of the angioplasty balloon. This assembly can be advanced through an 8-Fr sheath.

*Introduction of the Stent*

1. The exact localization of the lesion is marked with external metallic markers or with the road-mapping technique. Predilatation of the lesion may be necessary in tight lesions. The

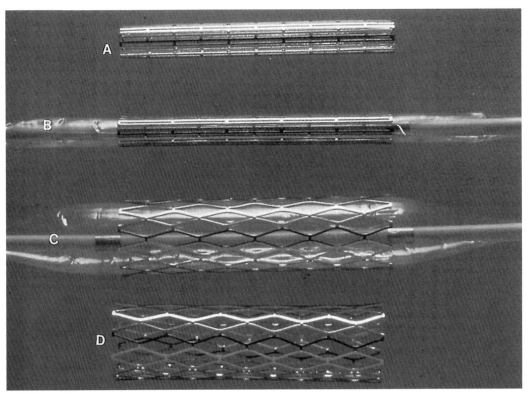

**FIG. 11.1. A:** The Palmaz stent. **B:** Mounted on an angioplasty balloon. **C:** Stent balloon expansion. **D:** Fully expanded stent. (Reprinted from Castañeda-Zúñiga WR. *Interventional radiology*, 3rd ed. Baltimore: Williams & Wilkins, 1997, with permission.)

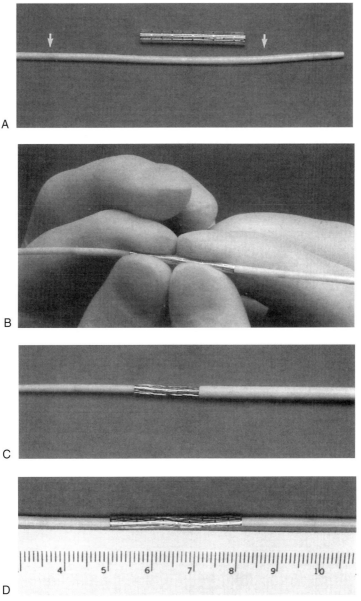

**FIG. 11.2.** Technique to mount the Palmaz stent over a balloon. **A:** The stent should be placed on the balloon between the markers. **B:** Crimping is accomplished by squeezing firmly inward, then rotating and squeezing in a new position until the stent is no longer compressible. **C:** Ideally, one should make sure that the stent mounted on the balloon fits easily into the nondiaphragmatic end of the sheath before placement. **D:** The leading end of the stent has been aligned with the 5-cm mark to demonstrate its withdrawal during inflation. (*Continued.*)

introducer sheath is advanced across the stenosis.

2. The stent is advanced through the sheath, and once the stent is centered in the lesion the sheath is slowly pulled back to uncover the stent and the balloon completely. Gentle contrast injections can be performed in small amounts to verify the location of the lesion.

3. Expansion of the stent is performed, inflating the balloon under fluoroscopic monitoring. The balloon is inflated to the recom-

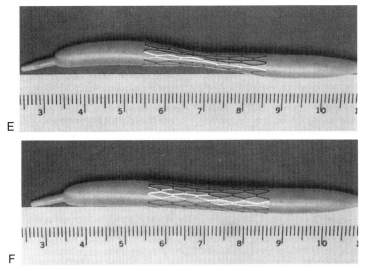

**FIG. 11.2.** *Continued.* **E:** After inflation to 4 atm, the partially inflated stent in the classic "hourglass" shape has withdrawn approximately 5 mm. **F:** The fully inflated stent has withdrawn 5.5 mm. (Reprinted from Bjarnason H, Hunter DW, Ferral H, et al. Placement of the Palmaz stent with use of an 8-F introducer sheath and Olbert balloons. *J Vasc Intervent Radiol* 1993;4:435–439, with permission.)

mended pressure (minimum 5 atm). Once the stent is expanded, the balloon is completely deflated and carefully rotated within the stent to refold the balloon; the deflated balloon is slowly removed, making sure that the balloon is free of the stent.

4. A control angiogram is performed, advancing a catheter over the guidewire and across the stent.

### *Potential Problems*

- Placement of the Palmaz in areas of flexion—that is, at the groin—should be avoided to prevent collapse of the stent. Also, placement of Palmaz stents in tortuous arteries is not recommended.
- Ruptured balloon can also occur and can be recognized by loss of the inflation pressure and aspiration of blood into the balloon. If the balloon ruptures with the stent partially expanded, the balloon is deflated and rotated two to three times inside the stent. The balloon is removed and exchanged for a new balloon, which, once positioned inside the stent, will complete the stent expansion. If the stent is not anchored securely in the arterial wall it will move out of the target when the ruptured balloon is removed. To stabilize the stent, the sheath tip is advanced to the stent while the balloon is withdrawn into the sheath. If the stent migrates out of the target when the tethered balloon is removed, the sheath is used to stabilize the stent, and then a new balloon is advanced into the stent and partially inflated to catch the stent. The assembly can then be advanced very carefully forward into the lesion. If any resistance is met with this maneuver, the migrated stent is deployed and a second stent used to cover the lesion.

### Strecker Stent

The Strecker stent (MEDITECH/Boston Scientific Corp., Natick, MA) is made of a single 0.1-mm tantalum wire mesh. Because of the loose structure of connected wire loops, the stent offers a high degree of longitudinal and radial flexibility. This balloon-expandable stent is not approved for use in the United States but has been used extensively in Europe. To avoid dislocation of the stent, the ends are attached to the balloon by silicone cuffs. During balloon expansion the silicone sleeves slide back, releasing the

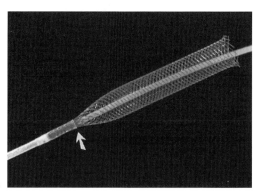 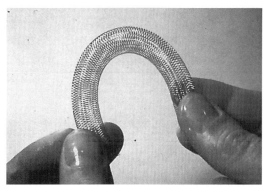

**FIG. 11.3.** The Wallstent. **A:** Withdrawal of the sheath allows progressive release of the stent (*arrow*) **B:** Note the flexibility of the stent. (Reprinted from Castañeda-Zúñiga WR. *Interventional radiology*, 3rd ed. Baltimore: Williams & Wilkins, 1997, with permission.)

stent. The stent is available in 4 to 12 mm, with a maximum length of 8 cm. The stent is mounted on a 5-Fr (for stents 4 to 8 mm in diameter) or 5.8-Fr (for stents 9 to 12 mm in diameter) Ultrathin ST balloon (MEDITECH/Boston Scientific Corp.).

The 4- to 8-mm stents are introduced through a 7-Fr sheath, the 9- to 11-mm stents through a 9-Fr, and the 12-mm stents through a 10-Fr sheath. The stent has excellent radioopacity and is flexible. The deployment of the stent is similar to that for the Palmaz stent. Predilatation of the lesion is recommended, and a metallic loading tool is used to advance the stent through the sheath. The stent is deployed by inflating the balloon. Several balloon inflations are sometimes required to obtain complete stent expansion. Because the stent is made of a single-knitted filament, the stent can be more easily removed with a forceps. The main disadvantage is the lack of hoop strength; lesser forces are necessary to cause eccentric deformation of the stent.

## SELF-EXPANDING STENTS

### Wallstent

The Wallstent (MEDITECH/Boston Scientific Corp.) is composed of Elgiloy, a surgical metallic alloy woven in a criss-cross pattern to form a tubular braid configuration.

The stent comes in a delivery system called *Unistep*. It consists of an exterior tube that constrains the stent until retracted during delivery; this tube is connected to a stopcock for priming the stent. There are radioopaque markers in the leading and trailing ends of the stent. An interior tube of the coaxial system contains a lumen that accommodates 0.035-in. or 0.038-in. guidewires. The stent is very flexible and pliable (Fig. 11.3). Stents up to 12 mm are mounted in a 7-Fr delivery system. Disadvantages of the stent are the poor fluoroscopic visibility (especially in obese patients) and the significant shortening, which makes precise placement difficult.

### *Preparation of the Stent*

Choose the appropriate stent for the patient; factors such as lesion length, possible lesion progression, and shortening of the stent must be considered. Usually, a stent that is longer than the minimum length that will provide adequate covering of the lesion is used. Predilatation of the lesion is recommended, especially in tight lesions, to facilitate stent advancement and subsequent dilatation. During deployment, the stent shortens by approximately 20% to 40% of the original constrained length.

1. Sterile saline is injected between the interior and exterior tubes via the stopcock to pro-

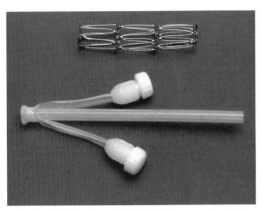

**FIG. 11.4.** The three-body Gianturco stent. The peel-away introducer is also shown.

vide lubrication until a few drops of saline are seen to emerge at the tip of the catheter.

2. The tip of the delivery system is inspected to verify that the leading end of the stent is covered by the exterior tube.

3. The delivery system is advanced over the guidewire.

4. Once the radioopaque markers are positioned in the lesion, the stent is deployed by slowly retracting the exterior tube while the metallic stainless steel tube is held stationary. The progression of the deployment is verified with fluoroscopy. As long as less than 75% of the stent is deployed, the stent with the delivery system can be pulled back as a unit and removed through the sheath if necessary. A reconstrainable delivery system is now available. With this new system, 83% of the stent can be deployed and can still be pulled back into the delivery system. To ensure complete deployment of the stent, the external tube is pulled all the way back into the metallic tube.

5. Once the stent is completely deployed, the delivery system is removed over the guidewire. When removing the delivery system or during balloon dilatation, make sure that the ends of the stent are not "caught" with the catheter; this may result in migration of the stent. If multiple stents are necessary, a 1- to 2-cm overlapping is suggested, with deployment of the stent at the leading end first.

## Gianturco Stent

The Gianturco stent is made of stainless steel wire bent in a zigzag pattern (Cook, Inc., Bloomington, IN). Multiple stents are connected by a wire strut (Gianturco type) or by a nylon monofilament (Gianturco-Rosch type). The expansile force of the stent increases proportionally to the number of leg bends and the diameter of the wire. Oversizing the stent by 25% to 50% also greatly increases the expansile force. The stent is compressed and introduced through a Teflon catheter of 8 to 12 Fr, depending on the caliber of the wire and the stent diameter.

The stent is deployed at the desired location with the help of a coaxial pusher while the pusher is held stationary; the Teflon catheter is pulled back, uncovering the stent that will expand. Sometimes the stent springs out when deployed, resulting in migration. This is more common with the single-body stent. Multiple-body stents are less likely to migrate during deployment (Fig. 11.4). The stent is rigid in the longitudinal axis. The main use in the vascular system is in large veins, such as the superior vena cava (SVC) and inferior vena cava (IVC). A slight modification of the stent position is possible while the stent is only partially extruded from the sheath. After deployment, the stent usually expands only to 50% to 60% and then continues to expand over a period of time.

## Cragg Stent

Nitinol is a nickel-titanium alloy with a unique thermal recovery property. When the nitinol is constrained to a desired shape and annealed at 500°C, it will memorize that shape. When the nitinol stent is cooled in iced water, it can be compressed and deformed. When the stent is warmed by the blood to its transitional temperature, it will resume its original shape. The Cragg stent is made of nitinol wire, 0.27 mm in diameter, bent in a zigzag configuration and tied with a 7-0 polypropylene suture in a spiral to form a tube. This stent is not approved by the U.S. Food and Drug Administration, but it has been used in Europe, Mexico, and South America. It is available in different sizes and has a 7% rate of shortening. The stent is deployed through a long (45-cm) 8.5-Fr introducer sheath. The constrained stent is preloaded in a Teflon capsule and is loaded into the sheath and pushed with a blunt catheter. Introduction of the stent is facilitated by means of infusion of cool saline solution, but this is not necessary in most cases. The sheath is initially advanced past the lesion to be treated; the stent is advanced under fluoroscopic control until it is across the lesion. To achieve stent deployment, the stent is held stationary with the pusher while the introducer sheath is pulled back. The stent has minimal shortening, good longitudinal flexibility, and greater expansion rate due to its thermal memory.

## B. ARTERIAL STENTING

### STENT PLACEMENT IN SUPRAAORTIC VESSELS

The initial experience of stent placement in supraaortic vessels is encouraging, but the long-term follow-up is not available. Patients usually present with arm claudication or vertebrobasilar or hemispheric vascular insufficiency (subclavian steal syndrome). High technical success rates have been reported with angioplasty alone, even in cases of proximal subclavian occlusions.

### Indications

- Suboptimal results after angioplasty, including recurrent or significant residual stenosis (more than 30% of lumen diameter) and residual peak systolic arterial pressure gradient (more than 10 mm Hg).
- Treatment of postangioplasty complications, including flow-limiting dissection and acute thrombosis.
- Primary stent placement is recommended in ulcerative lesions with evidence of distal embolization and markedly eccentric stenosis. Primary stent placement of ostial lesions is recommended by some authors.

### Contraindications

- Lesions involving the origin of the vertebral artery (second segment of the subclavian artery)
- Lesions of the origin of the right subclavian artery where placement of a stent extends across the origin of the carotid artery

### Technique

1. The procedure can be performed from the femoral or the axillary approach.
2. Cerebral angiography is necessary to define the contribution to cerebral flow by the affected vessels and to define the collateral flow. Pressure gradients are obtained and compared with bilateral systolic brachial cuff pressures.
3. The Palmaz stent is preferred in areas in which precise placement is critical. The Wallstent is the choice for curved or tortuous vessels.
4. The stent is deployed using a 90-cm-long 8-Fr guiding catheter (Cordis Corp.) or a long introducer sheath (Daig, Minnetonka, MN).
5. In subclavian stent placement, the origin of the vertebral arteries is considered, trying not to cover its origin with the stent. Also, the origin of the right common carotid artery is evaluated in right subclavian stent placement.
6. After subclavian angioplasty and stent placement, a delayed (20- to 45-second) restoration of the normal flow direction in the ipsilateral vertebral artery will occur. This delay may provide protection of embolic events, but the risk of neurologic complications may exist immediately after angioplasty (Fig. 11.5).
7. In lesions of the origin of the vessels, the ideal position of the stent is approximately 2 mm protruding into the aortic arch, with the rest of the stent within the branch artery.

### Postprocedure Care

- A bolus of 5,000 units of intravenous heparin is administered before the stent deployment. A short course (24 hours) of anticoagulation

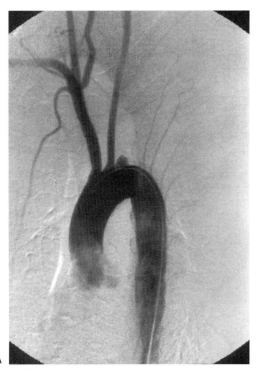
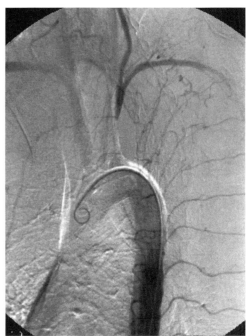
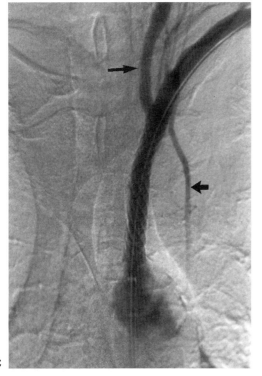

**FIG. 11.5.** Subclavian artery stent placement. Patient with severe claudication of the left arm. **A:** Arch arteriogram shows a 5-cm occlusion of the origin of the left subclavian artery. **B:** Delayed film shows opacification of the distal left subclavian artery via the inverted flow through the left vertebral artery. **C:** Film after guidewire recanalization and Palmaz stent placement shows excellent flow through the recanalized segment. Note the antegrade flow in the left vertebral artery (*arrow*) and left internal mammary artery (lima) (*arrowhead*).

is recommended. Antiplatelet agents such as aspirin, ticlopidine, or both are then continued orally.
- A baseline duplex scan should be obtained and followed by serial duplex and brachial cuff pressure measurements.

### Results

Technical success rates are more than 90%. Most technical failures are related to subclavian occlusions in which recanalization with a guidewire is not possible. Neurologic complications are rare, even when angioplasty is performed at the origin of the vertebral artery. Most complications are access related (2%). Long-term results are not yet available; cases of recurrent stenosis have been reported and successfully treated with additional angioplasty/stent placement.

## RENAL ARTERY STENT PLACEMENT

### Indications

- Renal artery stenosis (RAS) of 70% or more with inadequate response to angioplasty (residual stenosis of more than 30%, extensive dissection or pressure gradient after angioplasty of more than 10 mm Hg).
- Recurrent lesions after angioplasty.
- Ostial lesions.
- Lesion limited to the main renal artery not extending beyond 20 mm of the aortorenal border.
- Renal artery size of 4 to 7 mm.
- Although the experience is more limited, stents are useful for patients with fibromuscular dysplasia, Takayasu's arteritis, renal transplant stenosis, and aortorenal bypass stenosis after unsuccessful angioplasty.

### Contraindications

- Patient with normal blood pressure or with satisfactory control of hypertension.
- Lesion extends more than 20 mm from the aortorenal border or multiple lesions are present (placing a stent too distally into the renal artery will not allow a surgical bypass to be performed if needed).
- Advanced renal disease with creatinine level of more than 3.0 mg per dL in unilateral RAS or more than 4 mg per dL in bilateral RAS and proteinuria of more than 2 g per 24 hour.
- Kidney length of less than 7 cm.
- Renal artery diameter of less than 4 mm.
- Abdominal aortic aneurysm of more than 4 cm in diameter (relative contraindication; can be used as an adjunct to surgery).
- Contraindications to aspirin, heparin, or intravenous contrast medium.
- Unstable patient unable to sustain a surgical procedure in case of complications.
- Vessel rupture after angioplasty.

### Procedure Preparation

- Patient selection is critical; determining whether the RAS is responsible for the renal failure or hypertension is sometimes impossible.
- Aggressive hydration is recommended.
- Aspirin, 325 mg orally, starts the day before the procedure. Aspirin is continued indefinitely.
- Placement of a Foley catheter is recommended in bilateral stent placement.
- Modify antihypertensive medications as required.

### Technique

1. After performing a nonselective aortogram, the best approach is selected. In the majority of the cases, the procedure is performed from the femoral approach; the brachial or axillary approach is used occasionally in caudally oriented renal arteries.

2. A bolus of 3,000 to 5,000 units of heparin is administered once the access is obtained. Keep the activated clotting time at more than 300 seconds. Renal catheterization is performed with a cobra or a Simmons 1 catheter.

3. After crossing the lesion, 100 to 200 µg of nitroglycerin or 2.5 mg of verapamil are injected intraarterially to prevent spasm. The guidewire should not be placed too distally within the segmental branches to prevent vascular spasm.

4. It is essential that the renal artery is imaged in the proper oblique so that the lesion is exactly in profile. Usually, the left renal artery is best demonstrated in a 20- to 30-degree left anterior oblique (LAO) view and the right renal artery in an anteroposterior or 40-degree (LAO) view.

5. The stenosis is crossed with careful manipulation using a high-torque flexible-tip guidewire, such as the Wholey guidewire, or a steerable tapered guidewire TAD II (Peripheral System Group, Mountain View, CA). The angiographic catheter is advanced past the lesion and the pressure is measured. Exchange for a stiffer guidewire, such as the Rosen exchange length wire (Cook, Inc.), is then performed. Angioplasty of the lesion is performed; the guidewire is kept across the lesion at all times.

6. The Palmaz 154 is most commonly used in ostial lesions and can be dilated to 4 to 9 mm. To place a Palmaz stent, a long 7-Fr sheath or an angled 8-Fr "hockey stick" guiding catheter (Cordis Corp.) can be used. After the angioplasty, the stent is crimped over the balloon (usually 5-Fr shaft, 2 cm long). The guiding catheter is advanced across the lesion. The stent is then carefully advanced within the guiding catheter until the desired stent position is achieved. Once the adequate position of the stent is obtained, the guiding catheter is retracted, with the balloon kept immobile, uncovering half of the stent. Digital subtraction angiography is then performed through the guiding catheter and the position of the stent adjusted. The stent is completely uncovered, and after assessing final position with digital subtraction angiography, the balloon is inflated to obtain stent expansion. The balloon is then carefully removed after complete deflation and control angiogram performed through the guiding catheter. If necessary, the stent can be further dilated with larger balloons. Intravascular ultrasound is very useful to assess stent expansion.

7. In ostial lesions, the stent can protrude into the aortic lumen no more then 2 mm; the proximal end of the stent can be overdilated with a larger balloon to expand the stent struts against the plaque (Fig. 11.6).

## Postprocedural Care

- The femoral sheath is removed when the activated clotting time is less than 200 seconds.
- Vigorous hydration is recommended before and after the procedure to reduce the risk of contrast induced nephropathy.
- Blood pressure, fluid balance, and renal function are monitored closely.
- Intravenous heparin may be started 4 hours after the procedure, and then oral anticoagulation with warfarin (Coumadin) takes place for 4 weeks.
- Long-term acetylsalicylic acid and ticlopidine (250 mg per day) for 1 month are adequate alternatives.
- Close follow-up of blood pressure and the renal function are mandatory. A follow-up angiogram should be obtained, if possible, 6 and 12 months after stent placement.

## Results

The technical success ranges from 75% to 100%. Primary and secondary patency rates are 92% and 98% at 1 year and 79% and 92% at 2 years, respectively. Improvement or cure of hypertension has been noted in 67% to 100%.

In patients with renal insufficiency, improvement in renal function is seen in 34% of patients, stability in 39%, and deterioration in 27%.

## Complications

The incidence of complications is similar to the angioplasty related complications (approximately 13%).

- Thrombus formation around the ostium of the renal artery containing the stent is a complication that can be caused by a significant protrusion of the stent into the aortic lumen or by the presence of significant residual stenosis.
- Migration and embolization of the stents can also occur.
- Branches covered by the stent can be occluded.
- Contrast-induced nephrotoxicity can occur but can be partially prevented with hydration and reducing the amount of contrast.

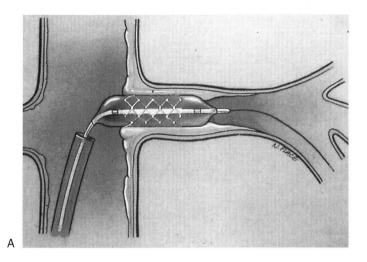

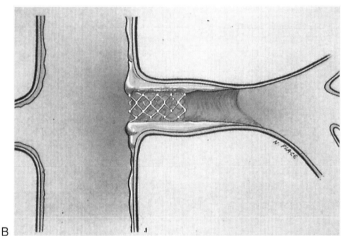

**FIG. 11.6. A:** Diagram of placement of Palmaz-Schatz stent in ostial renal artery stenosis. The balloon is fully inflated in order to expand the stent. **B:** After removal of the balloon, the stent acts as scaffolding to keep the renal artery ostium open. (Reprinted from Castañeda-Zúñiga WR. *Interventional radiology*, 3rd ed. Baltimore: Williams & Wilkins, 1997, with permission.)

- Renal failure can also be caused by distal embolism and cholesterol embolism.
- Early and late thrombosis of the renal artery can also occur.
- Stent re-stenosis due to neointimal hyperplasia is noted in 20% of patients at 1 year and is usually asymptomatic; when the re-stenosis is significant (more than 50%), it can be treated with angioplasty.

## AORTO-ILIAC AND PERIPHERAL ARTERIES

Currently, only the Palmaz stent and the Wallstent are approved by the Food and Drug Administration for use in the iliac arteries.

### Iliac Arteries

*Indications*

1. Failed angioplasty defined as
   a. Intimal dissection longer than the angioplasty site
   b. Residual stenosis of 30% or more
   c. Maximum transstenotic mean pressure gradient of 10 mm Hg or more postangioplasty
2. Re-stenosis within 90 days of a previous angioplasty or other endovascular procedure
3. Recanalization of total iliac occlusions
4. Ulcerative plaque
5. Leriche syndrome

## Contraindications

### Absolute

- Perforation or aneurysm at the angioplasty site
- Systemic infectious status (sepsis)

### Relative

- Lesions longer than 10 cm
- Dense, extensive calcifications of the iliac arteries
- Iliac artery aneurysm
- Diffuse, long-segment disease in small-caliber external iliac artery
- Severe, diffuse stenoses of the common femoral artery with poor (one vessel) distal runoff
- Severe hypertension
- Impaired pain sensation

## Technical Considerations

- Aspirin is given 1 day before the procedure and continued indefinitely.
- An ipsilateral approach is recommended as a first option. After the lesion has been crossed with the guidewire, heparin, 3,000 to 5,000 units, is administered intravenously to keep the activated clotting time at above 300 seconds.
- The lesion is predilated with an angioplasty balloon. A control angiogram is performed after the angioplasty to rule out rupture.
- The use of a stent with 10% to 15% greater diameter than the native vessel is critical to achieve adequate strut embedment.
- In heavily calcified lesions, the Palmaz stent is preferred because it provides better expansile force. When a contralateral approach is used or when placement of the stent must extend into an area of motion (at the groin), a flexible stent (Wallstent) is recommended.
- If multiple stents are used, the leading end of the lesion is stented first. One millimeter of overlapping is suggested for the Palmaz, and 10 mm of overlapping is recommended for the Wallstent.
- Sometimes, the origin of the internal iliac artery is covered by the stent. Although most of the time this is not clinically significant, gradual occlusion or severe ostial stenosis will develop in approximately 40% of the cases.
- In the aortic bifurcation, the kissing technique is used to treat lesions in the proximal common iliac artery and the distal aorta.

## Stent Placement of Complete Iliac Occlusions

Successful recanalization and stent placement of complete iliac occlusions are possible. The technical success rate ranges from 81% to 92%. Recanalization is performed with a straight or angled guidewire (MEDITECH/Boston Scientific Corp.). Frequent injections of contrast medium are required to verify intravascular position of the catheter and guidewire. An ipsilateral approach is more effective, but in some cases a contralateral approach using a sidewinder catheter is effective. After recanalization is achieved, primary stent placement without predilatation is recommended to decrease the risk of distal embolization. Other authors recommend overnight infusion of thrombolytics to reduce the thrombotic mass and facilitate the recanalization with the guidewire. The rate of distal embolization with thrombolytics is similar to that of guidewire recanalization followed by stent placement. Primary stent placement is not recommended in iliac occlusions less than 3 months old.

## Follow-Up

The follow-up includes clinical examination and ankle-brachial index at 1, 6, and 12 months. Angiographic follow-up is recommended at 6 and 12 months after stent placement. Aspirin (300 mg per day) is given 24 hours before stent implantation and then continued indefinitely.

## Results

The long-term patency of iliac stents is affected by the quality of the distal runoff, the size of the treated artery, the number of stents used, and the presence of diabetes mellitus and tobacco smoking. Patency rates are lower for occlusions compared with stenoses. With the Palmaz stent, the clinical success is 98.9% immediately after the procedure and 86.2% at 48 months. The angiographic patency rate is 87.5%. With the Strecker stent, primary patency rates at 3 and 5 years are 85% and 79%, respectively. With the Wallstent, the primary patency is 81% at 1 year, 71% at 2 years, and 61% at 4 years. The secondary clinical patency rates are 91%, 86%, and 86% at 1, 2, and 4 years, respectively.

The primary patency in primary stent placement for chronic iliac artery occlusions is 87% after 1 year, 83% after 2 years, and 78% after 4 years; secondary patency rates are 94%, 90%, and 88% at 1 year, 2 years, and 4 years, respectively.

### *Complications*

- Overall complication rate is 4% to 12%.
- Most complications are access-related, including groin hematoma (1.4%), groin pseudoaneurysm (0.7%), angioplasty site rupture (0.7%), and contrast-induced renal failure (0.7%).
- The incidence of distal embolization is 2% to 9%. The incidence of thrombosis of the stent is 5% to 12%.
- Most complications can be treated percutaneously. The incidence of stent thrombosis is higher when the stent is placed in an occluded (15%) segment than in a stenotic nonoccluded segment (2.7%). The incidence of thrombosis is also higher with stents in the external iliac artery (12.8%) compared to the common iliac arteries (1.1%). Longer-stented stenosis has a higher thrombosis rate (16.7%) than shorter ones (0.5%). Stented areas with poor runoff have higher (14%) incidence of thrombosis than segments with good runoff (2.1%).
- In stent placement of complete iliac occlusions, the overall rate of complications is 11.6% to 15.0%, with an incidence of major complications of 5.8%, including a 4.8% to 7% rate of distal embolization.

## Aortic Stents

The experience of stent placement of the abdominal aorta is limited; initial results are encouraging, but the indications and the role of stents versus angioplasty are not yet well defined. Total aortic reconstruction can be performed using stents, but until longer follow-up is available this treatment should be reserved for patients in whom surgery is contraindicated.

### *Indications*

- Focal stenosis of the abdominal aorta that failed to angioplasty due to significant residual stenoses (systolic pressure gradient of more than 20 mm Hg, 50% or more residual stenosis) or significant dissection (Fig. 11.7)
- Recanalization of total occlusions
- Graft anastomotic stenosis

### *Contraindications*

- Severe aorto-iliac disease that makes advancement of the delivery system impossible
- Acute aortic occlusions
- Stenosis close to the origin of the inferior mesenteric artery when the other mesenteric vessels are diseased
- Abdominal aortic aneurysm (stent-grafts are indicated)

### *Technique*

1. Access is obtained from the groin.
2. Angioplasty of the stenosis is performed using a single 10- to 12-mm balloon or 2 smaller (5- to 6-mm) balloons using the kissing technique; underdilatation of the aorta is recommended to prevent rupture.
3. In aorto-iliac lesions, the aorta and the iliacs are reconstructed using kissing stents. It is important to deploy the stents precisely to completely cover the aorto-iliac junction. Slight protrusion of the stent into the aortic lumen is preferred to cover the entire lesion and to ensure that the ostium of the iliac artery is widely open.
4. The 12- to 16-mm Wallstent can be advanced through a 9-Fr sheath.
5. The 308 Palmaz stent can be accurately placed and dilated in an expansion range of 8 to 12 mm. In an extreme case, this stent can be

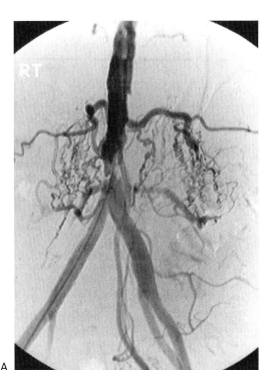
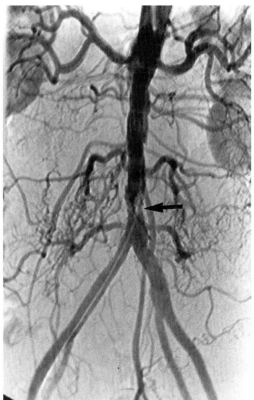
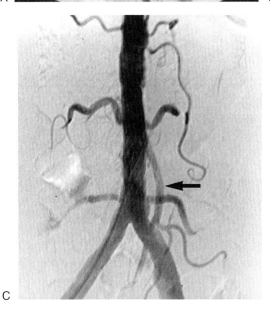

**FIG. 11.7.** Aortic stent case. A 69-year-old female with severe bilateral buttock and thigh claudication. **A:** Aortogram shows severe stricture of the distal abdominal aorta right above the iliac bifurcation. Note the extensive collateral network. **B:** After angioplasty, there is some improvement in the aortic lumen; however, a small filling defect persists (*arrow*). **C:** Angiogram after Palmaz stent placement shows a widely patent lumen without residual stenosis. Note the decrease in collateral opacification and patent inferior mesenteric artery (*arrow*).

dilated up to 15 mm, but considerable stent retraction is to be expected.

6. Balloon dilatation of the aorta should be performed with extreme care; low dilation pressures are recommended (5 atm).

7. If a single access is used, a large balloon used to dilate the aorta should not extend into the bifurcation, as overdilating the aorto-iliac junction can result in rupture of the iliac artery.

## *Complications*

Complications are very rare if careful technique is observed. Rupture of the aorta or the iliac artery is possible. Peripheral and mesenteric embolization may also occur.

## **Femoropopliteal Arteries**

Due to the small size of the vessels in the femoropopliteal region, placement of stents in this area is performed only in selected patients.

## *Indications*

- Stenosis after failed angioplasty (residual stenosis of more than 30% or peak systolic gradient of more than 5 mm Hg)
- Angioplasty-induced acute dissection resulting in flow impairment
- Recurrence of the lesion after angioplasty

## *Technique*

1. For lesions in the proximal and mid portion of the superficial femoral artery (SFA), a contralateral approach is usually adequate. Placement of an up-and-over Balkin sheath (Cook) is recommended.

2. For lesions in the mid to distal SFA, an antegrade approach is used, with placement of a 7-Fr introducer sheath to perform road mapping.

3. Occasionally, a popliteal approach can be used when the other techniques fail.

4. The lesions must be carefully negotiated with a guidewire. Angiography is performed to rule out distal embolization and to localize the lesion. Vessel spasm must be treated aggressively with intraarterial nitroglycerin or verapamil.

5. Stents 5 to 7 mm in diameter are used. The most commonly used stents are the Palmaz and Wallstent. The Strecker stent has been extensively used in Europe.

6. After the stent placement, a runoff is performed to rule out distal embolization.

7. Systemic anticoagulation is indicated, especially in vessels smaller than 6 mm. Anticoagulation is continued for 6 months.

## *Contraindications*

- Lesions that involve the femoral bifurcation should be approached carefully to prevent occlusion of the profunda femoral artery with the angioplasty or the stent.
- Rigid stents such as the Palmaz should not be placed in the common femoral artery to avoid deformation with the flexion of the extremity.

## *Results*

Lower patency rates have been reported with occlusions and vessel diameter of less than 5 mm. The initial success is high (99.7%). Using the Palmaz stent, the 4-year primary and secondary patency rates have been 82% and 95%, respectively, for the proximal and mid SFA. For the lower SFA and the popliteal artery, the 4-year primary and secondary patency rates have been 44% and 82%, respectively. The 4-year primary patency rate has been 80% for stenoses and 39% for occlusions. With repeated angioplasty, the secondary patency rates are 94% for stenoses and 86% for occlusions.

With the Wallstent, the reported primary patency rate is 49% at 12 months. The secondary patency rate is 67% at 12 months and 56% at 18 months. The overall rate of reocclusion is 43%.

In Germany, Dr. Strecker, using the Strecker stent, reported an overall 4-year patency of 48%, with 72% for stenoses and 26% for occlusions.

## *Complications*

The rate of major complications is 1% to 8%. Most complications are related to the access and are more common with an antegrade puncture. Most frequent complications include groin hematoma (7.4%), puncture site pseudoaneurysm (1.6%), and distal embolization (4% for stenoses and 14.1% for occlusions). The incidence of early stent thrombosis is 9.9%. Acute thrombosis can be usually treated by thromboaspiration or fibrinolysis. Secondary thrombosis is more common in the mid (11%) and distal SFA (17%). Patients with total occlusions have a higher risk of secondary thrombosis than patients with stenosis (20.6% versus 1.6%). The rate of restenoses due to intimal hyperplasia at 6 months is 11% in the SFA and 20% in the popliteal artery.

## C. VENOUS STENTS

Venous stenosis and occlusion are manifested clinically by pain, swelling, ulceration, and limb claudication. Management is difficult. Angioplasty in the venous system is associated with a high re-stenosis rate due to the presence of extrinsic compression, fibrosis, and elastic recoil. Both balloon-expandable and self-expandable stents have been successfully used in the venous system. The choice of the stent depends on the anatomic location of the lesion. Venous stent placement requires a special knowledge of balloon catheters, stents, and thrombolytic therapy.

### TECHNIQUE

1. The vein chosen for access is usually distant to the lesion. The right internal jugular approach is preferred for lesions in the IVC, iliac veins, or femoral veins. A femoral approach is usually selected to treat SVC lesions.
2. Venography is performed to evaluate the anatomy, measure the venous diameter, and obtain the pressure gradients.
3. If thrombus is demonstrated, urokinase is administered at a dose of 2,000 to 3,000 units per kg per hour for 12 to 36 hours to lyse the thrombus.
4. The lesion is predilated with a low-pressure angioplasty balloon.
5. Constant hemodynamic monitoring for cardiac arrhythmias is necessary when dilating the SVC.
6. A stent with 15% to 25% larger diameter than the adjacent normal vein is chosen.
7. After the procedure, patients are kept on systemic anticoagulation with heparin to keep the partial thromboplastin time at 75 to 100 seconds; then oral anticoagulation is continued for 3 to 6 months.
8. Follow-up includes clinical examination and duplex ultrasound.
9. Most commonly used stents are the Gianturco-Rosch stent and Wallstent.

### INDICATIONS

#### Malignant Venous Obstructions

SVC syndrome is usually caused by lung and mediastinal tumors. The syndrome is manifested as venous congestion and edema of the upper part of the body. Patients can present with brain edema and airway compression.

Primary or metastatic liver tumors or retroperitoneal tumors can cause IVC syndrome with incapacitating symptoms related to severe edema of the lower abdominal wall, lower extremities, and, occasionally, the scrotum (Fig. 11.8).

Response to angioplasty is usually poor in these cases, and stent placement is indicated in the following situations:

- Palliation of patients with severe symptoms that do not respond to conventional treatment with chemotherapy and radiotherapy
- Compressed SVC and IVC with more than 75% to 80% reduction in the vessel lumen
- Caval pressure of more than 22 mm Hg at the peripheral end of the caval stenosis

#### *Contraindications*

- Active systemic infectious process.
- Significant associated thrombosis (thrombolysis should be performed before stent placement to prevent pulmonary embolism).
- Tumor invasion of the caval wall is a relative contraindication; a stent graft can be used.

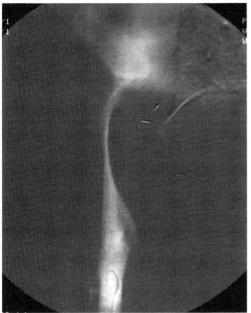 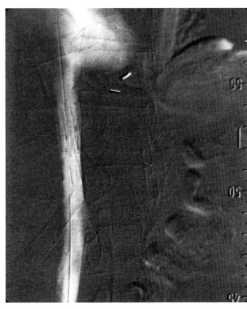

**FIG. 11.8.** Inferior vena cava stent placement. A 58-year-old man with extensive liver metastases secondary to prostatic carcinoma and inferior vena cava compression. The patient presented with severe inferior vena cava syndrome. **A:** Inferior vena cavagram shows severe extrinsic compression of the inferior vena cava in its intrahepatic segment. The pressure gradient across the obstruction was 17 mm Hg. **B:** Cavagram after placement of a two-body 22-mm Gianturco stent. Excellent venographic and hemodynamic result. The inferior vena cava syndrome improved markedly until the patient's death 3 months later.

- Poor cardiac function (these patients may not tolerate a temporary cardiac volume overload).

### Complications

The reported incidence of complications is 18%.

1. Fatal complications (infrequent)
   a. Pulmonary embolism—related to clot migration after stent placement
   b. Congestive heart failure—caused by sudden increase in preload after the recanalization
2. Nonfatal complications
   a. Vessel rupture—a rare complication
   b. Stent migration—more common with the single-body Z stent.
   c. Sepsis
   d. Pain
   e. Seizures—some authors recommend to perform the procedure under general anesthesia due to the reported seizure activities

### Results

Symptomatic improvement occurs in 74% to 100%. Excellent palliation is obtained in most patients.

## Benign Venous Obstructions

### Budd-Chiari Syndrome

Budd-Chiari syndrome is caused by the obstruction of the venous outflow of the liver. It is associated with hypercoagulable states, trauma, oral contraceptives, membranes or webs in the venous outflow, venoocclusive disease of the liver, and obstruction of the IVC by tumors. In many instances, no underlying cause is found. Budd-Chiari syndrome presents with abdominal pain, hepatomegaly, and ascites. The sequelae of portal hypertension are treated with portosystemic shunts. In advanced cases, orthotopic liver transplantation is the therapy of choice. In patients with obstructions of the hepatic veins or IVC caused by a membrane or web with a segment of less than 1 cm in length, angioplasty is the treatment of choice, with patency 80% to 100% in 1 year and 50% patency at 2 years. In longer segmental obstructions of the IVC, angioplasty with stent placement is indicated.

### *Technical Considerations*

- Venography of the IVC and hepatic veins is performed with pressure measurements.
- If associated thrombosis is identified, perform thrombolysis or percutaneous aspiration thrombectomy.
- In cases of complete obstruction, recanalization of the IVC can be performed using a Colapinto or a Brockenbrough needle (Cook, Inc.). The lesion is predilated with 15- to 18-mm angioplasty balloons or using the double-balloon technique.
- The Gianturco stent is most commonly used. In focal lesions of the hepatic veins, the Palmaz stent or the Wallstent can also be used.
- Stent placement into the right atrium should be avoided, as it will interfere with subsequent hepatic transplantation.

### *Results*

Technical success is 95% to 100%. Long-term follow-up is not available yet. The incidence of re-stenosis due to intimal hyperplasia is 12% after stent placement and 50% after angioplasty at 2 years.

### *Complications*

- Fever of unknown etiology—usually resolves within 72 hours after stent placement (if fever persists, blood cultures and antibiotic therapy should be started)
- Migration of the stent
- Pulmonary embolism—potentially fatal complication
- Disseminated intravascular coagulation—unknown etiology, potentially fatal

## Hemodialysis-Related Venous Stenoses

### *Central Vein Occlusion*

Central venous stenosis and occlusion are frequent complications seen in dialysis patients. These lesions are usually related to previously existing dialysis catheters. In patients with functioning grafts or arteriovenous fistulas, central venous occlusion manifests with symptoms of severe venous hypertension (painful, incapacitating extremity swelling) or early access failure. Recanalization of central venous obstructions followed by stent placement improves the performance of the dialysis graft and decreases the symptoms of upper extremity venous hypertension. Close follow-up and aggressive reintervention are necessary to maintain stent patency.

### *Dialysis Grafts*

Stenosis of the venous anastomosis is the main cause of hemodialysis graft failure. Aggressive treatment of venous stenoses with angioplasty and stent placement may improve the potency of hemodialysis fistulas and shunts. This is a controversial topic, as some authors do not advocate stenting of these lesions due to the very low primary patency rate (30% to 35% at 6 months).

### *Indications*

- Symptomatic venous stenoses that do not respond to angioplasty [more than 50% residual stenosis or lesions that recur after short intervals (less than 2 months)]
- Sealing of dissections or circumscribed perforations
- Chronic venous occlusions

### *Contraindications*

- Active infection
- Inability to dilate the lesion with high-pressure balloons

- Immature shunt or recent surgical anastomosis (high risk of rupture)
- Insufficient venous outflow due to small draining veins

### Technical Considerations: Central Vein Occlusions

- The procedure can be performed from the femoral venous approach or through the graft. Sheaths larger than 8 Fr should not be placed into the graft. A through-and-through access with an exchange guidewire communicating the brachial and femoral approaches facilitates the recanalization of chronic occlusions.
- The lesion should be predilated with a high-pressure balloon; if fresh thrombus is demonstrated, perform urokinase infusion.
- Crossing the origin of the internal jugular vein and brachiocephalic should be avoided if possible to preserve the contralateral venous access and in case that axillary–jugular vein bypass is required in the future.

### Technical Considerations: Dialysis Grafts

- Stents should not be placed in puncture areas in the dialysis graft.
- The Wallstent is more commonly used due to its large size and flexibility.
- The Gianturco and Palmaz stents can be used in focal lesions. The Palmaz stents should not be placed in mobile areas or areas with possible external compression.
- A flexible stent (Wallstent) should also be used in the elbow.

### Results

The technical success is very high (96% to 100%). Re-stenosis due to intimal hyperplasia is very common. The cumulative primary patency rate is 22% to 57% at 1 year and 45% at 2 and 3 years. With aggressive reintervention (including angioplasty, atherectomy, or both), the assisted patency rate is 86% at 1 year, 77% at 2 years, and 70% at 3 years.

### Complications

- The reported complication rate is 10% to 15%; complications include pseudoaneurysm and migration of the stent into the pulmonary artery.
- Patients and referring physicians should be warned against placement of ipsilateral central lines, which can result in stent migration.
- Early thrombosis can occur in 10% and usually can be treated with urokinase.

## Other Benign Venous Lesions

The initial experience with venous stent placement in other benign venous lesions is limited; possible indications for venous stent placement after failed angioplasty include the following:

- Fibrotic postsurgical scarring
- May-Thurner syndrome
- Post–liver transplant
- Venous spurs
- Post–thrombotic stenosis

# SUGGESTED READINGS

## A. Overview

Flueckiger F, Sternthal H, Klein GE, et al. Strength, elasticity, and plasticity of expandable metal stents: in vitro studies with three types of stress. *J Vasc Intervent Radiol* 1994;5:745–750.

Zollikofer C, Antonucci F, Stuckmann G, et al. Historical overview on the developments and characteristics of stents and future outlooks. *Cardiovasc Intervent Radiol* 1992;15:272–278.

## B. Arterial Stenting

### Stent Placement in Supraaortic Vessels

Jaeger HJ, Mathias KD, Kempkes U. Bilateral subclavian steal syndrome: treatment with percutaneous transluminal angioplasty and stent placement. *Cardiovasc Intervent Radiol* 1994;17:328–332.

Lyon RD, Shonnard KM, McCarter DL, et al. Supra-aortic arterial stenoses: management with Palmaz balloon-expandable intraluminal stents. *J Vasc Intervent Radiol* 1996;7:825–835.

Queral LA, Criado FJ. The treatment of focal aortic arch branch lesions with Palmaz stents. *J Vasc Surg* 1996;23:368–375.

## *Renal Artery Stent Placement*

Boisclair C, Therasse E, Oliva VL, et al. Treatment of renal angioplasty failure by percutaneous renal artery stent placement with Palmaz stents: midterm technical and clinical results. *AJR Am J Roentgenol* 1997;167:245–251.

Hennequin LM, Joffre FG, Rousseau HP, et al. Renal artery stent placement: long term results with the Wallstent endoprosthesis. *Radiology* 1994;191:713–719.

Henry M, Amor M, Henry I, et al. Stent placement in the renal artery: three-year experience with the Palmaz stent. *J Vasc Intervent Radiol* 1996;7:343–350.

Trost D, Sos T. Complications of renal angioplasty and stent placement. *Semin Intervent Radiol* 1994;11:150–160.

## *Aorto-Iliac and Peripheral Arteries*

Hausegger KA, Cragg AH, Lammer J, et al. Iliac artery stent placement: clinical experience with a nitinol stent. *Radiology* 1994;190:199–202.

Henry M, Amor M, Ethevenot G, et al. Palmaz stent placement in iliac and femoropopliteal arteries: primary and secondary patency in 310 patients with 2-4-year follow-up. *Radiology* 1995;197:167–174.

Long Al, Gaux JC, Raynaud AC, et al. Infrarenal aortic stents: Initial clinical experience and angiographic follow-up. *Cardiovasc Intervent Radiol* 1993;16:203–208.

Martin E, Katzen B, Benenati JF, et al. Multicenter trial of the Wallstent in the iliac and femoral arteries. *J Vasc Intervent Radiol* 1995;6:843–849.

Murphy KD, Encarnacion CE, Le VA, Palmaz JC. Iliac artery stent placement with the Palmaz stent: follow up study. *J Vasc Intervent Radiol* 1995;6:321–329.

Sapoval MR, Chatellier G, Long AL, et al. Self-expandable stents for the treatment of iliac artery obstructive lesions: long-term success and prognostic factors. *AJR Am J Roentgenol* 1996;166:1173–1179.

Strecker EP, Boos IBL, Hagen B. Flexible Tantalum stents for the treatment of iliac artery lesions: long term patency, complications and risk factors. *Radiology* 1996;199:641–647.

Vorwerk D, Guenther RW, Schurmann KS, et al. Primary stent placement for chronic iliac artery occlusions: follow-up results in 103 patients. *Radiology* 1995;194:745–749.

## C. Venous Stents

Baijal SS, Roy S, Phadke RV, et al. Management of idiopathic Budd-Chiari syndrome with primary stent placement: early results. *J Vasc Intervent Radiol* 1996;7:545–553.

Furui S, Sawada S, Kuramoto K, et al. Gianturco stent placement in malignant caval obstructions: analysis of factors for predicting the outcome. *Radiology* 1995;195:147–152.

Gray RJ, Horton KM, Dolmatch BL, et al. Use of Wallstents for hemodialysis access-related venous stenoses and occlusions untreatable with balloon angioplasty. *Radiology* 1995;195:479–484.

Nazarian GW, Austin WR, Wegryn SA, et al. Venous recanalization by metallic stents after failure of balloon angioplasty or surgery: four-year experience. *Cardiovasc Intervent Radiol* 1996;19:227–233.

Oudkerk M, Kuijpers TJA, Schmitz PIM, et al. Self-expanding metal stents for palliative treatment of superior vena cava syndrome. *Cardiovasc Intervent Radiol* 1996;19:146–151.

Park H, Chung JW, Han JK, Han MC. Interventional management of benign obstruction of the hepatic inferior vena cava. *J Vasc Intervent Radiol* 1994;5:403–409.

Venbrux AC, Mitchell SE, Sadaver SJ, et al. Long-term results with the use of metallic stents in the inferior vena cava for treatment of Budd-Chiari syndrome. *J Vasc Intervent Radiol* 1994;5:411–416.

Vorwerk D, Guenther RW, Mann H, et al. Venous stenosis and occlusion in hemodialysis shunts: follow-up results of stent placement in 65 patients. *Radiology* 1995;195:140–146.

# 12
# Stent-Graft Techniques

G. Michael Werdick, Michael S. Rosenberg, and Haraldur Bjarnason

The following discussion is a short introduction to the use of stent-grafts in vascular disease.

## A. USE OF STENT-GRAFTS FOR AORTO-ILIAC ANEURYSMAL DISEASE

G. Michael Werdick

The first publication on placement of an aorto-iliac stent-graft was in 1991. Since then, more than ten commercial systems have been developed and are undergoing clinical trials worldwide. At the time of this writing, only two systems have U.S. Food and Drug Administration (FDA) approval for use in the United States. Stent-grafts also may be built from readily available stents and surgical graft material, such as polytetrafluoroethylene (PTFE) or polyester [Dacron (Meadox Medicals, Inc., Oakland, NJ)], then used in an "off-label" fashion if the clinical situation indicates compassionate use.

### INDICATIONS

Stent-graft placement in the aorto-iliac circulation has been described for the treatment of aneurysm (both elective and emergent), iliac occlusive disease, pseudoaneurysm, and trauma. Use of stent-grafts in the thoracic aorta is anecdotal, and few cases have been described; however, extensive experience has been obtained in the treatment of infrarenal abdominal aortic aneurysm. Several clinical trials have been published comparing open surgery to stent-graft placement (or "endosurgery") in patients with abdominal aneurysmal disease. In this setting, three general categories of devices are used: bifurcated or tubular unibody, modular multicomponent, and aortomonoiliac grafts that are combined with a standard surgical femoral to femoral bypass graft.

### TECHNIQUES

Placement of stent-grafts in small vessels is similar to standard stent placement. Benchmade or prepackaged stents may be deployed over guidewires under fluoroscopic guidance, then dilated to size with angioplasty balloons.

In the case of abdominal aortic aneurysm, the procedure is much more involved. Preprocedure planning is crucial for successful stent-graft placement. Computed tomography (CT), magnetic resonance imaging, and angiography may be used exclusively or in combination. Three-dimensional reconstruction of helical CT and magnetic resonance imaging data is extremely helpful, if not essential, in planning. Calibrated catheters may be placed to assist in measurement. In preoperative planning, many precise measurements must be considered. The relationship of the aneurysm to the renal arteries, for example, is important to avoid covering them with the graft. Fifteen millimeters or more of normal aortic lumen is required inferior to the renal arteries to gain stable anchoring of an infrarenal portion of the stent-graft. The aortic diameter must be small enough to engage the graft and generally should be 28 mm or less. Because of the

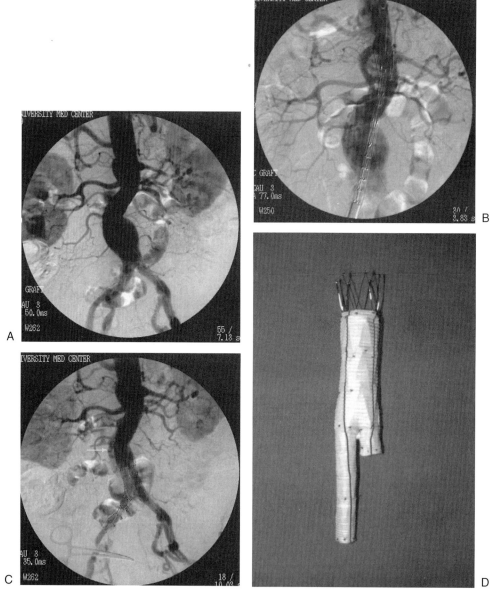

**FIG. 12.1. A:** Seventy-year-old man with an infrarenal abdominal aortic aneurysm that on computed tomography measured 6 cm in diameter. This angiogram demonstrates typical tortuosity with a long infrarenal neck. The ectasia/aneurysm extends to the common iliac arteries, and therefore a tube graft was not an option. **B:** A graft was made of Gianturco (Cook, Inc., Bloomington, IN) and a Dacron graft (Cooley Very-Soft, Meadox Medicals, Inc., Oakland, NJ) was sutured on to the stent skeleton. The graft was then delivered through a 21-Fr introducer from a right common femoral cutdown. The similarly constructed left limb is also deployed via a cutdown from the left common femoral artery on an 18-Fr introducer. The first portion of the stent-graft, which is an uncovered Gianturco stent, has been partially deployed across the renal arteries. **C:** Completed graft placement after the left limb has been placed and the aortic and the iliac artery limbs have been dilated. Note a small leak around the right margin (*white arrow*), which had disappeared at 1-month computed tomography follow-up. **D:** Stent made of Gianturco stents and Cooley Very-Soft graft at the site. This stent was used for this case. The left iliac component is missing from this picture.

size of the delivery system (20 to 28 Fr or 7 to 9 mm), a cutdown is made onto the common femoral arteries. This may change with the development of smaller delivery systems. Severe tortuosity of the iliac arteries may prohibit advancement of the delivery system into the abdominal aorta.

The stent-grafts were initially a straight tube if there was a good, normal aorta infrarenal and also proximal to the bifurcation where it could be fitted. Now bifurcated grafts have almost completely replaced the tubular grafts, as they require less precision in placement. The bifurcated stent-grafts are deployed in two steps. The first step is placement of a combined aortic stent-graft, which splits into two iliac components. The longer one is deployed into the common iliac artery on the right side. On the left side there is only a short stump of iliac graft, which is then accessed from the left common femoral artery, and a separate iliac component is placed into the short stump into the left common iliac vein (Fig. 12.1). Alternatively, a straight single-lumen stent-graft can be placed from the infrarenal aorta into either of the common iliac arteries. The contralateral common iliac artery is then occluded with a special closure device to prevent flow of blood retrograde into the aneurysm. A femoral to femoral bypass graft is then performed.

During the procedure, the patient receives antibiotics and is fully heparinized. Follow-up contrast-enhanced CT is performed at 1, 6, and 12 months postoperatively to ensure stable graft position, to check for continued expansion of the aneurysm, and to evaluate for flow of contrast-enhanced blood around the stent-graft, which would indicate leak into the aneurysm.

Aneurysms of the thoracic aorta have also been treated with endoluminal stent-grafts. The normal aorta has a larger diameter at that level, and the closeness to the neck vessels calls for more precision. The anterior spinal artery can also occlude during stent-graft placement in the thoracic aorta.

Thoracic stent-grafts have been custom made up to this point, but commercial products are waiting for FDA approval. Because the diameter is much larger, up to 40 mm, larger introducers are needed and the common femoral approach may not be adequate.

## COMPLICATIONS

Complications are listed in Table 12.1. The most common complication is continued blood flow into the aneurysm sac, referred to as an *endoleak*. The incidence of endoleak is reported between 7% and 34% and occurs most often around the fit at the infrarenal segment. The leak can also be at the iliac artery fit or at the junction of the iliac modules to the main body. Lumbar arteries and inferior mesenteric arteries can also be a source for such leak because of retrograde flow into the aneurysm sack. In early reports, leaks were aggressively pursued with extension of the grafts, embolization, or conversion to an open repair. Several studies, however, demonstrate that most of those leaks will seal within 1 month with a persistent endoleak rate of 10% or less. These may then be treated with the endovascular techniques as mentioned above. Preoperative embolization of lumbar arteries and inferior mesenteric arteries has been described, but this has been shown to be unnecessary and now is indicated only if there is filling of the aneurysmal sac at late follow-up.

The large size of the delivery system can cause thrombosis of the iliac arteries. Dissection, perforation, and pseudoaneurysm may also result from the force needed to advance the delivery system into the aorta across often very small and diseased iliac arteries. Case reports of bowel ischemia, ischemic neuropathy, and aor-

**TABLE 12.1.** *Potential complications of stent-graft repair of abdominal aortic aneurysm*

Endoleak
Graft migration
Continued expansion of the aneurysm
Ischemic bowel
Ischemic neuropathy
Aortoenteric fistula
Rupture
Iliac artery rupture or dissection
Renal embolization
Peripheral embolization
Sepsis

toduodenal fistulas have been published but appear also to be rare occurrences.

## OUTCOME

Stent-graft placement for the treatment of abdominal aortic aneurysm compares favorably to open surgery; however, there has yet to be a prospective randomized study to definitively compare the two. Mortality rates are 5% or less and not statistically different from matched comparisons.

Studies show technical failures to be higher in the stent-graft patients. Persistent endoleak or graft migration may require open repair or additional endovascular procedures. Results vary widely between studies, suggesting that the ideal system and technique have yet to be worked out. Generally, primary technical success is 80% to 90%, and secondary success after additional endoluminal procedures is 90% to 95%. Conversion to open repair is generally quite low, seen in 5% or less in most series.

Complications that prolong hospital stay are 20% for patients treated surgically. Studies have shown a marked reduction in the degree and number of postoperative complications in the stent-graft groups. Additionally, there appears to be a significant decrease in the need for blood transfusions in stent-graft patients.

## CONCLUSION

Stent-graft placement in the aorto-iliac circulation is technically possible for the treatment of occlusive and aneurysmal disease. The most extensive experience has been gained in the treatment of abdominal aortic aneurysm. Many different systems have been developed, and the technical details for the ideal system and technique have yet to be worked out. Early results suggest acceptable mortality rates and significant reduction in postoperative morbidity. Technical failure for stent-graft placement is higher than for open surgical repair. With improvement in delivery systems and operator experience, this is likely to improve. Long-term outcome has yet to be determined and, clearly, durability of this method needs to be proven before it can be considered an acceptable alternative to standard open surgical repair.

## B. STENT-GRAFTS FOR PERIPHERAL OCCLUSIVE DISEASE

### Michael S. Rosenberg

Endovascular treatment of peripheral occlusive disease in the iliac arteries and femoropopliteal circulation has continued to evolve. Stenting of atherosclerotic iliac lesions has yielded 5-year patency rates of up to nearly 90%, an improvement of approximatelmy 10% to 20% over angioplasty alone.

Femoropopliteal occlusions or stenoses have traditionally been more difficult to treat and to obtain durable results, especially if the treated blood vessel is less than 6 mm in diameter and the lesion length is greater than 10 cm. Furthermore, improvement in patency after femoropopliteal stent placement has not been established. Stent-grafts offer a means of "endovascular bypass" that may not dramatically affect the already high patency rates of iliac stenting but may decrease re-stenosis rates and improve patency in superficial femoral and popliteal artery interventions.

## INDICATIONS

The indications for placement of stent-grafts to treat peripheral occlusive disease are currently not well established. The most promising application for peripheral stent-grafts is in the treatment of long superficial femoral and popliteal artery occlusions or stenoses. This is especially true if the patient is a poor operative candidate for peripheral bypass surgery.

Although the potential impact on patency rates in the iliac arteries is less, there does remain some room for improvement beyond what is currently acceptable after conventional stenting. Stent-grafts may be indicated in complex long-segment iliac artery recanalizations. They may also be useful in treating ulcerated iliac lesions that could be a potential source of embolus after uncovered stent placement.

Stent-grafts have also been found to be very effective in treating pseudoaneurysms, acute vessel ruptures, and aneurysms in smaller vessels than the abdominal aorta. Several case reports support use of those in these circumstances.

## MATERIALS AND TECHNIQUES

The technique for peripheral endovascular stent-graft placement depends on the type of device used. Currently, no commercially available stent-grafts are available for peripheral use in the United States. Homemade stent-grafts have been developed and used in an "off-label" fashion, but mostly for the treatment of more urgent conditions, such as iliac artery pseudoaneurysm or rupture, and not for occlusive disease. These homemade versions have been constructed from a variety of commercially available balloon-expandable [Palmaz (Johnson & Johnson, Warren, NJ), Gianturco-Rosch Z stent (Cook, Inc., Bloomington, IN), Intrastent (Intratherapeutics Inc., St. Paul, MN)] or self-expanding [Wallstent (Boston Scientific Plymouth Technology Center, Minneapolis, MN)] stents, covered by a variety of commercially available surgical graft materials such as expanded PTFE (Impra, Tempe, AZ), Gore-Tex (W. L. Gore & Associates Inc., Flagstaff, AZ), or polyester (Dacron). The graft material is typically attached to one or both ends of the stent with a fine Prolene suture and is deployed in a similar manner as described elsewhere in this chapter (Fig. 12.2).

Two peripheral stent-grafts are currently in clinical trials in the United States. The Wallgraft endoprosthesis (Schneider, Minneapolis, MN) combines a self-expanding stent made of a supermetal alloy (Elgiloy) similar to the widely used Wallstent with a polyethylene terephthalate covering. This stent-graft design, in which the graft is on the outside of the stent, is referred to as an *endoskeleton*. The Hemobahn endovascular prosthesis (W. L. Gore and Associates) is an exoskeleton design that is composed of an expanded PTFE graft covered by a self-expanding nitinol stent (Fig. 12.3). Nitinol is a nonmagnetic nickel and titanium alloy that has a unique thermomemory property such that it will revert to its designed shape when in temperatures such as in the body. Both the Wallgraft and Hemobahn are fully supported stent-grafts, implying that the entire graft is supported by a stent framework, as compared with unsupported stent-graft designs, which incorporate a stent at only one or both ends of the graft.

The Cragg EndoPro System I (Mintec, Bahamas) stent-graft is widely used in Europe. This fully supported device consists of a nitinol stent endoskeleton covered by low-porosity woven polyethylene terephthalate graft material.

## OUTCOMES

Results of stent-graft placement in the superficial femoral artery are encouraging. Sixty-four percent primary and 73% secondary 5-year patency rates have been reported with the Cragg EndoPro System I. Secondary patency improved to 93% for lesions less than 15 cm in length. Results in the popliteal artery were less encouraging, with primary and secondary patency rates of only 33% and 50%, respectively.

Results with the Hemobahn endovascular prosthesis are even more encouraging. Feasibility of the Hemobahn endovascular prosthesis for treating occlusive disease was evaluated in a multicenter study in the United States and Europe. Overall technical success was very high (99%), with a primary femoral patency rate of 100% at 1 month, 91% at 3 months, and 80% at 6 months. This compares to 2-year patency rates of 43% to 79% after angioplasty and 1-year primary patency rates of 22% to 49% after uncovered stent placement in the superficial femoral artery. Re-stenosis or occlusion rates, or both, of up to 80% at 5 years are reported after angioplasty of superficial femoral artery lesions longer than 10 cm. Phase II clinical trials of the Hemobahn endovascular prosthesis are currently well under way in the United States.

Use of the Cragg EndoPro System I for treating iliac artery occlusions has also been success-

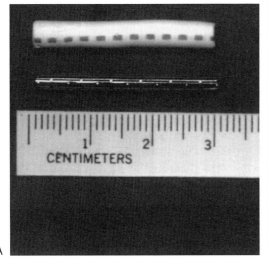
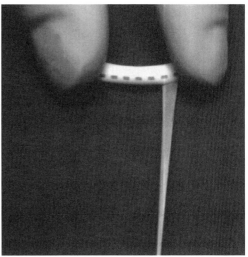
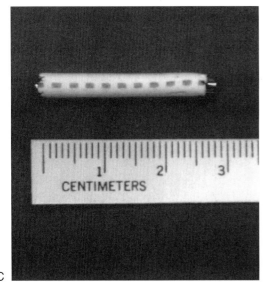

**FIG. 12.2. A:** A Palmaz P-308 stent is here next to a segment of a 3-mm thin-walled Gore-Tex graft. The graft is cut to the same length as the stent. **B:** The thin support membrane is peeled of the Gore-Tex graft. **C:** The graft material is placed on the stent and the stent sutured onto the stent using either a 5-0 or 6-0 suture. The stent-graft unit is then crimped onto a balloon and delivered as if it were a stent.

ful. Henry et al. (1994) report 97% primary and 100% secondary patency rates at 18 months, with essentially no decrease at 5 years (92% primary and 100% secondary patency).

Henry and Amor also report good results in the iliac artery from phase I trials of the Hemobahn endovascular prosthesis, citing 6-month primary and secondary patencies of 96%.

## COMPLICATIONS

Infection is the most dreaded complication, but there have been very few reports of infections. Fever and pain have been experienced in several cases. These have not been found to be related to infection but are characterized as immunologic response to the graft material. Macrophagic and giant cell accumulation around the graft has been seen. This complication is more commonly seen with PTE graft. The pain is confined to the site of the graft and can last for several weeks. The fever occurs within 12 hours of the procedure and resolves usually within 48 hours. Pain and fever together have been referred to as *postimplantation syndrome*. Access site complications are the same as for other interventions but more common because of the larger size of the introducers for stent-graft technology.

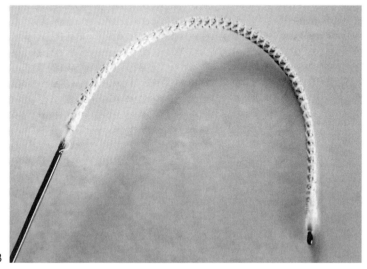

FIG. 12.3. **A:** The Hemobahn endovascular prosthesis constrained on the delivery catheter below and fully delivered stent above. **B:** Stent-graft on delivery system curved.

## CONCLUSION

Stent-grafts, although not available in the United States as commercial products, have been proven to be important tools in the treatment of small pseudoaneurysms, perforations, and shorter aneurysms of smaller vessels. Early results in the superficial femoral artery appear favorable; however, the use of stent-grafts in occlusive vascular disease still needs to be evaluated.

## C. USE OF COVERED METAL STENTS FOR TRANSJUGULAR-INTRAHEPATIC PORTOCAVAL SHUNTS

### Haraldur Bjarnason

Dr. Richter in Germany performed the first successful transjugular intrahepatic portocaval shunt (TIPS) procedure in humans using metal stents in 1988. The procedure has since then been shown to be effective in controlling variceal bleeding and, in patients with liver disease and portal hypertension, ascetic accumulation, with improvement in the quality of life. Extension of life expectancy has not been proven.

The TIPS has been plagued with a high incidence of stenosis causing recurrent variceal bleeding and reaccumulation of ascetic fluid. Spontaneous thrombosis of the shunts has also been relatively common early after the procedure. This has been thought to be partially due to communication with the biliary system. During the creation of the shunt, the puncture inadvertently is made through biliary radicals, which consequently causes thrombosis and intimal hyperplasia. Stents covered with PTFE or Dacron will exclude the biliary system from the blood, and several authors have advocated the use of covered stents for TIPS either primarily or secondarily when problems have occurred, such as thrombosis with or without demonstration of connection to the biliary system. Biliary communication is not always documented in shunts with intimal hyperplasia. There are still shunts in which intimal hyperplasia is a recurrent problem, and covered stents may still be helpful in those cases.

The reported rate of re-stenosis in TIPS ranges from 31% to 80%. Stenoses occur mainly in two locations: (a) at the hepatic vein end and (b) in the mid shunt. The hepatic vein end stenosis is probably caused by increased flow and shear forces at that point. The mid shunt also develops intimal hyperplasia, which is thought to be caused by exposure to bile from transected biliary ducts in some cases. This has also been thought to be the cause for acute thrombosis early after shunt placement.

## INDICATIONS

- Early thrombosis of TIPS where communication with the biliary system has been demonstrated
- Recurrent stenosis of the portocaval shunts after repeated angioplasty and additional stent placement (Communication with biliary ducts does not need to be documented.)
- Primary placement of covered stents in TIPS

## TECHNIQUES

Currently in the United States, covered stents for those purposes are made from regular vascular stents and vascular graft material. In TIPS, several combinations of stents and graft material have been used. The most popular fabric materials have been PTFE and Gore-Tex sutured onto different types of stents. The thin-walled Gore-Tex has a support membrane around it that needs to be peeled off before it can be expanded (Fig. 12.2). The Wallstent is the most commonly used stent for TIPS and has also commonly been used for covered stent assembly.

The technique for the Wallstent is roughly the following: A 4-mm Impra graft or thin-walled Gore-Tex graft is dilated with a 12-mm balloon and then cut to the measured tract length. The graft material needs to cover the length of the parenchymal portion of the tract and preferably not extend either into the hepatic or portal veins. The Wallstent is then partially opened (less than 1 cm) and the graft placed over the shaft and the distal end sutured loosely to the flared end of the Wallstent using 6-0 polypropylene sutures. The whole assembly is then placed inside a 10-Fr introducer sheath. The assembly inside the 10-Fr sheath is then carried over a wire into the shunt through a 12-Fr introducer sheath. The 10-Fr introducer is then pulled back, exposing the stent-graft, and the Wallstent is then deployed fully in that position. This can then be dilated to 8 to 12 mm.

The Gianturco stent (Cook, Inc.) covered with Impra has also been used for TIPS as well as for other vascular applications. The graft is treated the same way and sutured on a 12-mm, 6-cm Gianturco stent using the same 6-0 polypropylene suture. This assembly has to be delivered through a 12-Fr introducer. The same material can also be sutured onto a Palmaz and then brought through an introducer on a balloon catheter to the shunt. A 4-mm Impra or Gore-Tex graft is used but does not need to be predilated. It is sutured again with a 6-0 polypropylene suture onto both ends of the stent. The Palmaz P-308 stent is commonly used. It can be dilated to 8 to 12 mm as needed. If longer segments are treated, several Palmaz stents can be sutured inside the graft material or a longer Palmaz stent can be used.

## OUTCOME

Several animal studies have been conducted comparing covered stents with uncovered stents in TIPS. Haskal et al. found that the PTFE-covered Wallstent did significantly improve the patency rate of animal TIPS as compared with the bare Wallstent.

Bloch et al. covered Gianturco stents with Dacron, which then was covered with silicon.

Of six animals, the shunts were occluded in five by 3 weeks and in all six animals by 6 weeks. The thrombogenicity of the polyurethane coating and the Dacron were thought to be to blame for the poor results.

Otal et al. came to a similar conclusion with Dacron-covered stents without polyurethane covering. Only two of seven animals had patent shunts at 3 weeks. The occlusions were due to pseudointimal hyperplasia within the Dacron-covered shunt.

Finally, Tanihata et al. placed silicon-covered Wallstents in 14 animals with TIPS. At 3 weeks, 12 had occluded, and by 6 weeks all had occluded. This was thought to be caused by high thrombogenicity and a foreign-body reaction to the silicon.

A few human studies and case reports have been published on treatment of early TIPS occlusions with covered stents. Also, a few studies have looked at primary placement of covered stents in TIPS.

In general, PTFE has been shown to give excellent results for repair of failing or failed TIPS. This has especially been true for shunts with proven communication with the biliary system. Several case reports demonstrate good outcomes in individual cases with early TIPS failure or repeated TIPS stenosis. Some of those cases have had a documented connection to the biliary system and others have not. Most of those have in common severe pseudointimal hyperplasia. Placement of the PTFE graft seems to halt the process and secure a more permanent patency.

Ferral et al. used the Cragg EndoPro System primarily for TIPS in 13 patients. The stent-graft consists of a self-expandable nitinol Cragg stent covered with a low-porosity, fine polyester material. During a 2- to 3-month follow-up, two shunts occluded and a hepatic vein end stenosis occurred in one case (23% malfunction). The two occlusions were thought to be due to technical problems. Those results are similar to what would be expected with uncovered metal stents such as the Wallstent.

## CONCLUSION

Covered metal stents or stent-grafts appear to be a good solution to early failure of TIPS, especially those with biliary fistulas, but also in patients with recurrent shunt stenosis. Several combinations of materials have been used, but it appears that PTFE clothing is better suited than Dacron and silicon. Primary placement of covered stents in TIPS has not been shown definitely to be beneficial but there are now studies under way comparing use of bare stents with that of covered stents.

## SUGGESTED READINGS

Beheshti MV, Dolmatch BL, Jones MP. Technical considerations in covering and deploying a Wallstent endoprosthesis for the salvage of a failing transjugular intrahepatic portosystemic shunt. J Vasc Intervent Radiol 1998;9:289–293.

Bloch R, Pavcnik D, Uchida BT, et al. Polyurethane-coated Dacron-covered stent-grafts for TIPS: results in swine. Cardiovasc Intervent Radiol 1998;21:497–500.

Bosch JL, Hunink MGM. Meta-analysis of the results of percutaneous transluminal angioplasty and stent placement for aortoiliac occlusive disease. Radiology 1997; 204:87–96.

Cejna M, Thurnher S, Pidlich J, et al. Primary implantation of polyester-covered stent-grafts for transjugular intrahepatic portosystemic stent shunts (TIPSS): a pilot study. Cardiovasc Intervent Radiol 1999;22:305–310.

Cohen GS, Young HY, Ball DS. Stent-graft as treatment for TIPS-biliary fistula. J Vasc Intervent Radiol 1996;7:665–668.

Cragg AH, Dake MD. Percutaneous femoropopliteal graft placement. J Vasc Intervent Radiol 1993;4:455–463.

Cragg AH, Lund G, Rysavy J, et al. Percutaneous arterial grafting. Radiology 1984;150:45–49.

Cuypers P, Nevelsteen A, Buth J, et al. Complications in the endovascular repair of abdominal aortic aneurysms: a risk factor analysis. Eur J Vasc Endovasc Surg 1999;18:245–252.

DiSalle RS, Dolmatch BL. Treatment of TIPS stenosis with ePTFE graft-covered stents. Cardiovasc Intervent Radiol 1998;21:172–175.

Dolmatch BL, Blum U, eds. Stent-grafts: current clinical practice. Thieme Medical Publishers, New York, 2000.

Ferral H, Alcantara-Peraza A, Kimura Y, Castañeda-Zúñiga WR. Creation of transjugular intrahepatic portosystemic shunts with use of the Cragg Endopro System I. J Vasc Intervent Radiol 1998;9:283–287.

Gray BH, Sullivan TM, Childs MB, et al. High incidence of restenosis/reocclusion of stents in the percutaneous treatment of long-segment superficial femoral artery disease after suboptimal angioplasty. J Vasc Surg 1997;25:74–83.

Haskal ZJ, Davis A, McAllister A, Furth EE. PTFE-encapsulated endovascular stent-graft for transjugular intrahepatic portosystemic shunts: experimental evaluation. Radiology 1997;205:682–688.

Henry M, Amor M. Covered stents. Paper presented at the Eleventh Annual International Symposium on Endovascular Therapy, January 1999, Miami.

Henry M, Amor M, Cragg A, et al. Occlusive and aneurysmal peripheral arterial disease: assessment of a stent-graft system. Radiology 1996;201:717–724.

Henry M, Amor M, Ethevenot G, et al. Initial experience with the Cragg EndoPro System 1 for intraluminal treatment of peripheral vascular disease. J Endovasc Surg 1994;1:31–43.

Kawai N, Sato M, Nakai M, et al. Treatment of early TIPS occlusion with a polytetrafluoroethylene (PTFE)-covered zigzag stent [letter]. Cardiovasc Intervent Radiol 1999; 22:264–265.

May J, White GH, Waugh R, et al. Endovascular treatment of abdominal aortic aneurysms. Cardiovasc Surg 1999;7:484–490.

May J, Woodburn K, White G. Endovascular treatment of infrarenal abdominal aortic aneurysms. Ann Vasc Surg 1998;12:391–395.

Nazarian GK, Bjarnason H, Dietz CA Jr, et al. Refractory ascites: midterm results of treatment with a transjugular intrahepatic portosystemic shunt. Radiology 1997; 205:173–180.

Nazarian GK, Ferral H, Bjarnason H, et al. Effect of transjugular intrahepatic portosystemic shunt on quality of life. AJR Am J Roentgenol 1996;176:963–969.

Nazarian GK, Ferral H, Castañeda-Zúñiga WR, et al. Development of stenoses in transjugular intrahepatic portosystemic shunts. Radiology 1994;192:231–234.

Nolthenius RP, Berg JC, Biasi GM, et al. Endoluminal repair of infrarenal abdominal aortic aneurysms using a modular stent-graft: one-year clinical results from a European multicentre trial. Cardiovasc Surg 1999; 7:503–507.

Otal P, Rousseau H, Vinel JP, et al. High occlusion rate in experimental transjugular intrahepatic portosystemic shunt created with a Dacron-covered nitinol stent. J Vasc Intervent Radiol 1999;10(2 Pt 1):183–188.

Parodi JC, Barone A, Piraino R, Schonholz C. Endovascular treatment of abdominal aortic aneurysms: lessons learned. J Endovasc Surg 1997;4:102–110.

Sapoval MR, Long AL, Raynaud AC, et al. Femoropopliteal stent placement: long-term results. Radiology 1992;184: 833–839.

Saxon RR, Mendel-Hartvig J, Corless CL, et al. Bile duct injury as a major cause of stenosis and occlusion in transjugular intrahepatic portosystemic shunts: comparative histopathologic analysis in humans and swine. J Vasc Intervent Radiol 1996;7:487–497.

Saxon RR, Timmermans HA, Uchida BT, et al. Stent-grafts for revision of TIPS stenoses and occlusions: a clinical pilot study. J Vasc Intervent Radiol 1997;8:539–548.

Sze DY, Vestring T, Liddell RP, et al. Recurrent TIPS failure associated with biliary fistulae: treatment with PTFE-covered stents. Cardiovasc Intervent Radiol 1999;22: 298–304.

Tanihata H, Saxon RR, Kubota Y, et al. Transjugular intrahepatic portosystemic shunt with silicone-covered Wallstents: results in a swine model. Radiology 1997;205:181–184.

# 13
# Diagnostic Techniques

## A. NONINVASIVE VASCULAR EVALUATION OF PERIPHERAL VASCULAR DISEASE

Mark Wofford and Thomas R. Beidle

Currently, angiography is accepted as the gold standard for detecting arterial occlusive disease. Noninvasive methods play a major role in the initial evaluation of suspected peripheral arterial disease and include ankle-brachial indices (ABIs), segmental limb pressures, transcutaneous oxygen pressure (tcPO$_2$) measurements, color duplex ultrasound, and magnetic resonance angiography (MRA).

ABI is a simple test used to screen for hemodynamically significant obstructive disease. The ABI is measured at each ankle. A blood pressure cuff is inflated across the calf and Doppler used to detect the first sign of flow either in the anterior or posterior tibial artery at the ankle as the cuff is slowly deflated. The ABI is then obtained by dividing this pressure by the higher of the two brachial systolic pressures (measured in the same way), using a standard cuff. ABIs between 0.92 and 1.1 are normal, but values of less than 0.92 can be seen in patients with claudication. An ABI of less than 0.5 is associated with severe disease and rest pain, and tissue loss is seen with an ABI of less than 0.3. Precaution must be taken in patients with diabetes or other conditions in which vessels can be calcified. The vessels may then not be compressible by the blood pressure cuff, and a falsely high ABI is measured.

Toe pressure measurements are obtained with a small pneumatic cuff and a photoplethysmograph. A normal toe pressure is greater than 60% of ankle pressure. Measurements below 30 mm Hg correlate with poor prognosis for healing of toe ulcer.

Segmental limb pressure examination is designed to detect the level of arterial obstruction. Pressure cuffs are used to obtain ratios to brachial systolic pressure, similar to ABIs. Normal ratios are 1.35 for high thigh, 1.25 above the knee, 1.18 in the calf, and 1.1 in the ankle. A drop between levels of more than 30 mm Hg indicates significant stenosis. The accuracy of this test is relatively low. Moneta et al. found complete agreement between segmental pressures and angiography in only 34% of 151 limbs examined. In addition, segmental pressures cannot distinguish between occlusion and multiple or severe stenoses. This test gives good information with regard to clinical significant disease and can help to plan both angiographic planing and surgery or angioplasty.

tcPO$_2$ measurements assess tissue metabolism as a function of perfusion and are used to measure the degree of extremity ischemia. The test is often used to determine the likelihood that pedal ulcers or amputations will heal. Normal patients have no gradient in tcPO$_2$ values between the proximal and distal extremity. tcPO$_2$ values of less than 38 mm Hg are associated with nonhealing and delayed healing. A failure to increase the tcPO$_2$ value in the foot by more than 15 mm Hg by moving the patient from supine to standing position portends a poor prognosis for arterial reconstruction.

Color duplex sonography (CDS) is now accepted as an accurate method for evaluating arterial disease in the lower extremity. Sensitivity and specificity for detection of significant stenosis and occlusion are in the 90% to 100% range. A complete examination using color flow technology can be completed within 30 to 45 minutes.

A sector or curved array transducer (2 to 7 MHz) is used to evaluate the aorta and iliac vessels. Linear array transducers (5 to 12 MHz) are used to evaluate the common femoral, superficial femoral, popliteal, and trifurcation arteries. An occluded segment will demonstrate no flow with color and pulsed Doppler. Color Doppler is used to identify collaterals and the level of reconstitution of native arteries. The color Doppler beam can usually be electronically steered 20 degrees toward the direction vessel flow, and the linear transducer should be tilted approximately 10 degrees relative to the skin surface. Therefore, a color and pulsed Doppler angle of 60 degrees or below can be used to obtain accurate velocities and good color filling. The color scale should be set so that the high velocities in the center of a normal arterial segment are just below the aliasing threshold. As a result, focal stenosis will often demonstrate aliasing on the color Doppler image so that the pulsed Doppler gate can be directed to these sites. A doubling of peak systolic velocity within a 2-cm segment indicates hemodynamically significant stenosis of greater than 50% diameter reduction. An additional criterion for prediction of failure in a bypass graft includes peak systolic velocity of less than 45 cm per second in a normal segment of graft.

Analysis of pulsed Doppler waveforms obtained every 5 cm increases accuracy of the examination by suggesting proximal or distal occlusive disease and helping to quantify the severity of these lesions. Normal peripheral artery waveforms have a sharp, immediate upstroke and little or no flow throughout most of diastole (Fig. 13.1). A delayed, rounded systolic upstroke (pulsus tardus and pulsus parvus) and diastolic flow that is persistently elevated throughout diastole indicate a proximal occlusion or stenosis. A low-velocity, short-lived waveform is often preocclusive and indicates distal occlusion or severe stenosis. Abnormalities detected with waveform analysis can prompt evaluation of vessels in a more proximal or distal location to detect additional stenosis or occlusion (Fig. 13.2). In

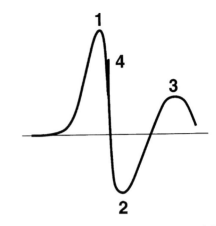

**FIG. 13.1.** Normal peripheral artery triphasic waveform. [*1*, positive wave peak of systolic segment; *2*, negative wave of early diastolic segment (dicrotic notch); *3*, positive wave of late diastolic segment; *4*, incisura.]

addition, intensive Doppler evaluation of problem areas, such as the distal superficial femoral artery in the adductor canal or the iliac arteries, could be prompted by findings of waveform analysis proximal or distal to these sites, thus increasing the accuracy of the color duplex examination.

High diastolic flow is also seen proximal to an arteriovenous fistula (AVF). Pulsatile, high-velocity venous flow is also seen proximal to an AVF. The AVF is often easily identified using color Doppler. A pseudoaneurysm (PSA) appears as a perivascular cystic structure using gray-scale ultrasound. Color Doppler shows color filling of the lumen with color signal, and the neck is also easily identified. "To-and-fro" flow in the neck using pulsed Doppler confirms the diagnosis of PSA. Ultrasound can be used to guide therapeutic compression or percutaneous thrombin injection in order to ablate the PSA.

MRA has been investigated as a possible alternative to conventional angiography. Methods used include two-dimensional time-of-flight, phase contrast, and gadolinium-enhanced. An extremity coil should be used for examining the lower leg. Studies have shown the accuracy of MRA to be between 57% and 99%, depending on which arterial segments were imaged and

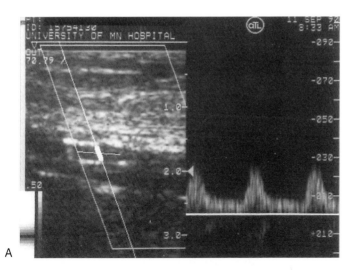

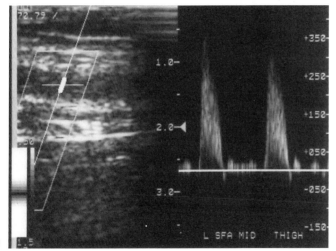

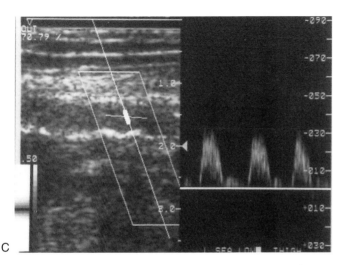

**FIG. 13.2.** Severe stenosis of the superficial femoral artery. **A:** Peak systolic velocity proximal to the stenosis is 25 cm per second. **B:** Peak systolic velocity at the site of stenosis is 300 cm per second, corresponding to velocity ratio of 12. **C:** Beyond the stenosis, peak systolic velocity is 30 cm per second.

which MRA technique was used. Although MRA is still not widely used for evaluation of the peripheral arteries, decreasing cost and examination time and increased availability could make this test a viable alternative test for evaluation of peripheral arterial disease in the future.

## B. EVALUATION OF DEEP VENOUS THROMBOSIS

### Mark Wofford and Thomas R. Beidle

One-third of the 20 million cases of deep venous thrombosis (DVT), if not treated properly, will develop pulmonary embolism (PE). One-half to two-thirds will eventually develop postthrombotic syndrome. Doppler ultrasound is the technique of choice for the diagnosis of femoropopliteal DVT. The sensitivity in symptomatic patients in detecting DVT is 89% to 100%, and the specificity is greater than 90%.

The three risk factors associated with DVT are generally regarded as the following: (a) venous stasis, (b) endothelial injury, and (c) hypercoagulability.

The veins of the lower extremity are divided into the deep and superficial system. Most lower extremity DVTs begin in the deep venous system of the calf. Ninety percent of PEs are secondary to proximal propagation of calf thrombus, but very few clinically significant PEs arise from localized calf thrombus. It is estimated that 40% of calf thrombi resolve spontaneously, and another 40% become organized or recanalize without extension, and the remaining 20% are believed to propagate proximally into the thigh or pelvic veins.

Ultrasound examination is routinely performed using B-mode pulsed Doppler and color Doppler. Compression is used to evaluate the deep venous structures using a linear high-frequency (5- to 10-MHz) transducer whenever possible. With compression, the lumen of a normal vein will completely collapse and the walls will touch each other. If thrombus is present, this will not happen (Fig. 13.3). It can be difficult to see if the lumen is filled with blood or fresh thrombus. It is therefore important to see whether the walls actually touch each other. Normal vessels above the calf should show spontaneous, phasic, and augmentable venous waveforms, along with complete filling on color Doppler examination (Fig. 13.4). Pulsed Doppler waveforms from both common femoral veins should

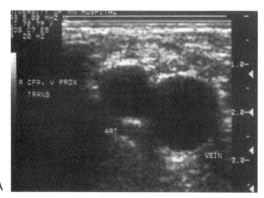
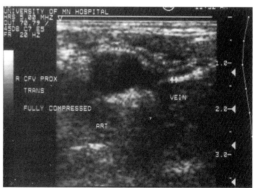

**FIG. 13.3. A:** The normal sonographic appearance of the common femoral vein. The vein is larger than the artery as a general rule. **B:** The vein is completely compressed by the transducer (*arrows*). The artery is just slightly compressed.

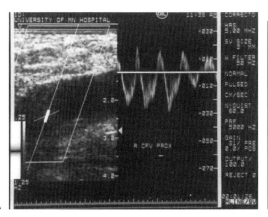 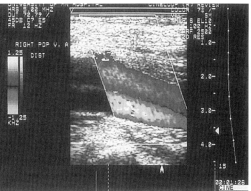

FIG. 13.4. **A:** Normal waveform of the common femoral vein, showing atrial contraction waves and respiration changes (phasicity). **B:** Normal longitudinal view of the popliteal artery and vein. Note complete filling of the vessels with color. Transverse view is also needed.

be obtained, even in unilateral examinations; thrombosis of nonvisualized portions of the iliac veins can be suggested if pulsatility and phasicity are dampened on one side.

One of the great advantages of ultrasound evaluation of the lower extremities is the ability to detect ancillary findings to account for the patient's symptoms (approximately 8% of patients). Ultrasound can detect lymphoceles, pseudoaneurysms, Baker cyst, enlarged lymph nodes, and saphenous vein clots.

## DEEP VENOUS THROMBOSIS OF THE UPPER EXTREMITY

The incidence of DVT in the upper extremity has been underrecognized. Central venous access and dialysis catheters are two of the most frequent predisposing conditions to the development of thrombosis of the brachiocephalic veins. Effort-induced thrombosis and extrinsic venous compression are additional causes.

It has been reported that 28% of patients with subclavian catheters developed thrombus of the subclavian vein. Venous thrombosis of the upper extremity is associated with PE in up to 16% of cases. These complications can occur even after anticoagulation has been started.

The main complications of upper extremity DVT include pain, swelling, PE, septic thrombophlebitis, and superior vena cava (SVC) syndrome. However, upper extremity DVT can be asymptomatic due to the rich collateral venous network in the upper extremity.

Sonography of the thoracic inlet veins requires knowledge of the regional anatomy. A complete ultrasound examination of the upper extremity and the thoracic inlet veins includes evaluation of the internal jugular veins (IJVs), innominate veins, subclavian veins, and axillary veins. The cephalad portion of the SVC can be visualized in some patients.

Gray-scale, color Doppler, and pulsed Doppler imaging are all essential for sonographic evaluation of the upper extremity veins. A high-frequency (7- to 10-MHz) linear transducer is usually optimal for visualizing the peripherally located portions of the subclavian veins and IJV, as well as the axillary veins and veins of the upper arm. A small footprint phased-array or sector transducer aids in visualization of the inferior IJV, the medial half of the subclavian veins, the innominate veins, and the SVC. A small footprint 7-MHz transducer can be used in thin patients to view the superior portion of the innominate veins. A 5-MHz small footprint transducer is preferred in large patients and for evaluating the lower innominate veins and SVC.

Pulsed Doppler samples and color Doppler images should be obtained from all venous segments. The compression technique can be used in the evaluation of the IJV, the axillary veins, and

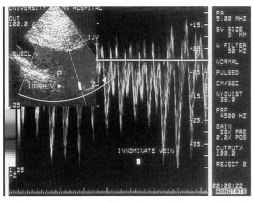

FIG. 13.5. Normal Doppler and conventional Doppler sonography of the innominate vein at the junction with the subclavian vein. Note the very sharp spikes representing the right atrial contractures.

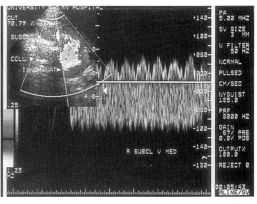

FIG. 13.6. Abnormal subclavian vein color flow Doppler sonogram. The Doppler waveform is obtained from the distal subclavian vein. The flow is not phasic, and there is spectral broadening. This would suggest central stenosis or occlusion.

the arm veins. As in the lower extremity, a normal vein should completely collapse, and a thrombosed vein will be incompressible. It is sometimes possible to directly visualize variably echogenic thrombus in regions in which venous compression is not possible. The echogenicity of the thrombus is not a reliable sign of the thrombus age. Color Doppler and Doppler spectral waveform analysis is critical in evaluating venous segments not amenable to compression.

The entire lumen of a normal upper extremity vein is often anechoic. Thrombus is sometimes apparent as echogenic material within the lumen. However, thrombus can be relatively anechoic or difficult to appreciate in obese patients due to decreased resolution and increased acoustic scatter. A normal color Doppler examination should demonstrate color completely filling the venous lumen. A nonobstructing thrombosis is seen as a filling defect within the color-filled lumen. An obstructing thrombus demonstrates an absence of color in the thrombosed segment. Color Doppler can also help identify collateral veins in the soft tissues surrounding a thrombosed venous segment.

Doppler waveforms from the IJVs, proximal subclavian veins, innominate veins, and SVC are normally pulsatile. This reflects right-sided cardiac pressure changes (Fig. 13.5). These veins also generally demonstrate phasic respiratory variation. Pulsatility and phasicity tend to decrease within more peripherally located veins (Fig. 13.6).

It is helpful to be familiar with the range of normal-appearing upper extremity venous waveforms and flow velocities when evaluating for proximal, nonvisualized thrombus. Color Doppler or spectral waveform Doppler venous waveforms from both sides should be compared. A side-to-side variation in the waveforms or velocities may be secondary to a proximal nonvisualized thrombus.

Determination of the accuracy of ultrasound of the thoracic inlet and upper extremity veins is complicated because of the variety of sonographic methods used to diagnose venous obstruction. Koksoy et al. used color and duplex ultrasound and contrast venography to study 44 patients with subclavian venous catheters. They concluded that the most useful combination of parameters included visualization of thrombus, absence of spontaneous flow, and absence of respiratory phasicity. In their study, sonography had a sensitivity of 94% and a specificity of 96% for depiction of thrombus; its positive predictive value was 94%, and its negative predictive value was 96%. Because of the overlying osseous structures, upper extremity venous ultrasound is a technically demanding study and can be difficult to interpret. Reliance on waveform analysis is necessary to suggest a centrally located obstruction.

Frequently, ultrasound is able to help confirm or rule out the existence of thrombosis, and no further evaluation is needed. Even if the innominate veins are not completely visualized, central venous outflow obstruction can still be suggested based on pulsed Doppler waveform analysis. Conventional venography remains the gold standard in equivocal cases of DVT. No other modality matches the noninvasiveness, portability, and low cost of ultrasonography, while still providing high accuracy and the ability to survey the surrounding soft tissues.

## C. OTHER DOPPLER APPLICATIONS

### Mark Wofford and Thomas R. Beidle

This section presents a brief overview of the salient duplex Doppler sonographic findings in evaluating the hepatic and portal vasculature, renal arteries, and organ transplants and in evaluating transjugular intrahepatic shunts (TIPS).

When evaluating abdominal structures, lower-frequency transducers tend to be used. If the vessels in question are not located beneath overlying ribs, a curved array transducer (transducer frequency between 3 and 7 MHz depending on the size of the patient) is preferred due to superior lateral resolution for deeper structures. A sector or vector transducer is used to scan between ribs. When trying to detect flow using color, power, or pulsed Doppler, pulse repetition frequency and wall filter should be lowered as much as possible. Note, however, that heart motion and breathing motion can limit optimization of Doppler parameters.

The portal, splenic, and mesenteric veins normally demonstrate antegrade flow, with only slight phasicity and pulsatility. The hepatic veins are normally extremely phasic and pulsatile. The hepatic, celiac, and renal arteries (including transplants) normally demonstrate low-resistance, monophasic flow. As expected, flow in the arterial and portal circulations varies between fasting and nonfasting states.

Portal hypertension generally results in decreased flow in the main portal vein. One exception occurs when the paraumbilical vein is recanalized, in which case flow in the right portal vein is often decreased or reversed. Hepatofugal flow in the main portal vein is a definite sign of portal hypertension, and a search for left gastric, splenorenal, and retroperitoneal varices should be performed using color Doppler. Portal vein thrombosis is easily evaluated with CDS. Complete color filling of the portal vein with waveform documentation of flow virtually excludes the diagnosis. Sometimes, flow may be severely diminished in the portal vein due to high resistance and may then be difficult to detect. Color Doppler is also helpful for detection of cavernous transformation of the portal vein. Occlusion of the hepatic veins (Budd-Chiari syndrome) gives rise to intrahepatic collaterals, loss of phasic flow, and decreased or reversed portal venous flow. CDS can also be used to determine patency of surgical portosystemic shunts.

CDS has been shown to be highly accurate for predicting TIPS malfunction. Kanterman et al. showed that CDS has a sensitivity of 92% and a specificity of 72% for detecting TIPS malfunction when multiple parameters are used in combination. The shunt is easily visualized in the liver parenchyma on gray-scale imaging (Fig. 13.7). An occluded shunt will demonstrate no flow on CDS examination. Due to the orientation of the shunt, it can be difficult to obtain a Doppler angle of 60 degrees or less in some patients. Often, the sonographer must attempt imaging from multiple different angles before reliable color flow and pulsed Doppler signals can be obtained. In addition, it is important to recognize that the stent used to create the portosystemic shunt often extends into the right or main, or both, portal veins a significant distance. Therefore, flow detected within the stent that is within the portal vein is actually portal

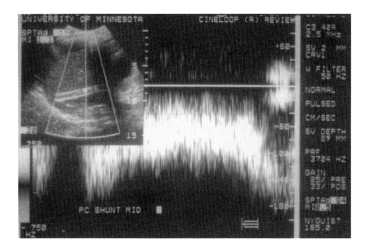

**FIG. 13.7.** Typical normal ultrasound Doppler of a transjugular intrahepatic shunt. The stent can be seen as bright lines, and the gate is in the mid shunt. The velocity is within normal range.

venous flow and not flow within the shunt (that portion of the stent that lies within the liver parenchyma). This distinction is important to avoid a false-negative conclusion of shunt patency, based on erroneous labeling of flow in the portal stent segment as flow in the shunt. In addition, velocity readings obtained from the portal venous segment of the stent should not be labeled as shunt velocities because, as described below, the reading within the portal circulation will be invariably low compared with a normal shunt velocity, and therefore shunt malfunction may be erroneously diagnosed. To avoid this pitfall, the sonographer must realize that the portal vein end of the shunt begins beyond the point at which the right portal vein or branches of the right portal vein separate from the stent.

Other parameters are used to detect critical narrowing of the shunt so that revision can be undertaken before shunt occlusion. A surveillance program including ultrasound at 3- to 6-month intervals with revision can result in assisted primary shunt patency of several years. Parameters for malfunctioning TIPS include peak shunt velocity of less than 90 or more than 190 cm per second, increase by more than 40 cm per second or decrease by more than 60 cm per second in peak shunt velocity between studies, main portal vein velocity of less than 30 cm per second, and reversal of right and left portal veins from hepatofugal to hepatopetal between studies. Each one of these parameters alone is relatively inaccurate, but when used in combination and with other sonographic findings such as amount of ascites and splenic size, maximum sensitivity and specificity are possible.

The role of CDS in evaluating the renal arteries for stenosis is controversial. Although some investigators have reported good results, others have been unable to duplicate these results. Two methods of detecting renal artery stenosis include insonation of the main renal artery and evaluation of intrarenal artery waveforms. The first method is challenging due to the difficulty in insonating the entire main renal artery bilaterally, particularly in large patients. In addition, identification of accessory renal arteries is difficult. Regardless, sensitivities of 84% to 98% and specificities of 97% to 99% have been reported. A peak systolic velocity of 180 to 200 cm per second correlates with stenosis of greater than 60% diameter reduction.

Evaluation of intrarenal waveforms is less technically demanding. This method uses acceleration indices and evaluation of systolic upstroke and early systolic peak to infer stenosis upstream, in the main renal artery. Therefore, this method can potentially detect stenosis of an accessory renal artery. Stavros et al. reported a sensitivity of 95% and a specificity of 97% for detection of renal artery stenosis. Others have not been able to duplicate these results.

Doppler evaluation of renal transplants includes evaluation of the main renal artery and vein, including their anastomotic sites, as well as the intrarenal arcuate arteries. A peak systolic velocity of greater than 100 cm per second indicates stenosis of the main renal artery. Renal vein thrombosis occurs in as many as 5% of transplants. Findings include lack of flow in the renal vein and high-resistance flow in the main renal artery, often with reversal of diastolic flow. PSA and AVF are complications of biopsy that are detected on the color Doppler examination. A resistive index above 0.70 in the arcuate arteries is a nonspecific finding that can indicate acute tubular necrosis, renal vein thrombosis, graft infection, obstructive hydronephrosis, and acute or chronic graft rejection.

CDS plays an important role in evaluation of liver transplants. Portal vein thrombosis and thrombosis of the inferior vena cava can be detected with nearly 100% accuracy, but thrombosis of the hepatic artery is more difficult to detect.

## D. CARBON DIOXIDE DIGITAL SUBTRACTION ANGIOGRAPHY

James G. Caridi, Irvin F. Hawkins, and Bret N. Wiechmann

The use of carbon dioxide as an imaging agent dates back to 1914, when it was originally used for the visualization of the abdominal viscera. It was subsequently used in the evaluation of the retroperitoneum and for hepatic veins the diagnosis of pericardial effusion. In the 1970s, the intraarterial use of carbon dioxide was pioneered by Hawkins. With the development of digital subtraction angiography, stacking software, tilting tables, and reliable delivery systems, it became viable as an angiographic imaging agent.

### UNIQUE PROPERTIES OF CARBON DIOXIDE

Carbon dioxide is a nontoxic, invisible gas that is highly compressible, nonviscous, and buoyant. Most important, carbon dioxide, as an intravascular imaging agent, lacks both allergic potential and renal toxicity. It is 20 times more soluble than oxygen and is rapidly dissolved in the blood.

Unlike iodinated contrast, carbon dioxide does not mix with blood but must displace it to render an image. Also, the buoyancy of carbon dioxide causes it to rise to the anterior, nondependent portion of the vessel. Therefore, in larger vessels (aorta and iliac arteries), if an insufficient volume is injected, there will be incomplete displacement of blood, resulting in diminished contrast and, potentially, a spurious image. Normal vessels may appear smaller than their true caliber. To overcome this phenomenon, either a larger amount of carbon dioxide must be administered or, using the buoyancy principle, the area of interest should be placed in the nondependent position.

### INDICATIONS

Carbon dioxide can be injected as a contrast agent in any luminal structure (arterial, venous, biliary tree, urinary tract, abscess cavity, fistula). We previously used carbon dioxide primarily in patients with iodinated contrast allergy and renal failure. However, its gaseous characteristics can occasionally provide additional information otherwise unattainable. Its very low viscosity permits detection of arterial bleeding, visualization of the portal system by hepatic parenchymal injection for TIPS procedures, visualization of small collaterals in ischemic disease, and AV shunting in tumors. The lack of viscosity also allows delivery via very small catheters and injections between the guidewire and the needle or catheter, making it ideal for interventional procedures such as angioplasty and stent placement. Furthermore, because of its rapid dissolution and elimination from the lungs, there is no maximum dose if less than 100 cc is injected every 2 minutes.

This is of great benefit in complex interventional procedures in which carbon dioxide can be used in combination with iodinated contrast to minimize the risk of renal compromise.

## CONTRAINDICATIONS

Our studies with rats suggest that the safety of cerebral carbon dioxide is questionable. We therefore avoid any arterial injections above the diaphragm and never administer carbon dioxide with the patient's head in an elevated position.

Carbon dioxide digital subtraction angiography has not been a problem in patients with chronic obstructive airway disease. However, in these patients, we do attempt to reduce the volume and allow more time between each injection. An evaluation by our laboratory using swine included the direct IVC administration of carbon dioxide at different volumes. This resulted in no change in either the arterial oxygen saturation or pulmonary artery, central venous, or systemic arterial pressure at an injection of 1.6 mL per kg. This is well below the individual dose required for diagnostic purposes.

## POTENTIAL COMPLICATIONS AND PRECAUTIONS

Because carbon dioxide is invisible, it is susceptible to contamination without detection. Our initial studies revealed water, rust, and particulate matter within reusable sources. Therefore, a pure medical-grade source and disposable cylinder (CMD, Gainesville, FL) are mandatory. Furthermore, a closed delivery system is imperative to eliminate the additional possibility of room air contamination. Because of diffusivity, an open syringe containing carbon dioxide can be replaced with less soluble room air in approximately 72 minutes. In addition, a system using stopcocks can be easily contaminated if they are inadvertently malpositioned or loose. In a closed system, one-way "check" valves and glued stopcocks can be used to reduce this possibility.

Another rare yet potential complication is "trapping." This occurs when an excessive volume of carbon dioxide is delivered or the blood-gas interface is reduced and interferes with normal dissolution. As a result, a bolus of gas can cause a "vapor lock" that can restrict blood flow and potentially cause ischemia. Abdominal aortic aneurysms (AAA); pulmonary outflow tract; and celiac, superior, and inferior mesenteric arteries are most susceptible because of their nondependent location.

If trapping does occur, it can be reduced by positional maneuvers. For example, if trapping during an inadvertent excessive, large-volume injection occurs in the pulmonary artery, bradycardia, hypotension, and coronary ischemia can result. By placing the patient in the left lateral decubitus position, carbon dioxide migrates to the nondependent portion of both the pulmonary artery and the right atrium. This allows blood flow to be reestablished beneath the residual carbon dioxide. Similarly, trapping in an AAA can be reduced by rolling the patient, first to one decubitus position and then to the other. As a precaution for trapping, fluoroscopy of susceptible sites can be performed between carbon dioxide injections. If persistent gas is visualized, positional changes can be instituted. For venous injections, fluoroscopy of the pulmonary artery will demonstrate dissolution of the gas within 10 to 30 seconds. If the gas remains longer, the possibility of room air contamination must be considered.

Injection of excessive volumes (more than 400 cc) is the most dangerous potential complication. Excessive doses are first and foremost avoided by ensuring that the carbon dioxide cylinder is never connected directly to the catheter. A carbon dioxide cylinder usually contains 3 million cc of pressurized gas and can flood the low-resistance circulatory system if a stopcock is inadvertently malpositioned. Also, because it is compressible, a syringe loaded under pressure will have an indeterminate volume of carbon dioxide and potentially result in an excessive dose. It is suggested that a noncompressed, known volume (usually 100 cc or less, depending on the site of evaluation) be administered via a dedicated injector or closed-plastic-bag system. Purging the catheter of saline or blood with a small volume of carbon dioxide should be performed before injection to eliminate compressed carbon dioxide and explosive delivery. We have also found that the elimination of explosive delivery reduces the subjective discomfort of pain, nausea, and the urge to defecate. Moreover, if using carbon dioxide to evaluate permanent dialysis access, great care should be taken to avoid explosive delivery and reflux into the cerebral circulation.

Carbon dioxide should be used cautiously with nitrous oxide anesthesia. In theory, nitrous oxide may diffuse from the soft tissue into the carbon dioxide "gas bubble" and cause a five- to sixfold increase in the occlusive effect. An innocuous 100-cc carbon dioxide injection may have the effect of 500 to 600 cc of gas and result in a vapor lock condition.

## DELIVERY

Currently, there are two safe delivery mechanisms: dedicated injectors and the closed-plastic-bag hand-delivery system. Because a dedicated carbon dioxide injector is not currently available in the United States, the closed-bag system can be used (Fig. 13.8). It consists of a 1,500-cc plastic-bag reservoir, extension tubing, one-way check valves with glued fittings, and a delivery purge syringe. Using a pure source, the bag is filled with carbon dioxide and flushed three times to purge any residual air. After this, the bag should be left flaccid to avoid any carbon dioxide compression. Next, the bag is connected to the delivery fitting. The delivery system is similarly flushed to eliminate room air before injection. It is then connected to the angiographic catheter, which is subsequently relieved of any residual blood or saline by forcefully injecting 3 to 5 cc of carbon dioxide. A controlled, nonexplosive delivery of known volume can then be performed. The check valves primarily prevent reflux of blood into the catheter and permit rapid injections without stopcock manipulation. No additional connecting tubes or stopcocks should be added to the system. All ports should be occupied and syringes attached to prevent any possibility of air contamination.

### General Delivery Principles

1. Use a closed system—that is, the plastic bag or a dedicated carbon dioxide injector.
   a. Never connect the catheter directly to the carbon dioxide cylinder. This avoids the potential inadvertent delivery of excessive and possibly lethal volumes.
   b. Malpositioned stopcocks can result in room air contamination and air embolus.
2. Avoid explosive delivery. Purging fluid (blood or saline) from the angiographic catheter results in a more consistent delivery with less discomfort.
3. Initially, inject small volumes of carbon dioxide. Increase or decrease volume as required for specific anatomy.
4. Wait 2 to 3 minutes between injections to allow any potentially trapped carbon dioxide to dissolve.
5. Elevate area of interest in poor flow conditions (feet, 10 to 15 degrees; renal artery, 30 to 45 degrees).
6. Vasodilators (nitroglycerin, 100 to 150 µg intraarterial) can be used to improve filling.
7. Delivery catheter.
   a. Use radioopaque-tipped catheter.
   b. At least one side hole is recommended for safety.
   c. Any flush catheter is acceptable.
8. Digital subtraction angiography imaging.

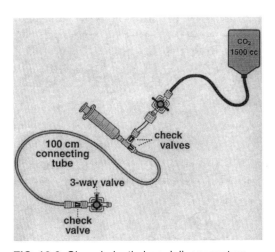

**FIG. 13.8.** Closed-plastic-bag delivery system.

a. Three to four frames per second using a 60-millisecond pulse width with adequate penetration.
b. When the carbon dioxide bolus is "broken up" (fragmented), use image stacking if available.
c. If imaging is consistently poor, consult an equipment applications specialist to optimize acquisition.

**Specific Procedure**

1. Runoff.
   a. Initially, obtain both leg runoffs with the catheter in the distal aorta. Perform aortogram after the runoff.
   b. Inject 20 to 40 cc in 1 second.
   c. Elevate the feet 10 to 15 degrees for optimal filling and obtain images of pelvis, thigh, knee, lower legs, and feet.
   d. If there is no stacking program, a longer injection (approximately 60 cc over 2 to 3 seconds) is necessary.
   e. Problem: Poor filling of the lower leg and feet.
      - Perform a selective injection of the common femoral or more distal arteries. Most runoff exams are currently examined in this fashion.
      - With stacking, inject 10 to 20 cc in 2 seconds. If filling remains poor, inject 20 to 40 cc over 3 to 4 seconds.
      - Without stacking, begin with 20 to 40 cc over 3 to 4 seconds.
      - Intraarterial nitroglycerin, 100 to 150 µg before injection.
      - When large volumes are required, discomfort may occur, precipitating patient motion and distorting images.
2. Aortogram.
   a. Usually performed after the runoff. We believe this allows the patient to become acclimated to carbon dioxide and, as a result, less discomfort and nausea are experienced with larger aortic injections.
   b. Attempt to obtain the aortogram without glucagon. Our experience is that carbon dioxide and glucagon may cause nausea.
   c. Higher flow rates may be necessary (25 to 50 cc in 1 to 2 seconds).
   d. The left renal is more difficult to image and may be better visualized by elevating that side. If necessary, a selective injection with a shepherd hook catheter (10 to 20 cc carbon dioxide in 1 second) can be performed. The ostium is usually apparent secondary to carbon dioxide reflux.
   e. Selective injections of the visceral arteries commonly require 10 to 30 cc in 1 to 2 seconds.
3. Venous system: Always image the pulmonary artery after the first injection to rule out air contamination (persistent gas). Normally, carbon dioxide should disappear after 10 to 30 seconds.
   a. SVC and IVC: 20 to 60 cc in 1 to 2 seconds.
   b. Subclavian: 20 to 40 cc in 1 to 2 seconds.
   c. Peripheral veins: 15 to 25 cc in 4 to 8 seconds. Rapid injection precipitates pain.
4. Interventional procedures.
   a. Using a Touhey-Borst fitting, carbon dioxide can be injected between the guidewire and needle or catheter. Wires without coil wrap are better (Glidewire, 0.018 torque wire).
   b. Use a 20- to 40-cc Luer-locked syringe. With a smaller syringe, carbon dioxide will simply compress without injecting.
   c. Wait 20 to 30 seconds for carbon dioxide to exit the catheter. Carbon dioxide will compress; purge fluid from the catheter and inject.
   d. After purging, subsequent injections require less pressure and delay.
5. TIPS.
   a. Using any needle, inject 20 cc of carbon dioxide into the hepatic parenchyma for visualization of the portal vein.
   b. With the guidewire in place, carbon dioxide can be used to verify the needle entry site and determine stent positioning.

# E. INTRAVASCULAR ULTRASOUND

## Wilfrido R. Castañeda-Zúñiga

Conventional, or digitally subtracted, angiography provides a two-dimensional representation of the vessel lumen, but it does not provide transmural information, nor can it give a real picture of the complex three-dimensional morphology vascular disease.

Intravascular ultrasound (IVUS) adds a new dimension to atherosclerotic and nonatherosclerotic vascular disease evaluation. In contrast to angiography, images obtained by IVUS depict the full cross-sectional thickness of the vascular wall. Furthermore, intravascular sonography provides a real-time method of assessing the effects of vascular interventional procedures, such as percutaneous transluminal angioplasty, percutaneous atherectomy, vascular stenting, and TIPS.

## TECHNICAL CONSIDERATIONS

Two types of systems have been developed for intravascular applications:

1. Mechanically rotating devices
2. Electronic array devices

Typically, IVUS catheters use a frequency ranging from 20 to 40 MHz. These high frequencies allow greater lateral and axial resolution—at the expense of lesser tissue penetration due to energy dissipation at the higher frequencies.

## MECHANICAL DEVICES

There are three types of catheter configurations:

1. The transducer is set at a 10-degree angle to the catheter long axis. Images are obtained by rotating the transducer to allow side viewing of the vessel wall (Fig. 13.9).
2. An acoustic reflector mounted at a 45-degree angle rotates around the long axis of the catheter while the transducer remains stationary.
3. The transducer and reflecting mirror relationship is fixed while the catheter drive shaft rotates.

The advantages of mechanical devices include the following:

- The presence of an acoustic reflector within the system means that ring artifacts occur within the catheter so that they are minimized, with no need for computer image enhancement.
- The drive shaft rotates at speeds varying from 1,200 to 1,800 rpm, providing a 20 to 30 frame-per-second rate.

A disadvantage is that the drive shaft is enclosed inside of a plastic catheter to protect the vascular wall, which increases the diameter of the device.

Three types of catheter design exist: over-the-wire, monorail, and fixed wire tip. The catheter diameter varies from 3.5 to 6.0 Fr in the 20- to 40-MHz range. Lower-frequency catheters in the 10- to 15-MHz range have larger diameters of up to 10 Fr.

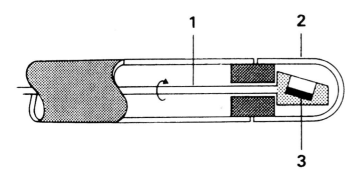

FIG. 13.9. Mechanical intravascular ultrasound device. The ultrasound element is mounted at the tip of a rotating flexible drive shaft (*1*); a sonolucent dome isolates the ultrasound element from the flowing blood (*2*); ultrasound transducer (*3*).

## ELECTRONIC ARRAY DEVICES

Available designs of electronic array devices use 64 or more transducer elements, with multiple integrated circuits to digitize, amplify, and multiply the signal for transmission to the system's computers (Fig. 13.10).

Advantages of the electronic design include the absence of rotating parts, allowing a smaller diameter, greater flexibility, and a true over-the-wire capability.

Disadvantages include a lower framing rate (10 frames per second) and ring down near artifacts that require computer image enhancement to eliminate them, with the consequent image degradation.

A more significant drawback of IVUS is the fact that the device is not able to acquire information about segments of the vascular wall located forward of the catheter tip. Ideally, an IVUS catheter should provide not only lateral information, but also information on lesion morphology, degree of obstruction, and length of the lesion. This information could then be used to help the interventionist cross the site of obstruction and select the most adequate interventional therapy for the lesion present. Three-dimensional reconstruction addresses the inability of two-dimensional ultrasound images to provide the spatial relationship of the segments of interest, by providing an image display in which any vascular segment is simultaneously projected in longitudinal relationship to the proximal and distal segments.

To further increase the information obtained with the IVUS, a Doppler probe has been placed on the ultrasound probe. This gives hemodynamic information, which can be very important when dealing with vascular stenosis.

## INTRAVASCULAR ULTRASOUND EVALUATION OF VASCULAR ANATOMY

Because of its ability to provide transmural information, IVUS is uniquely capable of providing insight into the structure and composition of the normal arterial wall. Significant information can be gained with IVUS with regard to the following:

- Normal arterial wall
- Plaque morphology
- Characterization of thrombi
- Mechanism of angioplasty

### Normal Arterial Wall

The three anatomic layers of the arterial wall provide distinctive ultrasound images (Fig. 13.11) referred to as the *three-layered image*. The inner bright acoustic reflection in normal adult arteries is produced by the interface of blood with the intima and mainly with the internal elastic lamina. The intimal layer is typically not thick enough to generate a distinct ultrasonic signal. The second, outer bright interface is produced by the external elastic lamina and adventitia. The central echolucent zone is the result of the media, which is sonolucent because of its homogeneous structure.

### Plaque Morphology

IVUS can help delineate the type of vascular disease present (atherosclerotic or nonatherosclerotic), its severity, and the degree of luminal obstruction. Typically, lipid collections are

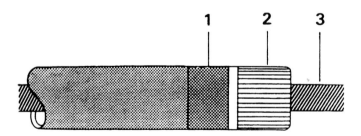

**FIG. 13.10.** Electronic array intravascular ultrasound catheter. Electronically switched, phased-array catheter with integrated circuitry for reduction of the number of wires (*1*); transducer elements (*2*); guidewires (*3*).

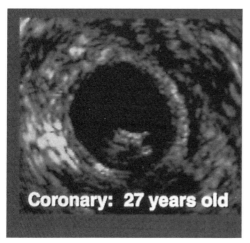 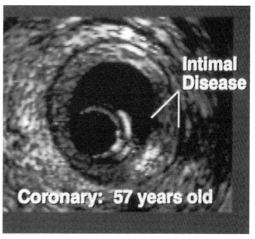

**FIG. 13.11. A:** Intravascular ultrasound image obtained in a healthy 27-year-old. Note that the intimal layer is not thick enough to generate a distinctive ultrasound signal. The media is sonolucent, and the second outer bright interface is produced by the external elastic lamina and adventitia. **B:** Three-layer appearance in a 57-year-old with extensive intimal thickening due to atherosclerotic disease. The media is seen as an echolucent ring around the intima. The second, outer bright ultrasonic interface seen outside the media represents the ultrasound signal produced by the external elastic lamina and the adventitia.

echolucent and can be easily identified within an atheroma; fibrous tissue and collagen are strong echo reflectors and are therefore visualized as echogenic, speckled to homogeneous bright structures. Calcium, which is the strongest echo reflector, produces intense, bright areas within the plaque, associated with a dropout of peripheral echoes due to the intense reflection of the ultrasound signal.

The extent of involvement of the arterial lumen is well demonstrated by angiography in orthogonal planes. However, if diffuse involvement is present, the severity of involvement might be underestimated by assuming that the adjacent segments are normal. IVUS with three-dimensional reconstruction greatly enhances our ability to assess the overall extent of atherosclerotic vascular disease present, as well as its severity, helping us to separate hemodynamically significant from nonsignificant lesions.

Alfonso et al. described the following with regard to IVUS and atherosclerosis:

- The atherosclerotic involvement of the arterial wall extends beyond the angiographic lesion site.
- IVUS frequently reveals the presence of atherosclerotic plaque at sites with a normal angiographic appearance.
- These plaques usually have a semilunar shape and do not disrupt the luminal contour.

IVUS and magnetic resonance imaging have excellent correspondence both for structural identification within the vessel wall and the histologic arterial structures when compared.

### Characterization of Thrombi

The ability to detect thrombi with IVUS is directly related to the thrombus composition. Thrombi consisting solely of platelet-rich plasma cannot be detected. As the content of red cells within the thrombus increases, there is a concomitant increase in the echogenicity. The intensity of reflection is brighter for a more recent thrombus than for organized older clots (Fig. 13.12). This speckled ultrasonic appearance is not specific to thrombus; it can also be seen with soft plaques composed mainly of loose connective or fibromuscular tissue.

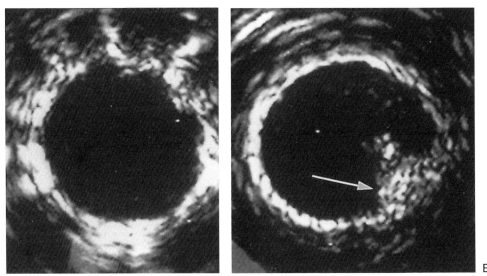

**FIG. 13.12.** Patient with successful thrombolysis of thrombosed intravascular stent. Intravascular ultrasound proximally (**A**) and distally (**B**) shows residual distal mural thrombi (*arrow*) after a successful lysis of a completely occluded stent site. (Courtesy of B.T. Katzen, M.D., of the Miami Vascular Institute.)

## CLINICAL APPLICATIONS

The applications of IVUS can be divided into four groups: (a) diagnostic adjunct, (b) therapeutic adjunct, (c) real-time therapeutic monitoring, and (d) therapeutic tool.

### Diagnostic Adjunct

As a diagnostic adjunct, IVUS increases accuracy in measuring the arterial diameter and circumference, lumen volume, percentage of stenosis, and arterial wall compliance. Qualitative information obtained by IVUS of interest for interventional vascular procedures includes location of vessel bifurcations, plaque evaluation, tissue characterization, vessel wall analysis, and perivascular tissue analysis. IVUS provides transmural quantitative information and therefore enhances our ability to select the optimal treatment for the existing pathology and thus obtain a better outcome.

IVUS has been found useful in the diagnostic evaluation of aortic pathology, mainly aortic dissection, traumatic rupture, and coarctation of the aorta. This is based on small studies, and the practicality of using IVUS for those applications has not been demonstrated. IVUS has been found to have comparable diagnostic accuracy to conventional angiography with regard to aortic dissection; however, IVUS does not provide additional information not already available from angiography. Angiography demonstrated in one study the entry site better, whereas IVUS was better at demonstrating the reentry site, circumferential involvement, the contents of the false lumen, and the presence of intramural hematoma in the aortic wall. In comparison with transesophageal echocardiography, IVUS was better at demonstrating distal involvement, whereas transesophageal echocardiography was better at demonstrating proximal changes.

The application of IVUS to the diagnosis of acute traumatic rupture can be of benefit, whereas the aortography demonstrated only a subtle contour irregularity of the proximal descending aorta. The IVUS can demonstrate a mural flap and a small, contained hematoma. Typically, a diagnosis can be established on the basis of the aortographic findings; occasionally a prominent aortic bump can simulate a traumatic PSA, and a plaque or artifact can simulate a mural flap. IVUS can distinguish a mural flap, subintimal hematoma, or a PSA from an atherosclerotic plaque or an aortic bump; however, the routine use of IVUS is unlikely in these unstable patients.

IVUS can give added information in the diagnosis of coarctation of the aorta. It defines better the full extent of narrowing, the position of the narrowest segment, and the extent of the collateral system.

### Therapeutic Adjunct

It has been demonstrated that the use of IVUS before vascular interventions can change the therapeutic approach in up to 44% of the cases.

## TRANSLUMINAL ANGIOPLASTY

IVUS can determine, on the basis of the type of plaque dissection present, whether further therapy is necessary to maintain vessel patency. That is, if a complete circumferential plaque dissection/fracture is present, severely compromising the vessel lumen, intravascular stent deployment may be necessary to avoid acute occlusion. IVUS is more sensitive than angiography in assessing postintervention lesion characteristics, particularly lesion composition, topography, and extension of dissection.

## INTRAVASCULAR ULTRASOUND IN DIRECTIONAL ATHERECTOMY

Debulking techniques such as lasers and atherectomy were developed to address the high incidence of re-stenosis after interventional vascular procedures. One reason why the use of these devices did not improve the long-term results is that they do not remove as much atherosclerotic plaque as expected.

Among the causative factors is the inability to accurately orient the cutting chamber toward the residual plaque. The atherectomy catheter can then be placed at the desired location, and the necessary amount of clockwise or counterclockwise rotation is applied, as determined by the IVUS study. Obviously, the most efficient way to remove plaque with the directional atherectomy catheter would be to include an ultrasound crystal within the atherectomy device to provide real-time imaging during plaque removal. Another factor that influences the results of directional atherectomy is the presence of plaque calcification. Calcification is present in 50% to 70% of lesions evaluated by IVUS. The IVUS can detect the presence of superficial calcium and change the therapeutic approach to another modality that would provide a better outcome.

## INTRAVASCULAR ULTRASOUND AND INTRAVASCULAR STENTS

Traditionally, angiography has been used to assess the vessel dimensions for selection of the stent diameter to be used. IVUS has been shown to be better than angiography for estimating vessel diameter before and after intervention. IVUS has also been shown to be better than angiography in determining underexpansion of the stent, asymmetric stent expansion, and residual narrowing after stent deployment.

A final and more challenging application of IVUS is in the deployment of intravascular stented grafts. IVUS can provide more reliable information in these cases than angiography regarding the site of placement, diameter, topography, and branch location. After device deployment, IVUS can help assess whether full stent and graft expansion has been achieved. IVUS findings can also lead to further intervention after deployment (e.g., further balloon expansion of the stented graft) or to the deployment of additional stents to seal a leaking stented graft site.

## CONCLUSION

The exact role of IVUS as a diagnostic and interventional tool needs to be further defined, particularly with regard to its cost-effectiveness.

# SUGGESTED READINGS

## A. Noninvasive Vascular Evaluation of Peripheral Vascular Disease

Carpenter JP, Owens RS, Holland GA, et al. Magnetic resonance angiography of the aorta, iliac and demoral arteries. *Surgery* 1994;116:17–23.

Coley BD, Roberts AC, Fellmeth BD, et al. Postangiographic femoral artery pseudoaneurysms: further experience with US-guided compression repair. *Radiology* 1995;194:307–311.

Crossman DV, Ellison JE, Willis H, et al. Comparison of contrast arteriography to arterial mapping with color-flow duplex imaging in the lower extremities. *J Vasc Surg* 1989;10:522–528.

Fletcher FP, Kershaw LZ, Chan A, et al. Noninvasive imaging of the superficial femoral artery using ultrasound duplex scanning. *J Cardiovasc Surg* 1990;31:364–367.

Hatsukami TS, Primozich JF, Zierler E, et al. Color Doppler imaging of infrainguinal arterial occlusive disease. *J Vas Surg* 1992;16:527–533.

Igidbashian VN, Mitchell DG, Middleton WD, et al. Iatrogenic femoral arteriovenous fistula: diagnosis with color Doppler imaging. *Radiology* 1989;170:749–752.

Kang SS, Labropoulos N, Mansour MA, Baker WH. Percutaneous ultrasound guided thrombin injection: a new method for treating postcatheterization femoral pseudoaneurysms. *J Vasc Surg* 1998;27:1032–1038.

Kempczinski R, Yas J. *Practical noninvasive vascular diagnosis*. Chicago: Year Book Medical Publishers, 1983:99.

Lacy JH, Box JM, Connors D, et al. Pseudoaneurysm: diagnosis with color Doppler ultrasound. *J Cardiovasc Surg* 1990;31:727–730.

Lynch TG, Hobson RW, Wright CB, et al. Interpretation of Doppler segmental pressures in peripheral vascular occlusive disease. *Arch Surg* 1984;119:465–467.

McCauley TR, Monib A, Dickey KW, et al. Peripheral vascular occlusive disease; accuracy and reliability of time of flight MR angiography. *Radiology* 1994;192:351–357.

Moneta GL, Yeager RA, Antonovic R, et al. Accuracy of lower extremity arterial duplex mapping. *J Vasc Surg* 1992;15:275–284.

Polak JF. Peripheral arterial diseases. In: Polak JF, ed. *Peripheral vascular sonography: a practical guide*. Baltimore: Williams & Wilkins, 1992:247–302.

Polak JF, Karmel MI, Mannick JA, et al. Determination of the extent of lower extremity peripheral arterial disease with color-assisted duplex sonography. *AJR Am J Roentgenol* 1990;144:1085–1089.

Quinn SF, Demlow TA, Hallin RW, Eidenmiller LR, Szumowski J. Femoral MR angiography versus conventional angiography: preliminary results. *Radiology* 1993;189:181–184.

Ramsey DE, Manke DA, Sumner DS. Toe blood pressures—a valuable adjunct to ankle pressure measurement for assessing peripheral arterial disease. *J Cardiovasc Surg* 1983;24:43–48.

Taylor PR, Tyrell MR, Cofton M, et al. Colour flow imaging in the detection of femoro-distal graft and native artery stenosis: improved criteria. *Eur J Vasc Surg* 1992;6:232–236.

Whelan JF, Barry MH, Moir JD. Color flow Doppler ultrasonography; comparison with peripheral arterial disease. *J Clin Ultrasound* 1992;20:369–374.

## B. Evaluation of Deep Venous Thrombosis

Cohen JR, Tyman R, Pillari G, Johnson H. Regional anatomical differences in the venographic occurrence of deep venous thrombosis and long-term follow-up. *J Cardiovasc Surg* 1988;29:547–551.

Coon WW, Willis PW. Deep venous thrombosis and pulmonary embolism—prediction, prevention and treatment. *Am J Cardiol* 1959;4:611–621.

Cronan JJ, Dorfman GS, Grusmark J. Lower extremity deep venous thrombosis: further experience with and refinements of US assessments. *Radiology* 1988;168:101–107.

Dorfman GS, Cronan JJ. Sonographic diagnosis of thrombosis of the lower extremity veins. *Semin Intervent Radiol* 1990;7:9–19.

Horattas MC, Wright DJ, Fenton AH, et al. Changing concepts of deep venous thrombosis of the upper extremity: report of a series and review of the literature. *Surgery* 1988;104:561–567.

Koksoy C, Kuzu A, Kutlay J, et al. The diagnostic value of color Doppler ultrasound in central venous catheter related thrombosis. *Clin Radiol* 1995;50:687–689.

McCarthy WJ, Vogelzang RL, Bergan JJ. Changing concepts and present-day etiology of upper extremity venous thrombosis. In: Bergan JJ, James JST, eds. *Venous disorders*. Philadelphia: WB Saunders, 1991: 407–420.

Monreal M, Raventos A, Lerma R, et al. Pulmonary embolism in patients with upper extremity DVT associated to venous central lines a prospective study. *Thromb Haemost* 1994;72:548–550.

Nazarian GK, Foshager MC. Color Doppler sonography of the thoracic inlet veins. *Radiographics* 1995;15:1357–1371.

Strandness DE, Langlois Y, Cramer M, et al. Long-term sequelae of acute venous thrombosis. *JAMA* 1983;250:1289–1298.

## C. Other Doppler Applications

Dada SH, Erley CM, Wakat JP, et al. Post transplant renal artery stenosis: evaluation with duplex sonography. *Eur J Radiol* 1993;16:95–101.

Dodd GD, Tublin ME, Shah A, et al. Imaging of vascular complications associated with renal transplants. *AJR Am J Roentgenol* 1991;157:449–459.

Foshager MC, Ferral H, Nazarian GK, et al. Duplex sonography after transjugular intrahepatic protosystemic shunt (TIPS): normal hemodynamic finding and efficacy in predicting shunt patency and stenosis. *AJR Am J Roentgenol* 1995;165:1–7.

Grant EG, Schiller VL, Millener P, et al. Color doppler imaging of the hepatic vasculature. *AJR Am J Roentgenol* 1992;159:943–950.

Hoffman U, Edwards JM, Carter S, et al. Role of duplex scanning for the detection of atherosclerotic renal artery disease. *Kidney Int* 1991;39:1232–1239.

Kanterman RY, Darcy MD, Middleton WD, et al. Doppler sonography findings associated with transjugular intrahepatic portosystemic shunt malfunction. *AJR Am J Roentgenol* 1997;268:467–472.

Kaveggia LP, Perrella RR, Grant EG, et al. Duplex Doppler sonography in renal allografts: the significance of reversed flow in diastole. *AJR Am J Roentgenol* 1990;155:295–298.

Kliewer MA, Tupler RH, Hertzberg BS, et al. Doppler evaluation of renal artery stenosis: interobserver agreement in the interpretation of waveform morphology. *AJR Am J Roentgenol* 1994;162:1371–1376.

Koslin DB, Mulligan SA, Berland LL. Duplex assessment of the portal venous system. *Semin Ultrasound CT MR* 1992;13:22–33.

Letourneau JG, Day DL, Ascher NL, et al. Abdominal sonography after hepatic transplantation: results in 36 patients. *AJR Am J Roentgenol* 1987;149:299–303.

Lomas DJ, Britton PD, Farman P, et al. Duplex Doppler ultrasound for the detection of vascular occlusion following liver transplantation in children. *Clin Radiol* 1992;46:38–42.

Millener P, Grant EG, Rose S, et al. Color Doppler imaging findings in patients with Budd-Chiari syndrome: correlation with venographic findings. *AJR Am J Roentgenol* 1993;161:207–312.

Olin JW, Piedmonte MR, Young JR, et al. The utility of duplex ultrasound scanning of the renal arteries for diagnosing significant renal artery stenosis. *Ann Intern Med* 1995;122:833–838.

Renowden SA, Blethyn J, Cochlin DL. Duplex and colour flow sonography in the diagnosis of post-biopsy arteriovenous fistulae in the transplant kidney. *Clin Radiol* 1992;45:233–237.

Stavros AT, Parker SH, Yakes WF, et al. Segmental stenosis of the renal artery: pattern recognition of tardus and parvus abnormalities with duplex sonography. *Radiology* 1992;184:487–492.

Sterling KM, Darcy MD. Stenosis of transjugular intrahepatic portosystemic shunts: presentation and management. *AJR Am J Roentgenol* 1997;168:239–244.

Taylor DC, Kettler MD, Moneta GL, et al. Duplex ultrasound scanning in the diagnosis of renal artery stenosis: a prospective evaluation. *J Vasc Surg* 1988;7:363–369.

Tessler FT, Gehring BJ, Gomes A, et al. Diagnosis of portal vein thrombosis: value of color doppler imaging. *AJR Am J Roentgenol* 1991;157:293–296.

## D. Carbon Dioxide Digital Subtraction Angiography

Coffey R, Quisling RG, Mickle JP, et al. The cerebrovascular effects of intra-arterial $CO_2$ on quantities required for diagnostic imaging. *Radiology* 1984;15:405–410.

Ehrman KO, Taber TE, Gaylord GM, et al. Comparison of diagnostic accuracy with carbon dioxide versus iodinated contrast material in the imaging of hemodialysis access fistulas. *J Vasc Intervent Radiol* 1994;5:771–775.

Hawkins IF. Carbon dioxide digital subtraction angiography. *AJR Am J Roentgenol* 1982;139:19–24.

Hawkins IF, Caridi JG, Kerns SR. Plastic bag delivery system for hand injection of carbon dioxide. *AJR Am J Roentgenol* 1995;165:1–3.

Paul RE, Durant TM, Oppenheiner MJ, Stauffer HM. Intravenous carbon dioxide for intracardiac gas contrast in the roentgen diagnosis of pericardial effusion and thickening. *AJR Am J Roentgenol* 1957;78:224–225.

Phillips JH, Burch GE, Hellinger R. The use of intracardiac carbon dioxide in the diagnosis of pericardial disease. *AJR Am J Roentgenol* 1966;97:342–349.

Rautenberg E. Rontgenphotographie der Leber, der Milz, und des Zwerchfells. *Deutsch Med Wschr* 1994;40:1205.

Rosenstein P. Pneumoradiology of kidney position-a new technique for the radiological representation of the kidneys and neighboring organs (suprarenal gland, spleen, liver). *J Urol* 1921;15:447.

Teffey EP, Johnson BH, Eger EI. Nitrous oxide intensifies the pulmonary arterial pressure response to venous injection of carbon dioxide in the dog. *Anesthesiology* 1980;52:52–55.

## E. Intravascular Ultrasound

Alfonso F, Macaya C, Goicolea J, et al. Intravascular ultrasound imaging of angiographically normal coronary segments in patients with coronary artery disease. *Am Heart J* 1994;127:536–544.

Benanati JF. Intravascular ultrasound: the role in diagnostic and therapeutic procedures. *Radiol Clin North Am* 1995;33:31–50.

DeGroff CG, Reller MD, Sahn DJ. Intravascular ultrasound can assist angiographic assessment of coarctation of the aorta. *Am Heart J* 1994;128:836–839.

De Scheerder I, De Man F, Herregods MC, et al. Intravascular ultrasound versus angiography for measurement of luminal diameters in normal and diseased coronary arteries. *Am Heart J* 1994;127:243–251.

Feltrin GP, Chiesura-Corona M, Miotto D, et al. Intravascular ultrasound evaluation for assessment and therapeutic decisions in aortic diseases. *Angiology* 1994;45:716.

Fisher RG, Hadlock F, Ben-Menachem Y. Laceration of the thoracic aorta and brachiocephalic arteries by blunt trauma. *Radiol Clin North Am* 1981;19:91–110.

Fitzgerald PJ, Sudhir K, Gupta M, Yock PG. Combined atherectomy/ultrasound imaging device reduces subintimal tissue injury. *J Am Coll Cardiol* 1992;17:223A.

Fitzgerald PJ, Yock PG. Mechanisms and outcomes of

angioplasty and atherectomy assessed by intravascular ultrasound imaging. *J Clin Ultrasound* 1993;21:579–588.

Frimerman A, Miller HI, Hallman M, et al. Intravascular ultrasound characterization of thrombi of different composition. *Am J Cardiol* 1994;73:1053–1057.

Froelich JJ, Hoppe M, Nahrstedt C, et al. The precise determination of vascular lumen and stent diameters: correlation among calibration angiography, intravascular ultrasound, and pressure-fixed specimen. *Cardiovasc Intervent Radiol* 1997;20:452–456.

Goldberg SL, Colombo A, Nakamura S, et al. Benefit of use of intracoronary ultrasound in the deployment of Palmaz-Schatz stents. *J Am Coll Cardiol* 1994;24:996–1003.

Gussenhoven EJ, Essed CE, Lancee CT, et al. Arterial wall characteristics determined by intravascular ultrasound imaging: an in vitro study. *J Am Coll Cardiol* 1989;14:947–952.

Hoppe M, Heverhagen JT, Froelich JJ, et al. Correlation of flow velocity measurements by magnetic resonance phase contrast imaging and intravascular Doppler ultrasound. *Invest Radiol* 1998;3:427–432.

Isner JM, Kaufman J, Rosenfield K, et al. Combined physiologic and anatomic assessment of percutaneous revascularization using a Doppler guidewire and ultrasound catheter. *Am J Cardiol* 1993;71:70D–86D.

Isner JM, Rosenfield K, Lorsodo DW, et al. Clinical experience with intravascular ultrasound as an adjunct to percutaneous revascularization. In: Tobis JM, Yock PG, eds. *Intravascular ultrasound imaging*. New York: Churchill Livingstone, 1992.

Kimura BJ, Fitzgerald PJ, Sudir K, et al. Guidance of directed coronary atherectomy by intravascular imaging. *Am Heart J* 1992;124:1365–1369.

Martin AF, Ryan LK, Gothlieb AI, et al. Arterial imaging: comparison of high resolution US and MR imaging with histologic correlation. *Radiographics* 1997;17:189–202.

Matar FA, Mintz GS, Douek P, et al. Coronary artery lumen volume measurement using three-dimensional intravascular ultrasound: validation of a new technique. *Cathet Cardiovasc Diagn* 1994;33:214–220.

Nakamura S, Colombo A, Galione A, et al. Coronary stenting guided by intravascular ultrasound [Abstract]. *Circulation* 1993;88:3211.

Pande A, Meier B, Fleisch M, et al. Intravascular ultrasound for diagnosis of aortic dissection. *Am J Cardiol* 1991;67:662–663.

Pandian NG, Kreis A, Weintrasub A, Kumar R. Intravascular ultrasound assessment of arterial dissection, intimal flaps and intraarterial thrombi. *Am J Cardiol Imaging* 1991;5:72–77.

Pichard AD, Mintz GS, Satler LF, et al. The influence of preintervention intravascular ultrasound imaging on subsequent transcatheter treatment strategies [Abstract]. *J Am Coll Cardiol* 1993;21:133A.

Siegel RJ, Chae JS, Maurer G, et al. Histopathologic correlation of the three-layered intravascular ultrasound appearance of normal adult human muscular arteries. *Am Heart J* 1993;126:872–878.

Stone GW, St Goar F, Klette MA, Linnemeier TJ. Initial clinical experience with a novel low-profile integrated coronary ultrasound angioplasty catheter: implications for routine use [Abstract]. *J Am Coll Cardiol* 1993;21:134A.

The GUIDE Trial Investigators. Impact of intravascular ultrasound on device selection and endpoint assessment of interventions: phase I of the GUIDE Trial [Abstract]. *J Am Coll Cardiol* 1993;21:134A.

Weintraub AR, Erbel R, Gorge G, et al. Intravascular ultrasound imaging in acute aortic dissection. *J Am Coll Cardiol* 1994;24:495–503.

White RA, Verbin C, Kopchok G, et al. The role of cine fluoroscopy and intravascular ultrasonography in evaluating the deployment of experimental endovascular prostheses. *J Vasc Surg* 1995;21:365–374.

Williams DM, Dake MD, Bolling SF, Deeb GM. The role of intravascular ultrasound in acute traumatic aortic rupture. *Semin Ultrasound CT MRI* 1993;14:85–90.

Williams DM, Simon HJ, Marx MV, Starkey TD. Acute traumatic aortic rupture: intravascular US findings. *Radiology* 1992;182:247–249.

# 14

# Fibrinolytic Therapy

Haraldur Bjarnason

Hemostasis is based on a complicated system composed of different entities such as platelets, the vascular wall, and multiple molecules with a variety of functions. There is a complex system of reactions that work in symphony to keep this vital system at bay. Failure of one of those functional units can lead to severe derangement of the entire system. It is important to have some basic knowledge of the system. Many of the drugs we use daily affect certain portions of the system, and we use them for exactly that reason.

## A. INTRODUCTION TO THE HEMOSTATIC CASCADE

### COAGULATION CASCADE

The final product of the coagulation cascade is fibrin. Fibrin is the product of conversion of fibrinogen by thrombin locally, at the site of injury.

The hemostatic system is divided into two pathways based on the nature of the initial stimulus and the first steps in the pathway. A short overview follows, but for a more in-depth discussion of this complex system we refer to hematology textbooks (Fig. 14.1).

The intrinsic pathway or contact activation pathway consists of several plasma proteins activated by contact with negatively charged surfaces within the bloodstream such as prekallikrein, high-molecular-weight kininogen, and factor XII (Hageman factor). Deficit of those factors is known but does not seem to affect coagulation. The importance of the contact pathway in the activation of coagulation therefore appears to be relatively unimportant *in vivo*.

The extrinsic pathway or tissue factor pathway is better understood, and deficits in components of this pathway as well as in the common pathway are well known. Factors V, VII, VIII, IX, and X all are known to cause bleeding problems if deficient or abnormal. Factor XI deficit, the first step of the intrinsic pathway, has also been described and demonstrated to cause bleeding problems during surgery. The extrinsic pathway is activated by tissue factor, a protein that is expressed on cells that are predominantly in the extravascular space when exposed to the bloodstream after vascular injury. On injury, tissue factor becomes exposed to factor VIIa and initiates the extrinsic pathway (Fig. 14.1). Thrombin activates platelets at the site of injury in addition to its action in converting soluble fibrinogen to insoluble fibrin.

Fibrin is finally stabilized by factor XIIIa, which is activated by thrombin and facilitates cross-linkage of the fibrin strands and makes the mesh tighter.

There are several factors whose function it is to inhibit the spontaneous progression of the cascade. Protein C is the best known, with its cofactor, protein S. Protein C inactivates factors VIIIa and Va. There are known inherited disorders of protein S and protein C that lead to hypercoagulability (Fig. 14.2).

Antithrombin III (ATIII) is a protein that circulates in the bloodstream and is the most important inhibitor of coagulation. It inhibits factors IIa, IXa, Xa, XIa, and XIIa. ATIII achieves this by binding to the factors and tying up the active site. ATIII action is increased 1,000-fold by the

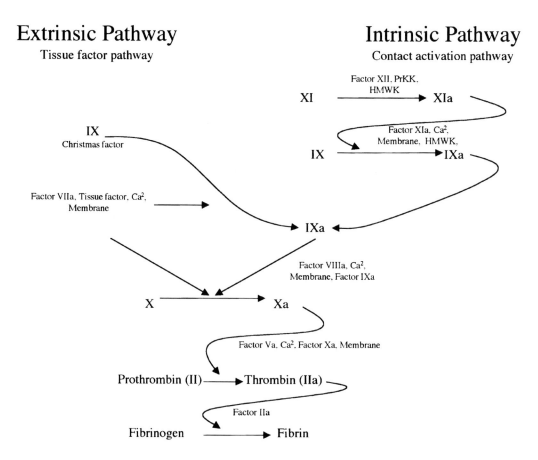

**FIG. 14.1.** Extrinsic and intrinsic coagulation cascades. Inactive clotting factors are denoted by Roman numerals; activated clotting factors are indicated by Roman numeral plus a. The identities of the clotting factors are listed below. All factors promote coagulation with the exceptions of proteins C and S. Protein C inhibits factors V and VIII, and protein S serves as a cofactor in this inhibition. These inhibitory functions are indicated by a minus sign. [I, fibrinogen; II, prothrombin; III, tissue factor; IV, calcium ($Ca^2$); V, proaccelerin (labile factor); VI, not assigned; VII, proconvertin (stable factor); VIII, antihemophilic factor; IX, Christmas factor; X, Stuart factor; XI, plasma thromboplastin antecedent; XII, Hageman factor; PrKK, prekallikrein; HMWK, high-molecular-weight kininogen.]

presence of heparin sulfate, and it is through this effect that heparin acts (Fig. 14.3).

More in-depth discussions of the complex interrelations between the coagulation pathways and the kallikrein and complement systems may be found in most standard textbooks of internal medicine.

Whatever the initial stimulus (intrinsic or extrinsic), the final common pathway of the two cascades is the conversion of soluble fibrinogen to fibrin by the action of the enzyme thrombin. Insoluble fibrin forms the matrix for hemostatic plugs in instances of vascular injury and the matrix that solidifies thrombi in the process of thrombosis. Fibrin also provides the framework for the reparative connective tissue. On this framework, healing proceeds, with fibroblastic proliferation and ingrowth of capillaries.

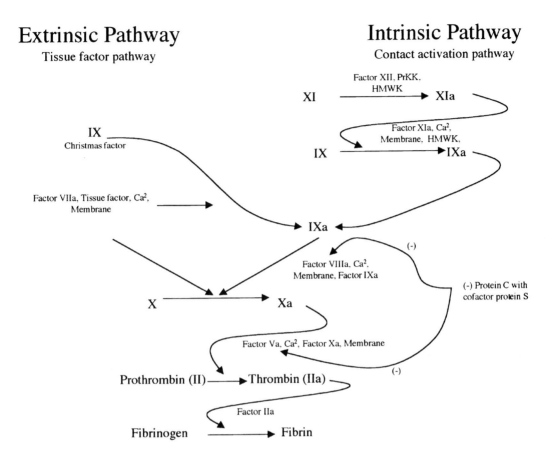

**FIG. 14.2.** The casket. Protein C and protein S have been added so one can put those into perspective. [I, fibrinogen; II, prothrombin; III, tissue factor; IV, calcium (Ca$^2$); V, proaccelerin (labile factor); VI, not assigned; VII, proconvertin (stable factor); VIII, antihemophilic factor; IX, Christmas factor; X, Stuart factor; XI, plasma thromboplastin antecedent; XII, Hageman factor; PrKK, prekallikrein; HMWK, high-molecular-weight kininogen.]

If the activated coagulation mechanisms were allowed to operate unchecked, the entire human circulation would coagulate after a single vascular injury. This is avoided, at least in part, through the action of various plasma coagulation factor inhibitors. These include $\alpha_2$-macroglobulin, $\alpha_2$-antiplasmin, ATIII (see above), $\alpha_1$-antitrypsin, and C1 inactivator. The coagulation mechanism may also be inhibited by various other pharmacologic means: (a) dicumarols, which are vitamin K antagonists that interfere with the hepatic synthesis of the vitamin K–dependent clotting factors (II, VII, IX, and X); (b) heparin, which blocks the action of thrombin on fibrinogen; (c) antiplatelet agents such as acetylsalicylic acid and dipyridamole, which decrease platelet function; (d) fibrinolytic agents such as urokinase and streptokinase; and (e) certain snake venoms, which cause fibrinogen depletion.

## B. ANTICOAGULATION AGENTS

### HEPARIN

Heparin is well known by all of us. It is a normal product of the human body found in many tissues and especially in mast cells. It is a proteoglycan with high molecular weight. The drug's effect is through its action on proteins known as *serpins*, the best known of which are ATIII and heparin cofactor II.

ATIII is a proteolytic inhibitor produced by the liver, but it is vitamin K independent. It has its effect by irreversibly neutralizing factors IXa, XIa, XIIa, and especially Xa and thrombin (factor IIa) (Fig. 14.3). Heparin increases the effect of ATIII by a factor of 1,000 to 10,000. Heparin is then released from the complexes and can affect other complexes in the same manner.

The anticoagulation effect of heparin is measured by coagulation tests. Usually, activated partial thromboplastin time is used, but activated clotting time can be used when high doses are being used as in cardiovascular intervention.

The biological half-life of heparin is not constant but dose dependent. If 25 IU per kg is given, the half-life is 30 minutes but will increase to 150 minutes if 400 IU per kg is given. Protamine sulfate can be used to reverse the heparin effect. One milligram of protamine will reverse 100 IU of heparin.

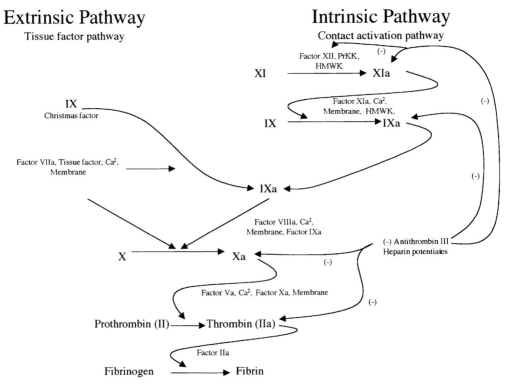

**FIG. 14.3.** The casket. This figure has antithrombin III (ATIII) included. Heparin acts through ATIII. The effect of ATIII is wide. [I, fibrinogen; II, prothrombin; III, tissue factor; IV, calcium ($Ca^2$); V, proaccelerin (labile factor); VI, not assigned; VII, proconvertin (stable factor); VIII, antihemophilic factor; IX, Christmas factor; X, Stuart factor; XI, plasma thromboplastin antecedent; XII, Hageman factor; PrKK, prekallikrein; HMWK, high-molecular-weight kininogen.]

**TABLE 14.1.** *Vitamin K–dependent coagulant factors produced by the liver*

| Factor | Half-life (hr) |
|---|---|
| Factor VII | 6 |
| Protein C | 8 |
| Factor IX | 24 |
| Protein S | 30 |
| Factor X | 36 |
| Prothrombin | 72 |

Low-molecular-weight heparin is now available. It is given subcutaneously and cannot be given intravenously. The doses are more standardized than for regular heparin, and routine monitoring is not required. Low-molecular-weight heparin does not influence the activated partial thromboplastin time, and if measures are needed, one has to depend on direct heparin activity measurements (usually the anti-Xa activity assay).

## COUMARIN DRUGS

The coumarin drugs have been in clinical use since the 1940s. Warfarin (coumarin) and dicumarol are the best known. The drug inhibits production of vitamin K–dependent coagulant factors produced by the liver. The factors are listed in Table 14.1, together with the half-lives of those factors. Note that protein C is also a vitamin K–dependent factor.

The drug is given orally, and the starting dose is usually 5 to 10 mg. The maintenance dose is usually between 5 to 20 mg per day.

Monitoring is done with measurements of international normalization ratio. If the international normalization ratio is high or if bleeding occurs, the effect can be reversed with infusion of fresh-frozen plasma or by giving vitamin K. One to 3 mg of vitamin K can be given subcutaneously. Vitamin K can also be given orally.

## C. THROMBOLYTIC AGENTS

Fibrin dissolution is a proteolytic process that results in the formation of soluble fibrin degradation products (fragments X, Y, D, and E). The process is mediated by plasmin, a relatively nonspecific plasma protease formed from the zymogen (inactive precursor) plasminogen (Fig. 14.4). Normally, plasminogen is found in the circulation in concentrations of 12 to 25 mg per dL, whereas plasmin is not detectable. The latter may, however, be detected in certain disease states. Several factors are responsible for controlling this process, such as plasminogen activator inhibitor and antiplasmin (Fig. 14.4). The process and control of fibrinolysis are complicated, and in-depth discussion is beyond the reach of this short introduction.

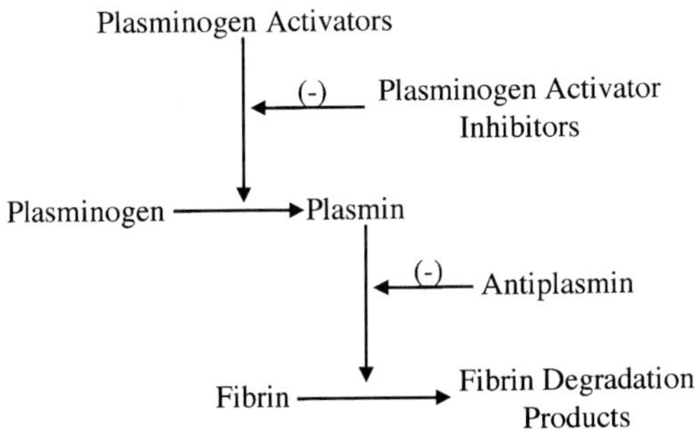

### STREPTOKINASE

Streptokinase is a single-chain protein with 414 amino acids produced by beta-hemolytic streptococci. Streptokinase has no fibrin specificity and causes systemic fibrinolysis. The drug's biological half-life is 1 to 2 hours. The drug may induce antibody formation and 6 months should elapse between administrations. The liver degrades the drug.

### UROKINASE

Urokinase is a protein product of the normal human kidney that is found in trace quantities in the circulation and is available commercially for pharmacologic administration as a fibrinolytic agent. Commercial laboratories derive the substance from cultures of human fetal kidney cells. Unlike streptokinase, urokinase does not form active complexes with plasminogen to produce its fibrinolytic effect. Rather, it acts

**FIG. 14.4.** The basic fibrinolytic casket.

directly on plasminogen to produce plasmin. Like streptokinase, it also activates circulating plasminogen molecules to produce plasmin, which in turn results in systemic fibrinolysis.

The drug was the most commonly used thrombolytic drug until 1999, when it was recalled from the market; it is currently not available here in the United States.

### TISSUE PLASMINOGEN ACTIVATOR

Tissue plasminogen activator (t-PA) is a protein that can be found in human tissue such as the endothelium. The native t-PA is a single-chain molecule. It is transformed into a double-chain form by plasmin. The catalytic efficiency of t-PA on plasminogen activation increases 1,000-fold in the presence of fibrin. This led people to believe that very little systemic lysis would occur with t-PA. Since then, we have learned that one form of the fibrin-degrading products, (DD)E, is almost as potent as fibrin as a stimulator for the t-PA catalyzed conversion of plasminogen to plasmin, which explains the fact that t-PA causes systemic effects. This drug has mainly been used for treatment of myocardial infarct and pulmonary emboli, but because urokinase is not available in the United States, t-PA is increasingly used for peripheral artery thrombolysis and venous thrombolysis. The half-life for the drug is 5 minutes, and the liver breaks down the drug.

### PRO-UROKINASE

Pro-urokinase (prUK) is a proenzyme with little enzymatic activity itself. The drug is a single-chain glycoprotein, which is split into a two-chain active molecule by plasmin or kallikrein. prUK has no fibrin affinity but causes little systemic fibrinolysis and has indirect fibrin specificity through plasminogen selection. The PURPOSE trial took a look at the dose range for prUK in the treatment of peripheral arterial thrombosis and compared three dose regimens of prUK to urokinase. None of the prUK doses did significantly better or worse than urokinase alone. The rate of bleeding complications was slightly higher, and the rate of lysis was slightly increased in the highest dose of prUK (8 mg per hour) as compared with the other two doses (2 mg and 4 mg per hour) and urokinase. There was also faster decrement in fibrinogen in that dose group (8 mg per hour). This drug is undergoing clinical trials and might be on the market in the not-too-distant future.

## D. HYPERCOAGULABLE STATE

Several conditions cause the blood to be especially prone to coagulation. Here, we mention only a few.

### PROTEIN C/PROTEIN S DEFICIENCY

Both protein C and protein S inhibit or slow down the coagulation cascade. Both proteins are vitamin K dependent. Protein C is a plasma protein with protease activity and is activated on contact with thrombin when it is bound by thrombomodulin, an endothelial receptor. Protein S is a membrane-bound cofactor of protein C. When protein C is activated it causes inactivation of factors Va and VIIIa, causing anticoagulation effect (Fig. 14.2). Activated protein C also promotes thrombolysis. Patients with low levels of protein C, protein S, or both have increased tendency to thrombosis, especially venous thrombosis.

### FACTOR V LEIDEN

Factor V Leiden is a condition that causes resistance to the inhibitory effect of protein Ca (also called *activated protein C resistance*) on factor Va. It is caused by a defect in factor V, causing significant increased risk of venous thrombosis. Factor V Leiden is the most common known inherited risk factor in the general population for venous thrombosis.

### ANTITHROMBIN III DEFICIENCY

ATIII is a proteolytic inhibitor produced by the liver that is vitamin K independent. It has its effect by irreversibly neutralizing factors IXa, Xa, XIa, and XIIa and thrombin. Heparin increases the effect of ATIII dramatically.

ATIII deficiency is quite uncommon in the general population, occurring in 1 in 2,500 to 1 in 5,000. Approximately 50% (15% to 100%) of patients with ATIII deficiency will develop deep vein thrombosis, but only 2% will experience arterial thrombosis.

ATIII concentrate can be given but is expensive and has to be given intravenously.

## HOMOCYSTINURIA

Patients with homocystinuria have been found to have an increased prevalence of venous thrombosis. Those patients are also at increased risk for peripheral vascular, cerebrovascular, and coronary artery disease. Sixty percent of patients with homozygous defect of cystathionine synthase develop thromboembolism by the age of 40 years. Homocystine decreases the function of thrombomodulin at the endothelial cell level and thereby decreases the conversion of protein C to activated protein C. It also seems to lead to decreased conversion of plasminogen to plasmin by decreasing the binding of t-PA to the endothelium. The treatment is high doses of vitamin $B_6$. More common in the general population is hyperhomocysteinemia, in which there is only a mild elevation in plasma homocysteine levels. This has also been found to be a mild risk factor for venous thromboembolism and premature atherosclerosis.

## ANTIPHOSPHOLIPID SYNDROME

Antiphospholipid syndrome is due to circulating autoantibodies to negatively charged phospholipids. Cardiolipin is the best known of these phospholipids. There is increased incidence of venous and arterial thrombosis and fetal abortion. The condition is seen in up to one-half of patients with lupus or lupuslike disorders. The autoantibodies are thought to either inhibit endothelial cell prostacyclin production or block protein Ca inactivation of factors Va and VIIIa.

# E. THROMBOLYSIS

It has been stated that the evolution of fibrinolytic therapy has come in generations. The first generation was characterized by infusion of relatively high doses of the fibrinolytic agent intravenously. The thrombi or thromboemboli that were intended to be lysed were usually remote from the site of infusion.

Streptokinase was the initial drug used for thrombolytic therapy in this era, but urokinase was later identified as the preferred drug for this use.

Most of the experience of using thrombolytic agents came from large, randomized studies in which thrombolytic agents were evaluated for treatment of pulmonary emboli. All these studies demonstrated a relatively high bleeding complication rate in the group of patients treated with fibrinolytic agents. Dr. Dotter then developed the second-generation therapy. His approach was characterized by infusion of streptokinase locally into the clot to be treated. Significantly smaller doses of the thrombolytic agent were needed.

## THROMBOLYSIS TECHNIQUES AND SPECIFIC AREAS

### Thrombolysis for Acute Lower Extremity Arterial Thrombosis

Thrombolysis for peripheral arterial thrombosis and visceral artery thrombosis is now performed using catheter-directed infusion of the drug. The infusion catheters have a segment with multiple side holes that are buried in the thrombus, and the agent is then infused for hours or days.

McNamara and Fischer proposed a regimen for peripheral arterial and graft occlusions. They used infusion of 4,000 IU per minute of urokinase until antegrade blood flow was reestablished and then 2,000 IU per minute until lysis was complete. They did report clinical improvement in 81% of the patients. This regimen was adapted by many interventionists and was one of the most commonly used protocols for urokinase use. McNamara and Fischer used systemic heparin infusion to keep the partial thromboplastin time (PTT) at approximately three times normal and were able to decrease the incidence of catheter-related thrombosis to 3%. Bleeding complications were not frequent.

The Thrombolysis and Peripheral Arterial Surgery trial compared three dose regimens for urokinase infusion and found that the most effective and safest regimen was an infusion of 240,000 IU per hour of urokinase for 4 hours and then 120,000 IU per hour for a maximum of 48 hours. This regimen has become very widely used.

t-PA has not been widely used in the United States for treatment of peripheral arterial thrombolysis. A standard dose regimen has not yet been established. Because urokinase is not available in the United States, there is increased interest in the use of t-PA. No dose study has been performed for t-PA, but currently the most common dose is 0.05 mg per kg per hour as a continuous infusion into the thrombus through an infusion catheter. Doses as small as 0.005 mg per kg per hour have been used with success, but these results have not yet been published.

### Pulsed-Spray Pharmacomechanical Thrombolysis

An enhanced method of thrombolysis was proposed by Bookstein and Valji. It consists of forceful spraying of a thrombolytic agent such as urokinase into the thrombus in short pulses along with systemic heparinization. The combination of mechanical disruption of the clot by the advancement of the infusion catheter and the spraying of the fibrinolytic agents into the thrombus reduced clot lysis time and decreased the risk of hemorrhage.

The method is based on use of a 5-Fr multiside-hole catheter specially designed for this purpose, with multiple, closely spaced side slits over its distal segment placed across the thrombosed segment. This segment comes in different lengths (Angiodynamics, Inc., Glens Falls, NY). The catheter tip is placed just proximal to the distal end of the thrombus to prevent embolization of distal thrombi at the start of the procedure. A 0.032-in. end-occluding wire is used to occlude the end hole. The solution consists of 250,000 IU of urokinase diluted in 9 mL of sterile, degassed water with 1 mL of heparin (5,000 IU in 1 mL) was used. This is now replaced with approximately 1 to 3 mg of t-PA. The entire thrombus is initially sprayed with pulses of 0.2 to 0.3 mL each, beginning with the distal portion of the thrombus, sparing the distal 2 cm of the thrombus plug. Two pulses are given every minute. If the thrombus is longer than the infusion segment of the catheter, the catheter is withdrawn every 5 to 10 minutes.

This method has not gained general acceptance for use in peripheral arterial thrombolysis except in cases in which rapid thrombolysis is needed and low doses of the drug are important. The method requires larger resources and long time in the angiographic suit, which has made it less appealing than traditional infusion treatment. The actual rate of thrombolysis has been debated and most experience came from treatment of clotted dialysis accesses, in which the method gained wide acceptance until mechanical devices replaced use of thrombolytic agents.

### Lacing

Lacing was proposed by Sullivan et al. to accelerate the rate of thrombolysis. The technique consists of the intrathrombus deposition of either low- or high-dose urokinase, using a coaxial method. A diagnostic 5- or 6-Fr catheter is advanced through the thrombus. Urokinase mixed with 10 to 20 mL of saline is injected into the thrombus as the catheter is being withdrawn.

Sullivan et al. found that by lacing the thrombus with a high dose (230,000 IU) of urokinase, the time to lysis was decreased significantly and the total dose required to achieve lysis is reduced significantly as well.

The incidence of major complications was 22.9%. The rapid lysis is attributed to the faster conversion of plasmin into plasminogen within the thrombus. The plasmin in the thrombus has a longer life than in the circulation because in the presence of fibrin it is protected from its inhibitor, $\alpha_2$-antiplasmin. This method has kept some loyal users but is not widely used.

### Grafts

It has long been recognized that fibrinolysis of bypass grafts is both faster and more frequent than is thrombolysis of native artery thrombosis. One of the reasons for this is that there are no collaterals where the fibrinolytic agents can egress away from the clot. The most recent experience indicates that the first line of treatment for occluded grafts is thrombolysis (Fig. 14.5).

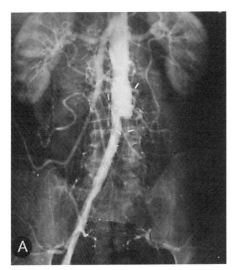
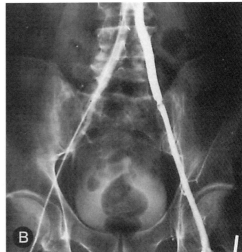
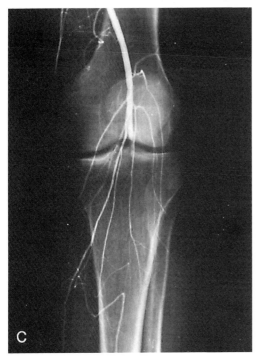

**FIG. 14.5.** This 56-year-old man with a 5-year-old aorto-iliac graft presented with new left lower extremity ischemia. **A:** Initial angiogram discloses complete occlusion of left limb of graft. A catheter was placed over the bifurcation into the occlusion, and streptokinase infusion at 5,000 IU per hour was started. **B:** Angiogram 19 hours after initiation of streptokinase shows patency of the graft limb and stenosis at the distal anastomoses. **C:** Angiogram of the popliteal area at the same time shows complete mid–popliteal artery embolic occlusion. (*Continued.*)

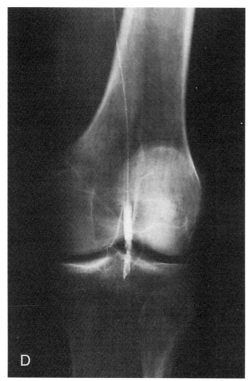
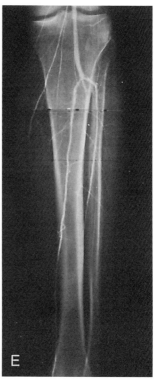

**FIG. 14.5.** *Continued.* **D:** The iliac artery stenosis was treated with balloon angioplasty, and a 3-Fr coaxial catheter was placed into the popliteal occlusion infusion continued there. **E:** Angiogram after another 24 hours shows resolution of the embolus.

The advantages of thrombolytic therapy include the following:

- It uncovers the underlying stenotic lesion at the anastomotic sites or in the native vessels that have contributed to the occlusion.
- The anatomic detail obtained by arteriography is helpful for surgical repair or during a percutaneous angioplasty.
- It is less traumatic compared with mechanical extraction by Fogarty catheters.
- Thrombi in small vessels distal to the graft anastomosis are better managed by thrombolytics compared to a surgical thrombectomy.
- Thrombolytic therapy converts an emergency to an elective procedure.
- The longtime mortality and morbidity are probably decreased for those patients who undergo fibrinolysis, even if a surgical procedure is also needed.

### Thrombolysis versus Surgery for Lower Extremity Occlusive Disease

Three papers in the recent past have specifically focused on the role of thrombolysis in acute to subacute (less than 6 months) thrombosis of native arteries or bypass grafts in the peripheral circulation (iliac, femoral, popliteal, and leg) as compared with surgical revision: the Rochester study (Ouriel), the STILE study, and the Thrombolysis and Peripheral Arterial Surgery trial study. Those studies came to the following conclusion: The number and magnitude of surgical procedures are reduced by 40% to 60% if thrombolytic treatment is given initially. One-year limb salvage is enhanced, and the recurrent ischemia is less in the cohort treated with thrombolytic therapy for acute thrombosis (less than 14 days). Two of the studies indicated that there were less mortality and morbidity from cardiovascular incidents at 30 days and 1 year in the group treated with thrombolysis initially (Rochester, STILE).

*Upper Extremity*

Most cases of hand ischemia are caused by atherosclerotic occlusion with superimposed thrombosis, emboli, or trauma, whereas only a small number are due to Raynaud's disease, connective tissue disorders, or thoracic outlet syndrome.

Approximately 4% to 12% of patients with thoracic outlet syndrome present with subclavian vein or subclavian artery thrombosis.

Although two-thirds of the cases of hand ischemia are due to occlusions proximal to the wrist, which are potentially amenable to operation, the remaining one-third are due to occlusion of the small arteries of the hand and fingers, a type in which surgical reconstruction is often difficult or impossible. Spasm is rarely an important factor, and cervicodorsal sympathectomy is of little value.

There is therefore ample room for fibrinolytic therapy in the treatment for acute upper extremity thrombosis, and there may be an indication for treatment of a more chronic arm ischemia with fibrinolytic agents.

Thrombolytic treatment is preceded by a careful anatomic evaluation. This should involve complete upper extremity angiography, including arch aortography, as an initial step. Pharmacoangiography using 25 to 50 mg of tolazoline, 50 mg of papaverine, 0.1 mg of nitroglycerin, or 50 to 100 mg of lidocaine is used to identify fixed vascular occlusions. Vasodilators are not needed when initial forearm angiography demonstrates intraluminal filling defects. A transfemoral approach should be used in most instances, even though some authors have advocated the use of an antegrade puncture of the brachial artery for placement of the infusion catheters. The rationale behind the recommendation of this approach was the occurrence of periprocedural transient ischemic attack, which was thought to be related to catheter crossing near the origin of the cerebral arteries. Most authors use the femoral approach for the infusion catheters.

The same infusion methods apply for the treatment of thrombus in the upper extremities. If the thrombus is in the hand and the smaller vessels, a microcatheter can be placed coaxial as close to or into the thrombus and the drug given directly into the thrombus. Vasodilators, heparin, or additional thrombolytic agent can be given via the outer guiding catheter upstream.

Lambiase et al. treated six patients with acute (less than 30 hours) symptoms of upper extremity thromboembolic disease. Five patients had complete symptomatic recovery and a very good technical success. In a patient who had fixed ischemic changes of several fingers with profound motor impairment before the procedure, the fingers had to be amputated, and there was persistent motor impairment.

*Mesenteric Arteries*

There are only few reports in the literature of fibrinolytic treatment for acute ischemic bowel, and therefore the role of this treatment for the condition is not well defined.

In most instances a common femoral approach has been used and a multi-side-hole infusion catheter has been placed across the thrombosed area. All three available thrombolytic agents have been successfully used.

Patients with findings suggestive of bowel infarction will require surgical intervention, and careful clinical evaluation of the patient must be conducted before thrombolytic therapy is initiated. Certain patients with ischemic bowel may respond to local fibrinolytic treatment.

*Renal Arteries*

Renal artery thrombosis has been successfully treated using catheter-directed infusion of fibrinolytic agents. The renal artery is an end artery in the sense that intrarenal collateral practically does not exist, and only after long-standing arterial stenosis or occlusion does collateral develop. The kidney is irreversibly damaged after 90 minutes of warm ischemia. A metaanalysis of 97 reported renal artery occlusions in the literature revealed that 50 kidneys in which complete main renal artery occlusion had been present for more than 3 hours all resulted in irreversible damage of the renal parenchyma. Only when the ischemia lasted less than 3 hours (three cases) was complete recovery of renal function without loss of renal parenchyma observed. Patients with partial occlusion of the main renal artery or segmental renal artery occlusion fared better than patients with complete main renal artery occlusions. The authors concluded that after complete blockage of the main renal artery for more than 90 minutes ("the ischemic tolerance of the renal parenchyma"), the renal function is partially or completely lost. Therefore, immediate surgical or medical treatment of acute complete or incomplete occlusion of the main renal artery or segmental branches does not improve the clinical outcome once the ischemic tolerance of the kidney has been exceeded.

## Venous

### Iliofemoral Venous Thrombolysis

Although acute morbidity from venous thrombosis, mainly pulmonary emboli, is quite significant, the chronic morbidity may be no less important, even though less obvious. Up to two-thirds of patients with acute deep vein thrombosis will develop one or more of the symptoms referred to as *postthrombotic syndrome* as time passes.

It has been demonstrated that normal valves in the distal superficial femoral vein, popliteal vein, and calf veins will prevent ulcer formation and that malfunction of those is the most important risk factor for development of postthrombotic syndrome.

Multiple, small studies have compared systemic thrombolytic therapy with conventional anticoagulation therapy for acute deep vein thrombosis of the lower extremity. All these studies have in common that the clinical symptoms have improved much earlier, and venographic resolution has also been more complete in the thrombolytic group than in the anticoagulation group. The bleeding risk is higher in patients treated with systemic thrombolytic therapy than in those treated with anticoagulation alone.

A small study by Arnesen and colleagues demonstrated impressive differences in clinical findings and venographic findings 6.5 years later.

Catheter-directed delivery has been used in the past to decrease the rate of bleeding complications.

### Inclusion Criteria

- Patients with iliofemoral deep vein thrombosis or isolated iliac or femoral vein thrombosis
- Age of thrombus of less than 10 days (this is not absolute, and much older thrombi have been successfully treated)

### Exclusion Criteria

- Any contraindication to anticoagulation.
- Intracranial malignancy or arteriovenous malformation.
- Recent intracranial or spinal surgery.
- Uncorrectable coagulation disorder (relative contraindications).
- Pregnancy or very recent delivery.
- Recent major surgery (relative contraindications).
- Complete contraindication in recent intrathoracic or intraabdominal surgery.
- Peripheral vascular or orthopedic surgery may be a less strong contraindication.

Either the popliteal or posterior tibial vein at the ankle level is used for access. Earlier access was gained via the internal jugular vein but this is now used only exceptionally.

Multi-side-hole infusion catheters are used. With the availability of long infusion catheters one attempts to infuse the entire length of the thrombus from the beginning. The dose of urokinase was approximately 2,000 IU per kg per hour, which usually comes to 150,000 units per hour. t-PA doses are 0.05 to 0.005 mg per kg per hour, which comes to 0.35 to 3.50 mg per hour in a 70-kg person. Full anticoagulation is given during the procedure and the PTT is kept at approximately 60 to 80 seconds.

Infusion is then continued in the intensive care unit or in a step-down unit. The patients are taken back to the radiology suite every 12 to 24 hours for contrast evaluation and readjustment of infusion catheters. At the same time, adjunctive therapy such as angioplasty, stent placement, or mechanical thrombectomy can be carried out.

Complete cleaning of thrombus is the ultimate objective, but 30% residual lumen narrowing is accepted if there are no collaterals and no pressure gradient across the area (acceptable pressure gradient is 2 to 3 mm Hg). Chronic thrombosis or narrowing caused by compression or tumor ingrowth is treated with endoluminal stents. Angioplasty is usually of little help in treating residual narrowing. Metal stents are used in the treatment of 45% to 50% of patients (Fig. 14.6).

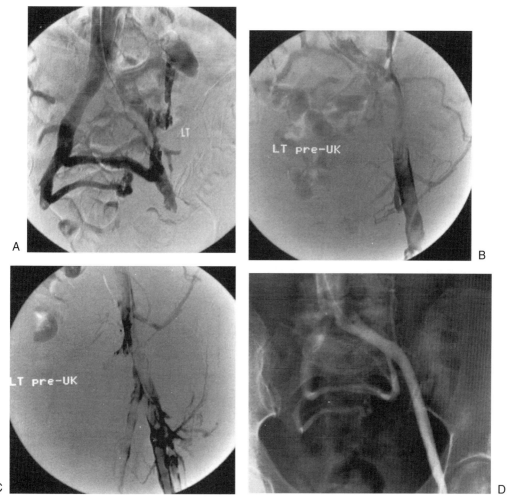

**FIG. 14.6.** This 65-year-old woman presented 1 week after arthroscopy on her left knee with swelling of the left lower extremity. **A:** A venogram of the common iliac vein, using a right internal jugular vein approach, reveals thrombosis of the vein. **B:** The catheter has now been advanced into the distal external iliac vein, which is also thrombosed. **C:** Thrombosis of the common femoral, superficial femoral, and greater saphenous veins is demonstrated. **D:** After infusion of urokinase (UK) over approximately 48 hours and balloon angioplasty of the proximal common iliac vein, the iliac and femoral veins were completely open and no thrombus was seen. There was no pressure gradient across the proximal common iliac vein where an impression caused by the right common iliac artery can be seen. (LT, left.)

A multicenter registry was conducted, and Mewissen et al. published the results. Four hundred seventy-three patients were included. Urokinase was used, with a mean dose of 7.8 million IU. The mean treatment time was 53.4 hours. Major bleeding complications were encountered in 11% of patients. Six patients (1%) developed pulmonary emboli. Two patients died as a consequence of the treatment from intracranial bleeding and pulmonary emboli. Overall, more than 50% lysis was achieved in 83%. The 1-year primary patency rate was 60%. The authors conclude by commenting that catheter-directed thrombolysis is safe and effective and that the information from this registry can help to select which patients would benefit from thrombolytic therapy.

### *Upper Extremity Deep Vein Thrombosis*

Upper extremity deep vein thrombosis is usually divided into two groups: primary thrombosis and secondary thrombosis.

Primary thrombosis, or idiopathic axillary and subclavian vein thrombosis, is also referred to as *thoracic outlet thrombosis* or *effort thrombosis*. The thrombosis is due to chronic injury to the vein as it passes over the first rib behind the anterior scalenus muscle, where it is compressed by the clavicular bone and by the costoclavicular ligament. The subclavius muscle may also play a role in the pathogenesis.

Secondary thrombosis has become more common because of increased use of central vein catheters. Other causes of secondary thrombosis are malignant disease with compression of the subclavian vein, pacemaker wires, and trauma.

The accepted treatment regimen has been anticoagulation treatment or fibrinolytic treatment. The reported rates of disability vary from 17% to 75% without thrombolytic treatment.

At the University of Minnesota, the treatment is based on catheter-directed thrombolysis. A catheter is placed from an upper arm vein if possible. The basilic vein or the brachial vein is usually selected, but the cephalic vein is less preferable, as it will enter the central veins too proximally. A multi-side-hole infusion catheter is placed and urokinase or t-PA is infused at the same dose as above. The patients are treated with anticoagulation concomitantly, keeping the PTT at approximately 60 to 80 seconds. Heparin is usually infused through a sidearm of the access introducer. Common femoral vein access can also be selected if for some reason an arm access is not applicable.

At the University of Minnesota, 67 thrombolytic treatments were given to 61 patients from 1984 to 1994. Sixty-four percent of the patients had catheter-related thrombosis, but 25% had primary thrombosis. The average treatment time was 37 hours. An arm access was used in 64% of the cases, and in 16% access was gained from the common femoral vein. In 20%, combined access was used. The overall success rate was 77.6%. If the thrombus was acute, the success rate was 92%; in chronic thrombosis the success rate was 69%. Stents were placed in only six patients. Angioplasty was routinely used.

Several smaller studies have been published on effort thrombosis, and the immediate thrombolytic success has varied between 40% and 100%.

There is consensus among surgeons that those patients with thoracic outlet syndrome and thrombosis should have the first rib resected and the anterior scalenus muscle removed after successful thrombolysis. A surgical venoplasty is also performed at the same time.

## F. LABORATORY MONITORING

### LABORATORY TESTS DURING THROMBOLYTIC TREATMENT

During thrombolytic therapy, certain laboratory parameters should be monitored. Hemoglobin, hematocrit, and platelet levels should be measured every 6 to 12 hours. Platelet levels may fall secondary to heparin-induced thrombocytopenia, and a fall in hemoglobin and hematocrit will signal clinically silent bleeding. During the infusion treatment, there will be some degradation of fibrinogen and many centers measure the serum fibrinogen every 12 to 24 hours. There are no studies to support this practice, but in the STILE study, patients with hemorrhagic complications had significantly lower fibrinogen levels at the end of treatment than those without bleeding complications. If the fibrinogen falls below 150 mg per dL, thrombolytic treatment can be halted until the fibrinogen level resolves or cryoglobulin or fresh frozen plasma can be given while continuing treatment.

## G. COMPLICATIONS

The complications for venous and arterial thrombolysis are similar: (a) hemorrhage, (b) allergic reaction, (c) embolization, (d) catheter-related thrombosis, (e) extravasation, (f) myoglobinuria, and (g) renal failure.

### HEMORRHAGIC COMPLICATIONS

Bleeding complications for arterial thrombolysis are reported to be from 8% to 30%. Serious complications, often defined as complications requiring blood transfusion, surgical intervention, or a major change in therapy, range from 8% to 13%.

Most complications are related to the introducer sheath. Most commonly there are subcutaneous hematomas at the puncture site, but occasionally, in the case of high puncture where the puncture has gone through the inguinal ligament, a retroperitoneal hematoma may result. Intracranial hemorrhage was quite significant in some of the systemic fibrinolytic literature but seems to be relatively rare with local therapy (0.1% to 1%).

### DISTAL EMBOLI

Distal embolization from an occluded native artery or arterial bypass has been estimated to be 8% to 15%. Most of the distal emboli are symptomatic. They can usually be treated with continuous urokinase infusion, and it is typical for a patient treated with a fibrinolytic agent for an arterial bypass to develop increasing pain 2 or 3 hours after the initiation of treatment. This is due to emboli breaking off the thrombus, migrating distally, and causing symptoms there. This is usually treated successfully by continuing fibrinolytic treatment, sometimes with larger doses.

Symptomatic pulmonary emboli are very rare with venous thrombolysis. A large venous registry and other smaller case studies have not reported on pulmonary emboli as a frequent complication.

### PERICATHETER THROMBOSIS

A 26% incidence of pericatheter thrombosis has been reported with catheter-directed fibrinolytic therapy. The incidence of this has decreased significantly with use of concomitant heparin administration. With the use of t-PA, there is a concern that cerebral bleed incidence may be increased with concomitant heparin. Many centers have therefore moved away from concomitant heparin infusion when using t-PA.

## SUGGESTED READINGS

Arnesen H, Heilo A, Jakobsen E, et al. A prospective study of streptokinase and heparin in the treatment of deep vein thrombosis. *Acta Med Scand* 1978;203:457–463.

Badiola CM, Scoppetta DJ. Rapid revascularization of an embolic superior mesenteric artery occlusion using pulse-spray pharmacomechanical thrombolysis with urokinase. *AJR Am J Roentgenol* 1997;169:55.

Becker GJ. Second-generation fibrinolytic therapy: state of the art. *Semin Intervent Radiol* 1985;2:409.

Belli A-M. Thrombolysis in the peripheral vascular system. *Cardiovasc Intervent Radiol* 1998;21:95.

Berridge DC, Gregson RHS, Makin GS, Hopkinson BR. Tissue plasminogen activator in peripheral arterial thrombolysis. *Br J Surg* 1990;77:179–181.

Bjarnason H, Kruse JR, Asinger DA, et al. Iliofemoral deep venous thrombosis: safety and efficacy outcome during 5 years of catheter-directed thrombolytic therapy. *J Vasc Intervent Radiol* 1997;8:405–418.

Blum U, Billmann P, Krause T, et al. Effect of local low-dose thrombolysis on clinical outcome in acute embolic renal artery occlusion. *Radiology* 1993;189:549–554.

Bookstein JJ, Valji J. Pulse-spray pharmacomechanical thrombolysis. *Cardiovasc Intervent Radiol* 1992;15:228–233.

Clause ME, Stokes KR, Perry LJ, Wheeler HG. Percutaneous intraarterial thrombolysis: analysis of factors affecting outcome. *J Vasc Intervent Radiol* 1994;5:93–100.

Comerota AJ, Aldridge SC. Thrombolytic therapy for deep venous thrombosis: a clinical review. Can J Surg 1993;36:359–364.

Diffin DC, Kandarpa K. Assessment of peripheral intraarterial thrombolysis versus surgical revascularization in acute lower-limb ischemia: a review of limb-salvage and mortality statistics. J Vasc Intervent Radiol 1996;7:57.

Dotter CT, Rosch J, Seaman AG. Selective clot lysis with low-dose streptokinase. Radiology 1974;111:31–37.

Galland RB, Earnshaw JJ, Baird RN, et al. Acute limb deterioration during intra-arterial thrombolysis. Br J Surg 1993;80:1118–1120.

Gore JM, Granger CB, Simoons ML, et al. Stroke after thrombolysis: mortality and functional outcomes in the GUSTO-I trial. J Vasc Intervent Radiol 1996;7:304.

The International Working Group. Thrombolysis in the management of lower limb peripheral arterial occlusion: a consensus document. Am J Cardiol 1998;81:207–218.

Kandarpa K, Chopra PS, Aruny JE, et al. Intraarterial thrombolysis of lower extremity occlusions: Prospective, randomized comparison of forced periodic infusion and conventional slow continuous infusion. Radiology 1993;188:861–867.

Katzen BT. Mechanism and clinical application of fibrinolysis. Proceedings of the Joint Meeting of the European and American Societies of Cardiovascular and Interventional Radiology 1987:10.

Katzen BT, van Breda A. Low dose streptokinase in the treatment of arterial occlusion. AJR Am J Roentgenol 1981;136:1171–1178.

Koeleman BPC, Reitsma PH, Allaarat CF, Bertina RM. Activated protein C resistance as an additional risk factor for thrombosis in protein C-deficient families. Blood 1994;84:1031–1035.

Lambiase RE, Paolella LP, Haas RA, Dorfman GS. Extensive thromboembolic disease of the hand and forearm: treatment with thrombolytic therapy. J Vasc Intervent Radiol 1991;2:201–208.

LeBlang SD, Becker GJ, Benenati JF, et al. Low-dose urokinase regimen for the treatment of lower extremity arterial and graft occlusion: experience in 132 cases. J Vasc Intervent Radiol 1992;3:475–483.

Lonsdale RJ, Berridge DC, Earnshaw JJ, et al. Recombinant tissue-type plasminogen activator is superior to streptokinase for local intra-arterial thrombolysis. Br J Surg 1992;79:272–275.

Loscalzo J, Schafer AI, eds. Thrombosis and hemorrhage, 2nd edition. Baltimore: Williams & Wilkins, 1998.

McNamara TO, Fischer JR. Thrombolysis of peripheral arterial and graft occlusions: improved results using high-dose urokinase. AJR Am J Roentgenol 1985;144(4):769–775.

Mewissen MW, Seabrook GR, Meissner MH, et al. Catheter-directed thrombolysis for lower extremity deep venous thrombosis: report of a national multicenter registry. Radiology 1999;211:39–49.

Meyerovitz MF, Goldhaber SZ, Reagan K, et al. Recombinant tissue-type plasminogen activator versus urokinase in peripheral arterial and graft occlusions: a randomized trial. Radiology 1990;175:75–78.

Nachman RL, Silverstein R. Hypercoagulable states. Ann Intern Med 1993;119:819–827.

O'Donnell TF Jr, Browse NL, Burnand KG, Thomas ML. The socioeconomic effects of an iliofemoral venous thrombosis. J Surg Res 1977;22:483–488.

Ouriel K, Kandarpa K, Schuerr DM, et al. Prourokinase versus urokinase for recanalization of peripheral occlusions, safety and efficacy: the PURPOSE trial. J Vasc Intervent Radiol 1999;10:1083–1091.

Ouriel K, Shortell CK, Azodo MVU, et al. Acute peripheral arterial occlusion: predictors of success in catheter-directed thrombolytic therapy. Radiology 1994;193:561–566.

Ouriel K, Shortell CK, DeWeese JA, et al. A comparison of thrombolytic therapy with operative revascularization. J Vasc Surg 1994;19:1021–1030.

Ouriel K, Veith FJ. Acute lower limb ischemia: determinants of outcome. Surgery 1998;124:336–342.

Ouriel K, Veith FJ, Sasahara AA for the TOPAS Investigators. Thrombolysis or peripheral arterial surgery. (TOPAS): phase I results. J Vasc Surg 1996;23:64–75.

Pillari G, Doscher W, Fierstein J, et al. Low-dose streptokinase in the treatment of celiac and superior mesenteric artery occlusion. Arch Surg 1983;118:1340.

Schweitzer J, Altmann E, Stosslein F, et al. Comparison of tissue plasminogen activator and urokinase in the local infiltration thrombolysis of peripheral arterial occlusions. Eur J Radiol 1996;22:129.

Sheeran SR, Hallisey MJ, Murphy TP, et al. Local thrombolytic therapy as part of a multidisciplinary approach to acute axillosubclavian vein thrombosis (Paget-Schroetter syndrome). J Vasc Intervent Radiol 1997;8:253.

Shortell CK, Ouriel K. Thrombolysis in acute peripheral arterial occlusion: predictors of immediate success. Ann Vasc Surg 1994;8:59–65.

Simo G, Echenagusia AJ, Camunez F, et al. Superior mesenteric arterial embolism: local fibrinolytic treatment with urokinase. Radiology 1997;204:775.

The STILE Investigators. Results of a prospective randomized trial evaluation surgery versus thrombolysis for ischemia of the lower extremity. The STILE trial. Ann Surg 1994;220:251–268.

Sullivan KL, Gardiner GA, Shapiro MJ, et al. Acceleration of thrombolysis with a high-dose bolus technique. Radiology 1989;173:805–808.

Valji K, Bookstein, JJ, Roberts AC, Sanchez RB. Occluded peripheral arteries and bypass grafts: lytic stagnation as an end point for pulse-spray pharmacomechanical thrombolysis. Radiology 1993;188:389–394.

Van Breda A, Graor RA, Katzen BT, et al. Relative cost-effectiveness of urokinase versus streptokinase in the treatment of peripheral vascular disease. J Vasc Intervent Radiol 1991;2:77–87.

Vannini P, Ciavarella A, Mustacchio A, Rossi C. Intra-arterial urokinase infusion in diabetic patients with rapidly progressive ischemic foot lesions. Diabetes Care 1991;14:925–927.

Weaver FA, Comerota AJ, Youngblood M, et al. Surgical revascularization versus thrombolysis for nonembolic lower extremity native artery occlusions: results of a prospective randomized trial. J Vasc Intervent Radiol 1996;24:513–523.

Weaver FA, Comerota AJ, Youngblood M, et al. Surgical revascularization versus thrombolysis for nonembolic lower extremity native artery occlusions: results of a prospective randomized trial [Abstract]. Radiology 1997;204:293.

# 15
# Mechanical Thrombectomy

### Zhong Qian, Hector Ferral, and Xiaoping Gu

Mechanical thrombectomy consists of the techniques designed to achieve mechanical dissolution, fragmentation, and aspiration of intravascular thrombosis. The technique of percutaneous mechanical thrombectomy was first described in 1989, with the goal of fast removal of thrombus and restoration of patency of thrombosed arteries, veins, bypass grafts, and stents. More than a dozen devices have been developed since then, but only a few of them have been extensively tested experimentally and clinically. This chapter includes the devices for which there is substantial clinical experience and that are commercially available in either the United States or Europe.

## DEVICE DESIGNS

The thrombectomy devices are classified into five major categories. Some devices are a hybrid form of more than one mechanism of thrombectomy. The hybrid devices are usually classified into the category to which its primary mechanism belongs.

1. Recirculation thrombectomy, in which the hydrodynamic vortex created by a high-speed spinning impeller macerates thrombus. Similar to a blender, the device fragments the thrombus and recirculates the particle, progressively reducing particle size. The Amplatz thrombectomy device (ATD) (Microvena, White Bear Lake, MN) and the Trac-Wright catheter (Kensey Nash, Exton, PA) belong in this category.

2. Hydraulic thrombectomy, which uses a high-velocity fluid jet to generate a negative pressure—the so-called Venturi effect, which not only macerates thrombus but also evacuates the fragmented thrombi out of the circulation. The AngioJet catheter (Possis Medical, Minneapolis, MN), the Hydrolyser (Cordis, Miami), and the Oasis catheter (Boston Scientific, Natick, MA) are in this category.

3. Aspiration thrombectomy, referring to thrombus removal through a catheter system using a steady suction applied with a large syringe with or without adjunctive clot maceration, such as the Günther catheter (custom-made) and the Tulip thrombectomy sheath (Schneider Europe, Zurich, Switzerland).

4. Nonrecirculation thrombectomy, which uses a low-speed rotational mechanism for maceration of thrombus, with or without concomitant use of thrombolytic agent. These devices include the Arrow-Trerotola percutaneous thrombolytic device (PTD) (Arrow International, Reading, PA), the Rotatable pigtail catheter (Cook Europe, Bjaeverskov, Denmark), the Castañeda thrombolytic brush catheter (Micro Therapeutics, Irvine, CA), and the transluminal extraction catheter (Interventional Technologies, San Diego, CA).

5. Special energy-assisted thrombectomy, in which ultrasound, laser, or radiofrequency is used for fragmentation of thrombus or thrombolysis, such as the Acolysis system (Angiosonic, Morrisville, NC).

Although most of these devices have been miniaturized and made more clinically practical after multiple refinements, none of them has achieved a complete clinical success, and each device is limited to having advantages in certain applications.

### Amplatz Thrombectomy Device

The ATD was first described in 1989 and has been extensively investigated experimentally and clinically in the United States and Europe. The device consists of an 8-Fr polyurethane catheter with an impeller mounted on the end of a drive shaft that is protected by a 1-cm-long, open-

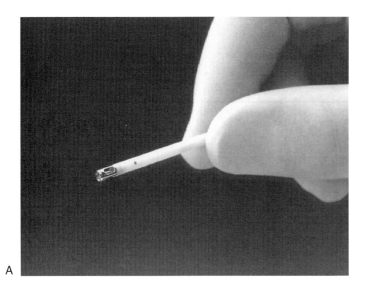
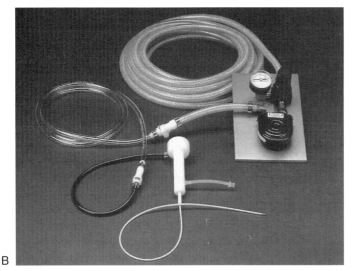

FIG. 15.1. A: Close-up of tip of the Amplatz thrombectomy device. Note an open-ended metal capsule with two side holes for flow recirculation during thrombectomy. B: The whole system consists of a thrombectomy catheter, a pressure gauge, and a foot petal.

ended metal capsule with two side holes used for flow recirculation (Fig. 15.1). The edges of the capsule are slightly rounded and the impeller is slightly recessed to reduce the potential vascular injury by the device. The impeller is driven by a compact air turbine at speeds of up to 150,000 rpm. The high rotational speed of the impeller permits the thrombus to be recirculated for subsequent fragmentation. At the proximal end of the catheter, the drive shaft extends through a Y-connector that allows the infusion of saline, contrast medium, or adjunctive fibrinolytic agents, any of which also serve to lubricate and cool the device. The ATD is commercially available in two sizes: 6 Fr and 8 Fr, with the lengths of 120 cm, 75 cm, and 50 cm. The *in vitro* experiment proved that the ATD liquefied 99.2% of 4-day-old thrombus and 98.8% of 10-day-old thrombus. However, the ATD has a detectable hemolytic effect, as observed clinically and experimentally, despite the absence of permanent consequences. Until recently, lack of steerability and torque control had been the major drawbacks of the ATD. The 6-Fr catheter shaft made with Pebax, a polyether block amide material, has come to the market, with a high torque control. The ATD has to be introduced into the target vessel through a guiding catheter or an introducer sheath in most cases; the over-the-wire system is under investigation. Although a wide range of clinical applications have been reported, including in dialysis grafts, pulmonary arteries, peripheral arteries and veins, vena cava, and transjugular intrahepatic portosystemic shunting (TIPS), the device is currently approved by the U.S. Food and Drug Administration (FDA) for the use of declotting hemodialysis grafts only.

### AngioJet

The AngioJet rheolytic thrombectomy device was initially described in 1992. After having undergone a major redesign, the current system is purportedly more practical and safe. The AngioJet system consists of an infusion drive unit delivering saline under high-pressure, a pump set, and a thrombectomy catheter. The catheter is fabricated from a double-lumen tube—a large lumen for effluent evacuation and wire passage and a small-caliber metal tubing supplying the pressurized fluid to be transformed into high-velocity jets. The metal tubing at the catheter tip is modeled as a 360-degree loop with three mini-orifices (50 μm in diameter), retrogradely facing the evacuation lumen (Fig. 15.2A). There are three additional side holes at the tip of the catheter, which are proximal to the "gap zone" between the metal loop and the large aspiration lumen. The jets from these side holes optimize the hydrodynamic vortex to augment device efficiency. When the catheter is activated, thrombus is fragmented and aspirated into the large lumen by a Venturi effect created by the fluid jets. The catheter is available in three models: (a) peripheral model—the AngioJet-F105 with 5-Fr catheter shaft, recently replaced by the newer version device, the so called e-Train 110; (b) arteriovenous graft model—the AngioJet AV60 with 5-Fr tip diameter; and (c) coronary model—the AngioJet-LF140 with a cap on the catheter tip designed for use in coronary arteries. Unlike the former two models, the AngioJet-LF140 creates six fluid jets and has no side holes. All models of these catheters can be introduced either over an 0.018-in. guidewire or through an 8-Fr guiding catheter. However, insertion of the device over the wire may compromise the efficiency of thrombectomy because the presence of the wire in the catheter lumen reduces the free lumen. The working length is 60 cm with AngioJet-AV60, 105 cm with AngioJet-F105, 110 cm with e-Train 110, and 140 cm with AngioJet-LF140. One should be aware that the dedicated pump drive unit is a one-time capital investment for use of this system. The device reportedly has been used in dissolution of acute and subacute thrombus in peripheral arteries and veins, the coronary arteries, dialysis fistulae, and TIPS. This device has been approved by the FDA for dialysis graft declotting.

### Hydrolyser

The Hydrolyser was introduced in 1993 as an alternative device. This 7-Fr nylon double-lumen catheter has a 6-mm oval side window at the distal portion of the catheter, in which adjacent thrombus is fragmented and aspirated by the Venturi effect created by a fluid jet from a small inflow lumen to a large outflow lumen (Fig. 15.3). Its closed round tip represents the

**FIG. 15.2. A:** AngioJet catheter used in the peripheral vascular system. **B:** AngioJet catheter used in the coronary arteries.

**FIG. 15.3.** The double-lumen Hydrolyser catheter. (Reprinted from van Ommen VG, van der Veen FH, Geskes GG, et al. Comparison of arterial wall reaction after passage of the Hydrolyser device versus a thrombectomy balloon in an animal model. *J Vasc Intervent Radiol* 1996;7:451–454, with permission.)

atraumatic design. The Hydrolyser is powered by a conventional contrast injector, with a rate of 3 mL per second at a pressure of 750 psi. Although the catheter accepts a 0.025-in. guidewire, the catheter proves to be much more efficient coupling with use of a 9-Fr guiding catheter. Its major drawbacks include lack of flexibility and difficulty to cross the aortic bifurcation over the wire. Its current clinical applications include peripheral arterial thrombosis and failing dialysis grafts. This catheter has been approved for use in the setting of dialysis shunts.

## Oasis Catheter

The experimental data on the Oasis catheter (previously known as the SET catheter) were first published in 1994. The catheter is currently manufactured by Boston Scientific Corp. The catheter shaft has three lumens. The inflow lumen is connected with a conventional angiographic power injector to establish a fluid jet across a metal nozzle on the tip of the catheter. The jet is directed backward into the second lumen, the outflow lumen, which is in line with the nozzle, creating a Venturi effect. Adjacent thrombus is pulled into the working area of the catheter tip by the pressure gradient generated by the Venturi effect and is fragmented by the strong fluid jet (Fig. 15.4). The resultant particles are evacuated through the outflow lumen into a plastic collecting bag. The third one is a guidewire lumen and accepts an 0.018-in. guidewire for catheter advancement. The catheter is available in four sizes: 5 Fr, 6 Fr, 8 Fr, and 10 Fr. The catheter seems to be very effective for removal of acute and subacute thrombus in peripheral artery, vein, or dialysis grafts. The suction action generated by Venturi effect may cause vessel collapse in vessels with small diameter during the activation procedure. Application in hemodialysis access grafts was approved by the FDA in 1999.

## Arrow-Trerotola Percutaneous Thrombolytic Device

The Arrow-Trerotola PTD was added to the family of mechanical thrombectomy devices in 1999. The PTD designed for declotting dialysis grafts consists of a rotational nitinol basket, drive cable, and handheld electric motor powered by an accompanying battery pack (Fig. 15.5). The basket and the cable are housed in a 5-Fr catheter that constrains the self-expanding basket in unexpanded form. The device is compatible with 5-Fr sheath. A commercial set comes with a 6-Fr sheath. As soon as the tip of the PTD is positioned at the target site, the basket will be opened by withdrawing the 5-Fr catheter and then rotated at the speed of 3,000 rpm by activating the motor. By slowly pulling the activated device through the thrombosed graft, the basket strips the clot from the graft walls and macerates the thrombus into fragments smaller than 3 mm in diameter; most of them are less than 1 mm. The resultant fragments can be aspirated through the side port of the 6-Fr sheath. The PTD is available for clinical use in the United States.

## TECHNICAL CONSIDERATIONS

Percutaneous mechanical thrombectomy has proved to be safe and effective if an appropriate

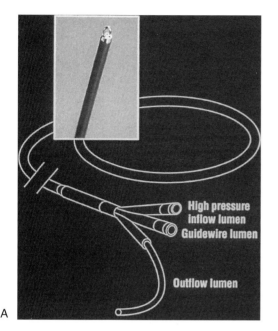
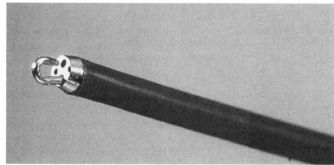

**FIG. 15.4.** The Oasis catheter. **A:** Catheter has three lumens: inflow lumen, guidewire lumen, and outflow lumen. **B:** Close-up of the tip of the catheter. (Reprinted from Sharafuddin MJ, Hicks ME. Current status of percutaneous mechanical thrombectomy. Part III. Present and future applications. *J Vasc Intervent Radiol* 1998;9:209–224, with permission.)

device is chosen and is used properly. However, several technical concerns related to use of these devices have gained a great amount of attention, based on the current clinical data and experimental results. These concerns are not just for a particular device but for all in general use. Familiarity with these issues may help in choosing optimal treatment and minimizing potential procedure-related complications.

## Hemolytic Effect

All mechanical thrombectomy devices can cause intravascular hemolysis, especially those using high-speed rotation or high-velocity fluid jet. Significant hemolysis may result in renal insufficiency due to the nephrotoxic effect of free hemoglobin and anemia secondary to massive consumption of red blood cells. Early clinical and experimental results with the ATD show that although consistent hemolysis is detected after the thrombectomy, it usually represents a transient process (within 24 hours) without adverse clinical consequences. However, more extensive hemolysis was found when the device was used in native vessels than in grafts and when used in the vena cava than the aorta. To minimize the potential risks to patients who undergo thrombectomy procedure, several measures have been suggested, including (a) use of high-speed rotation and fluid jet with caution in anemic and pediatric patients or in patients with

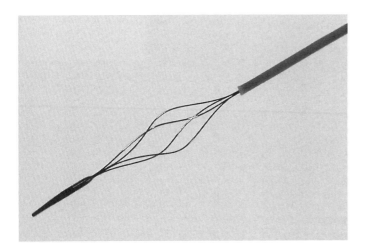

**FIG. 15.5.** Close-up of the tip of the percutaneous thrombolytic device. (Reprinted from Trerotola SO, Davidson DD, Filo RS, et al. Preclinical in vivo testing of a rotational mechanical thrombolytic device. *J Vasc Intervent Radiol* 1996;7:717–223, with permission.)

renal insufficiency; (b) prophylactic urine alkalinization to reduce risk of nephrotoxicity in the above-mentioned patient population; (c) close monitoring of the device's activation time (e.g., the activation time being limited to no more than 10 minutes with the ATD); and (d) measurement of the hematocrit and levels of plasma-free hemoglobin and haptoglobin in the selected patient population.

### Blood Loss

Use of aspiration or hydraulic devices is evidently associated with the potential for significant blood loss, which may be clinically relevant in anemic or pediatric patients. Strict control of the activation time and close supervision of the treated outcome should be exercised. Monitoring of the hematocrit level and determination of the output volume of the effluent are strongly recommended. Supplementary hydration or even blood transfusion should be considered in some selected cases.

### Endpoint of the Procedure

Improvement in lumen patency can usually be observed after two to three passes with most thrombectomy devices. However, incomplete results due to residual thromboses are not uncommon after use of the device according to the manufacturers' recommendations. It is sometimes difficult to determine the age of thrombus. In case of embolism, clots are sometimes formed over a period of time and contain different components representing varying ages of thrombi. Clots older than 14 days or organized thrombi usually do not respond well to most of the mechanical thrombectomy devices. Moreover, thrombotic vascular occlusions are often associated with severe underlying diseases, such as atherosclerosis, intimal hyperplasia, or dissection. These underlying lesions may complicate thrombus dissolution by mechanical thrombectomy. Occasionally, thrombus elongating to the distal segment of the vessel may not be reached by the device due to its relatively larger profile. In all these circumstances, if a complete thrombectomy cannot be achieved after initial application of the device in the suggested manner, other adjunctive techniques may be attempted to accomplish complete thrombus dissolution. These adjunctive modalities include pharmacologic thrombolysis, surgical balloon thrombectomy, surgical revision, or stent-graft placement. Balloon dilation or stent placement may also be used to treat the underlying diseases after the thrombectomy procedure. If an aggressive pursuit of complete mechanical thrombectomy is planned, one should always weigh the possible benefits gained from prolonged use of thrombectomy device against the potential adverse effects of the additional operation. The aspiration type of thrombectomy devices proved effective in removing embolized organized clots, whereas the mesh-basket device and thrombolytic brush catheter are more attractive in clearing residual wall-adherent thrombus.

## PERIPROCEDURAL MEDICATION AND MONITORING

The thrombectomy procedures are generally performed with patients under conscious sedation. Analgesia can be achieved by giving midazolam and fentanyl intravenously as needed. Prophylactic antibiotics are not needed as a routine measure, although antibiotics may be given to patients who are susceptible to infection. In case of arterial spasm, especially in the infrapopliteal arteries, 10 mg of nifedipine can be given sublingually to relieve it.

Intravenous administration of heparin should be initiated with a bolus of 3,000 to 5,000 units before the procedure and may be extended for additional 2 days as needed. To reduce the risk of distal microembolism during arterial thrombectomy, local low-dose urokinase infusion (50,000 to 100,000 units) or tissue plasminogen activator has been suggested, although its benefits have yet to be proved. If pharmacologic thrombolysis is considered as an adjunctive measure, one can start with a dose of 50,000 units of urokinase over a period of 10 minutes, followed by 25,000 units per hour up to a total dose of 250,000 to 300,000 units of urokinase. The dose of tissue plasminogen activator has not yet been defined.

Peripheral pulse oxygen saturation, electrocardiogram, respiratory rate, and hemodynamic state need to be continuously monitored during and immediately after the procedure.

## CLINICAL APPLICATIONS

Although the only application of mechanical thrombectomy currently approved by the FDA is declotting of hemodialysis access grafts, a number of additional applications have been reported with encouraging results:

- Hemodialysis access graft thrombosis
- Thrombosis in native artery
- Peripheral arterial bypass graft
- Coronary arterial thrombosis
- Proximal deep venous thrombosis
- Iliocaval thrombosis
- Central venous thrombosis in upper body
- Proximal pulmonary embolism
- Visceral arterial thrombosis
- TIPS and portal vein thrombosis

Because currently available devices vary widely regarding their ideal indications and efficacy, proper indications of each particular device should be identified based on the manufacturer's recommendations before the device is used. In general, mechanical thrombectomy is limited to acute or subacute thrombus (less than 2 weeks). The response of aged clots is less uniform and is usually poor. Aspiration catheters may be applicable, especially in smaller arteries, because they simply remove all material instead of macerate. The efficacy of the aspiration catheter is generally restricted by the limited inner diameter of the catheter. A higher short-term success rate of thrombectomy has been observed in synthetic grafts compared with native arteries. In dialysis patients, long-term success is not directly related to the ability of the device to properly declot the vessels but depends heavily on the underlying lesions in the native vessel and technical problems of the grafts created by surgeons.

## COMPLICATIONS

Complications associated with mechanical thrombectomy vary widely in terms of nature, frequency, and severity, depending on the individual device. Generally speaking, potential complications can be classified into two groups: procedure-related complications and device-related complications.

### Procedure-Related Complications

- Puncture site hematoma or pseudoaneurysm. These complications occur in 1% to 3% of cases; resultant consequences could be devastating if not detected immediately and treated effectively. The management of these complications is not different from management of the same complications occurring from other interventional procedures.

- Rethrombosis. This complication can occur due to a number of reasons, including the lack of anticoagulation, incomplete thrombectomy, vessel damage, and inflow or outflow underlying lesions. Therefore, adequate anticoagulation is mandatory during and immediately after the procedure. Antiplatelet-aggregation therapy is also recommended to minimize this potential problem. Residual thrombus can be further treated by adjunctive therapies such as thrombolysis. Inflow or outflow stenotic lesions should be corrected after the thrombectomy.
- Embolization. Microemboli are usually asymptomatic and carry low risk for ischemia in peripheral and pulmonary arteries, but they can be critical to the visceral arteries. One must be extremely cautious in those cases with a high potential for microembolization to the visceral arteries. If microembolization occurs, emergent thrombolytic therapy may be indicated. Macroemboli defined as angiographic visible fragments will lead to much more grave consequences in the arterial system than in the venous system. Experimental data and clinical experience suggest that the pulmonary circulation in general tolerates a small amount of emboli clinically, despite lack of knowledge of the long-term effects. Distal embolism in peripheral arteries could be a catastrophe to patients due to ischemic complications resulting from poor tissue perfusion. The management of distal embolism includes use of aspiration thrombectomy and emergency pharmacologic thrombolysis. Several strategies have been proposed to minimize the potential of downstream macroembolism: (a) avoiding traversal of the thrombus with the device before activation of the device; (b) use of a careful, gradual leading-to-trailing approach during the thrombectomy; and (c) performing the thrombectomy procedure before dilating underlying outflow stenosis.

### Thrombectomy Device–Related Complications

- Vessel wall damage: Most of the currently commercially available devices are safe and cause no or little vessel damage. Studies show that the thrombus itself can cause endothelial denudation, but it is a reversible process. Severe vessel-wall injuries have been reported, such as intimal flaps, contrast medium extravasation into the wall, and local or massive extravasation. Fortunately, most of vessel wall injuries are minor and self-limited. More severe damage can be treated with interventional techniques; extremely few cases need surgical intervention.
- Hemolysis: Intravascular hemolysis is common after a mechanical thrombectomy procedure. The vast majority of these cases is transient and of no clinical consequence. The details are described in "Technical Considerations."
- Blood loss: This complication is associated with macerating and aspirating devices. Prevention and management are described in "Technical Considerations."

## SUGGESTED READINGS

Bildsoe MC, Moradian GP, Hunter DW, et al. Mechanical clot dissolution: new concept. *Radiology* 1989;171: 231–233.

Bücker A, Schmitz-Rode T, Vorwerk D, Günther RW. Comparative in vitro study of two percutaneous hydrodynamic thrombectomy systems. *J Vasc Intervent Radiol* 1996;7:445–449.

Drasler WJ, Jenson ML, Wilson GJ, et al. Rheolytic catheter for percutaneous removal of thrombus. *Radiology* 1992;182:263–267.

Manicone JA, Eisenbud DE, Hertz SM, et al. The effect of thrombus on vascular endothelium of arterized veins grafts. *Am J Surg* 1996;172:163–167.

Nazarian GK, Qian Z, Coleman CC, et al. Hemolytic effect of the Amplatz thrombectomy device. *J Vasc Intervent Radiol* 1994;5:155–160.

Reekers JA, Kromhout JG, van der Waal K. Catheter for percutaneous thrombectomy: first clinical experience. *Radiology* 1993;188:871–874.

Rilinger N, Gorich J, Scharrer-Pamler R, et al. Short-term results with use of the Amplatz thrombectomy device in the treatment of acute lower limb occlusions. *J Vasc Intervent Radiol* 1997;8:343–348.

Sharafuddin MJ, Hicks ME. Current status of percutaneous mechanical thrombectomy. Part I. general principles. *J Vasc Intervent Radiol* 1998;8:911–921.

Sharafuddin MJ, Hicks ME. Current status of percutaneous mechanical thrombectomy. Part II. devices and mechanisms of action. *J Vasc Intervent Radiol* 1998;9:15–31.

Sharafuddin MJ, Hicks ME. Current status of percutaneous mechanical thrombectomy. Part III. present and future applications. *J Vasc Intervent Radiol* 1998;9:209–224.

Sharafuddin MJ, Hicks ME, Jenson ML, et al. Rheolytic thrombectomy with the AngioJet-F105 catheter: preclinical evaluation of safety. *J Vasc Intervent Radiol* 1997;8:939–945.

Stiegler H, Arbogast H, Ness H. Thrombectomy, lysis, or heparin treatment: concurrent therapies of deep vein thrombosis—therapy and experimental studies. *Semin Thromb Hemost* 1989;15:250–258.

Tadavarthy SM, Murray PD, Inampudi S, et al. Mechanical thrombectomy with the Amplatz device: human experience. *J Vasc Intervent Radiol* 1994;5:715–724.

Trerotola SO, Davidson DD, Filo RS, et al. Preclinical in vivo testing of a rotational mechanical thrombolytic device. *J Vasc Intervent Radiol* 1966;7:717–723.

Uflacker R. Mechanical thrombectomy in acute and subacute thrombosis with use of the Amplatz device: arterial and venous applications. *J Vasc Intervent Radiol* 1997; 8:923–932.

Vicol C, Dalichau H, Köhler J, et al. Performance of indirect embolectomy aided by a new developed flush-suction catheter system: forty-seven experimental embolectomy procedures in test animals. *J Cardiovasc Surg* 1994;35:193–200.

Wengrovitz M, Healy DA, Gifford RRM, et al. Thrombolytic therapy and balloon catheter thrombectomy in experimental femoral artery thrombosis: effect on arterial wall morphology. *J Vasc Intervent Radiol* 1995;6:205–210.

Yasui K, Qian Z, Nazarian GK, et al. Recirculation-type Amplatz clot macerator: determination of particle size and distribution. *J Vasc Intervent Radiol* 1993;4:275–278.

# 16
# Inferior Vena Cava Filters

Steven F. Millward and Randall V. Olsen

Deep vein thrombosis (DVT) occurs in approximately 2 million Americans each year. Pulmonary embolism (PE) is estimated to develop in 600,000 of these patients, and 60,000 die of PE. Most patients with DVT, PE, or both are initially treated with heparin for 5 to 10 days followed by oral warfarin. Warfarin therapy may be required for durations varying from 6 weeks to lifelong depending on the patient's risk factors for DVT. Only about 14% of all patients with a diagnosis of DVT, PE, or both require treatment with an inferior vena cava (IVC) filter. Prophylactic anticoagulation treatment of hospitalized patients in various categories of risk for the development of venous thromboembolism is well established so that filter placement is also not often required in this group of patients.

Surgical ligation and clipping of the IVC have been replaced over the past 25 years by placement of an intraluminal device. Early devices such as the Mobin-Uddin umbrella and the Hunter balloon resulted in a high rate of IVC thrombosis. The original 24-Fr stainless steel Greenfield filter (MEDITECH/Boston Scientific, Watertown, MA) was developed in 1973, and excellent results were achieved with this device, which was initially inserted while in an operating room via a venous cutdown and subsequently by percutaneous insertion. Newer filters designed for ease of percutaneous placement are now routinely inserted by interventional radiologists. Filters should be placed percutaneously while in an interventional radiology suite rather than surgically. Although no large controlled trials have been performed to demonstrate the effectiveness of IVC filters, published literature suggests that they are effective in preventing recurrent PE and that life-threatening complications from these devices are rare.

## INDICATIONS FOR PLACEMENT OF AN INFERIOR VENA CAVA FILTER

### Standard Indications

The most common indications for IVC filter placement are the following:

- A contraindication to anticoagulation, such as overt gastrointestinal bleeding, recent intracranial hemorrhage or surgery, or an underlying hemorrhagic state such as liver failure or thrombocytopenia, in patients with documented iliofemoral DVT and/or PE
- In patients with documented iliofemoral DVT or PE who develop bleeding while receiving appropriate anticoagulation therapy with coagulation studies in the therapeutic range
- A failure of anticoagulation, which is a less common but clear indication, when recurrent PE or substantial extension of DVT is documented to occur despite anticoagulant treatment, with coagulation studies in the therapeutic range (activated partial thromboplastin time 1.5 to 3.0 times mean of laboratory normal range for patients receiving heparin or international normalized ratio of less than 2.0 for patients receiving warfarin)

The presence of DVT, PE, or both should be documented by an objective test. Contrast venography and venous ultrasonography are appropriate modalities to confirm the presence of DVT. For the diagnosis of PE, pulmonary angiography is the most reliable modality. However, a high-probability ventilation-perfusion lung scan in a patient with a high or moderate clinical probability of PE is diagnostic of PE. A contrast-enhanced helical or electron-beam computed tomography scan is also an acceptable method of demonstrating that a patient has had a PE.

## Other Indications

Placing an IVC filter in a patient for an indication other than one of the standard indications discussed above requires careful consideration. Filters are probably indicated in patients with massive PE who have marginal cardiopulmonary function and in patients who undergo pulmonary embolectomy. Filters may also be indicated in patients with less definite contraindications to anticoagulation, such as recurrent falls and poor compliance with medications.

## Prophylactic Indications

Prophylactic filter placement remains controversial and should rarely be used to justify placement of a filter. Prophylactic indications can be divided into two categories. The first category includes patients who have DVT or PE but in whom there is no contraindication to anticoagulation. Examples in this category include the presence of a free-floating thrombus in the IVC. Filter placement has also been recommended by some as an alternative to anticoagulation therapy in patients with cancer; however, other studies have shown no benefit in treating patients with brain metastases with filters rather than anticoagulation. Filter placement rather than anticoagulation has also been recommended by some authors in patients older than age 65.

A second category of prophylactic indications is in patients who have no documented DVT or PE but are thought to be at high risk. There is no strong evidence to show that filters are superior to prophylactic anticoagulation in most situations. However, some authors have advocated the use of filters in high-risk trauma patients who cannot receive prophylactic anticoagulation or venous ultrasound for DVT surveillance. Several nonrandomized studies have been performed in trauma patients with neurologic injury and in patients with severe pelvic or lower extremity fractures. These have suggested that there is a decreased rate of fatal PE in patients treated with prophylactic filters compared with historical controls who did not receive a filter.

## CONTRAINDICATIONS TO FILTER PLACEMENT

Absolute contraindications to filter placement are uncommon: complete thrombosis of the IVC and a lack of available venous access. Severe blood coagulopathy can usually be corrected before filter insertion, but in some circumstances it may make surgical venotomy and closure preferable.

Many authors recommend avoiding wherever possible placement of filters in young patients with a long life expectancy because the longevity of filters in the body is not known.

The presence of advanced malignant disease is considered by some authors to be a relative contraindication to filter placement because it may give little survival benefit. However, other workers have failed to show a strong association between advanced malignancy and early mortality after filter placement and conclude that the presence of advanced malignant disease should not be a deterrent to filter placement.

## TECHNIQUE OF PLACEMENT

### Choice of Access Route

Percutaneous filter placement is now the preferred method over surgical placement in almost all patients. Preprocedural venous ultrasonography is useful in confirming that the proposed access vein for percutaneous placement is free of thrombus. The right common femoral vein is most commonly used. If this site is thrombosed, some authors will use a left common femoral vein approach. However, filter placement through this route may be more difficult because of the angle between the common iliac vein and the IVC. The right internal jugular vein offers a straighter access to the IVC and is the access of choice according to several authors. In patients in whom infrarenal IVC thrombosis is present or suspected, the right internal jugular vein is the preferred access route.

In unusual circumstances it may be appropriate to use the external jugular vein or subclavian vein for access. In these situations, in which angulation of the access route may be a problem, use of the Simon Nitinol filter (Nitinol Medical

Technologies, Woburn, MA) may be appropriate because this filter can even be inserted through an antecubital approach. Puncture of the femoral vein should be performed with an open-ended needle with continuous suction to ensure that the femoral vein is entered directly, avoiding the superficial femoral artery. Jugular vein punctures should also be performed with an open-ended needle, and puncture of the vein under direct ultrasound guidance may be preferable.

### Preprocedural Venacavography

Preprocedural venacavography is best performed immediately before filter insertion. The catheter should be positioned at the confluence of the iliac veins and contrast material injected through a power injector (20 mL per second for 2 seconds), with rapid (at least two frames per second) cut-film or digital imaging in the anteroposterior plane.

Renal vein and IVC variants that influence filter placement occur in approximately 18% of the population. The renal vein orifices should be identified either by inflow of unopacified blood or by reflux of contrast material (Fig. 16.1). If there is suggestion of an anomaly of the renal veins, selective renal venography should be performed. Some authors recommend the use of routine selective renal venography. The filter should usually be placed so that its superior aspect is immediately below the most inferior renal vein or other substantial sized venous collateral channel. Missing the presence of a circumaortic left renal vein and placing the filter between the two left renal veins could allow recurrent emboli to pass through the left renal vein ring.

The venacavogram should then be carefully scrutinized to ensure that there is no thrombus at the site of proposed filter placement. The diameter of the IVC at the site of proposed filter placement should then be measured either directly from cut-films (allowing for magnification) or from digital images by using a sizing catheter during venacavography. Current commercially available filters cannot be placed in an IVC with a diameter of greater than 28 mm, with the exception of the Bird's Nest filter (Cook, Inc., Bloomington, IN), which can be

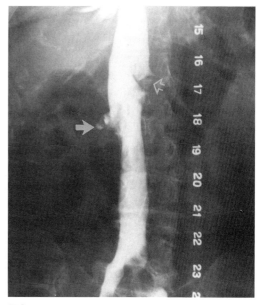

**FIG. 16.1.** Venacavogram performed by contrast injection into the right common iliac vein. Reflux into the left common iliac vein and right renal vein (*arrow*) is visualized. Inflow of nonopacified blood from the left renal vein (*open arrow*) identifies the site of this structure.

used when the IVC diameter is not greater than 40 mm. The distance between the most inferior renal vein and the iliac veins is measured to ensure that there is an adequate length of IVC for filter placement. This is particularly important with the Bird's Nest filter because it requires a 6- to 7-cm length for insertion. The iliac vein orifices should also be identified. In particular the left iliac vein orifice should be carefully looked for when venacavography is performed through the right femoral vein (Fig. 16.1) to avoid overlooking duplication of the IVC, which may require the placement of two filters. Other clues to the presence of IVC duplication include a small diameter to the infrarenal IVC and absence of inflow from the left renal vein. Selective contralateral iliac venography is usually unnecessary.

Adequate cavography is important to avoid placement complications, including disasters such as in deployment of the filter in the aorta. In patients with an absolute contraindication to the use of conventional contrast material, carbon dioxide digital subtraction venography or intravascular ultrasonography can be used.

### Oversized Inferior Vena Cava

IVC diameters greater than 28 mm occur in 1% to 3% of the population. The Greenfield, Simon Nitinol, and Vena Tech-LGM (B. Braun Medical, Evanston, IL) filters are not recommended for placement in venae cavae greater than 28 mm in diameter. In this situation most authors recommend either placement of one of these filters in each common iliac vein or placement of a Bird's Nest filter in the oversized IVC. If the IVC tapers inferiorly to a suitable length segment with a diameter of less than 28 mm, then placement of a Greenfield, Simon Nitinol, or Vena Tech-LGM filter in this segment may also be appropriate in some patients.

### Suprarenal Filter Placement

Placement of a filter in the suprarenal IVC may be required in up to 4% to 6% of patients, usually because of extensive infrarenal thrombus. However, suprarenal placement may also be appropriate in patients with renal transplants, in patients with renal cell carcinoma with renal vein tumor extension, and in either a pregnant woman or a young woman anticipating pregnancy if she requires a filter. Patients with a very small diameter infrarenal IVC may be at increased risk of IVC occlusion because of incomplete filter opening, and this may be an indication for a suprarenal filter.

### Filter Placement in the Superior Vena Cava

Placement of a filter in the superior vena cava has been described and may occasionally be appropriate for patients with proven upper extremity DVT who are considered to be at high risk for PE and have an indication for a filter.

## IMMEDIATE COMPLICATIONS OF FILTER PLACEMENT

Although the development of the new low-profile filters has resulted in devices with relatively simple placement techniques, immediate complications of placement still occur in up to 20% of patients. Some of these are due to operator error or inexperience with the device, and others can be due to difficult patient anatomy. These complications include misplacement, kinking of the introduction sheath, difficulty releasing the filter from the delivery sheath, difficulty anchoring the filter, incomplete opening, clustering of the legs, tilting, and acute penetration of the IVC.

Selection of an appropriate access site can facilitate filter placement. Tortuous access routes should be avoided if possible. A left femoral approach should be used only if the angle between the iliac vein and IVC is favorable. The importance of performing adequate venacavography has previously been discussed. Placement of the filter in a branch of the IVC, such as the right renal vein when a jugular approach is used (Fig. 16.2), can

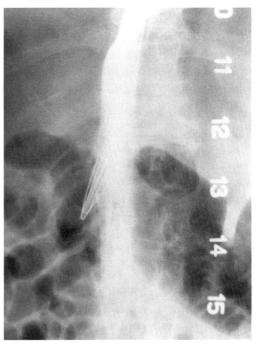

**FIG. 16.2.** Vena Tech-LGM filter of the original design has been deployed into the right renal vein from a jugular approach. This complication can be avoided by maintaining the guidewire in the inferior vena cava beyond the renal veins during placement of the introduction sheath. No attempt was made to remove this filter but a second filter was placed in the infrarenal inferior vena cava.

be avoided by ensuring that the guidewire remains in the IVC beyond the renal veins at all times during the procedure. Similarly, adequate venacavography will help ensure that filters are not misplaced superior to renal vein branches.

To avoid a placement complication, careful selection of an appropriate filter is required. The Bird's Nest filter should probably be used in most patients with an IVC diameter greater than 28 mm but is not appropriate in patients whose infrarenal IVC is less than 6 to 7 cm in length. A Simon Nitinol filter may be the most appropriate for use through a tortuous approach. Finally, great care must be taken to ensure that a jugular filter is chosen for a jugular approach and a femoral filter for a femoral approach.

Compression of the IVC from an adjacent tumor can make filter placement difficult. Incomplete opening of Greenfield, Simon Nitinol, or Vena Tech-LGM filters can occur. Filters may fail to open fully because of a small IVC. Delay during filter deployment, occurring because of technical difficulty or operator inexperience, could allow thrombus formation within the introduction sheath and can also cause incomplete filter opening. Continuous flushing of the sheath is recommended for the Simon Nitinol filter and is suggested for the titanium Greenfield filter (MEDITECH/Boston Scientific). Other causes of incomplete opening include deployment of the filter within thrombus and failure to stabilize the pusher or delivery device on deployment. Greenfield, Simon Nitinol, and Vena Tech-LGM filters may also tilt on deployment. Manipulation of incompletely opened or tilted filters or filters showing leg asymmetry can be performed, but this practice is of unproved benefit and should probably be avoided. In cases in which a filter shows severe tilting or incomplete opening, placement of a second filter may be appropriate, although the effect of these complications on filter function is not known.

Premature filter deployment, either inferior or superior to the desired location, can usually be avoided by careful technique. In cases in which a filter becomes stuck in the introduction sheath or other major technical difficulties occur, it is better to remove the sheath with the filter and start the procedure over again than to deploy the filter in an inappropriate position.

## POSTPLACEMENT MANAGEMENT

Immediately after filter placement, a radiograph should be obtained to document the position of the filter. A postplacement venacavogram should be obtained whenever malposition is suspected (Fig. 16.2). Anticoagulation treatment, when not contraindicated, should be used in conjunction with the filter. Patients should understand the importance of identifying to future health care providers that they have a filter so that complications occurring during venous procedures can be avoided. Guidewire entrapment has been described with both Greenfield and Vena Tech-LGM filters. Magnetic resonance imaging of patients with ferromagnetic filters such as the Bird's Nest filter, and to a lesser extent the stainless steel Greenfield filters, results in metallic artifacts that can degrade abdominal images. Although these filters do show deflection within a phantom, imaging appears to be safe as long as several weeks have passed after filter placement so that incorporation of the filter into the IVC wall has occurred. The Vena Tech-LGM, Simon Nitinol, and titanium Greenfield filters are nonmagnetic. Patients with filters should undergo follow-up abdominal radiographs and evaluation of the insertion vein and IVC to detect complications that may require treatment. Venous ultrasonography can be used for insertion vein and IVC imaging. Appropriate timing for this follow-up has not been determined but follow-up within the first month, then between 3 and 6 months, and then yearly is probably appropriate.

## LONG-TERM COMPLICATIONS OF FILTERS

### Insertion Vein Thrombosis

Rates of insertion vein thrombosis as high as 41% were reported after percutaneous insertion of the 24-Fr stainless steel Greenfield filter. However, placement of newer filters through small introduction sheaths is still associated with the development of nonocclusive, occlusive, and symptomatic insertion vein thrombosis in 25%, 10%, and 3% of patients, respectively. In addition, the development of occlusive thrombus appears to be more common in patients with extrinsic compression of or partially occlusive thrombus in the IVC or ipsilateral iliofemoral veins before filter placement.

### Inferior Vena Cava Occlusion

The reported rates of occlusive thrombus developing in the IVC in patients with filters vary considerably. These differences may be due to true differences in filter performance but may also be related to different methods of patient follow-up. Examples of reported occlusion rates are 5% to 7% for the 24-Fr stainless steel Greenfield filter, 2% for the titanium Greenfield filter, 2.9% for the Bird's Nest filter, 7% to 9% for the Simon Nitinol filter, and 8% for the Vena Tech-LGM filter. However, other reports indicate a 15% rate for the 24-Fr stainless steel Greenfield filter, 15% for the Bird's Nest filter, and 24% for the Vena Tech-LGM filter. If only patients receiving objective testing of the IVC are included in the analysis, then higher rates of 22% for the Bird's Nest filter and 28% for the Vena Tech-LGM filter are apparent. Crochet et al. estimated IVC occlusion rates for the Vena Tech-LGM filter of 8% at 2 years and 30% at 6 years. It is difficult to use these data to compare the Vena Tech-LGM filter to other filters, because other filters have not been subjected to such rigorous follow-up. Although IVC occlusion may not produce severe symptoms in small numbers of patients followed for short periods of time, IVC occlusion should not be regarded as a benign complication (Fig. 16.3). In one study in which 48 patients received Bird's Nest filters, seven developed symptomatic IVC thrombosis and of these four died, mainly of acute renal failure due to extension of thrombus into the renal veins. IVC occlusion may also result in severe venous stasis symptoms, although Crochet et al. found that the development of these symptoms after filter placement was related to the initial extent of DVT rather than the presence or absence of IVC occlusion.

### Recurrent Pulmonary Embolism

The true rate of recurrent PE in patients with IVC filters is unknown because studies in which all patients are followed with objective imaging of

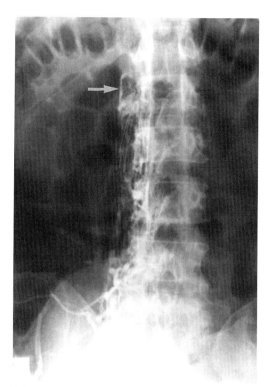

FIG. 16.3. Venacavogram demonstrates the presence of extensive occlusive thrombus in the inferior vena cava in a patient with a Vena Tech-LGM filter. A moderate-sized thrombus (*arrow*) is seen propagating on the cephalic aspect of the filter, which could result in recurrent pulmonary embolus.

the lungs have not been performed. Recurrent PE rates of 4% to 5% are commonly reported, with a mortality rate from recurrent PE of 2% to 3%. There appears to be little difference among the various available IVC filters in recurrent PE rates.

### Migration

Both caudal and cephalic migrations occur with all types of filters; however, clinically significant migration with resultant symptoms is rare. Cephalic migration of Bird's Nest filters was reported with the original design. However, the new modified Bird's Nest filter has been reported to migrate during massive embolus capture: in a single patient with inferior strut migration only and in two patients with migration of the entire filter to the right atrium. Two symptomatic migrations of Simon Nitinol filters into the pulmonary artery and right atrium, respectively, have been described. In the series of Ferris et al. one Simon Nitinol filter migrated to the pulmonary artery but did not produce symptoms.

### Inferior Vena Cava Penetration

Penetration of filter components through the wall of the IVC is frequently demonstrated on follow-up imaging and was documented in 9% of filters in the study of Ferris et al. None of these patients developed symptoms due to penetration, although four Simon Nitinol filter components penetrated adjacent organs. More recent work has suggested that some apparent strut penetrations may actually still be confined to the IVC wall. However, small bowel volvulus and hydronephrosis have been caused by penetration of the IVC by the struts of Greenfield filters in two patients. Although at present reports of serious consequences caused by IVC penetration are rare, the prevalence of this complication may increase as larger numbers of patients have indwelling filters for long periods of time.

### Fracture of Filter Components

Asymptomatic fracture of filter components has been reported with the 24-Fr stainless steel Greenfield filter, the Simon Nitinol filter, and the Vena Tech-LGM filter. There have been no reports to date of fractures occurring with the titanium Greenfield filter. In the study of Ferris et al., 2% of filters showed fracture of a strut or leg during follow-up. No traumatic event could be linked to this complication. Although none of these patients had symptoms, it is possible that a fracture of a filter could result in migration of a component or impairment of the device's ability to trap thrombi.

### Guidewire Entrapment

The potential for dislodgment of an IVC filter caused by entrapment of a guidewire introduced during a venous procedure, such as placement of a central venous catheter, has been investigated in an *in vitro* model. These results have been confirmed, with clinical reports of guidewire entrapment occurring with Vena Tech-LGM and Greenfield filters. The importance of fluoroscopically controlling venous procedures in patients with IVC filters should be stressed both to the patient and his or her physicians. It may be possible to free trapped wires with percutaneous interventional techniques.

## CHOICE OF INFERIOR VENA CAVA FILTER

Currently no ideal IVC filter exists. Several other filter types are available. There are three versions of the Greenfield filter: the original stainless steel filters, the titanium Greenfield filter, and the over-the-wire Greenfield filter. The other available filters are the Bird's Nest filter, the Simon Nitinol filter, and the Vena Tech-LGM filter, which are available for use in North America. Each of the currently available devices possess some advantages and some disadvantages. It is probably more appropriate to become familiar with the insertion of one or possibly two filters than to attempt to use all the available devices. Operator inexperience can result in serious complications. Choice of a device may depend on clinical circumstances, such as the length and diameter of the IVC, the available insertion route, and personal preference.

### 24-French Stainless Steel Greenfield Filter

The 24-Fr stainless steel Greenfield filter (Fig. 16.4) was developed in 1973, and excellent results have been reported with its use. Although it is still regarded as the gold standard for comparison of other filters, the high rate of insertion vein thrombosis occurring after percutaneous insertion has led to it being largely replaced by smaller-profile devices.

### Titanium Greenfield Filter

The titanium Greenfield filter is a modified titanium version of the original 24-Fr stainless steel Greenfield filter (Fig. 16.5). It is cone-shaped, has six limbs, and is 4.7 cm in height. The original design showed a high rate of IVC penetration, so that the hooks at the base were modified. With this modification a device with similar characteristics to the 24-Fr stainless steel Greenfield filter

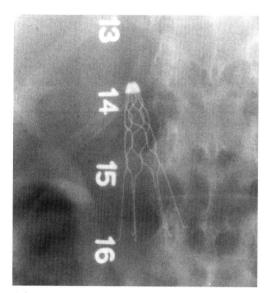

FIG. 16.4. The original stainless steel Greenfield filter.

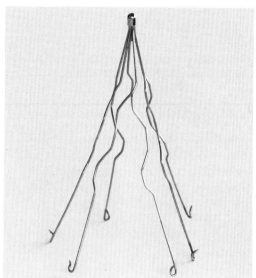

FIG. 16.5. Titanium Greenfield filter.

is considered to have been produced. Although it has a broader base than the 24-Fr stainless steel Greenfield filter, it is recommended for IVC diameters not exceeding 28 mm. The filter is preloaded in a 12-Fr carrier and is placed through a 14-Fr outer-diameter sheath. A reinforced sheath of 15-Fr outer diameter is available and could be used in patients with tortuous anatomy. Because of the relatively rigid carrier, the right common femoral vein and right internal jugular vein are recommended for placement in most situations. The left femoral approach can be used, although kinking of the introduction sheath may occur with this approach. A color-coded filter is available for femoral or jugular approaches. Release of the device is straightforward and uses a trigger mechanism. Unlike the original (24-Fr) stainless steel filter, the device is not deployed over a guidewire, and tilting of the filter (more than 15%) occurs in approximately 23% of placements. Poor distribution of filter legs with clustering occurs in up to 71%. Manipulation of the filter after deployment is feasible; however, the risks and benefits of doing so are unknown and this practice is probably not to be generally recommended. The IVC occlusion rate with the device appears to be less than 5% and the reported recurrent PE rates are also less than 5%. However, the titanium Greenfield filter has shown decreased clot-trapping efficiency compared with the Bird's Nest, Simon Nitinol, and Vena Tech-LGM filters in an *in vitro* model and poorer trapping efficiency than the Vena Tech-LGM filter in an *in vivo* model.

## 12-French Stainless Steel Greenfield Filter

The U.S. Food and Drug Administration has approved the 12-Fr stainless steel Greenfield filter. Like the original Greenfield filter, it is made of 316L stainless steel but the new design permits loading into a 12-Fr carrier. It is placed through a 12-Fr inner-diameter introduction sheath, passed through the sheath over a guidewire, and deployed with the guidewire in place. It is supplied preloaded in jugular or femoral carriers and has a similar release mechanism to the titanium Greenfield filter. A reinforced 15-Fr outer-diameter introduction sheath is also available. Preliminary satisfactory reports of the use of the device have been presented.

## Bird's Nest Filter

The Bird's Nest filter has two V-shaped struts at each end for venous fixation with four intervening wires that form a mesh on deployment (Fig.

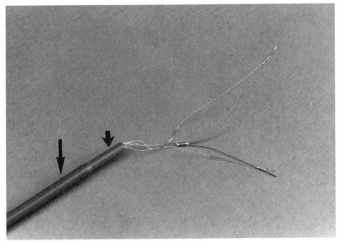

FIG. 16.6. **A:** A Bird's Nest filter has been partially extruded. The introduction sheath (*arrow*), filter catheter (*arrowhead*) and one of the V-shaped struts are shown. **B:** With the entire filter extruded, the guidewire pusher (*arrow*), both V-shaped struts, and four intervening filter wires are shown.

16.6). It is placed through a 12-Fr inner-diameter (14-Fr outer-diameter) introduction sheath. A device is produced both for femoral and jugular approaches, with a longer filter catheter and introduction sheath for the jugular approach. The filter is preloaded in the filter catheter. The Bird's Nest filter can be used when the IVC diameter is up to 40 mm but requires an IVC length of 6 to 7 cm for deployment. Placement involves initial deployment and anchoring of the first V-shaped strut, followed by formation of the four wires into a compact mesh, then deposition and fixation of the second V-shaped strut. Release of the filter from the guidewire pusher has been simplified with the development of a release button. Nevertheless, use of this device is technically more difficult than use of the titanium Greenfield filter, 12-Fr stainless steel Greenfield filter, Simon Nitinol filter, and Vena Tech-LGM filter. Wire prolapse can occur, particularly in patients with a short infrarenal IVC, but this is not generally considered to impair the filter's clinical effectiveness. The Bird's Nest filter is usually placed through a right common femoral or internal jugular vein approach but can be performed through a left femoral approach, although even with this flexible filter design, kinking of the introduction sheath has been described in patients with tortuous anatomy.

The initial design of the device allowed migration to occur. The filter has subsequently been modified, with stiffening of the V-shaped struts and addition of hooks. This has resulted in improved stability, although inferior strut migration has been described in a single patient and another report described two filters migrating to the heart.

The most commonly quoted IVC occlusion rate with the Bird's Nest filter is 2.9%, as reported by Roehm et al. However, Mohan et al. reported a 14.6% rate of symptomatic IVC thrombosis, and just over half of these patients died as a result of extension of thrombus into the renal veins, causing acute renal failure. The development of IVC thrombosis in these patients may have been related to filter wire prolapse.

The device shows excellent clot-trapping efficiency, and recurrent PE rates are less than

5%. The Bird's Nest filter is made of stainless steel, and although it is probably safe for a patient with a well-incorporated Bird's Nest filter to undergo magnetic resonance imaging, the device does produce artifacts that degrade abdominal images.

### Simon Nitinol Filter

The Simon Nitinol filter is 3.8 cm in height and is composed of six anchoring limbs with hooks and a dome (Fig. 16.7). This filter is made from nickel and titanium and is nonmagnetic. It cannot be used when the IVC diameter exceeds 28 mm. It is placed through a 7-Fr inner-diameter (9-Fr outer-diameter) introduction sheath. Infusion of ice-cold saline through the introducer sheath previously was recommended to facilitate the deployment of this filter. This recommendation was valid for the earlier version of the filter. More recently, with the development of a new delivery system, this practice is not necessary for filter deployment. Infusion of cold saline does facilitate the passage of this filter through the delivery system. The filter can then be advanced into and pushed to the end of the introduction sheath. The filter is deployed by withdrawing the introduction sheath. When placed through a femoral approach, the dome may retract caudally during deployment, and the manufacturer recommends allowing the filter dome to fully expand before deployment of the legs to facilitate accurate placement. The filter has a thermal memory and assumes its rigid complex shape when warmed to body temperature in the IVC. It is normally placed through a femoral or jugular vein approach, but the cooled filter is extremely flexible and the small introducer sheath allows placement through even an antecubital vein approach.

Symptomatic IVC occlusion has been reported in 9% of patients, with a total IVC occlusion rate of approximately 12%, although few of the patients in this series received objective imaging of the IVC. The recurrent PE rate is less than 5%, and the device shows excellent clot-trapping efficiency in *in vitro* experiments. However, in the series of Ferris et al., one Simon Nitinol filter migrated to the pulmonary artery and 19% penetrated the IVC, with strut penetration into adjacent structures, including the abdominal aorta, the iliac artery, and the duodenum. Two symptomatic migrations of Simon Nitinol filters into the right atrium and pulmonary artery have been described.

### Vena Tech-LGM Filter

The Vena Tech-LGM filter is a nonmagnetic, cone-shaped device constructed with six flat diagonal struts with stabilizing side rails, each of which has an upward-pointing fixation barb (Fig. 16.8). The original design of the filter had a longer cone than the side rails, which may have been the cause of incomplete opening that sometimes occurred particularly when the jugular approach was used. This seems to have been corrected with the new design of the filter, which has a cone and side rails of almost equal height. The new filter design is 3.8 cm in height.

The device is supplied loaded in a syringe carrier. Three delivery formats are available: femoral only, jugular only, and dual access. The delivery system is very simple, and thus this may be the easiest of the currently available devices to deploy. It should normally be placed through either the right femoral or right internal jugular vein

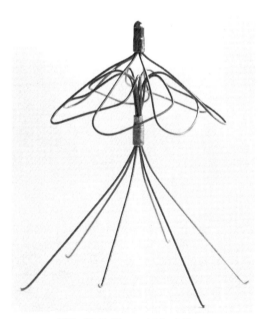

**FIG. 16.7.** Simon Nitinol filter.

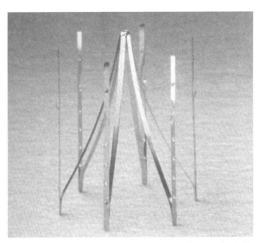

FIG. 16.8. The new, modified Vena Tech-LGM filter. (Courtesy of B. Braun Medical, Evanston, IL.)

## RETRIEVABLE FILTERS

Retrievable filters are permanent filters that have a design feature, usually a hook, to permit snaring and removal. There are currently no approved devices for use in North America. Potential problems with retrievable filters include technically difficult snaring and removal methods and incorporation of the device into the IVC. However, there may be a future role for a retrievable filter as an alternative to a permanent filter in a patient with a long life expectancy who has a short-term contraindication to anticoagulation. The Gunther Tulip filter (William Cook, Europe) is a retrievable filter currently undergoing evaluation in Europe.

## TEMPORARY FILTERS

Temporary filters differ from potentially retrievable permanent filters in that they are attached to a catheter or guidewire so that removal of the device is usually required. Because of the attached catheter or wire, removal may be technically simpler than snaring of a retrievable filter. However, some of the devices require an external catheter, which could predispose to filter infection, particularly through the femoral approach. A potential role for temporary filters may be in protecting patients who are undergoing lower-limb thrombolysis from PE. There may also be a role for temporary filters in young patients who require a filter in whom there is a short-term contraindication to anticoagulation. The Protect Infusion Catheter (C. R. Bard, Inc., Covington, GA) (Fig 16.9) and the Tempofilter (B. Braun Medical) are currently undergoing evaluation in the United States.

approach but can be placed through the left common femoral vein, the right external jugular vein, and the left internal jugular vein. Care should be taken to avoid kinking the introduction sheath when a tortuous approach is used. The filter is simply extruded from the carrier syringe into a 10-Fr inner-diameter introduction sheath. The outer diameter of the sheath has been increased to 13 Fr to stiffen it. The filter is pushed through the sheath using a pusher catheter and the filter is deployed by withdrawing the sheath to uncover the filter, which is held in position by the pusher catheter.

IVC occlusion rates of 4% to 8% are most commonly reported. However, a study in which all patients received objective follow-up imaging of the IVC estimated IVC occlusion rates of 8% at 2 years, increasing to 30% at 6 years. Recurrent PE rates of 2% to 6% are reported. The device shows good clot-trapping efficiency both in *in vitro* and in *in vivo* experiments.

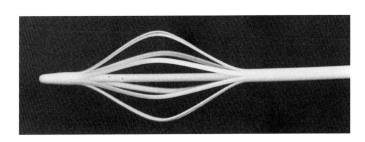

FIG. 16.9. The Protect Infusion Catheter.

## SUGGESTED READINGS

Becker DM, Philbrick JT, Bayne Selby J. Inferior vena cava filters. Indications, safety, effectiveness. *Arch Intern Med* 1992;152:1985–1994.

Crochet DP, Stora O, Ferry D, et al. Vena Tech-LGM filter: long-term results of a prospective study. *Radiology* 1993;188:857–860.

Ferris EJ, McCowan TC, Carver DK, McFarland DR. Percutaneous inferior vena caval filters: follow-up of seven designs in 320 patients. *Radiology* 1993;188:851–856.

Grassi CJ. Inferior vena caval filters: analysis of five currently available devices. *AJR Am J Roentgenol* 1991;156:813–821.

Greenfield LJ, Michna BA. Twelve-year clinical experience with the Greenfield vena caval filter. *Surgery* 1988;104:706–712.

Greenfield LJ, Proctor MC, Cho KJ, et al. Extended evaluation of the titanium Greenfield vena caval filter. *J Vasc Surg* 1994;20:458–465.

Hicks ME, Malden ES, Vesely TM, et al. Prospective anatomic study of the inferior vena cava and renal veins: comparison of selective renal venography with cavography and relevance in filter placement. *J Vasc Intervent Radiol* 1995;6:721–729.

Hirsh J, Hoak J. Management of deep vein thrombosis and pulmonary embolism. *Circulation* 1996;93:2212–2245.

Kaufman JA, Geller SC, Rivitz SM, Waltman AC. Operator errors durring percutaneous placement of vena cava filters. *AJR Am J Roentgenol* 1995;165:1281–1287.

Kaufman JA, Thomas JW, Geller SC, et al. Guide-wire entrapment by inferior vena caval filters: in vitro evaluation. *Radiology* 1996;198:71–76.

Mohan CR, Hoballah JJ, Sharp WJ, et al. Comparative efficacy and complications of vena caval filters. *J Vasc Surg* 1995;21:235–246.

Murphy TP, Dorfman GS, Yedlicka JW, et al. LGM vena cava filter: objective evaluation of early results. *J Vasc Intervent Radiol* 1991;2:107–115.

Neuerburg J, Gunther RW. Developments in inferior vena cava filters: a European viewpoint. *Semin Intervent Radiol* 1994;11:349–357.

Roehm JOF, Johnsrude IS, Barth MH. The Bird's Nest inferior vena cava filter: progress report. *Radiology* 1988;168:745–749.

Reed RA, Teitelbaum GP, Taylor FC, et al. Use of the Bird's Nest filter in oversized inferior venae cavae. *J Vasc Intervent Radiol* 1991;2:447–450.

Simon M, Athanasoulis CA, Kim D, et al. Simon Nitinol inferior vena cava filter: initial clinical experience. Work in progress. *Radiology* 1989;172:99–103.

Simon M, Rabkin DJ, Kleshinski S, et al. Comparative evaluation of clinically available inferior vena cava filters with an in vitro physiologic simulation of the vena cava. *Radiology* 1993;189:769–774.

Vesely TM. Technical problems and complications associated with inferior vena cava filters. *Semin Intervent Radiol* 1994;11:121–133.

# 17
# Central Venous Access

## Maria Rodrigues Gomes and Gwen K. Nazarian

Central venous catheters have become an important tool in modern medical management, enabling delivery of fluids and drugs as well as access for hemodialysis or hyperalimentation and for manometry. The interventional radiologist has become increasingly involved in the placement of these catheters, due to a combination of their knowledge of angiointerventional techniques and their experience in ultrasound- and fluoroscopic-guided access to the arterial and venous systems, both of which greatly assist in successful catheter placement. Experience is especially important when dealing with difficult cases in which the usual venous accesses are not available due to venous thrombosis or stenoses resulting from previous catheters. It is important for the interventional radiologist to be familiar with the wide variety of access sites, the available devices, and their care. The interventionist must also be ready to deal with any complications that can occur, both immediate and delayed.

## CENTRAL VENOUS CATHETERS

Central venous catheters can be broadly divided into two categories: short-term and long-term catheters.

The short-term catheters are intended for up to 3 weeks of use and are most often placed at the bedside. If the access is difficult and the patient's clinical status allows, the patient may be transported to the radiology suite to perform this procedure under fluoroscopic or ultrasound guidance. Examples of these catheters are triple-lumen and Swan-Ganz catheters.

Long-term central venous accesses are divided into external central venous catheters and totally implanted central venous catheters. They are intended to be in place for months.

The external catheters are inserted through a subcutaneous tunnel into the central venous system. These catheters are available in different designs and materials, and have a Dacron cuff attached to them, which is positioned within the tunnel and will, with time, become embedded in the subcutaneous tissues, securing the catheter position and acting as a barrier against infection. The two main types are the Hickman and the Groshong catheters (C. R. Bard, Inc., Covington, GA). The Hickman catheter is made of silicone rubber and has an end hole. The Groshong catheter is made of polyurethane and has a slit on the side near the tip that functions as a valve to prevent blood backflow into the catheter, while still allowing aspiration. This eliminates the need for heparin in the flush solution. Both catheter types come in different French sizes and as single- or multilumen catheters (Fig. 17.1).

Totally implanted central venous catheters can be in place for months to years. These devices consist of a catheter that is tunneled into the central vein just like the external catheters; however, they are connected peripherally to a reservoir that is implanted subcutaneously. The reservoir has a silicone membrane that is accessed percutaneously by special noncutting needles (Fig. 17.2).

Peripherally inserted central catheters (PICCs) have assumed a more prominent position. They consist of 3- to 7-Fr single- or double-lumen catheters placed through a peripheral vein in the arm. They can be left in place for several weeks (Fig. 17.3).

## PATIENT PREPARATION

When obtaining consent, it is important to inquire about relevant past medical and surgical

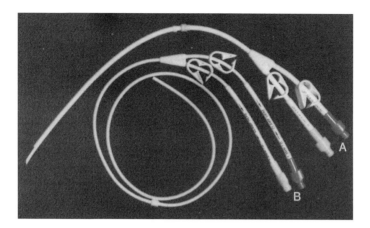

**FIG. 17.1. A:** A 13.5-Fr dialysis catheter with two cuffs. A distal Dacron cuff secures the catheter in position and prevents migration of bacteria around the catheter; the proximal cuff is impregnated with silver nitrate to prevent bacterial growth. **B:** A 10-Fr double-lumen Hickman catheter with only one cuff. (Reprinted from Castañeda-Zúñiga WR. *Interventional radiology*, 3rd ed. Baltimore: Williams & Wilkins, 1997, with permission.)

history such as previous surgery in the area (mastectomy, lymph node dissection, lung resections), radiation therapy to the area, and previous central venous catheters. Also, physical examination to search for signs of venous thrombosis (edema of the extremity, numerous superficial collaterals) should be performed. It is also important to inquire about the handedness of the patient to try and place the catheter in the contralateral side to facilitate care. As the catheters are preferably placed on the nondominant side, the patient should be asked about his or her handedness.

### RISKS

The risks of central venous catheter placement are explained to the patient. These include venous thrombosis, catheter failure, infection (the most frequent and serious complication), catheter breakage, arterial puncture, pneumothorax (rare with fluoroscopically guided placements), bleeding, and air embolism.

Renal function, coagulation, hemoglobin, and platelets are checked if indicated by the clinical history and physical examination. Platelet transfusions are used to increase the level to 50,000 per µL if possible. Coagulopathy is corrected as much as possible with fresh frozen plasma and cryoprecipitate.

An absolute contraindication to placement of a permanent central venous catheter is the presence of bacteremia. A temporary nontunneled catheter can be placed if necessary until blood cultures are negative.

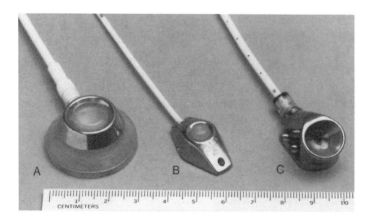

**FIG. 17.2. A:** Port-a-Cath for chest wall placement. **B:** P.A.S. port for placement in the upper arm or forearm. **C:** Cathlink (SIMS Deltec, Inc., St. Paul, MN) can be accessed with an intravenous catheter that can be advanced through the cone-shaped center of the metal housing. (Reprinted from Castañeda-Zúñiga WR. *Interventional radiology*, 3rd ed. Baltimore: Williams & Wilkins, 1997, with permission.)

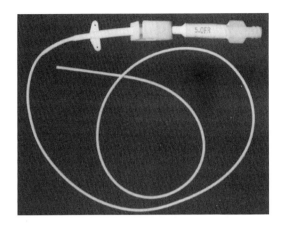

FIG. 17.3. A 5-Fr single-lumen peripherally inserted central catheter. (Reprinted from Castañeda-Zúñiga WR. *Interventional radiology*, 3rd ed. Baltimore: Williams & Wilkins, 1997, with permission.)

## SUBCLAVIAN AND INTERNAL JUGULAR ACCESS IN CATHETER PLACEMENT

### Subclavian Vein Access

#### *Anatomic Considerations*

The axillary vein becomes the subclavian vein at the lateral border of the first rib. The subclavian vein is slightly anterior and inferior to the artery. The thoracic duct enters the subclavian vein close to the junction of this vessel with the internal jugular vein. Structures to be aware of are the phrenic nerve and the internal mammary artery, which lie in contact with the subclavian vein. Also, the brachial plexus is superior and posterior to the subclavian vein.

An important, though somewhat unnoticed, structure is the subclavius muscle, which originates from the inferior surface of the mid-portion of the clavicle and radiates to the first rib, where it has insertions at the junction of the cartilage and the bone. If the catheter passes through this muscle, positional kinking of the catheter can occur, often preventing aspiration and infusion. This can also lead to chronic trauma and eventual breakage of the catheter at this site. This muscle can usually be avoided if percutaneous access is achieved more laterally than the typical surgical procedure.

#### *Fluoroscopic Guidance*

Access to the subclavian vein is often achieved by puncturing into the vein with fluoroscopic guidance and contrast opacification. Contrast medium is injected through a peripheral vein in the ipsilateral arm. The patient is positioned supine, with the head and shoulders in a neutral position and the arm slightly abducted by the side. The subclavian vein is entered as it passes anterior and superior to the first rib. Using this technique, the risk of puncturing the lung and causing a pneumothorax is minimal, as the needle will hit the first rib before it enters the chest. The skin is punctured approximately 3 cm lateral and slightly inferior to the proposed entry point into the vein. This makes the angle from the subcutaneous tunnel into the vein slightly less acute and minimizes the risk of placing the access through the subclavius muscle. A syringe attached to the needle is used to aspirate as the needle is advanced into the vein.

As the needle punctures the vein, one can see an indentation into the contrast opacified vein, and the needle tip can usually be seen to "pop" into the vein. Once blood has been aspirated, the needle is fixed in place. The syringe is removed, and a guidewire is passed through the needle into the central venous system. Should any difficulty be encountered in passing a wire into the right atrium, standard angiographic techniques can be used to identify any stenosis or occlusion and to manipulate a wire into an appropriate position.

#### *Ultrasound Guidance*

Ultrasound guidance may be useful, particularly in cases in which the patient has a history of allergy to contrast agents or abnormal renal function. A 5- to 10-MHz ultrasound probe is used. The subclavian and axillary veins are examined in the transverse plane with regard to compressibility and position with respect to the artery. The first rib can be seen as a shadow on the ultrasound image underneath the vein. The vein is then visualized in the longitudinal plane, and the needle is placed through the skin lateral to the probe. Using real-time visualization, the needle can be seen traversing the tissues, and a short forward thrust is made when the needle is seen indenting the superficial wall of the vein. Often, the needle tip needs to be advanced against—or even through—the posterior wall of the vein, always over the first rib.

### Internal Jugular Vein Access

#### Anatomic Considerations

The internal jugular vein originates in the jugular foramen at the base of the skull and then travels down the neck, until it is just posterior to the medial segment of the clavicle, where it joins the subclavian vein to form the innominate vein. The course of the vein is a straight line from the posterior aspect of the medial border of the clavicle to a point between the mastoid process and the ramus of the mandible. The vein is surrounded by a fascia, which also surrounds the carotid artery and the vagus nerve. The internal jugular vein is medial to the sternocleidomastoid muscle in its cephalad segment and it is between the two heads of the muscle in its middle portion.

#### Ultrasound Guidance

A 5- to 10-MHz probe is used to visualize the vein. The vein and the artery are first examined in the transverse plane using compression. The skin entry site is selected approximately three to four fingerbreadths above the clavicle; however, if a tunneled catheter is to be placed, a lower entry site may be chosen. For tunneled catheters, the puncture is usually made from a lateral approach to allow for a smooth curve and reduce the risk of kink formation. The same techniques are used as for the puncture of the subclavian vein (see above).

### Other Techniques

In complex cases in which stenoses or occlusions are present in the more central venous system, it is helpful to use a transfemoral approach. A guidewire and catheter are negotiated into the subclavian or jugular vein. The catheter is then replaced with an Amplatz gooseneck snare (Microvena, White Bear Lake, MN), and the snare is opened at that point where the occluded vein branch or collateral vein is to be punctured, and used as a target.

As an alternative to contrast opacification, a small guidewire can be placed, either from an arm vein or the femoral vein to the subclavian vein, and used as a target for puncture.

### Formation of a Subcutaneous Tunnel

When venous access has been secured, a subcutaneous tunnel is created. First, a 0.5-cm-long skin incision is made at the puncture site, and a 5-Fr short catheter or vessel dilator is then placed over the wire to replace the needle. As the wire is removed, the distance from the junction of the superior vena cava and the right atrium to the skin entry site is measured. The catheter is left in place, capped, and flushed. A subcutaneous tunnel is created from the puncture site to a second skin incision. For a subclavian or internal jugular access, the second incision is made medial on the chest, preferably presternal or parasternal on the ipsilateral side. The subcutaneous tissue is usually thinnest in this most comfortable location, and there is the least movement of the catheter during different postures. The tissues between both skin incisions are infiltrated with local anesthetic. The tunnel is made approximately 6 to 10 cm long.

To create the tunnel, an 18-gauge needle with a blunt-tipped stylet is passed under direct vision and palpation through the subcutaneous tissues between the two skin incisions. A wire is then advanced through the needle, and the needle is replaced with a peel-away sheath that fits the catheter to be used. The dilator from the peel-away sheath is removed and the catheter pulled through the peel-away sheath. The Dacron catheter cuff is positioned approximately 2 to 3 cm into the tunnel from the skin entry site. The same technique can be used for tunneling translumbar and transhepatic catheters. If the tunnel is long, as it is for translumbar catheters, it can be created in steps.

After the catheter is pulled through the subcutaneous tunnel, it is cut to the appropriate length by using the measures previously made from the right atrium to the skin. The 5-Fr dilator is replaced over a wire for a peel-away or cut-away sheath of adequate size for the catheter being used. The catheter is then placed through the peel-away sheath into the venous system. In the supine position, the tip of the catheter should be positioned in the upper right atrium because it will withdraw some when the patient changes to the upright position. The incision at the puncture site is closed with subcuticular absorbable sutures, and the skin incision at the catheter outlet is closed with monofilament cuticular sutures, which are also used to secure the catheter in position.

## PERCUTANEOUS TRANSLUMBAR CENTRAL VENOUS CATHETER PLACEMENT

Percutaneous translumbar central venous catheter placement is useful in patients whose subclavian and jugular veins are impossible to use due to their occlusion or due to occlusion of the superior vena cava.

### Technique

A 5-Fr pigtail catheter is placed into the inferior vena cava using a femoral vein access. The loop of the pigtail is placed just below the level of the renal veins or at the level of the third lumbar vertebra. The patient is then rolled over into a prone position, and the skin in the lumbar area and right flank is cleansed and draped. Under fluoroscopy and using cephalad angulation of the image intensifier, an entry site is selected above the iliac crest and 8 to 10 cm to the right of the midline. A second skin entry site is chosen on the right flank, preferably in the anterior axillary line and slightly more cephalic, to serve as the permanent catheter exit site, providing easy access to the line.

After a small incision is made (0.5 to 1.0 cm), a needle is advanced under fluoroscopy using a down-the-barrel puncture technique toward the target provided by the pigtail catheter loop. The position of the needle across the loop is confirmed by oblique fluoroscopy. Injection of contrast medium may also be done to confirm the intraluminal position of the needle. A guidewire is then passed through the needle and advanced into the high inferior vena cava. The needle and pigtail catheter are withdrawn. Subsequent tunnel creation and catheter placement are then similar to that for other accesses. The catheter tip is placed at the level of the hepatic veins or just slightly higher.

Ultrasound guidance can also be used to gain access to the inferior vena cava in thin patients and children (Fig. 17.4).

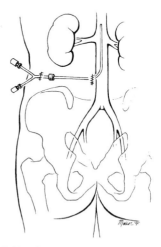

**FIG. 17.4.** For the translumbar catheter the tip is left at the level of the hepatic veins. (Reprinted from Castañeda-Zúñiga WR. *Interventional radiology*, 3rd ed. Baltimore: Williams & Wilkins, 1997, with permission.)

## PERCUTANEOUS TRANSHEPATIC CENTRAL VENOUS CATHETER PLACEMENT

In instances in which a translumbar approach to the inferior vena cava may not be possible or practical, an alternative is the use of percutaneous transhepatic access to the inferior vena cava.

Ultrasound is used to select the access site. An intercostal approach is avoided if possible. The middle hepatic vein is preferred because of its anterior course, which facilitates a subcostal approach and allows for a peripheral puncture. Once there is blood return, a wire is passed through the needle into the infrahepatic inferior vena cava and is manipulated across the right atrium into the superior vena cava to establish a stable wire position. The remainder of the procedure is similar to that described previously.

Proper position of the catheter tip is verified, preferably at the junction of the inferior vena cava and right atrium. In children, growth must be taken into account and an even higher position may be indicated (Fig. 17.5).

Transhepatic catheters are placed in an anterolateral approach rather than from an anterior abdominal approach due to the risk of dislodgment with abdominal distension or growth. Direct complications related to this or the translumbar catheter location are rare, and the catheters have been well tolerated.

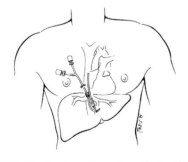

**FIG. 17.5.** For the transhepatic catheter the tip is left in the lower right atrium in adults. For children, however, it is preferably left in the upper portion of the right atrium because growth can cause the catheter to be pulled out into the tract. (Reprinted from Castañeda-Zúñiga WR. *Interventional radiology*, 3rd ed. Baltimore: Williams & Wilkins, 1997, with permission.)

## PERIPHERALLY INSERTED CENTRAL CATHETERS

Peripherally inserted central catheters (PICCs) were developed to downsize from the usual central venous catheters and simplify placement, while still being able to deliver drugs into high-volume central vessels. The main advantage of these catheters is that they are placed into a peripheral arm vein, which decreases the rate of complications. A peripheral vein, preferably in the nondominant arm, is accessed. The preferred veins are the basilic and brachial veins. The catheter is advanced via this vein to the central venous system and into the superior vena cava, preferably to the junction with the right atrium. The lumen of these catheters is small, however, and the catheters are long, so there is a relatively high resistance to flow, which may reduce their usefulness for infusion of blood products and blood drawing. Venous thrombosis is also a potential risk.

The procedure is performed in the angiographic suite, and fluoroscopy is used to verify correct position and for guidance if the venous anatomy is difficult or if obstructions are present. If an obvious superficial vein is not seen when a tourniquet is applied, ultrasound or venography may be used to direct access to the veins, preferably just above the antecubital fossa. After access is obtained, a peel-away sheath is placed and the catheter placed through it.

PICCs are an excellent alternative to centrally placed catheters such as Hickman and Port-a-Caths (C. R. Bard, Inc.) when intermediate to long-term central venous access is needed. The complication rate is relatively low and typically of no clinical significance. The lines can be used for long periods of time—up to several months. Placement is simple and significantly more accurate if the radiologic technique is used.

## TOTALLY IMPLANTED CENTRAL VENOUS CATHETERS

Totally implanted catheters are convenient and cosmetically appealing to patients. Between uses there is no external catheter and the ports can be accessed easily. The complication rate is similar to or even lower than that for other central venous catheters. It is therefore an attractive alternative for patients who need only intermittent access.

The ports can be used with all venous access sites, but the subclavian and internal jugular vein accesses are the most commonly used. The port is placed on the upper anterior chest wall in most cases, and the venous access technique is the same as for other long-term catheters. When access has been secured to the venous system, a skin incision large enough to accommodate the port base is made. Blunt dissections are made down onto the underlying pectoralis muscular fascia. A pocket just large enough to accommodate the port is then created in that plane. A subcutaneous tunnel is made in the same way as described for the other accesses. After the catheter has been pulled through the tunnel, it is connected to the port. A suture is often placed into the base of the pocket just below the incision and threaded through the holes at the base of the port to secure its position.

The pocket is closed by pulling the subcutaneous tissues together using absorbable sutures, and the skin is closed with subcuticular sutures. The silicone membrane should not be directly under the incision site. The port is then flushed using heparin solution and can be used immediately without increased risk of infections or complications.

Arm ports are placed by the interventional radiologist, either in the operating room or in the radiology suite. Strict sterile conditions are applied. The port is usually placed in the nondominant arm, and access is gained to the basilic, cephalic, brachial, or median antecubital vein. Ultrasound is used for access guidance if the vein cannot be accessed in the conventional way. A venogram can also be used to guide the access and to verify patency and continuity of the veins. Once venous access is gained, an incision is made at the skin entry site, and a peel-away sheath is placed. The polyurethane catheter is advanced into the central venous system through the peel-away sheath. A transverse skin incision large enough to accommodate the transverse diameter of the port is made approximately 2 to 3 cm distal to and slightly lateral from the access site. The port must be located either above or below the antecubital fossa because of the flexion at the elbow joint. A pocket is created using blunt dissection large enough to accommodate the entire port. The port is then sutured to the underlying tissues to prevent migration and twisting. The catheter is now cut close to the skin entry site and the end connected to the port. The pocket is closed in the same fashion as that described for the other ports.

Several reports have shown that the arm ports are comparable to other long-term access devices. Arm ports have all the advantages of a totally implantable system but without the need for puncture of a central vein, therefore eliminating some of the complications of central vein access. The infection rate is comparable to those of other centrally placed lines.

## COMPLICATIONS RELATED TO CENTRAL VENOUS CATHETERS

Central venous catheter–related complications can be divided into two main categories: acute and delayed. Acute complications include those directly related to the actual line placement, whereas delayed complications include complications that can be related to the catheter itself, such as infection, thrombosis, and vein stenosis.

### Acute Complications

Pneumothorax is the most common complication encountered when blind puncture of the subclavian vein is made, accounting for 30% of all complications. The incidence varies greatly. However, by using either fluoroscopic guidance or ultrasound guidance, the incidence of this complication can be almost eliminated (0% to 1.6%).

Subclavian artery puncture is the second most common complication with subclavian vein puncture, accounting for 20% of all complications, which may result in a large mediastinal hematoma. The incidence varies from 0% to 7% when blind puncture of the subclavian vein is performed. Puncture of the carotid artery also occurs. Blood can also enter the chest wall and cause an extrapleural hematoma or an acute or delayed hemothorax.

With inadvertent puncture of the subclavian or carotid arteries the needle should be withdrawn and pressure applied for 10 minutes or until bleeding has ceased.

A common complication after subclavian vein catheterization is line misplacement. Misplacement occurs most often during catheter placement, but it may also be delayed due to displacement and caused by shifting of the soft tissues and respiratory motion. One must also be aware of the possible presence of a left superior vena cava, a variation that occurs in approximately 0.3% of the general population and 4.3% of patients with congenital cardiac disease. This is important, as infusion of high-osmolar or otherwise irritating fluids should be made into high-volume veins.

Fluoroscopy is optimally used for line placement, but if not available, a chest film is obligatory after placement.

The ideal position of the catheter tip is at the junction of the superior vena cava with the right atrium. Too low a position can potentially cause atrial or even ventricular arrhythmias. The catheter can also rub against and damage the tricuspid valve. Too high a position can cause erosion through the wall of the superior vena cava, especially if a non-Silastic catheter is placed from the left subclavian vein.

A common problem during placement of tunneled catheters is kinking of the peel-away sheath. Kinks occur as the sheath enters the vein and also where it turns into the superior vena cava, particularly if the right subclavian vein approach is used. Advancing the catheter into the peel-away sheath toward the kink while retracting may help resolve this problem. A hydrophilic-coated wire inside the catheter protruding approximately 5 to 10 cm out will also help lead the catheter through the peel-away sheath. The wire will usually pass the kink easily, and access will then be secured even if the peel-away sheath comes out of the vein.

Catheter fragmentation with distal embolization into the central venous system and pulmonary artery is a well-known complication. This can occur either during placement or removal of the catheter. This was a well-known complication when it was a common practice to place the catheters through a needle and the catheter was

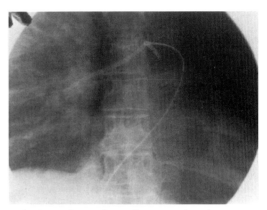

**FIG. 17.6.** The catheter fragment has been captured using an Amplatz gooseneck snare and is being pulled into the main pulmonary artery. (Reprinted from Castañeda-Zúñiga WR. *Interventional radiology*, 3rd ed. Baltimore: Williams & Wilkins, 1997, with permission.)

sheared as it was pulled back through the needle. Delayed pinch-off and catheter breaks have also been reported. Catheter fragments can usually be removed by the interventional radiologist, using a snare (Fig. 17.6).

Arrhythmias occur usually during the line placement secondary to irritation of the right atrium or right ventricle by the guidewires and catheters. Rarely is this arrhythmia symptomatic, but if the patient develops persistent supraventricular tachycardia, this can usually be converted with vagal stimulation. If vagal stimulation is unsuccessful, pharmacologic treatment needs to be used. Adenosine is the drug of choice. Rarely, electric conversion may be necessary.

The presence of other catheters or implants, such as inferior vena cava filters, venous stents, and pacemaker wires, may increase the risk of complications. The wire or the catheter may become caught on these devices. There are reports of dislodgment of such devices during wire manipulations, particularly during blind procedures.

### Delayed Complications

Catheter-related infections are the most important and most common delayed complications. Patients receiving chemotherapy have a higher rate of infections than those receiving parenteral nutrition. Low white blood counts under 2,000 per mL have been found to predispose to catheter-related infections, and counts above 20,000 per mL were also found to correlate with significantly increased risk of catheter-related infections.

Totally implanted vascular access devices have a significantly lower incidence of catheter-related infections than external ones, according to some studies. Prophylactic use of antibiotics has failed to demonstrate a definite decrease in the rate of infection. Multilumen catheters have been found to have a higher incidence of infection than those with a single lumen. The most important preventive factor has been found to be the establishment of intravenous care teams.

Depending on the causal agent and severity, catheter-related sepsis may often be treated without catheter removal. If infectious signs fail to improve in 48 to 72 hours of antibiotic treatment or if infection recurs after cessation of treatment, the catheter should be removed. Exit site infections are less likely to require catheter removal when compared with tunnel infection. Patients with septic thrombophlebitis or septicemia almost always require catheter removal.

Catheter-related thrombosis occurs in 5% to 10% of patients, but the incidence varies greatly among the reports, depending on the criteria used to define catheter-related thrombosis and the means used to detect this thrombosis. The symptoms can be quite variable, and many patients are asymptomatic. A definition of catheter-related thrombosis may include fibrin sheath surrounding the catheter, thrombosis of the catheter lumen, or venous thrombosis. All central venous catheters will develop a fibrin sheath that covers the entire extent of the catheter within weeks. In addition many patients requiring central venous catheters are hypercoagulable secondary to their underlying disease. The catheter materials also vary in their thrombogenicity. Polyvinyl chloride has been shown to be the most thrombogenic. Silastic or silicone rubber is the least thrombogenic.

Low-dose warfarin (1 mg per day) has also been used successfully to prevent thrombosis.

## CATHETER RESCUE

Nonfunctional or malpositioned catheters can often be salvaged by relatively simple measures. This is especially important for tunneled catheters or totally implanted catheters sparing a surgical procedure.

A malpositioned catheter can often be easily repositioned. A temporary catheter may simply be exchanged over a guidewire. By passing a wire through the lumen of a permanent malpositioned catheter and buckling the catheter and wire into the superior vena cava, such malpositions can often be corrected in a simple way. Hydrophilic guidewires are used for this purpose, primarily because of the high friction between regular guidewires and silicone catheters. A strict sterile protocol must be followed. If this is not successful, the rapid flushing of 20 mL of normal saline can sometimes cause the catheter to recoil into the correct position. This is especially successful during placement of the catheter but has not been found to be helpful in catheters that have been in place for a period of time.

If those measures are unsuccessful, a sidewinder catheter can be placed from a femoral approach and the loop of the catheter is then used to pull the catheter down. A pigtail catheter and a vascular snare can also be used for the same purpose.

If necessary, a tunneled catheter can be exchanged by using the same venous entry site but creating a new tunnel. A cutdown is performed on the catheter in the tunnel as it turns toward the vein. The central portion of the catheter is removed over a wire and a new tunnel is created. The peripheral portion of the catheter is pulled out through the skin entry site.

Occluded central venous catheters have been shown to be easily treated by the injection of urokinase or streptokinase into the line. The current protocol calls for the injection of 1 to 2 mL of a solution of urokinase (5,000 units per mL of urokinase) slowly into each lumen. The volume injected is determined by the estimated volume of each lumen. The catheter is clamped for 20 to 30 minutes. The catheter is then unclamped, and the lumen is aspirated of the drug and clot. If the catheter remains occluded, the procedure is repeated up to two or three times. Using this method, the majority of occluded catheters can be salvaged. If this therapy fails, infusion of urokinase over 12 hours can be attempted. Two hundred to 500 units per kg per hour of urokinase can be infused over 8 hours successfully. Also a soft-tipped wire can be advanced into the lumen of the catheter to try and break up the clot mechanically.

A fibrin sleeve stripping may also be performed percutaneously. The procedure consists of initially passing a 0.035-in. guidewire through the catheter into the right atrium or inferior vena cava. An Amplatz snare is then placed from the femoral approach and passed over the wire to the most peripheral portion of the catheter close to its entry site. The snare is then closed gently around the catheter and pulled carefully down to the end of the catheter. Because of the wire through the catheter, the snare can be easily passed over the catheter many times and the procedure repeated as often as needed. The effect can be monitored by injecting contrast through the catheter ports (Fig. 17.7).

The external portion of a tunneled catheter can tear after long use due to fatigue of the material or damage by sharp instruments. Such breaks can be easily repaired using available repair kits.

## CATHETER REMOVAL

For tunneled catheters, after cleansing and draping the skin exit site, local anesthetic is applied, and the external stitches are cut if still present. Using blunt dissection at the catheter skin exit site, the Dacron cuff is released from the subcutaneous tissues, with slight tension applied to the catheter. The catheter is pulled out through the subcutaneous tunnel and pressure applied to the vascular access site until hemostasis is obtained. Elevating the torso of the patient may be helpful as well. The catheter skin exit site can be closed with Steri-Strips to appose the skin edges.

For subcutaneous ports, a skin incision is made over the port, large enough for its removal, usually through the scar from placement. Blunt dissection is used to free the port, and a knife blade can be used to cut the fibrous capsule that forms around the port. The nonabsorbable sutures securing the port are cut and the port removed. The catheter is removed as described previously and the subcutaneous pocket is closed with absorbable sutures.

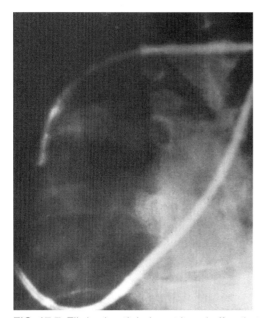 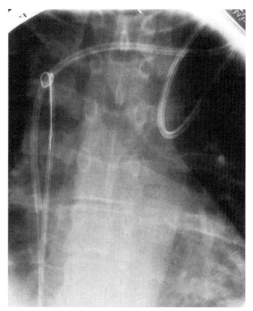

**FIG. 17.7.** Fibrin sheath being stripped off catheter tip using an Amplatz gooseneck snare.

## SUGGESTED READINGS

Akins EW, Hawkins IF Jr, Mucciolo P, et al. Percutaneous central venous catheter placement: use of the blunt needle for subcutaneous tract formation. *AJR Am J Roentgenol* 1992;158:881–882.

Bern MM, Lokich JJ, Wallach SR, et al. Very low doses of warfarin can prevent thrombosis in central venous catheters. *Ann Intern Med* 1990;112:423–428.

Bozzetti F, Scarpa D, Terno G, et al. Subclavian venous thrombosis due to indwelling catheters: a prospective study on 52 patients. *JPEN J Parenter Enteral Nutr* 1983;7:560–562.

Brant-Zawadzki M, Anthony M, Mercer E. Implantation of P.A.S. Port venous access device in the forearm under fluoroscopic guidance. *AJR Am J Roentgenol* 1993;160:1127–1128.

Brothers TE, Von Moll LK, Niederhuber JE, et al. Experience with subcutaneous infusion ports in three hundred patients. *Surg Gynecol Obstet* 1988;166:295–301.

Brown PWG, McBride KD, Gaines PA. Technical report: Hickman catheter rescue. *Clin Radiol* 1994;49:891–894.

Cardella JF, Fox PS, Lawler JB. Interventional radiologic placement of peripherally inserted central catheters. *J Vasc Intervent Radiol* 1993;4:653–660.

Cha EM, Khoury GH. Persistent left superior vena cava. Radiologic and clinical significance. *Radiology* 1972;103:375–381.

Crain MR, Mewissen MW, Paz-Fumagalli R, et al. Fibrin sleeve stripping for salvage of failing hemodialysis catheters: technique and initial results. *Radiology* 1996;198:41–44.

Crummy AB, Carlson P, McDermott JC, Andrews D. Percutaneous transhepatic placement of a Hickman catheter. *AJR Am J Roentgenol* 1989;153:1317–1318.

Denny DF. Placement and management of long-term central venous access catheters and ports. *AJR Am J Roentgenol* 1993;161:385–393.

Eastridge BJ, Lefor AT. Complications of indwelling venous access devices in cancer patients. *J Clin Oncol* 1995;13:233–238.

Grannan K, Taylor P. Early and late complications of totally implantable venous access devices. *J Surg Oncol* 1990;44:52–54.

Henriques H, Karmy-Jones R, Knoll S, et al. Avoiding complications of long-term venous access. *Am Surg* 1993;59:555–558.

Herbst CA. Indications, management, and complications of percutaneous subclavian catheters. *Arch Surg* 1978;113:1421–1425.

Jesseph JM, Conces DJ, Agustyn GT. Patient positioning for subclavian vein catheterization. *Arch Surg* 1987;122:1207–2109.

Kaye C, Smith D. Complications of central venous cannulation. *BMJ* 1988;297:572–573.

Keohane P, Attrill H, Northover J, et al. Effect of catheter tunnelling and a nutrition nurse on catheter sepsis during parenteral nutrition: a controlled trial. *Lancet* 1983;2:1390.

Laméris J, Post PJ, Zonderland HM, et al. Percutaneous placement of Hickman catheters: comparison of sonographically guided and blind techniques. *AJR Am J Roentgenol* 1990;155:1097–1099.

Lund GB, Lieberman RP, Haire WD, et al. Translumbar inferior vena cava catheters for long-term venous access. *Radiology* 1990;174:31–35.

Massumi R, Ross A. Atraumatic, nonsurgical technique for removal of broken catheters from cardiac cavities. *N Engl J Med* 1967;277:195–196.

Mauro M, Jaques P. Radiologic placement of long-term central venous catheters: a review. *J Vasc Intervent Radiol* 1993;4:127–137.

Mitchell SE, Clark RA. Complications of central venous catheterization. *AJR Am J Roentgenol* 1979;133:467–476.

Nazarian GK, Bjarnason H, Dietz CA, et al. Changes in tunneled catheter tip position when a patient is upright. *J Vasc Intervent Radiol* 1997;205(1):173–180.

Northsea C. Using urokinase to restore patency in double lumen catheters. *ANNA J* 1994;21:261–264.

Page AC, Evans RA, Kaczmarski R, et al. The insertion of chronic indwelling central venous catheters (Hickman lines) in interventional radiology suites. *Clin Radiol* 1990;42:105–109.

Ranson MR, Oppenheimer BA, Jackson A, et al. Double-blind placebo control study of vancomycin prophylaxis for central venous catheter insertion in cancer patients. *J Hosp Infect* 1990;15:95–102.

Selby JB, Tegtmeyer CJ, Amodeo C, et al. Insertion of subclavian hemodialysis catheters in difficult cases. *AJR Am J Roentgenol* 1989;152:641–643.

Takasugi J, O'Connell T. Prevention of complications in permanent central venous catheters. *Surg Gynecol Obstet* 1988;167:6–11.

Trigaux J, Goncette L, Van Beers B, et al. Radiologic findings of normal and compromised thoracic venous catheters. *J Thorac Imaging* 1994;9:246–254.

Urbaneja A, Fontaine A, Bruckner M, Spigos D. Evulsion of a Vena Tech filter during insertion of a central venous catheter. *J Vasc Intervent Radiol* 1994;5:783–785.

Yerdel MA, Karayalcin K, Aras N, et al. Mechanical complications of subclavian vein catheterization. *Int Surg* 1991;76:18–22.

# 18
# Foreign Body Retrieval

## Zhong Qian and Hector Ferral

With the increasing uses of indwelling arterial and venous catheters and endovascular devices and performance of interventional procedures, retained foreign bodies or iatrogenically placed devices are frequently encountered. These foreign bodies should be removed as quickly as possible to prevent such potential serious complications as thrombus formation, arterial embolization, myocardial damage, or septicemia. Percutaneous retrieval of intravascular foreign bodies has become a common practice in interventional radiology.

## TOOLS

### Loop Snare

A snare is usually made of guidewire-like material folded on itself and accompanied by an introducer catheter. The early snares were created by a loop formed with an 0.018-in. guidewire in parallel fashion in the same axis as the introducer catheter (Fig. 18.1), but in tubular structures such as the great vessels, a simple loop design was very difficult to engage the foreign body. Therefore, use of the snare technique was not prevalent until the introduction of the Amplatz gooseneck snare (Microvena, White Bear Lake, MN) in 1991. The Amplatz gooseneck snare is constructed of a nickel-titanium (nitinol) cable, featuring a highly radioopaque loop at a 90-degree angle to the cable (Fig. 18.2). The snare loop is available in a variety of sizes ranging from 2 mm to 35 mm in diameter. The nitinol cable has high tensile strength of approximately 18 to 20 lbs (8.1 to 9.0 kg), comparing favorably with the 2 to 3 lbs (0.9 to 1.4 kg) of stainless steel. Thin-walled Teflon shrink-wrap covers the cable shaft to reduce friction. The snare set comes with a polyethylene guiding catheter for facilitating snare manipulation. Use of the Amplatz gooseneck snare has simplified and accelerated the process of retrieving foreign bodies. The Amplatz snare is considered the first choice for such procedures. In general, use of snares is indicated only when the foreign body presents with a free-floating end or double-over segment that can be encircled. Under some circumstances, an object may have to be moved into a more favorable location with other tools such as forceps, tip-deflecting wire, or pigtail catheter before it is feasible to retrieve it with a snare. Snares have proved to be very safe. The only concern is blood loss through the guiding catheter during prolonged procedures. Blood loss can be minimized by using a Check-Flo sheath (Cook, Inc., Bloomington, IN), which has a sidearm attached to the hub of the sheath and a check valve. After the snare loop is passed through the valve, the sidearm assembly allows for flushing of the catheter or injection of contrast medium (Fig. 18.3). It eliminates bleeding during snare manipulation.

### Forceps

Early forceps have been abandoned for intravascular foreign body removal because of their nature of rigidity, which has the potential for damage to vascular and cardiac wall damage. A variety of flexible endoscopic grasping forceps have been adapted and proved to be practical devices for retrieval of intravascular foreign bodies. Today, the most commonly used devices include the Boren-McKinney retriever (Cook Urological Systems, Spencer, IN) and Cook retrieval forceps (Cook Urological Systems). The former device consists of a stainless steel, bivalve, round-edged, flexible forceps (Fig. 18.4). The newer Cook retrieval forceps are composed of single- ("rat-tooth") or multiple-

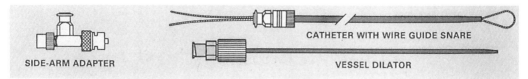

**FIG. 18.1.** Curry intravascular retriever set: snare loop with radioopaque sheath and sidearm adapter to allow flushing during manipulations and at the same time control bleeding along wires. (Courtesy of Cook, Inc., Bloomington, IN.) (Reprinted from Castañeda-Zúñiga WR. *Interventional radiology*, 3rd ed. Baltimore: Williams & Wilkins, 1997, with permission.)

("alligator") tooth forceps mounted on a flexible 0.038-in. shaft (Fig. 18.5). The forceps can be introduced through any catheter that accepts a 0.038-in. guidewire. If the course of retrieval is more tortuous, a slightly larger catheter such as 7- or 8-Fr size may be needed to minimize the friction around bends.

A major advantage of forceps over other techniques is the capability to grasp an object at its middle portion. Therefore, repositioning of a foreign body to a more favorable orientation for snaring as a preliminary step of the procedure can be eliminated. Forceps are particularly useful for retrieval of a foreign body without a free end. The chief concern of using forceps is the potential damage to vessel wall by their teeth, especially in the vascular tree.

### Baskets

Percutaneous retrieval of foreign bodies in the vascular system has been reported with baskets. The most commonly used basket is the Dotter retrieval set (Cook, Inc.), which consists of an 8-Fr Teflon

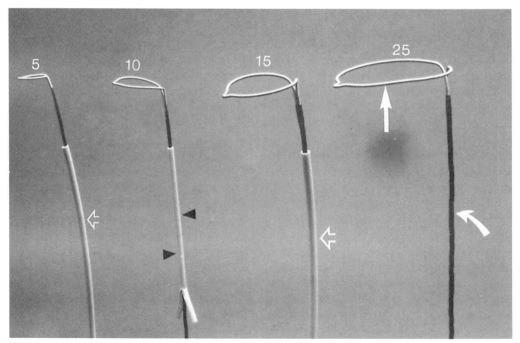

**FIG. 18.2.** Nitinol gooseneck snares (5, 10, 15, and 25 mm). Guiding catheters: 4-Fr (*small open arrow*), 6-Fr (*large open arrow*). "Cheater" for introduction of snare into catheter (*arrowheads*). Teflon shrink-wrap around cable (*curved arrow*). Gold-plated snare loop (*arrow*). (Reprinted from Castañeda-Zúñiga WR. *Interventional radiology*, 3rd ed. Baltimore: Williams & Wilkins, 1997, with permission.)

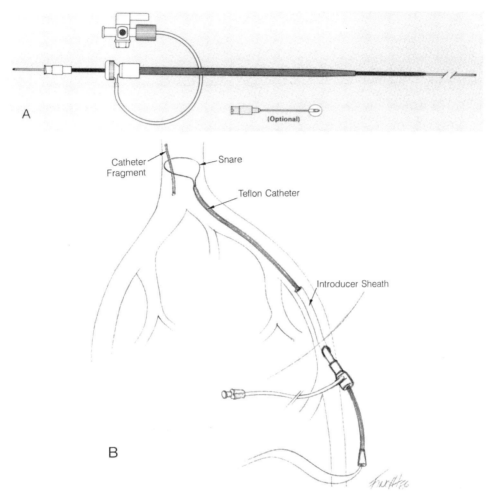

**FIG. 18.3.** Check-Flo sheath. **A:** Rubber seal within the hub prevents blood leakage during manipulations; a sidearm fitting allows flushing around the catheter. (Courtesy of Cook, Inc., Bloomington, IN.) **B:** Diagram of snare loop introduced through a Check-Flo sheath to entrap fragment of catheter in the aorta. (Reprinted from Castañeda-Zúñiga WR. *Interventional radiology*, 3rd ed. Baltimore: Williams & Wilkins, 1997, with permission.)

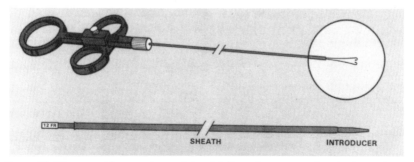

**FIG. 18.4.** Boren-McKinney retriever set consists of 5-Fr radioopaque sheath; three-ring handle with thumbscrew; stainless steel double-lobe retriever, which opens to 2.5 cm long and 3.3 mm diameter; and 12-Fr radioopaque polyethylene introducer with radioopaque Teflon sheath. (Courtesy of Cook Urological Systems, Spencer, IN.) (Reprinted from Castañeda-Zúñiga WR. *Interventional radiology*, 3rd ed. Baltimore: Williams & Wilkins, 1997, with permission.)

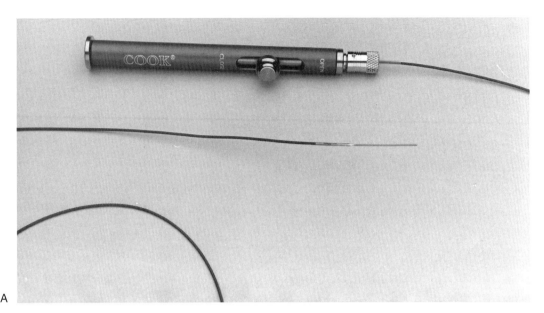
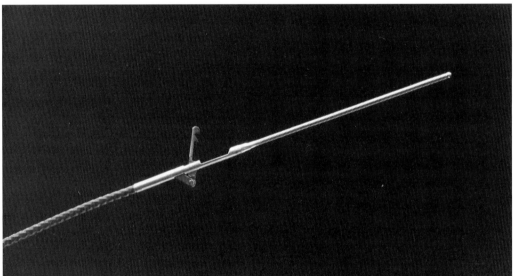

**FIG. 18.5.** Retrieval forceps. **A:** The forceps on a flexible 0.038-in. stainless steel shaft are controlled by a handle. **B:** Close-up of the "alligator" forceps.

catheter and a helical loop basket (Fig. 18.6). The basket is opened and closed by sliding it in and out of the outside catheter. As soon as the foreign body is trapped, the basket and catheter are withdrawn together. The use of a basket is indicated if a foreign body is a nonlinear object such as a bullet or a steel coil. It is useful when the plane of catheter fragment cannot be seen well or the broken catheter is nonopaque. The major drawback of using the Dotter basket is the risk of vascular injury due to the rigid basket tip. Better designed baskets (Cook Urological Systems) specifically for intravascular application are available now. The improved basket has a long, floppy tip (Fig. 18.7) to prevent vascular or cardiac perforation (Fig. 18.8).

**FIG. 18.6.** Dotter intravascular retrieval set consists of catheter with helical loop basket and introducer sheath. (Courtesy of Cook, Inc., Bloomington, IN.) (Reprinted from Castañeda-Zúñiga WR. *Interventional radiology*, 3rd ed. Baltimore: Williams & Wilkins, 1997, with permission.)

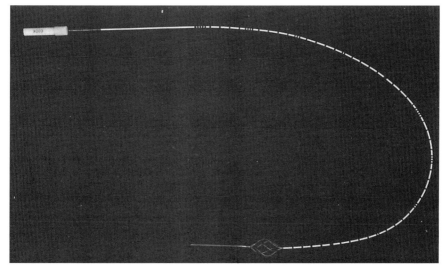

**FIG. 18.7.** Five-wire helical stone extractor with 5-cm filiform tip and 5-Fr radioopaque introducer catheter. (Courtesy of Cook Urological Systems, Spencer, IN.) (Reprinted from Castañeda-Zúñiga WR. *Interventional radiology*, 3rd ed. Baltimore: Williams & Wilkins, 1997, with permission.)

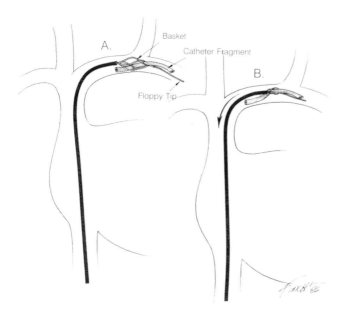

**FIG. 18.8.** Diagram showing removal of catheter fragment from innominate subclavian veins using helical loop basket with floppy tip. **A:** Fragment has been embraced by open basket. **B:** Basket is closed around the fragment, which is slowly withdrawn. (Reprinted from Castañeda-Zúñiga WR. *Interventional radiology*, 3rd ed. Baltimore: Williams & Wilkins, 1997, with permission.)

## Pigtail Catheter

The pigtail catheter can be used to facilitate retrieval of intravascular foreign bodies, especially those in the pulmonary trunk and heart chambers. A foreign body, such as a catheter or wire fragment, can be entangled by rotating a pigtail catheter around the object under fluoroscopy and then pulling it to a place where it is closer to the access sheath. Due to the lack of a tightly grasping mechanism, it is often impossible to remove a foreign body directly through an access sheath by simply pulling the pigtail catheter. This technique is usually used in combination with a snare, which enables removal of the foreign body through the access sheath. The use of the pigtail catheter is particularly useful if a catheter or wire fragment does not have a free end to engage a snare around.

## Balloon Catheter

Balloon catheters (Fogarty angioplasty balloons, Baxter Healthcare Corp., Irvine, CA) have been used for retrieval of catheter fragments from peripheral vessel or as an adjunct to mobilize the fragment into a better location for snaring. However, balloon catheters have no snaring ability and are of little value in the large vessels, such as the vena cava, main pulmonary artery, and heart.

## CLINICAL APPLICATIONS OF PERCUTANEOUS RETRIEVAL TECHNIQUES

Since percutaneous retrieval of intravascular foreign bodies was first reported in 1964, the nature of the foreign bodies and retrieval techniques have evolved with development of contemporary interventional practice. Recent reports describe a large scope of intravascular foreign bodies found in clinical setting, including embolization coils, vena cava filters, and metallic stents, as opposed to broken catheter or wire fragments as mainly encountered foreign bodies in the past.

## Retrieval of Misplaced Coils

A deployed coil without a free end constitutes a challenge for removal with conventional tools. Successful snaring (Fig. 18.9) or basketing of the coil allowed percutaneous extraction in most cases. In general, the forceps offer an advantage in removal of a circular-formed coil in the vessels. However, a flexible device is required to retrieve the coil in more complicated vascular structures such as the pulmonary arteries. In case that a coilon (stainless steel coil with an attached 8-mm polyvinyl alcohol sponge plug) is misplaced, it can be removed from the venous system through an 18-Fr Teflon coaxial dilator introduced percutaneously over the 8-Fr catheter of the snare loop. Such a large lumen is needed for removal of the embolic device because the reexpanded Ivalon plug precludes retrieval through the puncture site or through an introducer sheath. As an alternative to using a large sheath, a smaller-diameter sheath can be used. To allow extraction of a large-diameter foreign body, the sheath is slit longitudinally before introduction. The sheath will then open up or expand along the slit to accomodate the large-diameter object, and both sheath and foreign body are removed simultaneously.

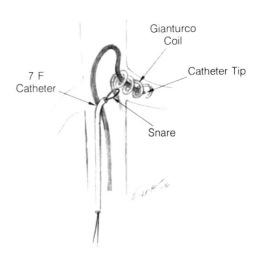

**FIG. 18.9.** Diagram of misplaced Gianturco spring coil (Cook, Inc., Bloomington, IN) protruding into the lumen of the aorta from the left renal artery and held in place by introducing catheter. A snare loop is being used to grab one end of the spring coil for percutaneous removal. (Reprinted from Castañeda-Zúñiga WR. *Interventional radiology*, 3rd ed. Baltimore: Williams & Wilkins, 1997, with permission.)

## Management of Misplaced Vena Cava Filter in the Heart

Although the long-term outcome of misplacement or migration resulting from incomplete opening of vena cava filters in the heart remains unclear, their potential risks of pericardial tamponade or acute myocardial infarction may require retrieval of these filters whenever it is possible. For the misplaced filter in the right atrium, several techniques have been described to achieve successful retrieval. The filter can be retrieved by using a 0.038-in. J-guidewire or the Amplatz snare to grasp the filter and then advancing a 24-Fr Coons-Amplatz sheath (Fig. 18.10) over the filter. If the Amplatz snare is used, the coaxial orientation of the loop to the vena cava permits the snare to be brought down over the filter apex, minimizing the risk of dislodgment and containing the entire filter. With this technique, retrieval of misplaced intracardiac Greenfield filters (Cook, Inc.) has been very successful. The major advantage of this technique is decreased risks of damage to the cardiac and vessel wall by containing filter hooks inside the sheath before retrieval. A basket catheter can also be used to pull the filter into a chest tube with a 45-degree-angle cut tip and then remove it from the right atrium. Repositioning of filters misplaced in the right atrium, instead of retrieval, has been proposed if retrieval is not possible. With this technique, the filter is encircled, collapsed, withdrawn into the inferior vena cava and then allowed to reexpand by using a snare through a femoral vein access. Technically, the repositioning is relatively easy to be accomplished, but unprotected filter struts could cause vascular damage during the procedure.

## Intravascular Bullet Removal

Although intravascular embolization of bullets is relatively uncommon, it has been reported with growing frequency due to the increasing incidence of civilian gunshot wounds. A successful retrieval of a bullet lodged in the venous system can be achieved by using a percutaneous technique. A multipurpose basket (MPB/25) (Boston Scientific, Natick, MA) is used to retract a bullet up to 0.25 caliber (equivalent to a 19-Fr diameter). Once it is trapped by the basket, the bullet is extracted through the femoral vein with a 24-Fr Amplatz dilator (Microvena, White Bear Lake, MN) with a longitudinal split at the distal tip of the tapered portion of the dilator. This distal, slightly opened, wedge-shaped configuration has allowed the bullet to be easily pulled into the sheath (Fig. 18.11). A combination of a balloon catheter and basket has been reported for retrieval of intravascular bullets. With this technique, a 7-Fr occlusion balloon catheter is placed proximal to the bullet to prevent central migration during retrieval of venous bullet emboli. The bullet is then trapped and extracted by a retrieval basket through the femoral access. The large caliber required for these procedures precludes their applications in the arterial system. An arteriotomy would be necessary in such case.

## Removal of Dislodged Catheter from Implanted Venous Access Ports

The use of implanted venous Port-a-Cath (C. R. Bard, Inc., Covington, GA) prevails for maintaining prolonged intravascular infusions or delivering chemotherapeutic agents. Although the port catheter dislodgment or fracture is uncommon, it may cause potentially serious complications due to drug leakage and embolized fragment. Catheter dislodgment should be suspected if there is no blood return from the Port-a-Cath and can further be confirmed by a chest radiograph (Fig. 18.12). In addition to early detection of dislodgment, immediate removal of the dislodged catheter should be attempted whenever possible to prevent further complications. Although there are no statistical data available on modality and mortality resulting from such events, a 71% incidence of serious complications and death was reported in cases in which no attempt was made to remove intravascular iatrogenic foreign bodies.

Most of the dislodged catheter can be easily retrieved with an Amplatz gooseneck snare if it is located in the peripheral veins, the vena cava, or the main pulmonary artery (Fig. 18.13). In case the catheter has lodged in the cardiac walls or the pulmonary arteries, a tip-deflecting wire (Cook, Inc.) may be helpful in

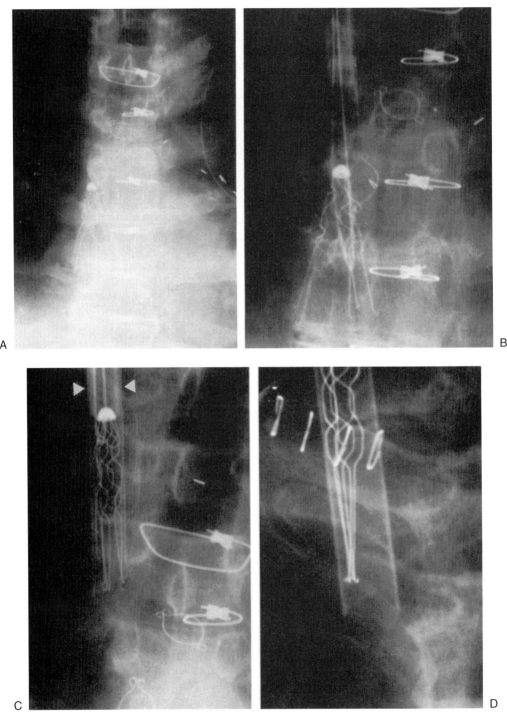

**FIG. 18.10.** Intravascular retrieval of misplaced Greenfield filter. **A:** A misplaced Greenfield filter is found in the right atrium on chest radiograph. **B:** One filter strut is grabbed by the looped guidewire. **C:** The filter is partially captured with the retrieval sheath (*arrowheads*) by advancing the sheath over the filter. **D:** The filter is completely pulled into the retrieval sheath and is ready to be removed from the vascular system. (Reprinted from Malden ES, Darcy MD, Hicks ME, et al. Transvenous retrieval of misplaced stainless steel Greenfield filters. *J Vasc Intervent Radiol* 1992;3:703–708, with permission.)

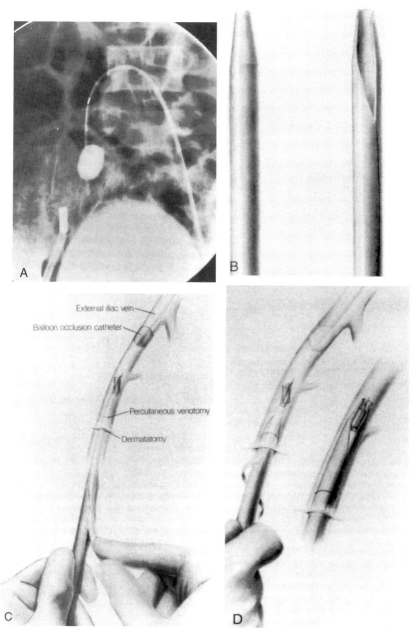

FIG. 18.11. **A:** Bullet retracted to the right external iliac vein; balloon occlusion is used to prevent recurrent embolization. **B:** Amplatz dilator before and after modification. Note the longitudinal split ending in a cut-out wedge at the top of the dilator. **C:** Catheter is fed into the split sheath. **D:** The sheath is rotated over the catheter into the femoral vein. The bullet was then pulled into the sheath. (Reprinted from Gaylord GM, Johnsrude IS. Split 24 Fr Amplatz dilator for percutaneous extraction of an intravascular bullet: case report and technical note. *Radiology* 1989;170:888, with permission.)

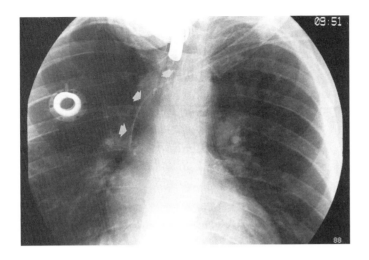

**FIG. 18.12.** Chest radiograph showing a catheter separating from the Port-a-Cath (C. R. Bard, Inc., Covington, GA).

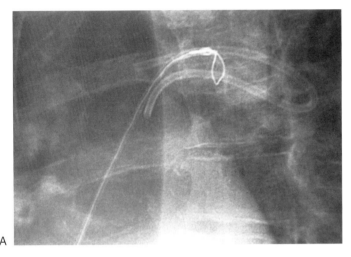

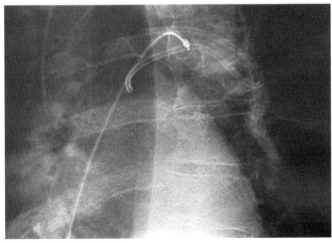

**FIG. 18.13.** A detached catheter migrated into the pulmonary trunk. **A:** A 10-mm snare is advanced over the foreign body. **B:** The loop of the snare is tightened by advancing the snare delivery catheter.

making the catheter end free from the wall, so it can then be grabbed with a snare.

## Retrieval of the Misplaced Wallstent

With increasing placement of the metallic stents, stent dislodgment or migration of stents to the nontarget anatomic sites occasionally occurs. If they become the source of thrombosis or affect the vital organs, such as the heart, removal or relocation of these stents is indicated. Successful retrieval of an embolized Wallstent (Boston Scientific Plymouth Technology Center, Minneapolis, MN) during transjugular intrahepatic portosystemic shunting procedure has been reported. If a Wallstent dislodges to the right atrium, a 25-mm Amplatz gooseneck snare is advanced into the inferior vena cava through a 10-Fr sheath in the right femoral vein. A balloon catheter is introduced through the right jugular access and advanced through the stent over a wire. The stent is then moved up into the superior vena cava by pulling the inflated balloon, so that the stent can be encircled by the snare and retrieved through the 10-Fr sheath placed in the femoral vein (Fig. 18.14). Use of a tip-deflecting wire will facilitate snaring the stent by separating it from the vessel wall during the procedure. If a complete retrieval of the migrated stent is not possible for technical reasons, the stent can be repositioned into the peripheral vessel, such as the iliac vein, as a temporary solution.

## Transvenous Retrieval of Pacemaker Wires

Broken pacemaker electrodes may migrate and entail the risks of caval thrombosis, iliofemoral phlebitis, venous perforation, or pulmonary embolism secondary to thrombosis. Transvenous removal of these migrating wires should be considered either as an alternative to surgical removal or as a complement to operation in cases in which the wire cannot be detached from the myocardium by simply pulling. In these instances, grasping the wire close to the myocardial implant site with a snare will provide the safest traction. Percuta-

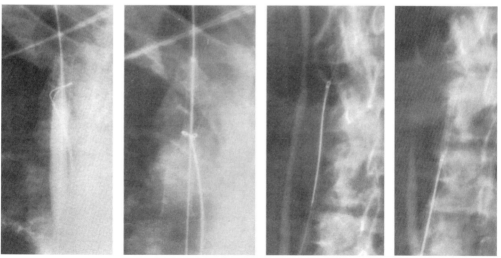

**FIG. 18.14.** Removal of the Wallstent in the superior vena cava. **A:** The Wallstent is surrounded by a snare. **B:** The mid portion of the stent is captured with the snare. **C:** Note that the mid portion of the stent is bent and the Wallstent is pulled down to the inferior vena cava with a snare. **D:** Wallstent is withdrawn into the sheath before removal. (Reprinted from Cekirge S, Foster RG, Weiss JP, McLean GK. Percutaneous removal of an embolized Wallstent during a transjugular intrahepatic portosystemic shunt procedure. *J Vasc Intervent Radiol* 1993;4:559–560, with permission.)

neous removal of pacemaker wires that have been in place for a long period of time (more than 6 months) is contraindicated. Most of those wires are embedded in myocardium, and removal attempts may result in severe myocardial or endocardial injuries.

## REASONS FOR FAILURE

The most common causes for failure of retrieval are as follows:

- No free ends are available for snaring; this problem can be overcome by manipulating the fragment into a more suitable location with a pigtail catheter or a hooked wire. A rotational maneuver with the catheter or wire may facilitate extraction of an entrapped object, such as a catheter or wire fragment.
- The catheter fragment is in a peripheral pulmonary artery branch, where manipulation of the snare onto the free end of fragment can be extremely difficult, particularly if biplane fluoroscopy is not used. However, the availability of the new small sizes (5 to 10 mm) of nitinol snares should facilitate foreign body retrieval from small peripheral vessels.
- Some draining catheter fragments are nonopaque. This problem can be solved by injecting contrast medium to locate the fragment and then performing blind manipulations. As an alternative, real-time ultrasound can be used for the entire procedure.
- One is unable to remove the catheter fragment intact due to the extreme friability of its walls. This complication usually arises only if a catheter is being reused. With improvement in the material technology, this predicament is extremely rare. If this problem is suspected after direct inspection of the proximal fragment, it is probably better either to remove the retained fragment surgically or to leave it alone because further attempts at percutaneous removal might create additional fragments.
- Thrombosis surrounding the foreign body, chronically lodged catheter, or guidewire may be extremely difficult to remove. This is encountered more frequently in the venous system and pulmonary arteries. If the dislodged body has been in a stable position for a long period of time (more than 6 months), there may be formation of clot around the catheter. Removal of these foreign bodies may be not possible.

## SUGGESTED READINGS

Auge JM, Oriol A, Serra C, Crexells C. The use of pigtail catheter for retrieval of foreign bodies from the cardiovascular system. *Cathet Cardiovasc Diagn* 1984;10:625–628.

Braun MA, Collins MB, Sarrafizadeh M, Koslow AR. Percutaneous retrieval of tandem right atrial Greenfield filters. *AJR Am J Roentgenol* 1991;176:872–874.

Cekirge S, Foster RG, Weiss JP, McLean GK. Percutaneous removal of an embolized Wallstent during a transhepatic portosystemic shunt procedure. *J Vasc Intervent Radiol* 1993;4:559–560.

Cekirge S, Weiss, JP, Foster RG, et al. Percutaneous retrieval of foreign bodies: experience with the nitinol Goose Neck snare. *J Vasc Intervent Radiol* 1993;4:805–810.

Chomyn JJ, Craven WM, Groves BM, Durham JJ. Percutaneous removal of a Gianturco coil from the pulmonary artery with use of flexible intravascular forceps. *J Vasc Intervent Radiol* 1991;2:105–106.

Cohen GS, Ball DS. Delayed Wallstent migration after a transjugular intrahepatic portosystemic shunt procedure: relocation with a loop snare. *J Vasc Intervent Radiol* 1993;4:561–563.

Duetsch LS. Percutaneous removal of intracardiac Greenfield vena caval filter. *AJR Am J Roentgenol* 1988;151:677–679.

Egglin TK, Dickey KW, Rosenblatt M, Pollak JS. Retrieval of intravascular foreign bodies: experience in 32 cases. *AJR Am J Roentgenol* 1995;164:1259–1264.

Fisher RG, Ferreyro R. Evaluation of current techniques for nonsurgical removal of intravascular iatrogenic foreign bodies. *AJR Am J Roentgenol* 1978;130:541–548.

Gaylord GM, Johnsrude IS. Split 24 Fr Amplatz dilator for percutaneous extraction of an intravascular bullet: case report and technical note. *Radiology* 1989;170:888.

Lassers BW, Pickering D. Removal of iatrogenic foreign body from aorta by means of ureteric stone catheter. *Am Heart J* 1967;73:375.

Lybecker C, Andersen C, Hansen MK. Transvenous retrieval of intracardiac catheter fragments. *Acta Anaesthesiol Scand* 1989;33:565–567.

Malden ES, Darcy MD, Hicks ME, et al. Transvenous retrieval of misplaced stainless steel Greenfield filters. *J Vasc Intervent Radiol* 1992;3:703–708.

Schneider PA, Bednarkiewicz M. Percutanous retrieval of Kimray-Greenfield vena caval filter. *Radiology* 1985; 156:547.

Sclafani SJA, Shatzkes D, Scalea T. The removal of intravascular bullets by interventional radiology: the prevention of central migration by balloon occlusion—case report. *J Trauma* 1991;31:1423–1425.

Selby JB, Tegtmeyer CJ, Bittner GM. Experience with new retrieval forceps for foreign body removal in the vascular, urinary, and biliary systems. *Radiology* 1990; 176:535–538.

Thomas J, Sinclair SB, Bloomfield D, Davachi A. Nonsurgical retrieval of a broken segment of steel spring guide from right atrium and inferior vena cava. *Circulation* 1964;30:106–108.

Tsai FY, Myers TV, Ashraf A, Shah DC. Aberrant placement of a Kimray-Greenfield filter in the right atrium: percutaneous retrieval. *Radiology* 1988;167:423–424.

Uflacker R, Lima S, Melichar AC. Intravascular foreign bodies: percutaneous retrieval. *Radiology* 1986;160: 731–735.

Yake WF. Percutaneous retrieval of Kimray-Greenfield filter from right atrium and placement in inferior vena cava. *Radiology* 1988;169:849–851.

Yedlicka JW, Carlson JE, Hunter DW, et al. The nitinol "goose-neck" snare for foreign body removal: an experimental study and clinical evaluation. *AJR Am J Roentgenol* 1991;156:1007–1009.

# 19
# Vascular Interventional Procedures in the Pediatric Age Group

## Hector Ferral

Angiography in the pediatric age group has undergone important changes during the last decade. The development of digital subtraction angiography has been by far the most important change in diagnostic angiography; however, the most important innovations in intravascular catheter techniques have occurred in the therapeutic area. Currently, interventional radiology in children is considered as an alternative to surgery in many circumstances, such as angioplasty, embolization, and arterial infusion.

## DIAGNOSTIC ANGIOGRAPHY

Angiography in pediatric patients requires specific considerations:

1. Ensure adequate room and patient temperature.
2. Assess femoral and distal pulses. Puncture the groin with the strongest pulse.
3. Use local anesthesia even if the patient is under general anesthesia because small amounts of a local anesthetic will prevent vessel spasm.
4. Single-wall puncture must be attempted to reduce injury to the arterial wall.
5. Four- and 5-Fr catheters with high torque are available, and these are the catheters to be used in children.
6. Digital angiography provides images of good resolution, provided that the child can hold his or her breath or that he or she is under general anesthesia, and therefore a breath hold can be maintained.
7. Ensure adequate sedation. In some cases, general anesthesia is necessary.
8. Intraarterial digital subtraction angiography has several advantages:
   a. Decreased procedure time
   b. Reduced contrast load
   c. Road-mapping capability

## Percutaneous Revascularization Techniques

Percutaneous transluminal angioplasty (PTA) has limited indications in children due to the absence of atherosclerotic disease in the pediatric age group.

### Percutaneous Transluminal Angioplasty of the Renal Artery

Renal artery stenosis is one of the primary indications of PTA in the pediatric age group. Fibromuscular dysplasia is the most common cause of renovascular hypertension in children, but it can also be secondary to arteritis, trauma, or part of a general dysplasia, as seen in Williams syndrome.

Technical success rates range from 85% to 100%, with reduction in blood pressure in 80% to 90% of patients. Currently, PTA is the therapeutic method of choice for fibromuscular dysplasia.

### Percutaneous Transluminal Angioplasty of Renal Transplantation Stenosis

Arterial stenosis may occur as a complication of renal transplantation, with an incidence ranging from 0.6% to 16.0%, and may be seen in either the native or the allograft artery and vein. The histologic findings in the stenotic renal artery are endothelial proliferation or suture granuloma. Extrinsic compression may also produce arterial stenosis.

### Indications for Angiographic Evaluation

- Patients with hypertension after renal transplantation
- Presence of a bruit over the graft
- Elevated peripheral renin levels

Diagnostic angiography is very important in the evaluation of the vascular problems that follow renal transplantation, and it is important in planning the therapeutic approach. The success and complication rates are difficult to assess, as the series published are generally small and have relatively short follow-up. Barth et al. reported their experience with 18 children. Angioplasty was successful in 14 cases (71%); two of them were redilated due to recurrence. Two of the four failures were due to technical problems, and the other two to transplant rejection.

### Complications

- Intimal dissection
- Branch artery spasm
- Balloon rupture with dissection or perforation of the vessel
- Distal embolization
- Graft loss

We think that these procedures should be performed in an operating room with a surgical team standing by to intervene in case of complications.

## BALLOON DILATION OF MESOCAVAL SHUNTS

Portocaval or mesocaval shunt failure may lead to life-threatening complications. The most common is bleeding, which presents with high mortality rates—up to 50%. Shunt failure is caused by thrombosis due to slow flow through the shunt.

The experience is limited. PTA for portocaval shunt thrombosis in children was performed in two patients with congenital liver fibrosis. These patients required surgery because of bleeding esophageal varices; both underwent end-to-side portocaval shunt procedures. In the postoperative period, both had recurrent bleeding. Angiography showed partial thrombosis of the shunt in both cases. Shunt recanalization was achieved after balloon dilation with control of bleeding.

PTA should be adopted as an emergency therapeutic method in these cases because it is effective and avoids the risks of surgery.

## BUDD-CHIARI SYNDROME

The Budd-Chiari syndrome is characterized by structural and functional abnormalities of the liver caused by obstruction to the outflow of hepatic venous or inferior vena cava flow. The disease is manifested by ascites, variceal hemorrhage, and liver dysfunction secondary to necrosis and fibrosis caused by intrahepatic venous congestion.

Whichever the cause, symptoms and outcome mainly depend on the capacity of portocaval collaterals to provide hepatic and splanchnic perfusion. Depending on the cause and level of the hepatic outflow impairment, treatment options include the following:

- Thrombolysis
- Surgical portosystemic or meso atrial shunting
- Angioplasty
- Transjugular intrahepatic portosystemic shunting (TIPS)
- Stents
- Liver transplantation

The goals of treatment are relief of portal hypertension and preservation of hepatic function. PTA has been used in selected patients with Budd-Chiari syndrome and has shown encouraging results. Balloon angioplasty successfully enlarges the hepatic vein or inferior vena cava lumen, resulting in initial physiologic and symptomatic improvement. The stenosis recurs in most patients and may be caused by the following:

- Fibrotic nature of the obstructing band
- External compression by regenerating hepatic tissue

Endovascular stents have been used to treat this syndrome to maintain vascular patency and prevent re-stenosis after balloon dilation.

At our institution these cases are evaluated by a multidisciplinary team, including hepatologists, transplant surgeons, and interventional radiologists. The decision to proceed with an endovascular approach must be approved by the primary team and all consulting physicians.

## TRANSJUGULAR LIVER BIOPSY

Transjugular liver biopsy is widely used in adults with liver disease when transhepatic biopsy is contraindicated. This method has not been used extensively in children. The main indications are the following:

- Severe coagulopathy
- Massive ascites
- Diagnostic evaluation of portal hypertension

A good set is the liver access and biopsy set (Cook, Inc., Bloomington, IN). This set uses a 7-Fr needle with a 49-cm catheter length. The needle is advanced to the right hepatic vein through an 8-Fr catheter introducer sheath. Then the Tru-Cut stylet enters the hepatic parenchyma, and this is followed by the fast-forward advancement of the outer sheath to obtain the tissue sample. This is a safe method even for children, with a low morbidity rate and a high rate of diagnostic biopsy specimens. Eighteen- or 19-gauge tissue cores may be obtained with this system.

## PORTAL HYPERTENSION AND TRANSJUGULAR INTRAHEPATIC PORTOSYSTEMIC SHUNTING IN THE PEDIATRIC POPULATION

Among pediatric patients, the most common causes of portal hypertension are congenital hepatic fibrosis, biliary atresia, and alpha$_1$-antitrypsin deficiency. Other causes include Budd-Chiari syndrome, cystic fibrosis, and postviral chronic hepatitis. Portal vein thrombosis is a frequent cause of prehepatic portal hypertension in children and should always be investigated as the underlying cause of variceal bleed in the pediatric population. Although several different treatment modalities are available for the management of portal hypertension, there is still a high mortality rate associated with gastrointestinal bleeding due to esophageal varices. Of the new treatment modalities, TIPS has demonstrated some of the best results in preventing recurrent gastrointestinal bleeding.

At this time, no single large series has been published in the literature regarding the use of TIPS in the pediatric population. Most of the reports are based on a few patients or are case reports. There is consensus that the TIPS procedure is indicated in pediatric patients with severe hepatic disease who are on the waiting list for liver transplantation.

The technique for TIPS creation in pediatric patients is very similar to the technique used in adults. In general, the Rösch-Uchida needle system (Cook, Inc., Bloomington, IN) is the preferred access system. Stent diameters will vary according to patients' anatomy. Stents as small as 5 mm in diameter may be used in pediatric TIPS.

After placement of the metallic stent, a follow-up portography is done, followed by pressure measurements of the portosystemic gradient (PSG). If the PSG is above 12 mm Hg, we perform redilation of the metallic endoprosthesis. Redilation is followed by measurement of the PSG to ensure a decrease below 12 mm Hg after treatment.

There is controversy regarding the indications for embolization of gastroesophageal varices after performing a TIPS procedure. Typically, embolization of gastroesophageal varices after TIPS is undertaken only in the following circumstances:

- Patients in whom the TIPS procedure was performed on emergency basis
- Patients with recurrent upper gastrointestinal bleeding
- Patients whose PSG remains above 12 mm Hg after adequate dilation of the shunt

The most frequent problems seen in the mid- and long-term follow-up of these patients include shunt dysfunction due to intimal hyperplasia, stenosis of the outflow hepatic vein, and shunt occlusion. These complications are usually treated with balloon dilation or by placement of a coaxial metallic prosthesis. These procedures are technically simple and are usually performed on an outpatient basis. This approach usually has a good secondary patency rate.

TIPS is an alternative treatment to obtain decompression of the portal system. The role of this procedure in the pediatric population remains to be determined.

## HEPATIC ARTERY STENOSIS AFTER TRANSPLANTATION

Liver transplantation is currently the treatment of choice for certain end-stage liver diseases. Stenoses of the hepatic artery and biliary anastomotic strictures are considered the most frequent complications. Stenoses of the hepatic artery may be due to trauma to the vessels during surgery, extrinsic compression, kinking, and angulation of the vessels. They can also be the result of fibrosis at the anastomotic site. Hepatic arterial anastomotic stenosis is shown clinically as a progressive loss of hepatic function. Some reports consider that PTA is the technique of choice to treat these lesions because it is safe and effective.

## TAKAYASU'S ARTERITIS

Takayasu's arteritis is a nonspecific inflammatory disease characterized by stenosis and occlusion of the aorta and its branches. The main symptoms of the disease are related to chronic arterial insufficiency.

Angioplasty has been used to treat stenosis of the abdominal aorta, carotid, subclavian, renal, and superior mesenteric arteries.

## METALLIC STENTS

Different types of metallic stents have been developed for endovascular applications. In the pediatric population, endovascular stents have been mainly used in the venous system, for pulmonary arteries, and for the creation of TIPS.

We think that self-expandable stents can be useful in treating symptomatic stenoses of the vena cava, especially in cases in which other treatments would represent higher risk to the patient (e.g., surgical correction) (Fig. 19.1). Careful patient selection and long-term follow-up will be extremely important to confirm their benefits.

## FIBRINOLYTIC THERAPY

The use of fibrinolytic agents in the pediatric age group has been limited. The main indications for use are complications related to umbilical artery catheterizations, thrombosed arteriovenous fistulas, and thrombosis resulting from angiographic or interventional procedures.

There are controversies regarding the fibrinolytic agent of choice, the protocols for administration, and the efficacy of this therapy. Urokinase was considered the thrombolytic agent of choice; however, this drug is no longer available, and now alternative thrombolytic drugs have been used—mainly recombinant tissue plasminogen activator.

The technique used to deliver the fibrinolytic agent is important. After diagnostic angiography, a 4- to 5-Fr angiographic catheter is introduced to the level of the occlusion. A 0.035-in. open-ended guidewire is passed coaxially through the angiographic catheter and advanced through the occlusion. Connecting the segment above the occlusion with the distal blood flow is considered an important factor in the success of fibrinolytic therapy. As the open-ended guidewire is withdrawn through the thrombus, the thrombolytic drug is injected. After this initial bolus, a continuous infusion of the drug is started. The patients are returned to the intensive care unit, where the thrombolytic agent is infused with a pump. Serial prothrombin time, activated partial thromboplastin time, C-fibrinogen, hemoglobin and hematocrit, and platelet count should be obtained to monitor treatment progress.

Fibrinolysis is an efficient method in thrombosis of the abdominal aorta and its main branches as a complication of umbilical artery catheterization. It is also valuable to reestablish patency in acutely occluded arteriovenous fistulas for hemodialysis and in the treatment of thrombosis occurring as a complication of angiography or interventional techniques.

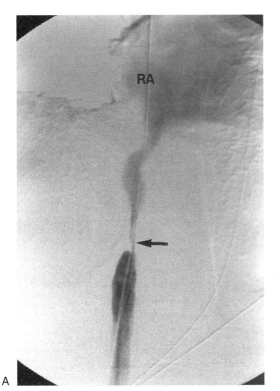
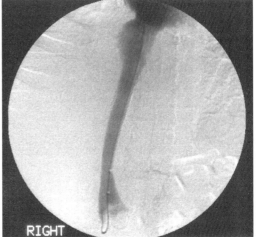
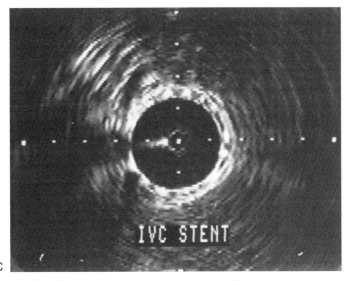

**FIG. 19.1.** A 14-year-old girl with history of recent contraceptive use who presented with massive swelling of both lower extremities. An abdominal computed tomography scan demonstrated caval thrombosis and renal vein thrombosis with extension of clot to both lower extremities. Thrombolytic therapy using low-dose urokinase (60,000 units per hour) was started via bilateral popliteal approach. **A:** Venogram performed 48 hours after urokinase infusion demonstrates a tight stricture of the intrahepatic inferior vena cava (IVC) (*arrow*). (RA, right atrium.) **B:** A balloon-expandable Palmaz stent (Johnson & Johnson, Warren, NJ) was placed with excellent results. **C:** Follow-up intravascular ultrasound 18 months after stent placement demonstrates a widely patent stent.

## EMBOLIZATION

Percutaneous transcatheter embolization is now a valuable procedure for the management and treatment of many vascular disorders in children. In many situations, embolotherapy is an alternative to the more conventional techniques of surgery, radiation, and drugs. Because embolotherapy involves risk, these procedures must be performed by personnel trained in managing the vascular disorders in the pediatric age group. Numerous embolic agents have been used for embolotherapy. The selection of the embolic material is dictated by the indication for treatment and the anatomy of the lesion to be treated. Temporary or permanent embolization agents are used for these procedures.

The embolic agents more frequently used in children are Gelfoam (surgical gelatin sponge), Ivalon (polyvinyl alcohol sponge), absolute ethanol, sodium tetradecyl sulfate (Sotradecol), detachable balloons, and stainless steel coils.

Gelfoam is an absorbable material that can be used for *temporary occlusion*. The occlusion lasts from a few days to a few weeks. It can also be used in combination with a nonabsorbable material for permanent vascular occlusion.

The remaining agents are considered permanent and can be divided into three categories: occlusion devices (stainless steel coils, detachable balloons), particles (Ivalon), and liquid agents (ethanol, Sotradecol, hot contrast medium). Coils should be used to occlude large or high-flow vessels. They usually do not occlude the lumen completely but induce thrombosis. The results of coil embolization are similar to those of surgical ligation, with development of collaterals distal to the site of coil obstruction. Coils are the agents of choice in the treatment of pulmonary arteriovenous malformations. Detachable balloons have the advantage of being carried by the blood to the desired site. They are useful for embolizations in which a high margin of safety is required, as in cranial or pulmonary arteriovenous malformations. The main problems of the detachable balloons are derived from their high cost and the possibility of deflation.

Ivalon is compressible when dry and re-expands to its original size in an aqueous medium such as blood. Particles range in size from 45 to 1,000 µm; thus, Ivalon is useful for permanent occlusion of small and medium-size arteries (Fig. 19.2). Careful dilution of the Ivalon particles in nonionic contrast is essential to avoid obstruction of the delivery catheter by the embolic agent.

Liquid sclerosing agents produce a deeper penetration and therefore a more thorough embolization. When these agents are used, a balloon catheter is necessary to avoid reflux proximal to the area being embolized.

A complete angiographic evaluation is essential before proceeding with embolization. The vessels feeding the lesion to be treated must be clearly identified before proceeding to avoid nontarget embolization. Digital subtraction angiography is of great help in these cases because it reduces the procedure time and reduces the amount of contrast medium used. In the pediatric population, the most common conditions necessitating embolization as a presurgical method include vascular malformations, uncontrollable gastrointestinal hemorrhage, tumors, and hypervascular masses.

## ARTERIOVENOUS MALFORMATIONS

Numerous methods have been used for the treatment of arteriovenous malformations: surgical excision, ligation of the feeding vessels, injection of sclerosing agents, and radiation. These methods, except total surgical excision, are successful in a small number of patients but commonly are followed by recurrence due to the development of collateral vessels. Therapeutic transcatheter embolization or direct injection of sclerosing agents probably offers the best results. Some authors have reported success in embolizing vascular malformations in different areas such as the head and neck, the abdomen, and the extremities.

The most important complications of arteriovenous malformations are pain, hemorrhage, ischemia, congestive heart failure, and platelet sequestration. The indications to treat these lesions are sometimes purely cosmetic. Before embolization, a careful angiographic study must

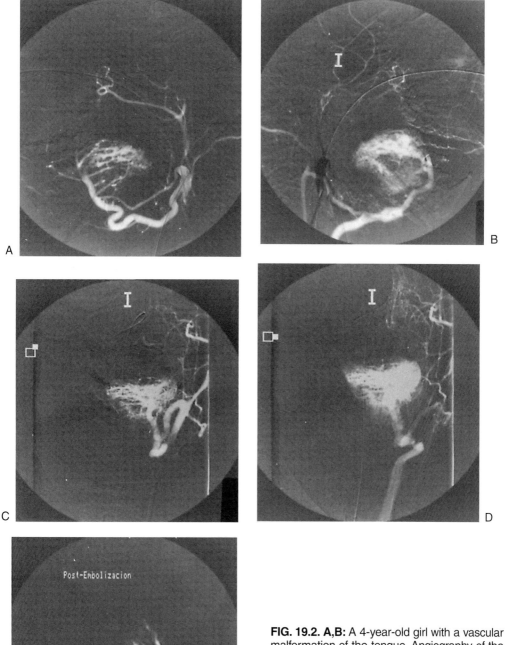

**FIG. 19.2. A,B:** A 4-year-old girl with a vascular malformation of the tongue. Angiography of the left external carotid artery shows an arteriovenous malformation with dilatated feeding artery and venous drainage. **C,D:** Lateral views of the left external carotid arteriogram. **E:** The arteriogram after embolization with Ivalon shows total occlusion of the arteriovenous malformation. (Reprinted from Castañeda-Zúñiga WR. *Interventional radiology*, 3rd ed. Baltimore: Williams & Wilkins, 1997, with permission.)

be carried out to demonstrate all the feeding arteries, the nidus, and the venous drainage. Occlusion of the proximal feeding arteries will result in recurrence in a short time due to the development of collaterals. Embolization must be performed as close to the nidus as possible, and the particles must occlude the peripheral vessels without passing through the malformation into the venous circulation. Large or very complex malformations have to be embolized in more than one session, and in these cases it is very important to keep patent the proximal vessels for subsequent procedures.

Ivalon is the ideal permanent occlusion agent for the treatment of arteriovenous malformations, considering that the size of the particles must be chosen carefully so they obstruct the vessels at the desired level. Because they occlude large feeding vessels, the occlusion devices are not the best embolic agents, although they can be used in arteriovenous malformations with high flow rates and in pulmonary arteriovenous malformations, which are generally fed by a single vessel.

Many patients experience symptoms of pain, nausea, vomiting, and fever after embolization. Pain may begin just after embolization and last for several days. Analgesics must be used, depending on individual patients, but in liberal amounts. Steroids do not reduce pain but can be useful to reduce edema and postembolic neuritis.

## HEMORRHAGE

Transcatheter embolization is the treatment of choice in bleeding originating from trauma and hemoptysis.

### Gastrointestinal Bleeding

The demonstration of extravasation of contrast medium is the only definitive sign of gastrointestinal bleeding, although some vascular abnormalities such as angiodysplasia, arteriovenous malformations, and tumor neovascularity may suggest the existence of a potential source of bleeding. Active bleeding of at least 0.5 mL per minute is necessary for detection by angiography. After identification of the bleeding site, the angiographic management includes pharmacologic agents (vasopressin) or the use of embolization techniques. Vasopressin (Pitressin) can be administered intravenously or intraarterially, selectively in the bleeding vessel. The selective infusion produces high regional levels of vasopressin, reducing the dose required for intravenous infusion and avoiding the side effects of high doses, such as vasoconstriction of other territories such as the myocardium. Vasopressin is supplied in 20-unit ampules. The infusion solution is prepared by diluting 100 units of vasopressin in 500 mL of normal saline (0.2 units per mL). The initial dose for bleeding control is 0.2 units per minute with an infusion pump. A control angiogram is performed 20 minutes after the initial infusion to evaluate vasospastic effect and control of bleeding. If a good spastic effect is achieved with successful control of bleeding, the dose is continued at this rate. If bleeding persists, the dose is increased to 0.4 units per minute. The patients must be followed closely in the intensive care unit. The vasopressin dose is tapered slowly on a 0.1-unit basis every 6 to 12 hours. If vasopressin is not effective in controlling the bleeding episode, more aggressive options must be considered—that is, embolization or surgery. Embolization is considered an alternative for the treatment of gastrointestinal bleeding. The treatment of transient lesions such as ulcers, erosions, and diverticula can be carried out with temporary materials. Gelfoam is useful in these cases because it is absorbed within a few days, permitting vessel recanalization once the acute period is over. Bleeding from lesions that will not heal spontaneously, such as arteriovenous malformations, tumors, or varices, requires permanent occlusion. In these cases, embolization is used to control the bleeding and to solve the underlying problem.

### Trauma

Uncontrollable hemorrhage from trauma can be a challenge for surgery. Localization and ligation of the injured vessels are often unsuccessful, especially in trauma involving the pelvis or retroperitoneum, resulting in further blood loss. Transcatheter embolization is a quick and effective method that can solve many of these life-threatening situations. Angiography can easily find the bleeding site, but one should always remember that more than one bleeding point may exist. A complete angiographic map of the involved area must be carried out before embolization. If the bleeding site is demonstrated, the most effective therapy is superselective catheterization and embolization with permanent agents. Subselective embolization with Gelfoam may be used in situations in which the bleeding site is not clearly documented. The purpose of this maneuver is to decrease the pressure of the area of highest suspicion. Embolization with Gelfoam is useful in these cases because it is a material that produces peripheral occlusion and permits posthealing recanalization. If rebleeding occurs after subselective Gelfoam embolization, a second look angiography is indicated to look for a specific bleeding site.

Occlusive devices can be used in vessels with high flow or in the case of massive bleeding in which an important arterial branch must be occluded as rapidly as possible.

### Hemoptysis

Cystic fibrosis is one of the most frequent causes of massive hemoptysis in the pediatric age group. Conservative methods are useful in treating minor bleeding, but in massive hemoptysis (300 mL within 24 hours or 50 mL per day for 3 consecutive days), surgery or transcatheter embolization is required. Bronchial artery embolization is an effective and widespread method for the treatment of massive or recurrent bronchial artery hemorrhage. A careful angiographic evaluation is imperative. The potential bleeding vessels must be identified, and in addition, when embolizing the bronchial circulation, we must remember that the spinal arteries frequently arise from the bronchial circulation. Embolization of a spinal artery during a bronchial artery embolization may have devastating consequences such as postembolization paralysis. Bronchoscopy may help locate the bleeding site, although in massive hemorrhage the blood makes localization impossible. The embolic agent of choice is Ivalon because it produces a permanent and distal occlusion. All significant vessels to the area must be occluded for effective hemostasis. It is important to keep the proximal vessels patent to allow repeated embolization because recurrent hemoptysis is frequent. These recurrences may be due to either recanalization of the previously embolized artery or revascularization by collateral circulation. The technique of choice is superselective catheterization of the bleeding vessels with coaxial microcatheter techniques.

## INTRALUMINAL FOREIGN BODY RETRIEVAL

The widespread use of central venous catheterization and percutaneous manipulations has resulted in more frequent embolization of catheter fragments from the peripheral vessels to the central venous system, the cardiac chambers, and the pulmonary artery. The potential risks of a foreign body include infection, thrombosis, and perforation. Therefore, percutaneous removal of centrally embolized foreign bodies has become the procedure of choice in the management of these events. The loop snare technique and the modified Dormia stone basket are the most widely used methods for foreign body retrieval. Snare loops are useful when one of the ends of the embolized fragments moves freely in the lumen of the vessel. In those cases in which no free end can be seen, the retrieval is better performed with the Dormia basket. Percutaneous manipulation in the pulmonary artery has some risks, including arrhythmias and, in some cases, cardiac perforations. The arrhythmias are more frequently induced during manipulations within the right heart than by contrast medium injection. Although severe arrhythmias appear in less than 1% of cases, extrasystoles and mild arrhythmias are quite frequent. The incidence of these events is even more common in those cases that require frequent exchange of guidewires and catheters.

We have performed selective catheterization of the pulmonary artery using a long introducer sheath with a curved tip, which provides us with a direct pathway to the pulmonary artery. The approach to the pulmonary artery is done through the right femoral vein with a long (70-cm) 8-Fr introducer sheath with a hemostatic valve. Initially, a catheter is advanced selectively into the pulmonary artery, and subsequently, the 8-Fr sheath may be advanced selectively into the pulmonary artery of choice.

After positioning the long introducer sheath in the pulmonary artery, catheters can be passed from the femoral vein to the pulmonary artery across the introducer sheath without the risk of causing arrhythmias. The technique can be used for foreign body removal and for embolization of pulmonary arteriovenous malformations.

## LONG-TERM CENTRAL VENOUS ACCESS

Tunneled catheters are widely used to provide long-term central venous access. Indications include administration of chemotherapy, long-term antibiotics, total parenteral nutrition, hemodialysis, and pheresis. These catheters have traditionally been placed surgically either by means of a cutdown on the cephalic, internal, or external jugular veins or by gaining percutaneous vascular access to the subclavian or internal jugular veins. Percutaneous tunneled catheter placements can be performed in children (ages ranging from 7 to 18) with local anesthesia, although some patients, due to their age or special psychological characteristics, require general anesthesia. Both internal jugular veins can be used, but the vein of choice is the right internal jugular vein. We recommend real-time ultrasonographic guidance for vessel puncture. This technique is simple and very safe. The incidence of complications such as arterial puncture and pneumothorax is reduced significantly. The use of the subclavian veins for access should be avoided because it is well documented that the subclavian vein access is associated with a higher incidence of complications, including central vein thrombosis and long-term venous strictures. Prophylactic antibiotics are administered, and extreme care is taken with skin preparation.

Two 1-cm skin incisions are made: The first is made 1 to 2 cm above the medial end of the clavicle at the venous access site, and the second is made 4 to 6 cm inferiorly on a point equidistant between the nipple and the midsternal line. With minimal dissection, a tunnel is created with a tunneler device from this point to the venous entry site. The required length of the catheter is measured fluoroscopically, and the catheter is passed through the tunnel to the venous access point. The dacron cuff is placed 1.5 cm deep to the inferior incision.

The percutaneous tract to the internal jugular vein is subsequently dilated and a peel-away sheath is introduced to facilitate passage of the catheter into the superior vena cava. Once the position of the catheter in the superior vena cava has been confirmed using fluoroscopy, the peel-away sheath and wire are removed and sutures are placed at the level of the surgical incision. The catheter is usually sutured in place and is flushed and locked with heparinized saline to prevent thrombosis. A chest x-ray is obtained to document catheter position. The duration of the procedure ranges between 15 and 30 minutes.

The percutaneous approach has proved to be effective for the insertion of long-term tunneled catheters and considerably reduces operative time.

Advantages of percutaneous access under fluoroscopic guidance include the following:

- Shorter procedure time
- Higher technical success rates
- No vessel ligation is performed
- Avoids catheter malposition

Percutaneous long-term tunneled venous catheter insertion can be expeditiously performed in the interventional radiology suite.

## SUGGESTED READINGS

Abad J, Hidalgo E, Cantarero J. Hepatic artery anastomotic stenosis after transplantation: treatment with percutaneous transluminal angioplasty. *Radiology* 1989;171:661–662.

Barth MO, Gagnadoux MF, Mareschal JL, et al. Angioplasty of renal transplant artery stenosis in children. *Pediatr Radiol* 1989;19:383–387.

Cohen AM, Doershuk CF, Sterm RC. Bronchial artery embolization to control hemoptysis in cystic fibrosis. *Radiology* 1990;175:401–405.

Coldwell DM, Stokes KR, Yakes WF. Embolotherapy: agents, clinical applications and techniques. *Radiographics* 1994;14:623–643.

Furuya KN, Burrows PE, Philips MJ, Roberts EA. Transjugular liver biopsy in children. *Hepatology* 1992;15:1036–1042.

Gomes AS. Embolization therapy of congenital arteriovenous malformations: use of alternate approaches. *Radiology* 1994;190:191–198.

Hackworth CA, Leef JA, Rosenblum JD, et al. Transjugular intrahepatic portosystemic shunt creation in children: initial clinical experience. *Radiology* 1998;206:109–114.

Hubbard AM, Fellows KE. Pediatric interventional radiology: current practice and innovations. *Cardiovasc Intervent Radiol* 1993;16:267–274.

Johnson SP, Leyendecker JR, Joseph FB, et al. Transjugular portosystemic shunts in pediatric patients awaiting liver transplantation. *Transplantation* 1996;62:1178–1181.

Nosher JL, Shami MM, Siegel RL, et al. Tunneled central venous access catheter placement in the pediatric population: comparison of radiologic and surgical results. *Radiology* 1994;192:265–268.

Park JH, Haw MC, Kim SH, et al. Takayasu arteritis: angiographic findings and results of angioplasty. *AJR Am J Roentgenol* 1989;153:1069–1074.

Yakes WF, Rossi P, Odink H. Arteriovenous malformation management. *Cardiovasc Intervent Radiol* 1996;19:65–71.

# 20

# Genitourinary Interventions

## Hector Ferral

## A. PERCUTANEOUS UROLOGIC TECHNIQUES

### RENAL ANATOMY

Knowledge of the anatomy of the kidneys is important before proceeding with percutaneous urologic techniques. From a practical standpoint, it is important to know the size and shape of the kidneys, calyceal anatomy, arterial anatomy, and the most common anatomic relations. Knowledge of the anatomy of the kidneys is the basis for prevention of complications during percutaneous interventional procedures. The human kidneys are retroperitoneal organs lying on each side of the vertebral column between the twelfth thoracic and the second to third lumbar vertebrae. The upper pole is more posteriorly located than the lower. The lateral margin of the kidney is more posteriorly positioned than the medial edge (Fig. 20.1). The surface of the kidney is covered by a fibrous nonadherent capsule. The organ is surrounded by the perirenal fat, which extends into the renal sinus. The fat is enclosed by the posterior and anterior layers of the renal fascia (Gerota's fascia) (Fig. 20.2).

Posteriorly, the twelfth rib crosses the kidney at a 45-degree angle. On the left, the lower half of the kidney usually extends below the pleural reflection, whereas on the right, approximately two-thirds of the kidney is located below this reflection (Fig. 20.3). These anatomic relations are important when considering the optimal percutaneous approach to the collecting system. Certain complications are associated with the intercostal approach: hydrothorax, pneumothorax, traversal of the lung parenchyma, catheter dislodgment (more common with an intercostal catheter), and postprocedural pain (caused by rubbing of the sheath or nephrostomy tube against the inferior costal margin, causing irritation of the intercostal nerve). If an intercostal approach is used, it is better to perform it between the eleventh and twelfth ribs.

Other anatomic structures that are in close relationship to the kidneys include the liver, spleen, and colon (Fig. 20.4). *Puncturing just off the lateral border of the paraspinal muscles (usually four fingerbreadths lateral to the spinous process) provides a consistently safe route into the renal collecting system by an infracostal route.* Punctures lateral to this reference are not recommended because the risk of traversing spleen, liver, or colon is much higher.

The adult kidney has an average of 8.7 calyces and 10.7 papillae. There is great variability in the positioning of the calyces. The optimal access should enter a posterior calyx on end; therefore, before puncturing, one should use multiple views on fluoroscopy to determine which calyx or calyces are posterior and better suited for access. The arterial anatomy of the kidney usually shows a main renal artery that divides into an anterior and posterior branch. Usually, the anterior division is the direct continuation of the main renal artery and divides into three or four branches, whereas the posterior division continues without significant branching. There are four anterior segmental arteries and one posterior segmental artery. These segmental arteries are end arteries that do not intercommunicate. The segmental arteries divide into 8 to 12 interlobar arteries. Any puncture and subsequent tract dilatation should avoid transection of any large artery. It is important to gain access to the collecting system in the most posterior calyx, as peripherally as possible. The needle should enter the calyx end-on rather than side-on to avoid transection of a major artery (Fig. 20.5).

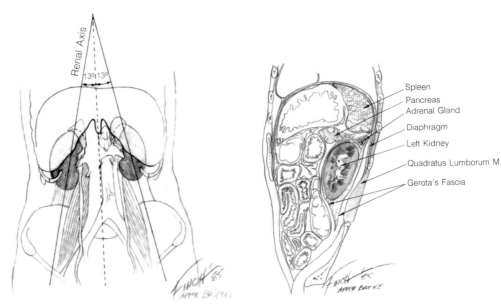

**FIG. 20.1.** The kidneys lie along the course of the psoas muscles, which tilt their axes approximately 13 degrees from the vertical so that the upper pole is more medial than the lower pole. (Reprinted from Castañeda-Zúñiga WR. *Interventional radiology*, 3rd ed. Baltimore: Williams & Wilkins, 1997, with permission.)

**FIG. 20.2.** Left sagittal section of the abdomen. Kidneys lie next to psoas muscles anteroposteriorly and follow the lordotic curve of the spine. Consequently, the upper poles are located more posteriorly than the lower poles. (M., muscle.) (Reprinted from Castañeda-Zúñiga WR. *Interventional radiology*, 3rd ed. Baltimore: Williams & Wilkins, 1997, with permission.)

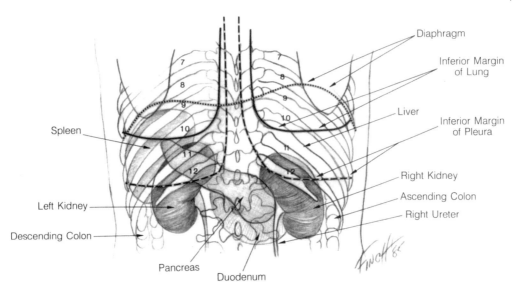

**FIG. 20.3.** Posterior view of the abdomen showing the relationship between the kidneys and the pleura. (Reprinted from Castañeda-Zúñiga WR. *Interventional radiology*, 3rd ed. Baltimore: Williams & Wilkins, 1997, with permission.)

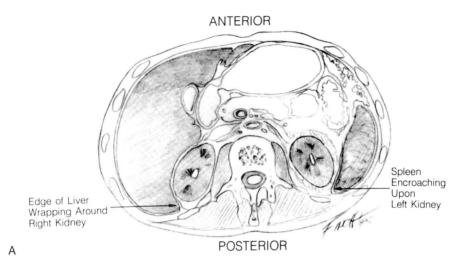
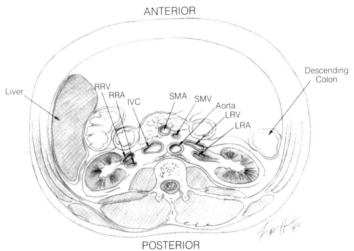

**FIG. 20.4. A:** The relationship between the kidneys and the liver and spleen. **B:** The relationship between the right kidney and the lower aspect of the liver and the left kidney with the descending colon. This emphasizes the importance of a medial puncture (four fingerbreadths lateral to the spinous process is a good rough reference). (IVC, inferior vena cava; LRA, left renal artery; LRV, left renal vein; RRA, right renal artery; RRV, right renal vein; SMA, superior mesenteric artery; SMV, superior mesenteric vein.) (Reprinted from Castañeda-Zúñiga WR. *Interventional radiology*, 3rd ed. Baltimore: Williams & Wilkins, 1997, with permission.)

## INDICATIONS FOR PERCUTANEOUS NEPHROSTOMY PUNCTURE

1. Urinary obstruction
   a. Operative ligation of the ureter
   b. Postpyeloplasty complications
   c. Impacted stone
   d. Fungal infection
   e. Retroperitoneal fibrosis
   f. Benign ureteral obstruction
   g. Malignancy
   h. Pyonephrosis
   i. Compression of ureter by the pregnant uterus
   j. Urodynamic studies (Whitaker test)
2. Urinary intervention
   a. Stone removal
   b. Ablation of calyceal diverticulum

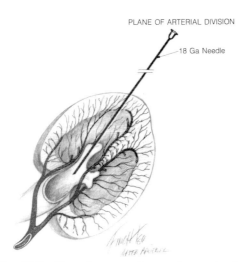

**FIG. 20.5.** The diagram shows the optimal course of the access needle into the collecting system through the plane that is relatively avascular to avoid damage to large vascular structures. (Reprinted from Castañeda-Zúñiga WR. *Interventional radiology*, 3rd ed. Baltimore: Williams & Wilkins, 1997, with permission.)

    c. Percutaneous pyeloplasty
    d. Nephroscopy to obtain biopsy
    e. Foreign body removal
    f. Infusion of solvents or drugs
    g. Treatment of urinary fistulae
    h. Placement of double-J stents
  3. Intervention in transplanted kidney
    a. Obstructions
    b. Fistulas or urinary leaks
    c. Stone removal
    d. Infusion of drugs

## CONTRAINDICATIONS TO NEPHROSTOMY PUNCTURE

The only major contraindication to nephrostomy puncture is bleeding diathesis. This can be corrected with the administration of fresh-frozen plasma, platelets, or specific blood components.

Urosepsis and pyonephrosis are not contraindications. In these conditions, a percutaneous nephrostomy tube placement may be a life-saving procedure. Prophylactic antibiotics are required and tube placement should be performed with minimal manipulation.

## PATIENT EVALUATION

1. Clinical history and physical examination.
2. Informed consent.
3. Coagulation profile.
4. Complete blood count (including platelets).
5. Creatinine and blood urea nitrogen.
6. Urinalysis.
7. Urine culture.
8. Review the radiologic studies. Usually these patients have an ultrasound, computed tomography scan, or intravenous pyelogram that may provide useful information to decide the best approach for access.

## PREPROCEDURAL PREPARATION

Discussion of the case with the referring physician is very useful. Access options, expected risks, and benefits should be discussed. This is especially important in patients in whom an interventional procedure is planned (e.g., stone removal, double-J placement, ureteral dilatation).

The expected risks, benefits, and planned interventions must be discussed with the patient as well. An informed consent should be obtained. A good intravenous line should be in place for administration of analgesics, sedatives, antibiotics, and fluids. Prophylactic antibiotics are indicated. The antibiotic of choice and dosage may be discussed with the patient's primary physician.

## PROCEDURAL CARE AND TECHNIQUES

The patient should be connected to a monitor for continued monitoring of vital signs (oxygen tension, pulse, blood pressure, and electrocardiograph). Oxygen supplement is highly recommended.

The patient is placed prone on the angiographic table.

### Ultrasound Guidance

We use the freehand technique for ultrasound guidance. This is a safe and simple way to obtain access into the collecting system. This approach is recommended for patients with hydronephrosis or patients with a transplanted kidney. The main advantage of using ultrasound guidance is the ability to select an optimal calyx for access. This technique is especially useful in transplanted kidneys.

## Fluoroscopic Guidance

A blind puncture technique may be attempted, especially if hydronephrosis is present. A puncture 1 cm lateral to the lateral process of L-1 or L-2 will usually provide access into the renal pelvis. A 22-gauge Chiba needle with a diamond tip is recommended. The needle is advanced 10 to 15 cm in a fast, forward thrust. Once the needle has been advanced, the trocar is removed and a connecting tube is connected to the needle hub. The needle is slowly withdrawn and continuous negative pressure is applied. Once urine is obtained, a small amount of contrast is injected to confirm needle location within the collecting system. At this point, a small amount of medical-degree carbon dioxide may be injected into the collecting system. The gas will migrate to the most posterior calyces (superiorly located when patient is in the prone position), and this provides an excellent target for definitive puncture.

## Fluoroscopy/Contrast

Contrast opacification of the collecting system may be achieved either by injecting intravenous contrast or direct injection through an open-ended catheter in the ureter placed via cystoscopic approach. We prefer the latter technique for percutaneous access for stone extraction cases. Localization of the optimal calyx is simple once the collecting system has been opacified.

## PERCUTANEOUS NEPHROSTOMY TUBE PLACEMENT

Once the optimal calyx is selected, a puncture is performed. We prefer to use a small needle system [Neff access set (Cook, Inc., Bloomington, IN) or Accustick (MEDITECH/Boston Scientific, Natick, MA)]; alternatively an 18-gauge, 15-cm diamond-tip needle may be used. To perform a successful puncture, simple steps should be followed:

1. Localize the selected calyx.
2. Angle the image intensifier so the selected calyx is seen on-end.
3. Mark the skin site and confirm the location of the puncture site. Make sure your puncture site does not look too lateral (ideal distance is four to five fingerbreadths lateral to the spinous process).
4. Inject 1% lidocaine for local anesthesia. A generous infiltration of the tissues is recommended to avoid a painful experience by the patient.
5. Perform a small skin nick with a No. 11 blade.
6. Traverse the skin with your needle.
7. Evaluate your needle position with fluoroscopy. You should see the selected calyx as a circle and your needle as a small dot. If you confirm this, the needle trajectory is optimal. At this point you may collimate and use magnification to restrict your radiation field. A plastic handle may be used to prevent direct radiation to your hands (Fig. 20.6). The use of this plastic handle involves a learning curve and some interventionists prefer not to use it.
8. Advance the needle (the patient will be breathing and your target may be moving under your needle. You should advance your needle in a fast, short, forward thrust when the target calyx is right under your needle).
9. Once you have advanced the needle far enough (usually 10 to 15 cm, depending on the patient's body girth), angle your image intensifier to the contralateral oblique view to assess the depth of your needle and confirm intracaliceal location.
10. If intracaliceal location is confirmed, a guidewire may be advanced into the collecting system. An 0.018-in. guidewire is advanced first if a skinny needle system is used. The small 5-Fr coaxial dilator is advanced over the 0.018-in. guidewire. The coaxial system is removed over the guidewire and the 5-Fr system is placed within the collecting system. Subsequently, a 0.035-in. guidewire is advanced into the collecting system to secure access. A 0.035-in. guidewire is used if an 18-gauge needle is used for puncture.
11. Secure adequate access by advancing the guidewire well into the collecting system so access is not lost during the dilatation maneuvers.

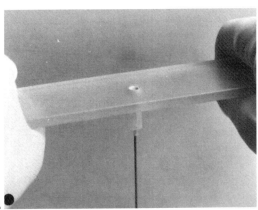

**FIG. 20.6. A,B:** Plastic handle to avoid direct hand radiation during percutaneous nephrostomy cases. (Reprinted from Castañeda-Zúñiga WR. *Interventional radiology*, 3rd ed. Baltimore: Williams & Wilkins, 1997, with permission.)

12. Dilate the tract with sequential dilators.

13. Advance the selected percutaneous nephrostomy tube (we prefer a multi-sidehole, self-locking pigtail system).

14. Secure the catheter by either stitching it to the skin or by the use of a retaining disk. We prefer the use of a regular stitch with 2-0 silk or 2-0 nonabsorbable nylon suture.

15. Connect the catheter to an external drainage bag.

## EMERGENCY BEDSIDE PERCUTANEOUS NEPHROSTOMY

Percutaneous nephrostomy tube placement may be performed at the bedside in critically ill patients. It may be performed using real-time sonographic guidance and a single-stick system with minimal to no guidewire manipulation [trocar set, Hawkins single-stick system, or DUAN catheter (Cook)]. The major disadvantage of this technique is the lack of anatomic reference and difficulty in determining the catheter's final position within the collecting system. For these reasons, we seldom perform percutaneous nephrostomy using this technique and prefer to bring the patient to the angiography suite.

## PERCUTANEOUS NEPHROSTOMY IN CHILDREN

Percutaneous nephrostomy is performed in infants for diagnostic and temporizing purposes. Important differences that should be remembered when performing percutaneous nephrostomies in children include the following:

- The kidneys of infants and young children lie lower than in adults.
- The kidneys are closer to the skin.
- The collecting system is usually severely dilated and less parenchyma is available.
- Heavy sedation or general anesthesia may be required. The patients are usually uncooperative.
- Small tubes and needles are available for pediatric percutaneous nephrostomy tube insertion.
- The most common indications include posterior urethral valves, congenital hydronephrosis, megaureter, postoperative strictures, pyonephrosis, and renal calculi.

## PERCUTANEOUS NEPHROSTOMY IN THE TRANSPLANT KIDNEY

Urologic complications occur in approximately 10% of all renal allograft recipients. Most of these

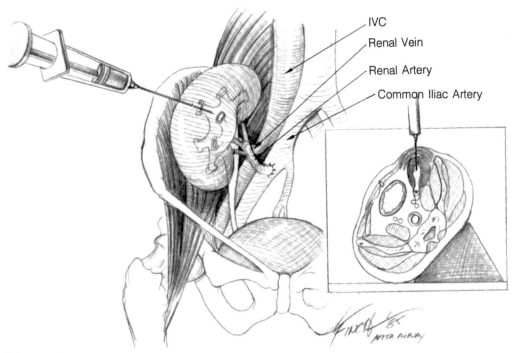

**FIG. 20.7.** Schematic drawing of anatomic relations of the transplanted kidney and its orientation in its extraperitoneal pelvic position. (IVC, inferior vena cava.) (Reproduced from Hunter DW, Castañeda-Zúñiga WR, Coleman CC, et al. Percutaneous techniques in the management of urological complications in renal transplant patients. *Radiology* 1983;148:407, with permission.)

complications involve either a urine leak, which usually occurs in the early postoperative period (within a month), or ureteric obstruction. The most common site of obstruction is the ureterovesical junction. This complication may be caused by kinks, fibrosis, rejection, ureteral ischemia, calculi, clots, fungus balls, and extrinsic compression (lymphocele, hematoma, urinoma, abscess, and bladder distention).

The diagnostic imaging evaluation of a renal transplant mainly includes ultrasound and renogram. If hydronephrosis is documented by ultrasound or the renogram suggests an obstructive problem or a urine leak, a percutaneous evaluation is indicated. A percutaneous antegrade pyelogram can be performed with a 21-gauge needle. This procedure is safe and relatively noninvasive. In addition, a urodynamic study (Whitaker test) may be performed in the same sitting. If the patient has increasing creatinine levels caused by obstruction, a percutaneous drainage is indicated.

In the usual situation, the transplanted kidney is extraperitoneal, with the renal pelvis on the posterolateral surface (Fig. 20.7). For patient preparation, prophylactic antibiotics should be given just before the procedure. The coagulation profile should be assessed and if abnormal coagulation is found, it should be corrected.

## TECHNIQUE

1. Real-time sonographic guidance is recommended.
2. Evaluate the graft sonographically and localize the most superficial calyx.
3. Once the best calyx is selected, check the site with fluoroscopy and determine if there are any bowel loops superimposed to your anticipated puncture site. In these cases, you must be careful and avoid puncture of loops of bowel.
4. Perform the puncture under real time sonography. We recommend a skinny needle system (Neff access set or Accustick set).

5. Once needle position within the collecting system has been documented, advance the 0.018-in. guidewire.

6. Advance the coaxial tract dilator over the 0.018-in. guidewire.

7. Remove the two inner dilators and leave the external dilator and the 0.018-in. guidewire (used as a security wire). You may inject contrast via the dilator using a K-50 over your 0.018-in. guidewire and determine the site and potential cause of obstruction or leak.

8. If an obstruction is documented, it will be important to attempt passage of the guidewire into the bladder. This maneuver should be postponed for a second session if the patient is septic at the time of the drainage procedure.

9. Once you have confirmed good position of your dilator within the collecting system, advance a 0.035-in. guidewire through the 5-Fr dilator.

10. If patient is not septic and internal/external drainage is feasible, attempt passage of your guidewire to the bladder. For this purpose, you may use a torque catheter such as a short multipurpose catheter.

11. If access into the bladder is successful, place the internal/external drainage catheter.

## COMPLICATIONS RELATED TO PERCUTANEOUS NEPHROSTOMY ACCESS

- Hemorrhage, which may be immediate or delayed. Most patients have hematuria after the percutaneous nephrostomy puncture but generally the urine clears within 24 to 36 hours. Severe bleeding or persistent bleeding must be addressed. Initially a larger tube may be placed to tamponade the tract. If this maneuver is not successful, angiography with embolization of vessel laceration should be performed.
- Urine leakage.
- Pain.
- Infection or sepsis, which is more common in patients with underlying pyonephrosis. High-pressure injections during diagnostic nephrostogram or catheter flushes should be avoided to prevent reflux of contaminated urine via the pyelovenous route into the systemic veins.
- Urinoma.
- Pneumothorax/hydrothorax.
- Retroperitoneal extravasation of urine.
- Catheter misplacement.

## B. SPECIAL TECHNIQUES

### WHITAKER TEST

The Whitaker test is indicated to assess the functional significance of a stenosis or to determine if hydronephrosis is due to partial obstruction. Access is obtained into the collecting system either by placement of a small catheter or double-needle technique. An infusion pump is hooked to the nephrostomy tube or needle and diluted contrast is injected into the collecting system at a rate of 15-20 mL per minute. Pressures in the collecting system are measured either with the use of a three-way stopcock connected to the drainage catheter or via the second needle located in the collecting system (Fig. 20.8). Pressures in the bladder are measured through a Foley catheter using a separate pressure setup (Fig. 20.8). A pressure gradient of 20 cm $H_2O$ between the pelvocalyceal system and the bladder is considered to be significant. A gradient below 15 cm $H_2O$ is normal, and between 15 and 20 cm $H_2O$ is equivocal.

### REPLACEMENT OF DISLODGED PERCUTANEOUS NEPHROSTOMY TUBE

If percutaneous nephrostomy tube dislodgment occurs, the nephrostomy tract can usually be reestablished with careful probing under fluoroscopic guidance. The longer the tube has been in place, the higher the chances of reestablishing access via the same tract. A mature tract has usually formed after 2 weeks. If dislodgment occurs, the patient must be instructed to report to the hospital as soon as possible. A mature tract will remain open for 6 to 12 hours in patients previously drained with 8- to 10-Fr catheters. Important aspects in the technique include the following:

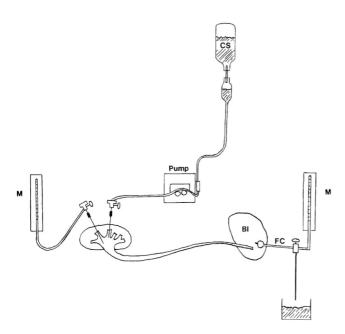

**FIG. 20.8.** Setup used for the Whitaker test. From reservoir, 20% to 30% contrast medium is pumped through 22-gauge Chiba needle into the collecting system. Simultaneously, the pressure in the renal pelvis is registered through a second Chiba needle in the collecting system. Pressure in the bladder is measured via a Foley catheter. For pressure measurements, two simple water manometers are sufficient. (BI, bladder; CS, contrast medium; FC, Foley catheter; M, manometer.) (Reprinted from Castañeda-Zúñiga WR. *Interventional radiology*, 3rd ed. Baltimore: Williams & Wilkins, 1997, with permission.)

- Do not attempt blind manipulations through the tract. A false tract may be created by blind manipulations. If a false tract is created, it will become very difficult to regain access.
- Do not infiltrate the tract with local anesthetic before attempting to catheterize it. Injections can traumatize the tract and obstruct it.
- Identify the orifice and inject a small amount of contrast through the orifice, using a K-50 connecting tube or a Christmas tree adapter.
- Once the tract is opacified, a floppy-tip 0.035-in. guidewire is gently manipulated through the tract. We recommend the use of a steerable multipurpose catheter or cobra catheter for wire support. Once the guidewire-catheter system has been advanced back into the collecting system, a new catheter can be placed.

If the patient is seen 24 hours after tube dislodgment or later, the tract is frequently closed and attempts to reopen will rarely succeed. In these patients, it is easier to create a new nephrostomy tract for drainage.

## REMOVAL OF OCCLUDED PERCUTANEOUS NEPHROSTOMY TUBES

Patients with long-term need for percutaneous nephrostomy tubes need to have their tubes changed at least every 6 months. Tube occlusion is frequent in patients who are prone to stone formation or in patients who keep their nephrostomy tubes without exchange for periods of time longer than 6 months. These patients will present to the hospital with a nonfunctional, nondraining tube. The technique for replacement of an occluded percutaneous nephrostomy tube is as follows:

1. Evaluate the patient, obtain informed consent, and discuss risks and benefits of the procedure.
2. Prepare the area in the usual sterile fashion.
3. Evaluate the catheter position with fluoroscopy. A nonfunctional catheter may be partially or completely dislodged.
4. If the catheter position appears to be satisfactory, attempt contrast injection and determine if any sideholes are still patent.
5. If a patent sidehole is identified and proper opacification of the collecting system is achieved, use a curved guidewire [0.035-in. Rosen wire (Cook, Inc.)]. The curved guidewire will exit through the first sidehole and access into the collecting system will be restored. If the nephrostomy tube is a self-locking tube, unlock the system before removing the tube or simply cut the tube with scissors to liberate the locking system.
6. If the tube is completely occluded you may try to negotiate a guidewire through the

occluded tube. If this maneuver is successful, you can proceed with catheter exchange.

7. If guidewire manipulation is unsuccessful you may use a peel-away sheath technique. First, cut the tube in close proximity to the hub. Slide a peel-away sheath 1-Fr larger than the existing tube (e.g., 9-Fr for an 8-Fr tube, 11-Fr for a 10-Fr tube) over the occluded tube under fluoroscopic guidance. Once you have completely advanced the peel-away sheath, pull the occluded nephrostomy tube. Your peel-away sheath should provide safe and immediate access into the collecting system for catheter replacement.

## DOUBLE-J CATHETER PLACEMENT

A double-J stent is indicated for temporary or long-term ureteral stenting. Once access into the collecting system has been obtained, the tract is dilated to 8 or 10 Fr. We usually place an intravascular sheath to secure access, facilitate catheter manipulation, and decrease the pain at the entry site. A security wire is advanced into the renal pelvis and a combination of angiographic catheter and guidewire is used to negotiate the ureter. In patients with long-standing occlusions, the course of the ureter may be extremely tortuous (Fig. 20.9). Once a guidewire has been successfully advanced into the bladder, we advance the angiographic catheter down to the bladder and inject a small amount of diluted contrast to confirm adequate location within the bladder. At this point, we proceed with double-J catheter placement.

1. Measure the distance between the bladder and the collecting system (Fig. 20.10).
2. Place an Amplatz 0.035-in. Superstiff guidewire (MEDITECH/Boston Scientific) down the bladder via the angiographic catheter. We prefer this wire for double-J placement because of its excellent support for catheter advancement.
3. Once the Amplatz wire has been advanced into the bladder, remove the angiographic catheter, and if a vascular sheath was used, remove the vascular sheath.
4. The double-J catheter, with the suture near the proximal J, is passed over the guidewire and advanced with the pusher (Fig. 20.11). (The sutures at the proximal end of the catheter may be removed before catheter insertion. This is up to the performing physician's preference.)
5. Once a good position has been reached within the bladder, the pusher is used to hold the stent in place while the silk suture is removed (Fig. 20.11).
6. The guidewire is then pulled back into the pusher.
7. When the wire leaves the stent lumen, the J will form (Fig. 20.11).
8. At this point, the pusher may be removed and double-J stent placement has been completed.
9. The security guidewire is used to advance an 8-Fr pigtail catheter into the renal pelvis. We usually leave the security percutaneous nephrostomy for 12 to 24 hours after double-J placement. The patient is brought back for a control nephrostogram 12 to 24 hours after double-J placement. Nephrostogram is performed via the percutaneous nephrostomy tube. If good function of the double-J is confirmed, the double-J is removed and complete internal drainage has been achieved.

## RETROGRADE CATHETERIZATION OF THE URETER WITHOUT ENDOSCOPIC ASSISTANCE

Babel and Witerkorn described a method for retrograde catheterization of the ureter without cystoscopic assistance (Fig. 20.12).

1. Procedure is performed in the angiographic suite with the patient under conscious sedation.
2. The external genitalia are sterilely prepped and draped.
3. Access to the bladder is obtained through the urethra using a Foley catheter or an angiographic catheter-guidewire combination.
4. A hockey-stick catheter is inserted and a small amount of dilute contrast is injected.
5. A T-shaped area of decreased opacity seen on fluoroscopy with the bladder minimally distended correlates to the anatomic bladder trigone. The upper margin of the T, usually slightly convex inferiorly, represents the transureteric ridge.
6. Using the hockey-stick catheter and a regular, soft, angled Glidewire (MEDITECH/Bos-

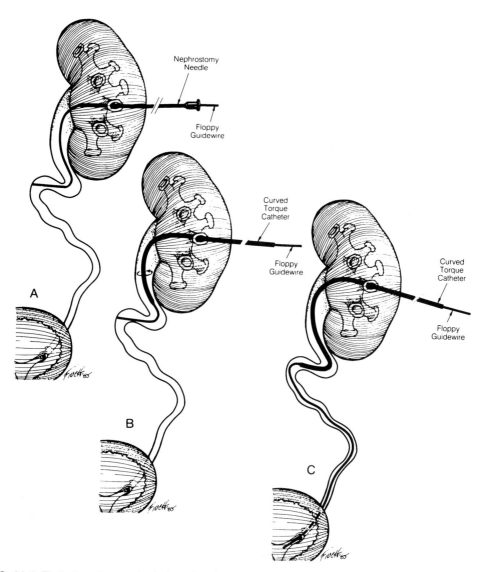

**FIG. 20.9.** Technique for manipulation of catheters and wire through tortuous, dilated ureters, especially those with severe obstructions. **A:** A floppy guidewire is advanced into the collecting system. **B:** After removal of the needle, an angiographic torque control catheter has been introduced over the wire. **C:** By gentle manipulation of the angiographic catheter and guidewire, the tortuous ureter has been negotiated, and the tip of the wire has been advanced into the bladder. (*Continued.*)

ton Scientific), it is possible to identify the ureter orifice and perform retrograde catheterization.

This technique is useful especially in patients who do not tolerate the endoscopic procedure or patients in whom percutaneous access to the kidney has failed or is contraindicated for other reasons.

## FLUOROSCOPY-GUIDED DOUBLE-J STENT EXCHANGE

The patency of double-J stents varies from 2 to 24 months. We recommend double-J stent exchange every 6 months for patients who need chronic double-J stenting. Double-J stents may be changed under direct cystoscopic visualiza-

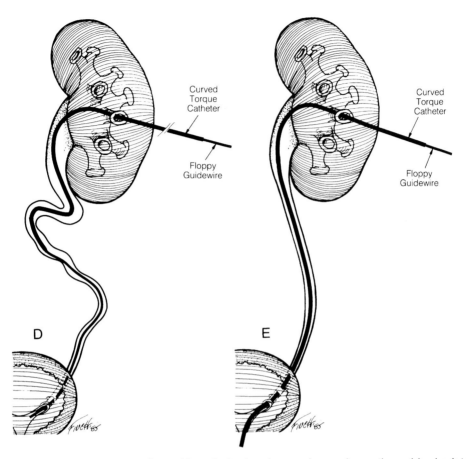

**FIG. 20.9.** *Continued.* **D:** The angiographic catheter has been advanced over the guidewire into the bladder. **E:** By gentle traction on guidewire and catheter, the ureter has been straightened. (Reprinted from Castañeda-Zúñiga WR. *Interventional radiology*, 3rd ed. Baltimore: Williams & Wilkins, 1997, with permission.)

tion or under fluoroscopic guidance. Important aspects of the technique include the following:

- Physical evaluation (investigate presence of active urinary tract infection).
- Intravenous line for administration of sedatives and antibiotics.
- The patient is placed supine on the angiographic table in a lithotomy position and the external genitalia prepped in sterile fashion.
- A small Foley catheter is inserted and 100 to 200 mL of dilute contrast (20% contrast; 80% saline) is infused into the bladder to distend the bladder and increase space for catheter manipulation.

- The Foley catheter is removed over a floppy-tip 0.035-in. guidewire.
- The catheter from the Amplatz Goose Neck Snares (Microvena, White Bear Lake, MN) is advanced into the bladder (Fig. 20.13). We recommend using a 15-mm or 25-mm Amplatz snare for double-J retrieval.
- The snare is advanced and the double-J is captured and retrieved.
- Once the tip of the double-J has been successfully withdrawn, a guidewire is placed through the double-J and advanced into the renal collecting system.
- An angiographic catheter may then be advanced over the wire, urine samples for cultures or

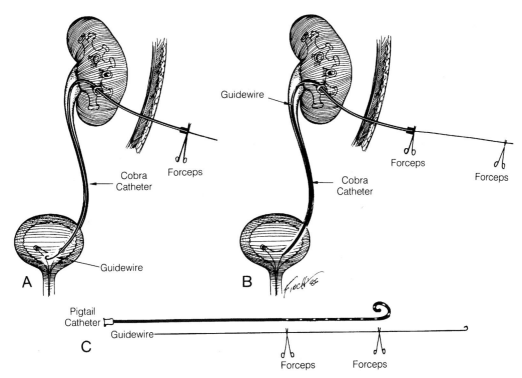

**FIG. 20.10.** Technique for measurement of distance from the ureterovesical junction to the renal pelvis. **A:** Through an angiographic catheter passed into the bladder, a guidewire has been introduced and passed beyond the tip of the angiographic catheter. Forceps have been placed on the guidewire as it exits from the hub of the catheter. **B:** While the angiographic catheter is kept in the same position, the guidewire has been pulled back until its tip lies within the renal pelvis; at this point, a second forceps has been applied over the guidewire, which is then removed. **C:** The distance between the forceps represents the distance from the bladder to the renal pelvis; corresponding sideholes have been created in the drainage catheter. (Reprinted from Castañeda-Zúñiga WR. *Interventional radiology*, 3rd ed. Baltimore: Williams & Wilkins, 1997, with permission.)

cytology may be obtained, and a nephrostogram may be performed.
- An Amplatz Superstiff guidewire is then advanced and coiled within the collecting system and retrograde double-J catheter placement can be performed.

## URETERAL STRICTURE DILATATION AND METALLIC STENT PLACEMENT

Dilatation of ureteral strictures with angioplasty balloons has been performed. The technical success is high (more than 95%). Fresh strictures without evidence of vascular compromise may be treated successfully and their patency rates may be as high as 91% at 1 year. If the stricture is older than 3 months, the patency rates drop to 53%. Strictures with evidence of vascular compromise have low patency rates—around 20% at 1 year. Important aspects of these procedures include the following:

- Evaluate the patient with a complete history and physical examination.
- Insert an intravenous line for administration of fluids, sedatives, and antibiotics.
- The optimal approach for ureteral dilatation requires a tract through a posterior calyx of the superior or middle calyceal group.
- Catheterize the ureter with a cobra or multipurpose catheter and negotiate the stricture with careful guidewire and catheter manipulation.

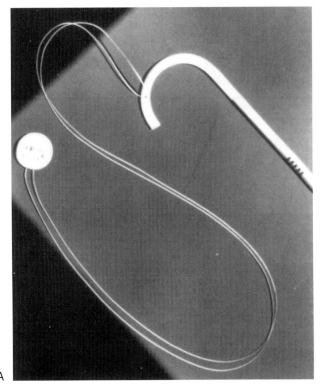
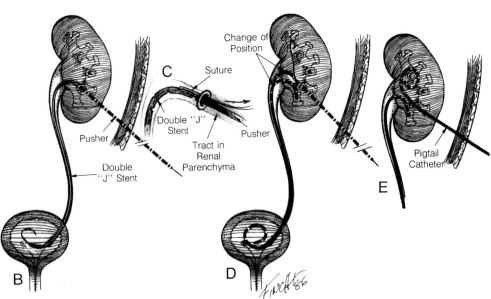

**FIG. 20.11.** Percuflex (MEDITECH/Boston Scientific) double-J stent insertion. **A:** Close-up view of proximal end of Percuflex double-J stent with sutures through a sidehole for repositioning purposes during introduction. **B:** Advancement of double-J stent over guidewire passed beyond site of obstruction. **C:** While the end of the stent is held in place with the pusher, the suture is pulled back. **D:** The stent has doubled back into the lower-pole infundibulum. **E:** The pusher has been replaced by a pigtail catheter for temporary external drainage. (Reprinted from Castañeda-Zúñiga WR. *Interventional radiology*, 3rd ed. Baltimore: Williams & Wilkins, 1997, with permission.)

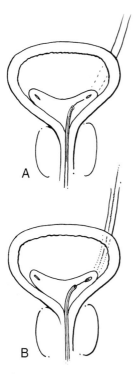

**FIG. 20.12. A:** A JB-1 catheter and a guidewire have been placed at the level of the bladder trigone. **B:** By gentle manipulation, the guidewire has been advanced into the ureter. (Reprinted from Castañeda-Zúñiga WR. *Interventional radiology*, 3rd ed. Baltimore: Williams & Wilkins, 1997, with permission.)

- Advance the guidewire into the bladder or ileal loop, remove angiographic catheter over the wire, advance an intravascular sheath to secure access, and place a security wire if necessary.
- Stricture dilatation with angioplasty balloons (4- to 8-mm diameter balloons are used for ureteric strictures; 8- to 10-mm for strictures at ureteroileostomy sites). We prefer to use high-pressure balloons [Blue Max (Boston Scientific) or Powerflex Plus [Cordis Endovascular, Miami]).
- The angioplasty balloon is inflated at high pressure (15 to 17 atm) and kept inflated for 3 to 5 minutes. The insufflation is repeated two to three times.
- After dilatation, an 8- to 10-Fr nephroureteral stent is placed for 4 to 6 weeks.
- Redilatation can be performed at 4 to 6 weeks. The patient is followed for 3 to 4 months in this manner, with up to three dilatations.

The strictures that respond favorably to balloon dilatation include the following:

- Strictures after recent ureteroneocystostomy and ureterolithotomy
- Recent ureteropelvic strictures after pyeloplasty
- Strictures after accidental ligation of the ureter without evidence of vascular compromise

The strictures that have unfavorable responses include the following:

- Strictures with evidence of vascular compromise
- Ureteroileostomy strictures in patients with a life expectancy of greater than 1 year
- Strictures associated with fistulae after radical hysterectomy

Metallic stent application in the ureter has been limited. The majority of patients in whom metallic stents have been used had ureteral obstruction due to malignancy. A smaller number of patients had metallic stents placed in benign strictures who had failed multiple percutaneous or endourologic attempts at treatment. The patency rates are variable, ranging from 25% at 3 months to 83% at 7 months. Continued investigation will be necessary to determine the efficacy of metallic stents in ureteral obstructions. At this point this mode of therapy is not advocated as the standard treatment for ureteral strictures.

## C. PERCUTANEOUS NEPHROSTOMY ACCESS FOR STONE REMOVAL

Percutaneous stone removal remains important in the management of urolithiasis. Close cooperation between the interventional radiologist and the urologist is essential for a successful procedure. The overall success rate for removal of pelvic stones is 96% to 98%. Percutaneous removal of urinary stones is applicable when external shock wave lithotripsy is not available or has

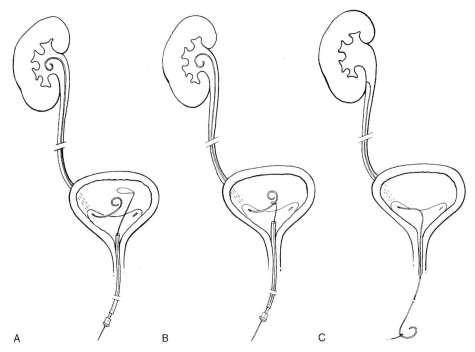

FIG. 20.13. **A:** A torque control and a vascular gooseneck snare have been placed into the bladder. Double-J is in place. **B:** The double-J catheter has been trapped by the vascular snare. **C:** The double-J catheter has been pulled out through the urethra. (Reprinted from Castañeda-Zúñiga WR. *Interventional radiology*, 3rd ed. Baltimore: Williams & Wilkins, 1997, with permission.)

been unsuccessful. The two relative contraindications for percutaneous nephrolithotripsy are active infection and bleeding diathesis; both are correctable.

## PATIENT EVALUATION

- Complete blood cell count
- Blood typing and cross-matching of two units
- Urinalysis and culture with sensitivity studies of any organisms so that infection can be treated with specific antibiotics
- Blood urea nitrogen and creatinine to evaluate renal function

## RADIOLOGIC STUDIES

- Intravenous pyelogram with oblique views (*extremely useful*) depicts the anatomy of calyceal system and stone(s) location.
- Computed tomography scan is sometimes useful—for example, in cases in which the kidney is malrotated or in patients with hepatosplenomegaly or anatomic variants such as horseshoe kidney.

## PREOPERATIVE AND OPERATIVE MANAGEMENT

In our institution, percutaneous nephrostomy access for stone removal is performed under general anesthesia, either in the angiography suite or in the operating room. We attempt a one-session removal if possible.

1. A retrograde open-ended catheter is placed cystoscopically by the urology team for opacification of the collecting system. This procedure is performed in the urology suite.

2. The patient is then transported to either the angiography suite or the operating room, where the anesthesiologist intubates the patient and puts the patient under general anesthesia.

3. The patient is then placed in the prone position. A 60-mL syringe with dilute hydrosol-

uble contrast is connected to the open-ended catheter in preparation for retrograde nephrostogram, before the patient is prepped.

4. The area of interest is sterilely prepped and draped.

5. A retrograde nephrostogram is performed and selection of the most suitable calyx for access is discussed with the urology team. We try to use the most posteriorly located calyx that will provide a straight access to the stone. Fluoroscopic evaluation with different oblique views is essential for calyceal selection.

## ACCESS TECHNIQUE AND TRACT DILATATION

1. Evaluate the retrograde nephrostogram and select the most appropriate calyx for access.

2. Undertake percutaneous puncture with either a small needle system or the 18-gauge trocar needle as described previously.

3. Once access is obtained, we make all efforts to advance a guidewire down the urinary bladder with a catheter and guidewire combination (see previous description). Although this maneuver is occasionally time consuming, having a security guidewire down the bladder is extremely useful. The security wire provides continuity to the collecting system in case of pelvic or ureteral laceration or perforation during the stone removal maneuvers and facilitates double-J stent placement at procedure termination if this is necessary.

4. Once the guidewire has been advanced down the bladder, the tract is dilated with the coaxial Amplatz system and the 10-Fr blunt, long dilator is advanced (Fig. 20.14).

5. A security guidewire is advanced into the urinary bladder using the coaxial system. (Two guidewires will remain in the bladder; one is the working guidewire and the other is the security wire.) We isolate the security wire and clamp it in an area where no manipulation is being performed to avoid accidental guidewire removal or dislodgment.

6. Tract dilatation is performed. This can be achieved either with a balloon system (Fig. 20.15) [Tract-Master (MEDITECH/Boston Scientific)] or by dilating sequentially with the Amplatz dilators (Fig. 20.14). We prefer the balloon technique because it is performed in a single step, making it fast and safe.

7. The tip of the balloon is passed into the renal pelvis. A pressure gauge is attached to an inflation device, which is then connected to the balloon catheter. The balloon is inflated with dilute contrast to a pressure of 12 to 14 atm. Initially, a waist is seen in the balloon at the level of the renal capsule, the posterior perirenal fascia, and the subcutaneous tissue. These deformities will generally disappear in 20 to 30 seconds.

8. The 30-Fr working sheath is advanced into the renal pelvis over the inflated balloon. The dilatation system is then withdrawn and the working sheath left in place within the renal pelvis.

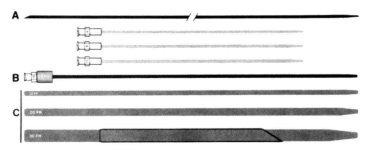

**FIG. 20.14.** Partial set of Amplatz renal dilators showing 8-Fr guiding dilator (**A**); 10-Fr blunt-ended Teflon dilator, which doubles as safety guidewire introducer (**B**); and 12-, 20-, and 30-Fr dilators (**C**). (Courtesy of Cook.) (Reprinted from Castañeda-Zúñiga WR. *Interventional radiology*, 3rd ed. Baltimore: Williams & Wilkins, 1997, with permission.)

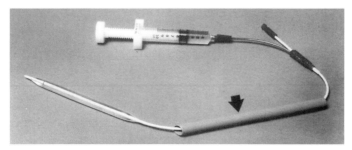

**FIG. 20.15.** Tract-Master one-step tract dilation system with sheath. Observe that a 10-mm-diameter, 15-cm-long balloon has been assembled with a 30-Fr (lumen) Teflon sheath (*arrow*). The sheath can be advanced easily over a fully inflated balloon into the renal pelvis. (Reprinted from Castañeda-Zúñiga WR. *Interventional radiology*, 3rd ed. Baltimore: Williams & Wilkins, 1997, with permission.)

9. Endoscopy is performed to confirm adequacy of the tract and visualization of the stone.

10. The urologist performs stone extraction. Different techniques for percutaneous removal of calculi exist. These are out of the scope of this publication and other text references are recommended.

## COMPLICATIONS

Complication rates associated with percutaneous extraction of renal and upper ureteral stones depend on the complexity of the stone, the renal anatomy, and the skill of the operator. Most complications occur with impacted caliceal and ureteropelvic junction stones, multiple stones, and staghorn calculi requiring extensive manipulations. Embedded ureteral stones were the chief cause of complications such as ureteral avulsion, perforation, and stricture. Most common complications include the following:

- Bleeding requiring transfusion (11%).
- Displacement of stone fragments.
- Pneumothorax or hydrothorax (3%).
- Urinoma.
- Retroperitoneal hematoma.
- Infection.
- Contrast medium reaction.
- Colon and duodenal perforation.
- Liver, spleen, and pancreas perforation.
- Death. The mortality rate is low (less than 0.1%). Most important causes include myocardial infarction and pulmonary embolism.

## D. PROSTATIC INTERVENTION

Balloon dilatation and stent placement of the prostatic urethra became available and somewhat popular in the late 1980s and early 1990s. The human experience with prostatic balloon dilatation and stent placement was found to be suboptimal and has been replaced by other endourologic procedures, mainly performed by urologists (i.e., laser surgery). Prostatic balloon dilatation and stent placement is rarely performed today.

### PROSTATIC BIOPSY

Probably the most common procedure in the prostate gland performed under direct imaging visualization is sonographic-guided prostatic biopsy. In most cases, ultrasound-guided transrectal biopsy is done on an outpatient basis. Biopsy is contraindicated in patients with coagulopathy or acute prostatitis. Sedation is usually not required, but broad-spectrum antibiotics should be started before and continued for 5 to 7 days after the procedure. Preprocedure enemas help decrease air or fecal artifacts. The patient is placed in a lateral decubitus, knee-chest, or lithotomy position and scanned. Biplanar imaging helps optimize the sensitivity of the procedure (Fig. 20.16). Once the area to be biopsied is localized, the needle is advanced through the probe channel

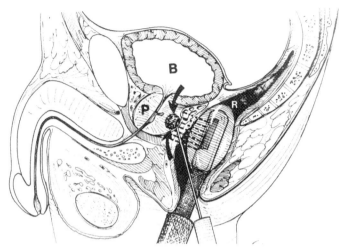

**FIG. 20.16.** A transrectal ultrasound transducer in position during real-time biopsy of prostatic nodule (*arrows*). (B, bladder; P, prostate; R, rectum.) (Reprinted from Castañeda-Zúñiga WR. *Interventional radiology*, 3rd ed. Baltimore: Williams & Wilkins, 1997, with permission.)

or biopsy guide. An automatic biopsy gun is recommended (Biopty) (C. R. Bard, Covington, GA). The procedure is well tolerated by the patient, and excellent core specimens are obtained. Hematuria, hemospermia, or mild rectal bleeding might occur during the next few days.

## RESULTS

In patients with clinical or laboratory abnormalities, the sensitivity of ultrasound-guided transrectal prostatic biopsy is high. Good tissue specimens are usually obtained with the automatic biopsy gun with minimal complications.

## SUGGESTED READINGS

Babel S, Winterkorn K. Retrograde catheterization of the ureter without cystoscopic assistance: preliminary experience. *Radiology* 1993;187:547–549.

Bjarnason H, Ferral H, Stackhouse DJ, et al. Complications related to percutaneous nephrolithotripsy. *Semin Intervent Radiol* 1994;11:213–225.

Coleman CC, Castañeda-Zúñiga WR, Kimura Y, et al. A systematic approach to puncture site selection for percutaneous urinary tract stone removal. *Semin Intervent Radiol* 1984;1:42.

Coleman CC, Kimura Y, Castañeda-Zúñiga WR, et al. Dilation of nephrostomy tracts for percutaneous stone removal. *Semin Intervent Radiol* 1984;1:50.

DeBaere T, Denys A, Pappas P, et al. Ureteral stents: exchange under fluoroscopic control as an effective alternative to cystoscopy. *Radiology* 1994;190:887.

Ferral H, Stackhouse DJ, Bjarnason H, et al. Complications of percutaneous nephrostomy tube placement. *Semin Intervent Radiol* 1994;11:198–206.

Irving HC, Arthur RJ, Thomas DF. Percutaneous nephrostomy in pediatrics. *Clin Radiol* 1987;38:245.

Pfister RC, Newhouse JH, Hendren WH. Percutaneous pyeloureteral urodynamics. *Urol Clin North Am* 1982; 9:41.

Stables DP. Percutaneous nephrostomy: techniques, indications and results. *Urol Clin North Am* 1982;9:15.

Swierzewski S, Konnak J, Ellis J. Treatment of renal transplant ureteral complications by percutaneous techniques. *J Urol* 1993;149:986.

Van Sonnenberg E, Casola G, et al. Symptomatic renal obstruction or urosepsis during pregnancy: treatment by sonographically guided percutaneous nephrostomy. *AJR Am J Roentgenol* 1991;158:91.

# 21
# Biliary Tract Intervention

Pablo A. Gamboa and Haraldur Bjarnason

## A. INTERVENTIONAL TECHNIQUES IN THE HEPATOBILIARY SYSTEM

### FINE NEEDLE PERCUTANEOUS TRANSHEPATIC CHOLANGIOGRAPHY

The differential diagnosis of jaundice caused by parenchymal liver disease or biliary obstruction is difficult if only the clinical and laboratory findings are considered. Such a distinction is critical to provide proper therapy. Ultrasound (US), computed tomography (CT), magnetic resonance imaging (MRI), and nuclear scanning, have improved the evaluation of hepatic disease and can determine the presence and level of obstruction. However, the demonstration of small lesions, partial obstructions, and biliary tree anatomy is beyond the capabilities of US, CT, and MRI. In addition, 10% to 20% of patients with surgically correctable obstructing lesions will not have dilated ducts demonstrable by imaging. In these cases, fine needle percutaneous transhepatic cholangiography (PTC) is invaluable.

### Indications and Contraindications

#### Indications

- Differentiation of medically treatable from surgically treatable jaundice
- Diagnosis of calculus disease
- Diagnosis of congenital abnormalities such as biliary atresia and choledochal cysts
- Evaluation of biliary-enteric anastomoses in patients with abdominal pain and jaundice
- Documentation of intrahepatic abscesses communicating with the biliary radicles
- Evaluation of physiology and anatomy of the biliary system by means of pressure-flow studies
- Evaluation before interventions such as biopsy and placement of drainage catheters

#### Absolute Contraindications

- Uncorrectable bleeding disorders after administration of vitamin K or blood products; in this situation, endoscopic retrograde cholangiopancreatography (ERCP) is the preferred procedure. Target international normalized ratio is more than 1.3 and platelet count is more than 75,000/µL.
- A history of a life-threatening reaction to iodinated contrast medium.
- Known hepatic vascular tumors or malformations.

### Preparation of the Patient

An initial careful review of the clinical history and any noninvasive methods (US, CT, chest x-rays) that aid in planning the best approach are performed. An intravenous (IV) line is started for the administration of fluids and drugs before and during the procedure. Laboratory studies to be evaluated include coagulation profile, complete blood count, and measurement of blood urea nitrogen and creatinine levels. IV antibiotics are started 6 to 12 hours before the procedure and continued for 48 to 72 hours afterward because the incidence of infected bile is 25% to 36% in malignant biliary obstruction and 71% to 90% in lithiasis. The most frequent organisms causing

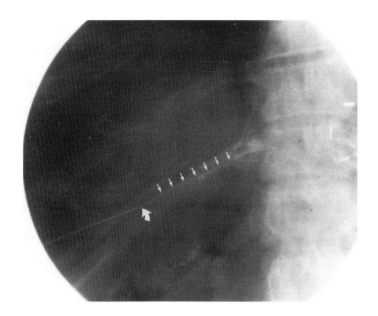

FIG. 21.1. Chiba needle tract opacified with contrast medium (*small arrows*). Curved arrow points to needle. (Reprinted from Castañeda-Zúñiga WR. *Interventional radiology*, 3rd ed. Baltimore: Williams & Wilkins, 1997, with permission.)

biliary sepsis include *Escherichia coli*, *Enterobacter aerogenes*, and *Streptococcus faecalis*. Antibiotics, such as ampicillin and gentamicin, or those with a similar broad spectrum of activity, such as second- and third-generation cephalosporins, which penetrate effectively into bile, can be used. Even with prophylaxis, these patients should be monitored because bloodstream infection can still occur. Midazolam and short-acting narcotics such as fentanyl are administered IV as premedication and during the procedure as needed, with vital signs and oxygen saturation carefully monitored.

## Procedure

The procedure is performed with the patient in the supine position, preferably with rotational fluoroscopy. The puncture site on the skin is selected using the visible liver as a guide, and the tract is planned so that it traverses safely the maximum amount of liver substance. Liver size and position and the interposition of bowel loops are evaluated during the initial fluoroscopic examination. The usual tract parallels the caudal edge of the liver, approximately one-third of the way up from the caudal edge to the dome of the diaphragm. Thus, the puncture is usually made from the seventh to the tenth intercostal space, slightly anterior to the mid-axillary line and below the costophrenic angle. The pleural reflection laterally reaches to the level of the tenth rib. Therefore, most punctures above the tenth rib will be transpleural (across the costophrenic angle), although not necessarily transpulmonary, because the lung commonly lies at a higher level during shallow normal breathing, reaching down into the costophrenic recess only during deep inspiration. To avoid the intercostal vessels, the puncture must be made close to the top edge of the lower rib. After sterile preparation, the puncture site is infiltrated with lidocaine down to the liver capsule. A 15-cm-long, flexible 21- to 23-gauge needle is used for the puncture. The needle is advanced under fluoroscopic control parallel to the tabletop until the tip reaches the midline at the level of the twelfth vertebral body. A syringe filled with dilute (1 to 1) contrast medium is attached to clear flexible polyethylene tubing, and this is connected to the hub of the needle. Under continuous fluoroscopic observation, contrast medium is steadily and slowly injected while the needle is slowly withdrawn, creating a thin, continuous tract of contrast behind the retreating needle, until a duct is entered (Fig. 21.1). As long as the needle tip is in the liver parenchyma, the contrast medium will stay close to the tip and have a faint, cloudlike

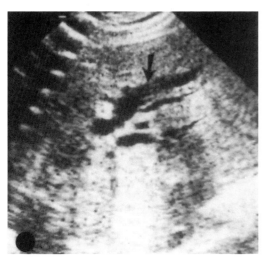 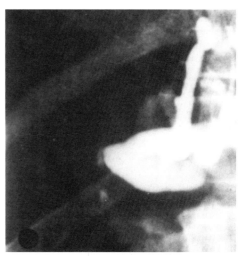

**FIG. 21.2.** Puncture under ultrasound guidance. **A:** Ultrasound shows marked dilatation of left biliary radicles (*arrow*). **B:** Puncture of left biliary radicles was performed with a 22-gauge Chiba needle. Contrast medium shows marked dilatation of left biliary radicles not communicating with the right biliary tree. (Reprinted from Mueller PR, Harbin WR, Ferrucci JT Jr., et al. Fine-needle transhepatic cholangiography: reflections after 450 cases. *AJR Am J Roentgenol* 1981;136:85–90, with the permission of the American Roentgen Ray Society.)

appearance; as soon as the tip enters a duct or vessel, the contrast will flow away from the tip. Thus, injection of slightly larger amounts of contrast permits identification of the structure. Branching of the ducts should quickly be obvious. If a duct is not found on the first puncture, subsequent needle passages should fan out from the first tract, first in a cephalad direction, then anterior, then posterior, and finally caudad. In more caudal passages, care should be taken not to advance the needle too far medially, not only because of the risk of puncturing adjacent structures but also because passage of the needle through the liver capsule is very painful. In patients with dilated bile ducts, a 99% to 100% success rate can be achieved with four to six needle passages, although as many as 14 passages have been used routinely by some authors without increasing the complication rate. Patients with nondilated bile ducts may require more punctures. In these cases, a direct puncture of the larger extrahepatic ducts, either blindly with the aid of bony landmarks or under US guidance, can be attempted (Fig. 21.2). No complications such as bile leak or hemorrhage have occurred as a result of this technique. Other branching structures within the liver may be entered during needle passages, such as branches of the hepatic vein, portal vein, or hepatic artery, and lymphatics. Most of the time, their recognition causes no difficulty. Bile ducts are recognized by the slow flow of puddled contrast medium, which may flow in a peripheral direction initially but eventually will flow centrally. With the patient supine, contrast medium, which has a higher specific gravity than bile, will flow to the most dependent posterior ducts. Unless an anterior duct has been entered, this preferential flow may cause difficulties with opacification of the central ducts and almost always causes some difficulty with opacification of the left ducts, which are located more anteriorly than those on the right. Opacification of left-sided ducts is important if an epigastric approach is desired for drainage (Fig. 21.3). Filling the ducts completely may require changing the patient's or table's position. Forceful attempts to opacify the entire biliary tree should be avoided, inasmuch as this increases the chances of bacteremia and sepsis by forcing bacteria from the duct's lumen into the bloodstream. At the completion of the procedure the needle is withdrawn and the patient is observed for complications such as sepsis and hemorrhage. Usually, bed rest for 4 to 6 hours is required.

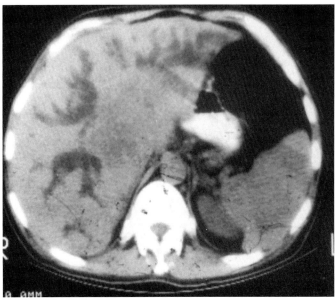

**FIG. 21.3.** Computed tomography scan through liver shows enormous dilatation of the biliary tree due to mass at the porta hepatis obstructing the common hepatic duct. Observe short distance from skin to left biliary radicles because of their anterior location. (Reprinted from Castañeda-Zúñiga WR. *Interventional radiology,* 3rd ed. Baltimore: Williams & Wilkins, 1997, with permission.)

### Percutaneous Transhepatic Cholangiography versus Endoscopic Retrograde Cholangiopancreatography

ERCP is useful to study the biliary and pancreatic ducts, as well as the pancreas. However, the 90% to 99% success rate of PTC in demonstrating the bile ducts compares favorably with ERCP's success rate of 70% in patients with suspected biliary obstruction. Additionally, PTC is less expensive than ERCP. There is no statistical difference between the PTC morbidity (3%) and mortality (0.1%) rates and the ERCP morbidity (3%) and mortality (0.3%) rates.

### Percutaneous Transhepatic Cholangiography in Patients with Suspected Biliary Disease and Nondilated Bile Ducts

PTC will determine the site and etiology of bile duct obstruction in 90% to 95% of patients with dilated bile ducts. However, the success rate decreases if ducts are nondilated. ERCP should be the initial invasive procedure in patients with nondilated ducts and suspected biliary pathology. If ERCP is unsuccessful or anatomically difficult secondary to surgically altered anatomy, PTC should be attempted, and if normal or inconclusive, a core liver biopsy will frequently provide a definitive diagnosis in these patients.

### Complications

A multicenter study found a complication rate of 3.28% in 3,596 patients who underwent PTC with a fine needle. The complications of PTC include death (0.14%), sepsis (1.8%), hemorrhage (0.28%), bile leakage (1%), and miscellaneous (pneumothorax, contrast reactions, hepatic arteriovenous fistulae, and vasovagal reactions).

## PERCUTANEOUS BILIARY DRAINAGE PROCEDURES

The techniques used for percutaneous transhepatic biliary drainage (PTBD) offer palliative decompression of the bile ducts in either benign or malignant disease with low morbidity and mortality rates. The traditional therapy for malignant biliary obstruction has been the surgical creation of a biliary-enteric anastomosis, but this technique has significant morbidity and carries an operative mortality rate of as much as 15% to 60%, depending on the severity of the liver disease. The principal application of PTBD has been for nonoperative drainage of malignant biliary obstruction. Patients with high obstruction at the hilum; patients with lesions producing noncommunicating obstruction of the right and left central ducts; patients with multiple branch obstructions or with a short life expectancy; and patients who are high operative risks because of cardiac, pulmonary, or renal complications are candidates for PTBD. Patients with lower common duct or periampullary obstructions who present in stable hemodynamic condition earlier in the course of their disease will typically undergo surgical palliation. Patients often will derive benefit from PTBD only when they are symptomatic or when their life expectancy may be shortened by recurrent cholangitis, progressive hepatic dysfunction, or nutritional deficiencies secondary to bile flow obstruction. In most patients with biliary obstruction secondary to malignancy, death occurs in 6 to 14 months. The net results of PTBD in improving the quality of life and the length of survival have to be weighed against the possible complications. Instrumentation and imaging advances have extended the use of the percutaneous approach to the management of biliary duct strictures and fistulae, biopsies, and intracanalicular radiation therapy.

## Instrumentation for Biliary Drainage Procedures

### Catheters

Drainage catheters can be divided into two large groups:

1. Exteriorized drainage catheters, which include catheters solely for external drainage and those that have an external portion but extend past the area of obstruction and permit internal drainage (internal/external stents)
2. Internal drainage catheters or stents, which include only catheters that have no external portion

These can be classified according to the material from which they are made as well as according to any alterations in shape or design incorporated to inhibit migration. Materials in common use for biliary stents include Percuflex, Teflon, C-Flex, Silastic, and metal alloys.

### Exteriorized Catheters

Among the most widely used are the VTC biliary catheter (MEDITECH/Boston Scientific, Natick, MA), Ultrathane drainage catheter (Cook, Bloomington, IN), and the Cope Catheter (Cook). Their design is similar. These catheters can function as internal/external stents or as internal stents if capped and their pigtail loop is formed in the duodenum. They come in different diameters and have multiple sideholes along the shaft for antegrade drainage and a retention suture locking pigtail. They are available with hydrophilic coating, which produces a more slippery surface than an uncoated surface, ideal to cross through tight stenoses.

### Internal Stents

The percutaneous placement of plastic internal stents has decreased due to their tendency to migration, obstruction, and difficult removal. Metal internal stents are discussed separately.

### Needles

Among the most widely used needles are the 21- to 23-gauge Chiba needles (several manufacturers) and the 18-gauge nephrostomy diamond-tip nonsheathed needle (Cook). Several introducer access systems, including the Accustick system (MEDITECH/Boston Scientific) and the Cope system (Cook), permit the placement of a 6-Fr short catheter from the initial access of a 21-gauge diamond-tip needle (Accustick) or 22-gauge needle (Cope) and the subsequent passage of a 0.035- or 0.038-in. guidewire.

## Indications and Contraindications for Biliary Drainage

### Indications

- Palliation for unresectable primary or metastatic malignancy

- Benign strictures, particularly stenotic biliary-enteric anastomoses
- Sepsis accompanying biliary obstruction
- Preoperative decompression
- Prophylactic decompression after PTC or ERCP
- Ancillary therapeutic maneuvers

### *Contraindications*

The only true and absolute contraindication to PTBD is the presence of an uncorrectable bleeding diathesis. Large ascites makes PTBD technically difficult, although not impossible. Combined extrahepatic and intrahepatic or multiple intrahepatic obstructions are also relative contraindications to PTBD. Even though extrahepatic obstruction exists, the intrahepatic ducts may not be dilated because of the multicentric mass effect of intrahepatic tumors. Also, the introduction of catheters into the biliary system invariably invites colonization by bacteria, and the stagnant bile in any nondrained segments of liver can rapidly become infected. Therefore, it is vital to perform high-quality PTC to evaluate the entire biliary duct anatomy in cases of intrahepatic tumors before placement of catheters.

## Patient Evaluation, Preparation, and Monitoring

Patient evaluation and preparation are similar as for PTC. Careful patient monitoring throughout interventional biliary procedures is critical. Sedatives and narcotics are given for painful, prolonged, or difficult procedures. Vital signs (pulse, blood pressure, breathing) are monitored continuously. Electrocardiographic and oximetric monitoring are generally recommended in all patients. Most biliary interventional procedures can be performed under local anesthesia, giving adequate amounts down to and including the liver capsule with the help of sedatives and analgesics. Several combinations have proved safe, mainly midazolam with fentanyl. Some centers use general anesthesia for their initial PTC and drainage procedures.

## Biliary Duct Demonstration

### *Ultrasound*

A good method in the patient with obstructed bile ducts is real-time US demonstration of the dilated left biliary radicles and US guidance to puncture a duct with either a skinny or an 18-gauge needle. Approach from the right side is more difficult because of the ribs. Systems that have been described for US guidance include transducer needle attachment guides and freehand techniques, which require greater experience by the operator. Once a duct is entered under US guidance, the procedure continues under continuous fluoroscopy in the usual fashion (Fig. 21.4).

### *Fluoroscopy*

The technique of fluoroscopic PTC is detailed above. Most authors advocate a right lateral approach, and this is also our initial choice.

### *Imaging of the Left-Sided Ducts*

In some situations the left ducts cannot be opacified from a puncture into a right-sided duct, despite several maneuvers. Most often, the cause is obstruction in the region of the porta hepatis impeding free flow of contrast from one side to the other. If there is a compelling indication for catheterization or examination of the left ducts, a left PTC must then be performed. This may be achieved from the right lateral approach as in the conventional technique but with a deeper pass of the needle, which is directed anteriorly and medially into the left lobe. Alternatively, a subxiphoid approach may be taken: A 22-gauge Chiba needle punctures the skin in the midepigastrium, just beneath the xiphoid cartilage, and is aimed in the transverse plane at an angle of 30 to 40 degrees toward the patient's right side (Fig. 21.5). The needle is advanced approximately 8 cm in a blind fashion and then slowly withdrawn during continuous contrast injection to locate the bile duct, as in the conventional right-sided technique. Real-time US greatly facilitates locating the left ducts, enabling

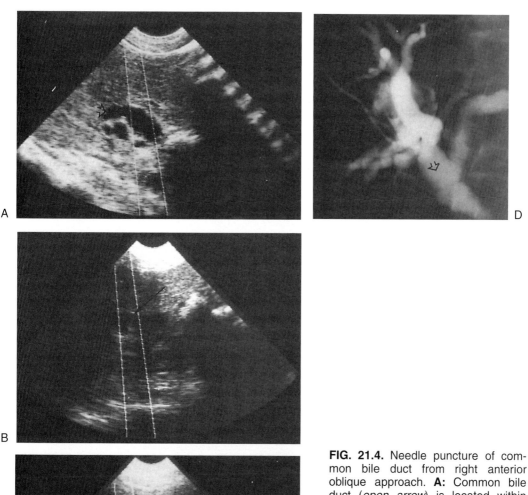

FIG. 21.4. Needle puncture of common bile duct from right anterior oblique approach. **A:** Common bile duct (*open arrow*) is located within puncture calipers. **B:** Echogenic needle (*arrow*) is advanced under real-time control. **C:** The needle is removed, and a Teflon sheath (*arrows*) is advanced into a dilated bile duct. **D:** Injection of contrast medium under fluoroscopic control demonstrates distended biliary system. The common bile duct is indicated by the open arrow. (Reprinted from Castañeda-Zúñiga WR. *Interventional radiology*, 3rd ed. Baltimore: Williams & Wilkins, 1997, with permission.)

entrance with a single pass in many cases. If a 21-gauge needle is placed under US guidance, a drainage catheter can be inserted via this same tract; otherwise, a second puncture is performed in the manner outlined below.

**Puncture Techniques**

*Single-Stick Techniques*

The materials used to access the biliary tree with single-stick techniques are described previously.

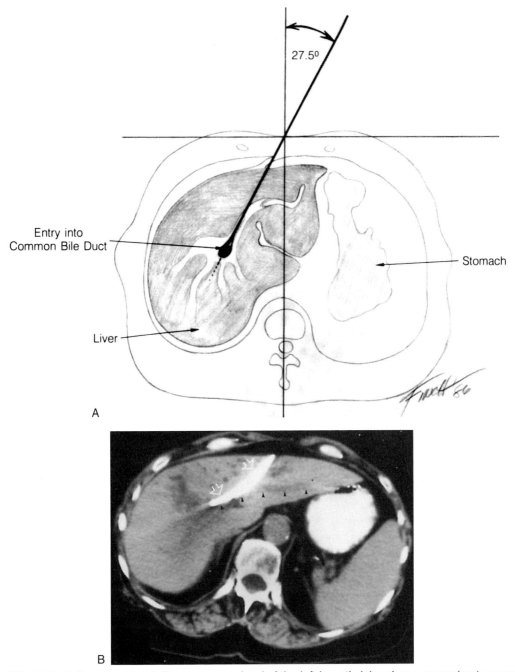

**FIG. 21.5. A:** Cross-sectional diagram across level of the left hepatic lobe shows approximate angulation needed to orient path of needle with long axis of the left hepatic duct. **B:** Computed tomography slice at the level of the biliary drainage catheter shows a favorable angle from the left hepatic duct (*arrowheads*) into common bile duct. Biliary drainage catheter is in place (*open arrows*). (Reprinted from Young AT, Cardella JF, Castañeda-Zúñiga WR, et al. The anterior approach to left biliary catheterization. *Semin Intervent Radiol* 1985;2:31–38, with the permission of Thieme.)

Basically, a 21- or 22-gauge needle enters the duct and is replaced over a 0.018-in. guidewire for a tapered 6-Fr coaxial sheath/dilator/cannula assembly (Accustick) or 6.3-Fr catheter with a large sidehole for passage of a larger wire (Cope). The most serious problem with any single-stick technique is that the skinny needle may enter a duct that is not suitable for biliary drainage because of its size, location, or the angle that it makes with the needle or with the common hepatic duct. This is obviated with the double-stick techniques.

### Double-Stick Techniques

As its name implies, a double-stick technique requires two punctures. The first puncture, for opacification of the ducts, is done with a 21-, 22-, or 23-gauge needle under ultrasonic or fluoroscopic guidance. The second, definitive puncture is done with a skinny needle or the 18-gauge diamond-tip nephrostomy nonsheathed needle, which accepts a 0.038-in. wire.

## Approaches

Although PTBD has traditionally been performed through a right lateral approach, other alternatives have been described, including anterior and posterior approaches.

### Right Lateral Approach

For the right lateral approach, after ductal opacification, a peripheral site in a right anterior duct is selected for the definitive puncture. The duct must be localized in two planes with the help of either biplane/rotational fluoroscopy or cross-table lateral films. The entry site should be below the tenth rib to avoid perforating the pleural reflection. Either a skinny needle or the 18-gauge needle will be used for the puncture, which is done under continuous fluoroscopy. The main disadvantage of the lateral approach is that the needle placement is done blindly: Even though fluoroscopy is used during needle advancement, it is usually in only the anteroposterior plane and therefore provides no depth perception.

### Anterior Approach

The anterior approach is technically simpler. The puncture is usually into an anterior left biliary radicle, although an anterior right duct also can be used. Guidance for left duct puncture can be ultrasonic or fluoroscopic. Rotational fluoroscopy is needed to minimize radiation exposure of the operator's hands. After opacification of the biliary radicles through a standard right lateral PTC, a peripheral entry site is selected in an anterior left biliary radicle. Puncture of a peripheral duct minimizes the size of the vessels at risk from inadvertent trauma and in addition places a sufficient number of drainage sideholes proximal to the stenosis. For this puncture technique, the skin entry site is aligned with the duct entry site in the center of the fluoroscopic field. The needle is advanced parallel to the x-ray beam so that the needle is seen as a dot, until indentation or displacement of the bile duct is seen, or until the operator feels that the appropriate depth has been reached. The multidirectional fluoroscope is rotated away into a new projection, and the relation of the needle tip to the biliary radicle is examined. The goal is to make a one-wall puncture, thus avoiding extravasation of contrast medium and false passage of the guidewire through a puncture hole in the back wall. Further manipulations are performed with the image intensifier angled 90 degrees from the entry tract to minimize the radiation exposure of the operator's hands. The tract should be angled approximately 30 degrees from vertical toward the patient's right side to create a favorable angle between the parenchymal tract and the bile duct for further manipulations. In most patients, the skin entry site for an anterior bile duct puncture is to the left of the midline near the costal margin and slightly below the xiphoid cartilage. When the left lobe is small, the puncture can be made on the right side of the epigastrium, beneath the right costal margin. The complications resulting from left-lobe puncture are fewer than with the right-lobe puncture. The left biliary duct puncture has several advantages:

- The left duct is more horizontal and superficial, making it an easier target.
- The pleural space can be avoided, as well as its related complications.

- There are shorter tracts and better angles for catheter and wire manipulations compared with right-sided approaches.
- Less pain (lack of ribs, no phrenic irritation), better tolerance, and easier catheter maintenance. The preferred access is through the left lobe in many centers, and US is the imaging method of choice for needle guidance in those.

### *Percutaneous Transjejunal Approach*

Access to the biliary tract can be gained in patients with biliary-enteric anastomoses who have previously undergone Roux-en-Y biliary surgery, bringing a limb of the Roux-en-Y to the skin and marking the site with surgical clips. Jejunal puncture is easily performed, is well tolerated by patients, and can be repeated as needed. Access from below is quite favorable for intrabiliary manipulations, particularly in patients with diffuse, chronic disease such as sclerosing cholangitis.

### *Discussion of the Different Approaches*

When an obstructing lesion in the region of the porta hepatis involves both the left and right hepatic ducts, the decision regarding where to place drainage catheters may be difficult. Drainage of both lobes separately reduces the risk of cholangitis secondary to persistent obstruction but leaves the patient with the morbidity of two external tubes. The right lobe is the larger, and therefore drainage of this lobe alone presumably provides better palliation than drainage of the left. Conversely, however, it has been said to be more likely that the radicles will be separately obstructed in the right lobe than in the left because of the anatomic differences between the two ductal systems, the right hepatic duct being shorter than the left with a shorter extrahepatic course. Right-sided drainage may therefore leave large areas undrained within the right lobe, requiring several catheters for adequate palliation in these cases.

## Technique of Catheterization

Once a duct is entered, a long floppy wire is passed as far as possible in the duct and the tract dilated to 6 Fr. The operator's hands are kept out of the primary beam by angulation of the C-arm or by rotating the patient if a conventional fluoroscopic unit is used. Next, a 5-Fr torque-control preshaped angiographic catheter is passed over the wire for manipulation within the biliary tree. This combination will often traverse a stricture, whether benign or malignant (Fig. 21.6). Otherwise, a hydrophilic wire used with care, alone or in combination with a regular or hydrophilic catheter, will negotiate almost all obstructions. Generally, once an 8-Fr drainage catheter has been passed beyond the stenosis or has been placed satisfactorily above it, the procedure is terminated to allow the parenchymal tract to mature, the biliary tree to decompress, and sepsis to defervesce (Fig. 21.7). Further manipulations or passage of larger catheters for long-term drainage are usually performed at a second sitting 2 to 5 days later. Samples of bile are aspirated and sent for culture, cytology study, or both. Transcatheter biopsy of a stricture may be performed preferably at a later date. The catheter is secured to the skin using suture, dressings, or any of the available fixation devices. If a pigtail catheter is to be introduced, it is advanced until the entire pigtail loop is seen to be well within the biliary tree. The wire is removed, and the position of the loop and sideholes within the biliary tree is checked fluoroscopically by injecting contrast medium in short 1- to 2-mL bursts. The loop should be coiled on itself and freely rotatable within the duct to prevent duct wall injury by the catheter tip. All the sideholes should lie within the biliary tree; otherwise, bleeding through the stent can be a problem, with the stent and bile ducts rapidly becoming obstructed by clots. For insertion of a drainage catheter, the 8- to 10-Fr catheter and the plastic or metal stiffening cannula are advanced together over the guidewire until the tip of the cannula is inside the duct. At this point, the cannula and guidewire are held together in a stationary position, and the catheter is advanced over the wire until the desired length of catheter is within the biliary tree. In some cases, advancement of the catheter over the wire through an area of severe obstruction is difficult, if not impossible. In these instances, several maneuvers are helpful: Use of a stiff wire and/or

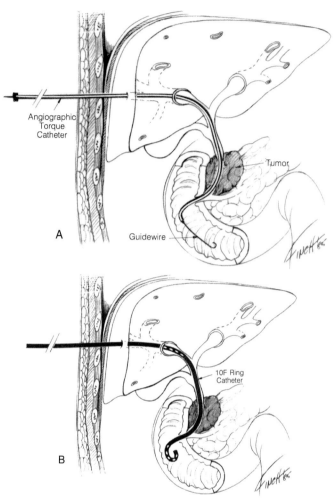

**FIG. 21.6.** Biliary drainage catheter placement. **A:** Through a right biliary radicle, a guidewire has been manipulated across the obstruction into the duodenum with the help of an angiographic torque control catheter. **B:** A 10-Fr pigtail catheter has been placed with the tip in the duodenum for internal drainage. (Reprinted from Castañeda-Zúñiga WR. *Interventional radiology*, 3rd ed. Baltimore: Williams & Wilkins, 1997, with permission.)

drainage catheters with hydrophilic coating or preshaping of the cannula and gentle advancement of it over the wire into a position at the level of the proximal common bile duct helps the subsequent advancement of the drainage catheter through the obstruction. Care should be taken not to perforate the distal wall of the duct with the tip of the catheter-cannula combination because a biloma may form. Once within the ducts, the catheter can be reshaped and locked using the suture string attached to it. In some patients with Cope loops, biliary salt encrustation on the thread blocks bile flow and prevents the loop from opening during catheter exchange at follow-up examination. In these cases, a hydrophilic (0.035-, 0.025-, or 0.018-in.) or small wire can usually be passed through the lumen of the catheter. If a guidewire cannot be passed, the catheter hub is cut, and while the monofilament threads are held, a peel-away sheath is advanced into the bile duct over the catheter until it reaches the loop. The catheter can then be pulled back forcefully against the sheath or dilator, cracking the encrusted loop open at the crossover point. The

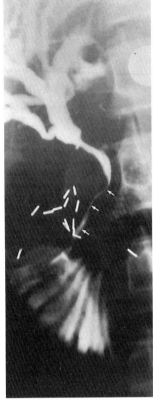
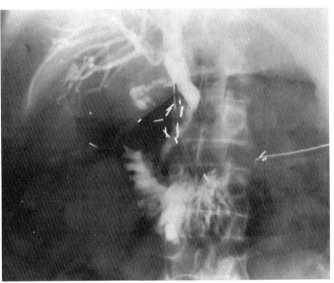

FIG. 21.7. **A:** Fine needle cholangiogram in patient with carcinoma of the pancreas causing obstruction of the distal common bile duct (*arrows*). Observe displacement and narrowing of the intrahepatic radicles. **B:** By an anterior epigastric approach, a locking drainage catheter has been placed across the obstruction with its tip in the duodenum for internal drainage with an exteriorized catheter. (Reprinted from Castañeda-Zúñiga WR. *Interventional radiology*, 3rd ed. Baltimore: Williams & Wilkins, 1997, with permission.)

loop is removed after advancing the sheath-dilator farther into the biliary tree. A new drainage catheter can then be inserted through the sheath or dilator.

### Catheter Placement for Internal-External Drainage

If the procedure goes smoothly, the patient remains stable, and the stenosis can be traversed without great difficulty, then passage of an internal drainage catheter to the distal common bile duct or duodenum can be accomplished at the first sitting. Often, however, the attempt to pass through the stenosis into the duodenum is postponed until a second sitting, when the biliary tree has been decompressed and the patient is more stable. In the second sitting, the skin and the previously placed external drainage catheter are sterile prepared. Contrast is injected through the catheter to locate the stenosis and opacify the ducts. A 0.035-in. floppy-tip guidewire is passed through the catheter, which is removed. Next, there are two suggested techniques. One technique uses a torque control angiographic catheter and the 0.035-in. floppy-tip wire. The catheter tip is used to direct the wire through the stenoses and around the bends in the tortuous, dilated ducts. The other technique relies for its directional control on a Glidewire (MEDITECH/Boston Scientific, Natick, MA). This is used to pass through stenoses and around bends in the biliary radicles into the duodenum, and the catheter follows. Once the guidewire has passed across the obstruction and into the duodenum, a stent can usually be passed over the wire. If the floppy-tip wire or the Glidewire passes through the obstruction but the 5-Fr catheter will not follow, an attempt is made to pass a smaller, 4-Fr catheter or a 5-Fr Glide-catheter (MEDITECH/Boston Scientific) through the lesion. If this is successful, the original wire is exchanged for a Superstiff wire, which usually allows the passage of larger catheters. Also, if the diagnostic catheter and wire pass through the lesion but the 8- or 10-Fr drainage catheter will not, we remove the drainage catheter and dilate the lesion with an appropriately sized balloon catheter. Once the obstruction is dilated, the drainage catheters are almost always advanced with ease through the ducts. In some cases, the ducts can easily be catheterized down to the point of the obstruction, but neither a catheter nor a wire will pass through the obstruction. In such cases, it is most helpful to decompress the biliary tree for 4 to 5 days with an external drain and then try again to traverse the obstruction. Decompressing the dilated bile ducts allows some time for edema at the obstruction site to resolve and the ducts to return to more normal dimensions. As a result, ductal tortuosity is decreased, which greatly facilitates catheterization. The segment of the catheter in which sideholes should be made is easily determined by the following technique: Once the guidewire is in the duodenum, a catheter is passed over the wire into the bowel and then pulled back to the level of the ampulla. A clamp or a bend is applied to the wire at the catheter hub. The wire is pulled back until the tip is well above the obstruction but inside the ducts, and a second clamp or bend is applied to the wire at the catheter hub. The wire is then removed. The length of wire between the clamps or bends represents the length of catheter in which sideholes have to be made.

## Complications

Cholangitis and biliary sepsis are the most frequent complications with any type of biliary drainage. The reported occurrence rates range from 3.3% to 67.0%. The larger the size of the catheter, the lesser the incidence of sepsis. Incomplete drainage of the biliary tree with obstruction of multiple segments carries a significantly higher risk of cholangitis (64%), compared with patients with a single stenosis (15%). Other complications include bleeding and biliary peritonitis (2% to 5%) and fluid and electrolyte disturbance.

## Dislodged Biliary Drainage Catheter

Accidental dislodgment of a drainage catheter, particularly a laterally placed one, commonly requires creation of a new percutaneous tract because attempts to catheterize the old tract usually fail. Anterior tracts tend to be easier to recatheterize because of their straighter and shorter courses. Two rules apply with these patients to avoid tract disruption: Do not manipulate the tract blindly and do not infiltrate local anesthetics within the tract before manipulation is performed. The technique for replacing a dislodged catheter begins with the patient in the supine position. The old entry site is prepared and draped. Under fluoroscopic control and filming, a Christmas tree adapter is inserted into the tract opening, and contrast medium is instilled to opacify the tract. Once the tract is opacified, a 0.035-in. floppy-tip guidewire is gently manipulated through the bends in the tract with the help of a torque control catheter until the wire reaches the biliary tree. Next, placement of a catheter or stent proceeds in the usual fashion. If the procedure is attempted more than 24 hours after accidental dislodgment, the tract is frequently closed, and attempts to reopen it more commonly result in false passages than in recatheterization.

## Drainage Dysfunction

The most common causes of drainage dysfunction are clogging by tumor overgrowth, biliary salts, blood clots, or intestinal contents or by displacement of the sideholes out of the biliary tree or below the obstruction. In most cases, an obstructed catheter can be easily cleaned by flushing or by passing a wire through it. Alternatively, it can be replaced with the aid of a guidewire.

## Patient Follow-Up

All patients should be evaluated periodically. Patients with external or internal/external drainage catheters should be evaluated every 3 months if asymptomatic or if at any time there is evidence of decreased bile output, an increase in serum bilirubin or liver enzymes, leakage of bile along the tract, or bleeding. As a rule, the catheter should be replaced each time the patient is evaluated for one of these reasons. Therefore, if the patient's life expectancy is short, placement of an internal stent is recommended. In some cases, the stent is patent, and the reason for the bilirubin elevation is a deterioration of liver function. These cases are probably diagnosed equally well by PTC or nuclear imaging. Nothing further needs to be done.

## Malignant Obstructive Jaundice

In a patient presenting for the first time with obstructive jaundice, the cause is likely to be malignant, as most patients with benign biliary obstruction will have a history of other disease or, more commonly, of surgical intervention. Often, US, CT, and MRI scans will not only provide the diagnosis of biliary obstruction but will indicate its level and may suggest the actual cause and determine resectability if a tumor is present. If the patient is unsuitable for curative surgery, a palliative procedure of choice can then be selected. Demonstration of resectability does not necessarily mean that a curable tumor is present, and tissue type should be established by fine needle biopsy, by ERCP, or after PTC. If the condition is considered appropriate for surgical management at this stage, the decision of whether to drain the biliary system preoperatively must to be made. In the incurable patient, the choice of method of palliation must be made. At present, there are two nonsurgical approaches to the problem of long-term palliation of malignant biliary obstruction: percutaneous transhepatic (internal-external and endoprosthesis) and endoscopic endoprosthesis, which may be preceded by transnasal catheter drainage. The aim of palliation is the extended relief of the symptoms of biliary obstruction, as the preterminal symptoms attributable to the cancer itself are not yet present. Plastic endoprosthesis has not been the answer to this problem either.

This is mainly due to migration and obstruction of the catheter. This requires repeat procedures to reestablish biliary flow, and although all patients experience dramatic relief of symptoms after stenting, their distress increases considerably with each occlusion and recurrence of jaundice necessitating a further procedure. For this reason, proponents of internal/external drainage argue that this technique is always preferable, as it allows regular irrigation of the catheter and simple catheter exchange when irrigation is no longer effective. However, not all patients tolerate these external catheters. Some cannot perform proper maintenance of their device. For others, the catheter is a constant reminder of their disease.

## Biliary Metal Stents

Several metallic endoprostheses are currently available for use in the biliary system. Metal stents can be divided into two groups: balloon-expandable and self-expandable stents. For further discussion of the individual stent, see Chapter 11. All available metal stents are approved for use in the biliary system. They are made of a variety of metal alloys, such as the Wallstent (MEDITECH/Boston Scientific Vascular) and Palmaz stent (Cordis, a Johnson & Johnson Company, Miami, FL), made of stainless steel, and nitinol stents such as the Smart stent (Cordis), Symphony stent (MEDITECH/Boston Scientific), and Memotherm stent (Bard, Covington, GA). Large series are needed to collect data regarding long-term outcomes for a variety of indications within the biliary tree.

Preliminary drainage for several days is preferable if there is appreciable bleeding into the bile ducts or biliary infection; however, if clear ducts are shown, immediate placement of the stent can follow. Because the expansible force of the self-expandable stents is less than that of balloon-expandable stents, it is recommended to predilate the stricture with a 7- to 10-mm angioplasty balloon over a guidewire. If the stent used is not long enough to provide adequate drainage, further stent(s) can easily be placed overlapping the first by the same technique. After placement of the stent, a temporary drainage catheter to allow for follow-up cholangiography after 24 hours may be placed above the stent, or all catheters may be removed. Migration of metal stents is uncommon. An advantage of the self-expandable metal stents over the balloon-expandable stents is their flexibility, permitting comparatively easy positioning where there is angulation of the tract and also allowing for stenting of curved ducts without deformity and with good stability of the stent.

### *Technique*

The maneuvers used to cross the stricture are similar to those described earlier. Either a metal stent or an internal/external catheter may be inserted immediately after crossing the lesion. Some authors believe that it is always better to use temporary drainage at this time and place a stent at a second session. It is also appropriate to wait if there is hemobilia and the bile ducts are full of clots, in which case temporary drainage until the bile is clear is advisable. This preference applies to the immediate use of an endoprosthesis irrespective of the material from which it is made.

### *Patient Follow-Up*

Patients with internal stents should be evaluated at the first sign of stent malfunction, as manifested by jaundice, abdominal pain, bilirubin elevation, or sepsis. Evaluation can be done noninvasively by hepatobiliary nuclear scans, which may confirm the suspicion of stent blockage, or, more invasively but more directly, by PTC, from which one can gain important anatomic information that is usually not clear from a nuclear scan.

### Comparison of Results

Plastic biliary endoprostheses have two main drawbacks: stent occlusion and migration. Because of these problems there has been great interest in the use of metallic biliary endoprostheses. Migration of metallic stents is unlikely, as they become incorporated and fixed in the wall of the common duct. Finally, the stent surface exposed to bile encrustation is minimal, which should diminish the risk of occlusion, which is in the range of 7% to 35% for metallic stents. Obstruction is due to neither foreign body reaction nor mucosal hyperplasia. In some series, tumor ingrowth and sludge formation have been causes of obstruction. The location of the lesion within the ductal system appears to influence the patency rates of the stent, as nonhilar lesions show better patency than hilar obstructions. Average time to reintervention after initial stent deployment is 5 to 9 months. Two small, randomized studies of plastic versus metallic stents indicated that in these study groups the overall costs of patient management were lower for the group treated with metallic stents, despite the higher initial cost of each single device. In addition, because of the relatively small size needed for percutaneous stent introduction, it is possible to insert the stent during the initial biliary drainage procedure and remove the temporary external drainage catheter after 1 to 3 days. This should lead to decreased patient discomfort and reduced length of hospital stay. Thus, primary stent placement can be another factor in reducing the cost of treating patients with metallic endoprostheses.

## BILIARY STRICTURES

Most bile duct strictures are secondary to surgical trauma. Other types, classified by cause, include the following:

- Malignant—primary or metastatic
- Benign—radiation fibrosis, congenital, infection, sclerosing cholangitis

Traditionally, patients with strictures of the biliary tract have been managed surgically, either a primary repair with an end-to-end anastomosis or creation of a biliary-enteric anastomosis. Surgical repair of biliary strictures is successful in 65% to 80% of patients. An alternative to operation is percutaneous ductal dilatation via either a transhepatic or endoscopic route. Data indicate that percutaneous dilatation of appropriate types of benign strictures generally yields good short-term results. This kind of therapy is offered to patients in the following circumstances:

- High surgical risk secondary to medical problems
- Strictures not amenable to corrective operation
- Previous attempt at surgical repair, which makes the second procedure extremely difficult
- Advanced liver disease

Percutaneous balloon dilatation is most successful in patients who have benign biliary strictures from previous choledochojejunostomy anastomosis, iatrogenic strictures, or, less often, sclerosing cholangitis. Unfortunately, studies reporting long-term results are not available, nor are the long-term effects of balloon trauma on the biliary tree known.

## Technique

Two principal steps are involved in the dilatation of strictures: the dilatation itself and long-term stenting. Because balloon dilators are easier to pass, they are preferred, and most strictures respond favorably. There is controversy about the best approaches for dilatation, which include a single inflation without stenting; a single inflation with short-term stenting; repeated short inflations with long-term stenting; and single or repeated long inflations with long-term stenting. The procedure is performed through a PTC access as described above and usually a 6- to 8-Fr introducer sheath is placed across the tract into the biliary system during the procedure. This preserves access to the biliary system and minimizes damage to the tract during the exchange of catheters and balloons. A 6- to 10-mm-diameter, 4-cm-long high-pressure balloon catheter is passed over the working wire into the stricture, and the balloon is inflated with an inflation device. An inflation device is recommended when prolonged balloon inflation is to be used because adjustments must be made during the 10 to 12 hours that the balloon remains inflated. A pressure of 4 to 6 atm is sufficient to dilate most strictures, but fibrotic strictures require much higher pressures, which is the reason for using a high-pressure balloon. At the beginning of inflation, a waist is seen in the balloon at the site of the stricture, but it disappears as the pressure is raised. For a short inflation, the balloon is deflated after 10 to 15 minutes and exchanged for an 8- to 10-Fr drainage catheter with sideholes proximal and distal to the dilated area. Generally, we dilate the lesion several times during the same sitting. The balloon catheter is then exchanged for a 10-Fr stent, which is left draining internally for 1 to 2 months, at which time follow-up cholangiography and repeat dilatation are performed. Cholangiography and repeat dilatation are performed as needed. A final 1-month follow-up pull-back cholangiogram is performed over a wire. If the dilatation site is widely patent, the stent is removed. If a high-grade stricture persists, the biliary stent is left in place. In equivocal cases, a biliary pressure study is performed. A pressure gradient greater than 15 cm $H_2O$ is considered significant, and the stent is left in place. In some patients, the stricture is so hard that it cannot be stretched by a routine balloon. A high-pressure balloon is used (20 atm pressure); Teflon coaxial dilators and the new flexible ureteral dilators (Cook, Inc., Bloomington, IN), which are tapered to the size of a 0.038-in. guidewire, have proved useful in these situations.

## Discussion

Primary surgical repair of biliary stricture is the traditional choice for most patients, with a success rate of 65% to 80%. Balloon dilatation has produced encouraging results in dilatation of choledochoenteroanastomosis (Fig. 21.8). In patients with sclerosing cholangitis, percutaneous dilatation of strictures, and long-term stenting have also been successful. In the authors' experience, balloon dilatation performed during three separate sessions over a 2- to 3-month period appear to help prevent stricture reformation, although this has yet to be proved by other investigators.

## Delayed Complications

The more common delayed complications include cholangitis, bile leakage, dislodgment of the catheter, and infection at the catheter site. They occur in 40% to 50% of patients and usually are mechanical problems related to catheter malfunction. The patient should be taught to recognize these complications and to report them immediately to the interventional radiologist for prompt correction. Less common delayed complications include peritoneal seeding of carcinoma and growth of tumor along the transhepatic catheter.

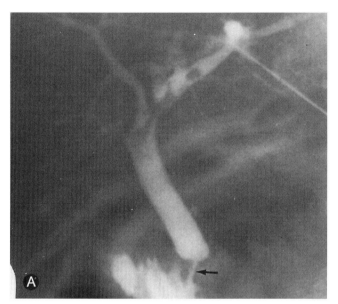
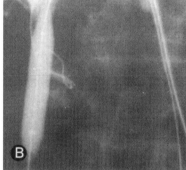
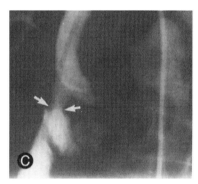

**FIG. 21.8.** Fever and jaundice 3 months after liver transplantation. Repeated cholangitis suspected. **A:** Fine needle cholangiography reveals mild dilatation of intrahepatic radicles and severe stenosis of the choledochojejunal anastomosis (*arrow*), although contrast medium is seen to pass into the jejunal loop. **B:** A left anterior epigastric approach was used to selectively catheterize the stenosed anastomosis, and a 10-mm angioplasty balloon catheter was passed and inflated at the site of stricture. **C:** Cholangiogram 2 months after dilatation before stent removal reveals a wide-open anastomosis (*arrows*). (Reprinted from Castañeda-Zúñiga WR. *Interventional radiology*, 3rd ed. Baltimore: Williams & Wilkins, 1997, with permission.)

## B. GALLBLADDER INTERVENTION

Laparoscopic cholecystectomy has become the operation of choice for symptomatic cholelithiasis. It is a safe procedure, with an incidence of bile duct injury ranging from 0.2% to 1.0% and a major cannulation rate of 1.6%. Therefore, any alternative therapy has to provide similar or better morbidity and mortality rates, as well as similar results, to substitute for surgical or laparoscopic cholecystectomy as the procedure of choice in the management of gallbladder disease.

### ANATOMIC RELATIONSHIPS OF THE GALLBLADDER

In a study based on CT scans of 100 patients with gallstones, four anatomic variations in the relationships of the gallbladder to the liver and anterior abdominal wall were reported. These variations are (a) completely intrahepatic gallbladder in 39%, (b) partially intrahepatic gallbladder in 35% (liver-free fundus), (c) gallbladder completely anterior to the liver in 17%, and (d) the gallbladder in a lateral position in 9%. In 51% of the patients, the colon was in direct contact with the gallbladder, but only 13% had the colon directly in front of the gallbladder. A transperitoneal approach is, however, possible in 65% of the patients using an inferosuperior oblique direction. In comparison, others have shown that up to 83% of patients require a transhepatic approach for cholecystostomy due to the above mentioned anatomic factors.

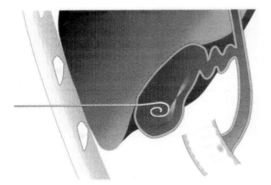

FIG. 21.9. Transhepatic gallbladder puncture. If leakage occurs, it will be extraperitoneal. (Reprinted from Castañeda-Zúñiga WR. *Interventional radiology*, 3rd ed. Baltimore: Williams & Wilkins, 1997, with permission.)

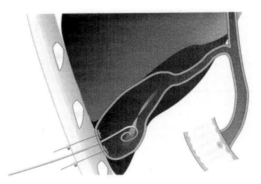

FIG. 21.10. Transperitoneal gallbladder puncture. Anchoring device has fixed the gallbladder fundus to the anterior abdominal wall. Catheter placement into the gallbladder fundus is through anchors. (Reprinted from Castañeda-Zúñiga WR. *Interventional radiology*, 3rd ed. Baltimore: Williams & Wilkins, 1997, with permission.)

## PERCUTANEOUS CHOLECYSTOSTOMY

Surgical cholecystostomy is of value in the management of both inflammatory and obstructive biliary tract disease in critically ill patients or when surgical removal of the gallbladder is technically difficult. It is viewed as a temporary solution to stabilize the patient before definitive surgical intervention. Because of the high-risk population involved, surgical cholecystectomy and cholecystostomy have a mortality rate, ranging from 10% to 36% and 5% to 30%, respectively. Percutaneous cholecystostomy can be performed at the bedside under local anesthesia in the intensive care unit. It allows postponement of the definitive procedure until the patient's sepsis is controlled. The technical success rate has been reported to be 98% to 100%. This procedure is an effective method to decompress the biliary system or an acutely inflamed gallbladder. Current indications of percutaneous cholecystostomy include (a) critically ill or elderly patients with calculus cholecystitis, (b) acute acalculous cholecystitis (AAC), (c) spontaneous gallbladder perforation and bile leakage, and (d) gallbladder sepsis following insertion of biliary stents. There are two basic approaches to the placement of a percutaneous cholecystostomy: the trocar catheter and the Seldinger technique. Trocar catheter placement is commonly performed under US guidance at bedside, providing safe, one-step percutaneous gallbladder drainage. The most common percutaneous approach used by investigators for placement of percutaneous tracks into the gallbladder is a transhepatic approach through the bare area or hepatic attachment of the gallbladder (Figs. 21.9 through 21.11). In theory, this approach would prevent biliary leakage from the gallbladder puncture site. However, there are no convincing data available so far to confirm this, and in studies in which both transhepatic and transperitoneal approaches are used, no significant difference has been observed in the incidence of biliary peritonitis with either technique.

### Transhepatic Technique of Percutaneous Cholecystostomy

A transhepatic route to the gallbladder is visualized with US, as close as possible to the fundus. Next, any of the above described introducer systems used for biliary duct access is used to enter the gallbladder and convert the tract to 6 Fr from the initial 21- or 22-gauge needle puncture. After a 0.035- or 0.038-in. guidewire is advanced into

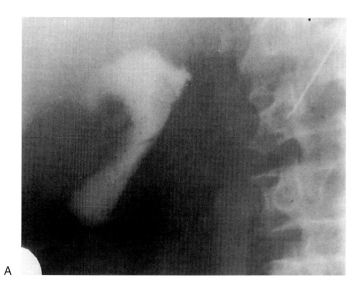

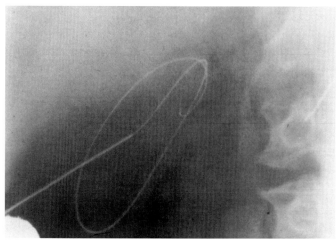

**FIG. 21.11. A:** Percutaneous cholecystogram in a patient in septic shock. A large amount of pus was aspirated. **B:** Guidewire looped in gallbladder. (Reprinted from Castañeda-Zúñiga WR. *Interventional radiology*, 3rd ed. Baltimore: Williams & Wilkins, 1997, with permission.)

the gallbladder, the tract is dilated to the desired size and a locking pigtail catheter is placed within the gallbladder. A gentle follow-up cholecystogram is performed and the catheter is secured to the skin and connected to external drainage.

### Transperitoneal Technique of Percutaneous Cholecystostomy

Transperitoneal technique of percutaneous cholecystostomy is preferred if the coagulation parameters are abnormal, cannot be corrected, or both. The technique can be performed without or with the use of suture anchors (Cope's technique). With Cope's technique, the gallbladder fundus is directly approached from the anterior abdominal wall and secured to it with removable suture anchors. The patient is placed in the supine position, and the gallbladder is localized under US and fluoroscopy. After the gallbladder is localized, it is punctured with a 17-gauge needle using an inferior to superior oblique approach as close as possible to the fundus. The puncture is performed between the right midclavicular and the anterior axillary line. The anchor is subsequently advanced through the needle to fix the gallbladder fundus against the anterior abdominal wall (Fig. 21.10) in a manner similar to performing percutaneous gastropexy.

The cholecystostomy tract is then dilated to allow a locking loop catheter to be inserted. The anchor's suture is then sewn to the skin under slight tension. The anchor device is not removed until a cholecystostomy tract is well formed, 10 to 15 days later, to prevent intraperitoneal bile spillage if the drain falls out accidentally. After the anchor's suture knot is severed, a 4-Fr catheter is inserted over the taut end of the anchor suture until it reaches the gallbladder, and the suture and needle are then pulled out. After the procedure, a fibrous reaction usually develops between the anterior gallbladder wall and the posterior surface of the anterior abdominal wall, beginning as early as 7 days after the percutaneous procedure. The rapid development of fibrous tissue around the drainage tract partially explains the low incidence of bile leakage after transperitoneal cholecystostomy. The key concern with the transperitoneal approach is guidewire dislodgment during the procedure if the gallbladder wall is relatively mobile or free. The single-puncture trocar technique can overcome this problem.

## Percutaneous Cholecystostomy for the Treatment of Cholecystitis

Percutaneous cholecystostomy is effective for treatment of both acalculous and calculous cholecystitis in selected patients. Because no chemical or imaging test is accurate in diagnosis of cholecystitis and the risk of sepsis is high, percutaneous cholecystostomy should be promptly performed if the diagnosis of AAC is suspected. Approximately 50% to 60% of patients will improve with percutaneous cholecystostomy; the improvement is usually dramatic and within 24 hours. For patients who do not respond, the gallbladder is eliminated as a source of clinical concern. The catheter is left in place for 2 to 3 weeks because the microbiologic results are often unreliable. This also ensures the formation of a mature fibrous tract to allow safe catheter removal without intraperitoneal bile leakage. Percutaneous cholecystostomy can be used as an immediate and definitive therapy for AAC, and cholecystectomy can be avoided in the majority of cases. In patients with acute calculus cholecystitis, percutaneous cholecystostomy is an effective temporary measure until cholecystectomy can be performed. It can also give lasting relief in patients who are considered inoperable; however, in these patients, percutaneous stone removal and gallbladder ablation should be considered.

## Percutaneous Cholecystostomy Catheter Care and Removal

Catheters are flushed with 5 mL sterile normal saline every 8 hours. If the follow-up cholecystograms show a patent cystic duct and the gallbladder to be free of stones or contrast extravasation, the catheter is removed after 2 to 3 weeks when a mature fibrous tract has developed from the skin to the gallbladder. Such a tract is necessary to prevent intraperitoneal leakage of bile when the tube is removed; leakage often occurs if the tube is removed sooner. Some patients have persistently immature tracts and need prolonged drainage (often more than 30 days) to allow complete tract maturation. Because it is difficult to predict which patients will have persistently immature tracts, routine tract evaluation should be performed before percutaneous cholecystostomy tube removal to prevent bile leaks. This can be performed by advancing a 6-Fr vascular sheath over a guidewire into the cholecystostomy tube tract. Pull-back contrast injection is then performed through the sidearm of the sheath as this is withdrawn; if there is leakage from the tract, the drain is left in place. For most patients all that is needed is more time to allow formation of a mature tract.

## Complications

In a review of 127 cases of diagnostic gallbladder puncture and percutaneous cholecystostomy, the overall complication rate was 12.6%. Major complications occurred in 8.7% of the cases; these included bile peritonitis, bleeding, vagal reactions, hypotension, catheter dislodgment, and acute respiratory distress. Minor complications were observed in 3.9% of the cases. The 30-day mortality rate was 3.1%; however, the deaths were related to the underlying diseases in these patients. Problems of direct transperitoneal puncture are bile leakage and peritonitis, loss of access secondary to decompression of the gallbladder, and transcolonic puncture. Delayed gallbladder rupture after percutaneous cholecystostomy may also occur when the gallbladder wall is necrosed.

## SUGGESTED READINGS

Boguth L, Tatalovic S, Antonucci F, et al. Malignant biliary obstruction: clinical and histopathologic correlation after treatment with self-expanding metal prostheses. *Radiology* 1994;192:669–674.

Boland GW, Lee MJ, Dawson SL, Mueller PR. Percutaneous cholecystostomy for acute cholecystitis in a critically ill patient. *AJR Am J Roentgenol* 1993;160:871–874.

Boland GW, Lee MJ, Mueller PR, et al. Gallstones in critically ill patients with acute calculus cholecystitis treated by percutaneous cholecystostomy: nonsurgical therapeutic options. *AJR Am J Roentgenol* 1994;162:1101–1103.

Bonnel DH, Liguory CL, Lefebvre JF, et al. Placement of metallic stents for treatment of postoperative biliary strictures: long-term outcome in 25 patients. *AJR Am J Roentgenol* 1997;169:1517.

Castañeda-Zúñiga WR, Tadavarthy SM, Laerum F, Amplatz K. Anterior approach for biliary duct drainage. *Radiology* 1981;139:746.

Chrisman HB, Tutton SM. Use of ultrasound guidance in hepatobiliary procedures. *Semin Intervent Radiol* 1997;14:421.

Clark CD, Picus D, Dunagan WC. Bloodstream infections after interventional procedures in the biliary tree. *Radiology* 1994;191:495–499.

Cope C. Percutaneous subhepatic cholecystostomy with removable anchor. *AJR Am J Roentgenol* 1988;151:1129–1132.

Culp WC, McCowan TC, Lieberman RP, et al. Biliary strictures in liver transplant recipients: treatment with metal stents. *Radiology* 1996;199:339.

Dick BW, Gordon RL, LaBerge JM, et al. Percutaneous transhepatic placement of biliary endoprostheses: result in 100 consecutive patients. *J Vasc Intervent Radiol* 1990;1:97–100.

Eschelman DJ, Shapiro MJ, Bonn J, et al. Malignant biliary duct obstruction: long-term experience with Gianturco stents and combined-modality radiation therapy. *Radiology* 1996;200:717.

Ferrucci JT, Wittenberg J. Refinements in Chiba needle transhepatic cholangiography. *AJR Am J Roentgenol* 1977;129:11.

Ferrucci JT Jr., Mueller PR, Harbin WP. Percutaneous transhepatic biliary drainage: technique, results and applications. *Radiology* 1980;135:1.

Garber SJ, Mathieson JR, Cooperberg PL, MacFarlane JK. Percutaneous cholecystostomy: safety of the transperitoneal route. *J Vasc Intervent Radiol* 1994;5:295–298.

Hamlin JA, Friedman M, Stein MG, Bray JF. Percutaneous biliary drainage: complications of 118 consecutive catheterizations. *Radiology* 1986;158:199.

Harbin WP, Mueller PR, Ferrucci JT Jr. THC: complications and use patterns of the fine-needle technique. *Radiology* 1980;135:15.

Huasegger KA, Kugler C, Uggowitzer M, et al. Benign biliary obstruction: is treatment with the Wallstent advisable? *Radiology* 1996;200:437.

Hausegger KA, Thurnher S, Bodendorfer G, et al. Treatment of malignant biliary obstruction with polyurethane-covered Wallstents. *AJR Am J Roentgenol* 1998;170:403.

Hayashi N, Sakai T, Kitqagawa M, et al. US-guided left-sided drainage: nine-year experience. *Radiology* 1997;204:119.

Hruby W, Urban M, Stackl W, et al. Stonebearing gallbladders: CT anatomy as the key to safe percutaneous lithotripsy. *Radiology* 1989;173:385–387.

Lammer J, Hausegger KA, Fluckiger F, et al. Common bile duct obstruction due to malignancy: treatment with plastic versus metal stents. *Radiology* 1996;201:167.

Lee ACN, Ho CS. Complications of percutaneous biliary drainage: benign vs malignant diseases. *AJR Am J Roentgenol* 1987;148:1207.

Martin EC, Laffey KJ, Bixon R. Percutaneous transjejunal approaches to the biliary system. *Radiology* 1989;172:1031.

Mathieson JR, McLoughlin RF, Cooperberg PL, et al. Malignant obstruction of the common bile duct: long-term results of Gianturco-Rosch metal stents used as initial treatment. *Radiology* 1994;192:663–667.

Morrison MC, Lee MJ, Saini S, et al. Percutaneous balloon dilatation of benign biliary strictures. *Radiol Clin North Am* 1990;28:1191–1201.

Mueller PR, Harbin WR, Ferrucci JT Jr., et al. Fine-needle transhepatic cholangiography: reflections after 450 cases. *AJR Am J Roentgenol* 1981;136:85–90.

Mueller PR, vanSonnenberg E, Ferrucci JT Jr. Percutaneous biliary drainage: technical and catheter related problems—experience with 200 cases. *AJR Am J Roentgenol* 1982;138:17.

Mueller PR, vanSonnenberg E, Ferrucci JT Jr., et al. Biliary stricture dilatation: multicenter review of clinical management in 73 patients. *Radiology* 1986;160:17.

Nealon WH, Urrutia F. Long-term follow-up after bilioenteric anastomosis for benign bile duct stricture [Abstract]. *Radiology* 1996;201:882.

Neff RA, Frankuchen EI, Cooperman AM, et al. The radiological management of malignant biliary obstruction. *Clin Radiol* 1983;34:143.

Nichols D, Cooperberg P, Golding R, Burhenne H. The safe intercostal approach? Pleural complications in abdominal interventional radiology. *AJR Am J Roentgenol* 1984;141:1013.

Petersen BD, Maxfield SR, Ivancev K, et al. Biliary strictures in hepatic transplantation: treatment with self-expanding Z-stents. *J Vasc Intervent Radiol* 1996;7:221.

Ring EJ, Kerlan RK. Interventional biliary radiology. *AJR Am J Roentgenol* 1984;142:31.

Rossi P, Bezzi M, Rossi M, et al. Metallic stents in malignant biliary obstruction: results of a multicenter European study of 240 patients. *J Vasc Intervent Radiol* 1994;5:279–285.

Rossi P, Bezzi M, Salvatori FM, et al. Clinical experience with covered Wallstents for biliary malignancies: 23-month follow-up. *Cardiovasc Intervent Radiol* 1997;20:441.

Salomonowitz E, Castañeda-Zúñiga WR, Lund G, et al. Balloon dilatation of benign biliary strictures. *AJR Am J Roentgenol* 1984;151:613.

Teplick SK. Diagnostic and therapeutic interventional gallbladder procedures. *AJR Am J Roentgenol* 1989; 152:913–916.

Teplick SK, Flick P, Brandon JC. Transhepatic cholangiography in patients with suspected biliary disease and nondilated ducts. *Gastrointest Radiol* 1991;16:193–197.

Tesdal JK, Adamus R, Poeckler C, et al. Therapy for biliary stenoses and occlusions with use of three different metallic stents: single-center experience. *J Vasc Intervent Radiol* 1997;8:869.

vanSonneberg E, D'Agostino HB, Goodacre BW, et al. Percutaneous gallbladder puncture and cholecystostomy: results, complications, and caveats for safety. *Radiology* 1992;183:167–170.

vanSonnenberg E, Wittich GR, Casola G, et al. Diagnostic and therapeutic percutaneous gallbladder procedures. *Radiology* 1986;160:23–26.

Vorwerk D, Kissinger G, Handt S, Günther RW. Long-term patency of Wallstent endoprostheses in benign biliary obstructions: experimental results. *J Vasc Intervent Radiol* 1993;4:625–634.

Warren LP, Kadir S, Dunnick NR. Percutaneous cholecystostomy: anatomic considerations. *Radiology* 1988;168: 615–616.

Wayne PM, Whelan JG. Susceptibility testing of biliary bacteria obtained before bile duct manipulation. *AJR Am J Roentgenol* 1983;140:1185.

Wu SM, Marchant LK, Haskal ZJ. Percutaneous intervention in the biliary tree. *Semin Roentgenol* 1997;32:228.

# 22

# Gastrointestinal Tract Intervention

Hector Ferral and Haraldur Bjarnason

## A. PERCUTANEOUS GASTROSTOMY AND JEJUNOSTOMY

### Haraldur Bjarnason

Enteric nutrition is the route of choice in patients who have normally functioning gastrointestinal tracts but who are unable to take in food orally. Enteric feeding can be given through a nasoenteric tube or via either transgastric or transjejunal enterostomy. The nasoenteric route can be used temporarily, but when long-term enteric feeding is foreseen, gastrostomy or jejunostomy should be considered.

Surgical gastrostomy was initially described in 1837. Less invasive techniques, such as percutaneous endoscopic gastrostomy and percutaneous gastrostomy under fluoroscopic guidance, have since been developed and are common practice. Enterostomies have also been created using laparoscopic approach.

### INDICATIONS AND CONTRAINDICATIONS OF BOTH PROCEDURES

#### Indications

- Inability to swallow
- Anatomic disorders, such as tumor; functional disorders, such as neurologic disorders
- Decompression
- After abdominal surgery
- Chronic bowel obstruction because of tumor or of other causes
- Support nutrition in patients who are unable for other reasons to take in adequate nutrition enterally

#### Contraindications

- Bleeding diathesis
- Splenomegaly or hepatomegaly causing interposition of the left liver lobe between the stomach and the skin
- Colon interposed in between the skin and the stomach
- Tumor ingrowth into the gastric wall is not in it self a contraindication to gastrostomy placement
- Ascites, which is only a relative contraindication and should not in general prevent one from proceeding with the procedure

### TECHNIQUE

#### Before the Procedure

- Check coagulation [(international normalized ratio (INR); platelet count more than 50,000)].
- Ultrasound of the left upper quadrant to mark the inferior margin of the liver.
- Intravenous access for administration of sedatives, pain medication, antibiotics, and glucagon.
- Nasogastric tube placement for insufflation of the stomach.
- Barium can be given orally 6 to 12 hours before the procedure to allow easier identification of the colon with the fluoroscope during the procedure.

#### Procedure

The skin of the left upper quadrant is sterilized and sterile drapes are applied around the procedure field. Many authors give antibiotics before and for 24 hours after the procedure, but this has not been proved to be beneficial. Typically a gram

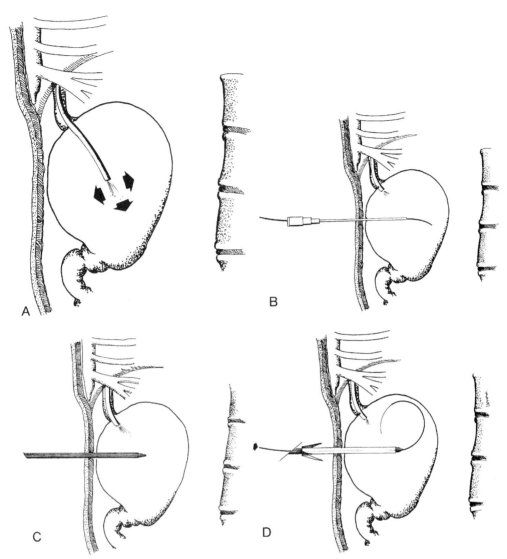

**FIG. 22.1. A:** The stomach is inflated with air through the nasogastric tube to bring the gastric wall close to the abdominal wall. **B:** The puncture is made with a 16- or 18-gauge needle pointing vertically down or toward the duodenum, and a 0.035-in. J-guidewire is advanced through it. **C,D:** Before inserting the peel-away introducer sheath, fascial dilators are advanced over the guidewire until an adequate diameter has been reached.

of cefazolin sodium is given. One-half to 1 mg of glucagon is then given intravenously just before the stomach is inflated with air to decrease air leak from the stomach into the duodenum. Gastric distension with air is important to the success of the procedure. An inflated stomach displaces the colon downward and brings the anterior gastric wall closer to the anterior abdominal wall (Fig. 22.1A). Rotational fluoroscopy is then performed to verify the anterior position of the stomach and to verify that there is no bowel in front of the stomach, between the stomach and the anterior abdominal wall.

The skin puncture site into the stomach is then selected. The anterior lower stomach body is punctured. One should bear in mind the gas-

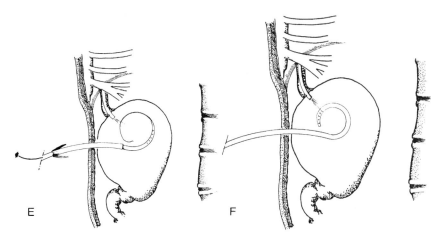

**FIG. 22.1.** *Continued.* **E:** The catheter is advanced over the guidewire through the peel-away sheath. **F:** After the catheter is in place, the guidewire is removed, and the sheath is peeled away. (Reprinted from Castañeda-Zúñiga WR. *Interventional radiology*, 3rd ed. Baltimore: Williams & Wilkins, 1997, with permission.)

troepiploic artery and the left liver lobe. Also, the inferior epigastric artery runs in the rectus sheath and can be inadvertently punctured. Lidocaine 1% (Xylocaine 1%) is used for skin and deep local analgesia down to the stomach.

Many authors recommend fixing the anterior stomach wall to the anterior abdominal wall before placement of the gastrostomy. Other authors do not use anchors for fixation and it is debated if the complication rate is any higher without fixation. The rationale for fixation is to prevent leakage around the gastrostomy into the peritoneal cavity and in case the tube comes accidentally out, the stomach stays attached to the anterior abdominal wall, preventing leakage and allowing for easier reentry into the stomach.

For fixation, specific anchors (Cope) from Cook (Cook, Inc., Bloomington, IN) (Fig. 22.2) and the so-called Brown-Mueller T-fastener set (MEDITECH/Boston Scientific, Natick, MA) are available (Fig. 22.3). The needle holding the fixation anchors is advanced into the stomach 1 cm from the site selected for the puncture. The anchors are then pushed out of the needle into the stomach and pulled up against the stomach wall. Two to four anchors are used with 1 cm in between, forming a rectangle if four anchors are used. The final access is then selected in between the anchors, and an 18-gauge needle is used for that purpose (Fig. 22.1B). This puncture is made in the direction of the antrum to ease conversion from a gastrostomy to a gastrojejunostomy later if needed. As soon as access is gained, a 0.035-in. guidewire is advanced into the stomach and an incision is made at the puncture site.

The tract is then dilated over the wire in the same manner for fixation or no fixation, either using fascia dilators (Fig. 22.1C) or angioplasty balloon (especially if larger-diameter tubes are to be used). The dilators are used in sequence, starting with a 6- or 8-Fr dilator and going up two French sizes each time. The target is a diameter 2 Fr larger than the tube itself (Fig. 22.1C). If an angioplasty balloon is used for the dilation, a balloon with a diameter approximately 2 Fr larger than the diameter of the tube is selected (1 mm diameter equals 3 Fr).

During this manipulation there is risk of intraperitoneal displacement of the guidewire and dilators. This can be overcome by keeping the stomach distended with air and by keeping the dilators lined up with the direction of the wires; fixation will prevent this from happening (Fig. 22.4).

Finally, a peel-away introducer sheath will be placed and the tube placed through it, preferably still over the guidewire. Some tube types will not require a peel-away sheath (Fig. 22.1D through F).

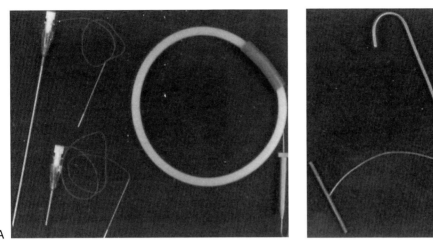

**FIG. 22.2.** Cope suture set. **A:** The anchor is preloaded in a 17-gauge needle. **B:** The anchor is released by advancing it with a guidewire. (Reprinted from Coleman CC, Coons HG, Cope C. Percutaneous enterostomy with the Cope suture anchor. *Radiology* 1990;174:889–891, with permission.)

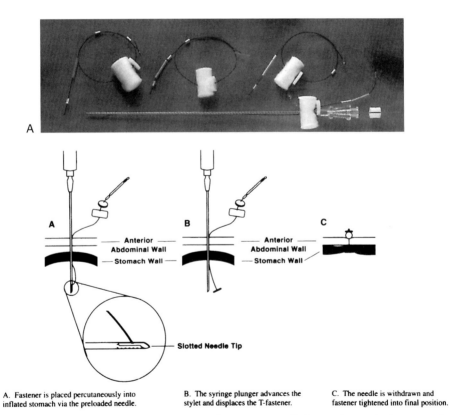

A. Fastener is placed percutaneously into inflated stomach via the preloaded needle.

B. The syringe plunger advances the stylet and displaces the T-fastener.

C. The needle is withdrawn and fastener tightened into final position.

**FIG. 22.3.** Brown-Mueller T-fastener. The T bar fits into a small slot at the end of the needle. The anchor is then pushed out with a pusher or a guidewire. (Courtesy of MEDITECH/Boston Scientific, Natick, MA.)

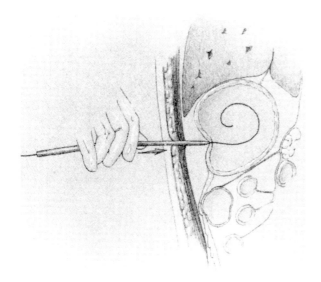

**FIG. 22.4.** If the gastric wall is not fixed to the anterior abdominal wall, it can easily be displaced backward by the advancing dilator or gastrostomy tube and can lead to loss of access. (Reproduced from Coleman CC, Coons HG, Cope C. Percutaneous enterostomy with the Cope suture anchor. *Radiology* 1990;174:889–891, with permission.)

If fixation is not used, the gastric lumen is entered in the same manner and a 0.035-in. guidewire is advanced through the needle into the stomach. The tract is then dilated in the same way. Alternatively, a single anchor can be placed through the initial puncture needle and the anchor left in during the dilatation for support. This can then be removed at the end of the procedure.

There are various types of feeding tubes available. They can be divided into gastric feeding tubes and gastrojejunal feeding tubes. The retention mechanism also varies. Some have an inflatable balloon inside the stomach such as Foley catheters (C.R. Bard, Covington, GA), the MIC gastrostomy button (Medical Innovations, Milpitas, CA) (Fig. 22.4), and the MIC gastrostomy tube (Medical Innovations) (Fig. 22.5). Others have a mushroom-like locking mechanism such as the Wills-Oglesby gastrostomy and gastrojejunostomy catheters (Cook, Inc.) and the Carey-Alzate-Coons gastrojejunostomy catheter (Cook, Inc.). Others have a pigtail lock keeping the tubes from coming out (MEDITECH/ Boston Scientific, Natick, MA). The balloon-inflatable tubes tend to have a larger outer diameter because of the material (silicon) and need for a balloon lumen in the body. The balloon will provide some seal from the inside, preventing stomach content from flowing to the skin surface.

Patients with esophageal reflux should have either a gastrojejunal tube or a transgastric jejunal tube placed to prevent aspiration. The procedure for gastrojejunal tube is practically the same as for regular gastrostomy. When access has been gained after placement of the anchors, an angiographic catheter, such as a multipurpose catheter, is passed through the tract into the duodenum and into the proximal jejunum passing the ligament of Treitz. Here, it is particularly useful to make the puncture at an oblique angle, pointing toward the pylorus to decrease the angle from the tract to the duodenum and to prevent the tube from kinking or buckling into the stomach.

The gastrostomy button (Fig. 22.6) is a simple skin-level device designed for long-term enteral feedings that has been used in children. Although migration of this device has been reported, the gastrostomy button usually avoids some of the problems associated with conventional gastrostomy tubes, including accidental removal, tissue reaction, discomfort, and psychological problems. Those tubes often have a port to the stomach for drainage or delivery of medication and a distal jejunal port for feeding (Fig. 22.7) (Table 22.1).

Occasionally, a nasogastric tube cannot be placed for the purpose of inflation of the stomach with air. In this case an initial puncture can be made with a 21-gauge needle under either fluoroscopic or computed tomographic guidance into an air pocket in the stomach. The stomach can then be distended through this needle.

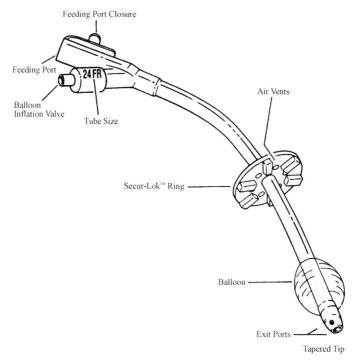

FIG. 22.5. MIC gastrostomy tube. (Reprinted from Castañeda-Zúñiga WR. *Interventional radiology*, 3rd ed. Baltimore: Williams & Wilkins, 1997, with permission.)

Gastrostomy placement in the pediatric population using this or similar techniques is also very successful. The procedure can often be performed with sedation, but general anesthesia may be needed.

Patients with ascites can still have the procedure performed. The ascites fluid should be drained before the procedure, and it is imperative to use fixation devices (anchor) for the procedure.

### After the Procedure

- After successful creation of gastrostomy and placement of a gastrostomy or gastrojejunostomy tube, a frontal and lateral film after contrast injection is obtained for verification of placement.
- Antibiotics are recommended by many for 24 hours after the procedure.
- The gastrostomy tube needs to be left to external drainage to prevent distension of the stomach and leak.
- The nasogastric tube can be removed after the procedure.
- Nurses' evaluation for fever and abdominal pain is requested.
- Pneumoperitoneum is a common and normal finding right after the procedure but should resolve within 24 to 72 hours.
- The following day, the radiologist needs to examine the patient for signs of peritonitis and bowel function. If there is no suspicion of peritonitis and bowel functions have resumed, feedings can be started. Feeding is started slowly with normal saline for 4 to 6 hours and then slow initiation of the enteric feeding. Nutritionists should be involved in this.

### OUTCOME

Technical success ranges from 99% to 100%. Major complications, including severe bleeding, peritonitis, and sepsis, have been reported in 1.5% to 6.0% of patients, and minor complications such as peritoneal irritation, local infec-

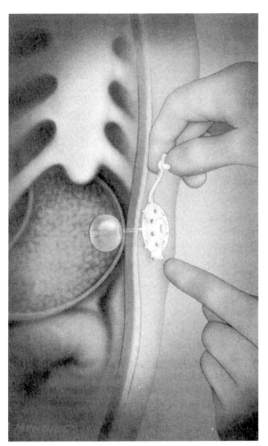

**FIG. 22.6.** MIC gastrostomy button in place. (Reprinted from Castañeda-Zúñiga WR. *Interventional radiology*, 3rd ed. Baltimore: Williams & Wilkins, 1997, with permission.)

tion, and tube migration occurred in 3.5% to 9.5% of cases.

The mortality rate from the procedure itself is low (0.7%). The postprocedural mortality at 30 and 90 days (13% and 50%, respectively) is primarily due to the underlying disease.

Pneumoperitoneum is commonly found after percutaneous gastrostomy, due to leakage of air into the peritoneum during the procedure. It is usually asymptomatic and no treatment is required.

The most common complications include occlusion and dislodgment of the feeding tube itself. Tube obstruction is more common in gastrojejunostomy tubes (58%) than gastrostomy tubes (8.3% to 15%) because of the smaller size of the first one. Patency rate is affected by tube diameter and length as well as the feeding formula used.

The key factor for prevention of tube obstruction is careful flushing with water or normal saline after each use. If tablets (medication) are given through the tubes, they must be ground into powder and dissolved carefully in fluid and the tube flushed carefully after. Most often the obstruction can be cleared either by flushing the tube with water or normal saline; if that does not work, a 0.035-in. Glidewire (MEDITECH/Boston Scientific Corp., Natick, MA) can be passed through the catheter lumen, followed by flushing. As an alternative, cola or a mixture of pancreatic enzymes (pancreozym) can dwell in the catheter for 20 minutes. Flushing can then be attempted again. If this is unsuccessful, the tube needs to be exchanged.

Catheter dislodgment can be a more serious problem, especially when it occurs soon (1 to 21 days) after creation of the gastrostomy tract. In these cases, one should initially attempt reestablishment of the immature tract using a 0.035-in. guidewire and an angiographic catheter such as the JB-1 catheter. If this is unsuccessful, a second puncture may be made at the same site. Then it is helpful to have the stomach fixed with anchors to the abdominal wall. If the tract is mature, tube replacement is usually simple and uncomplicated. The tract will close down very fast if nothing is left across the tract, so redilatation may be needed.

## CONCLUSION

Experience has demonstrated that percutaneous placement of gastrostomy or gastrojejunostomy tubes under fluoroscopic guidance is technically easy, effective, and safe. Improved nutritional status after placement of gastrostomy or gastrojejunostomy tubes is seen only in children and adults without malignant disease. Patients with malignancy do not improve their nutritional status or life expectancy with feeding tube placement.

## SMALL-BOWEL FEEDINGS

### Indications and Contraindications

#### Indications

Transpyloric enteric nutrition is indicated for patients who need enteric nutrition and have histories of the following:

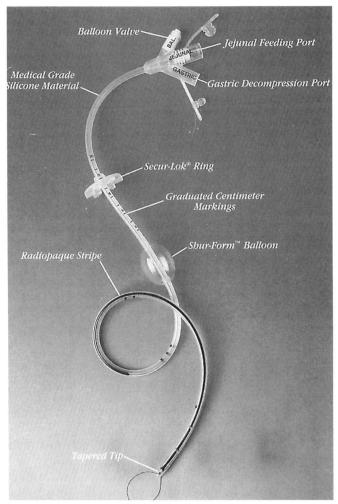

**FIG. 22.7.** MIC gastrojejunostomy tube. (Reprinted from Castañeda-Zúñiga WR. *Interventional radiology*, 3rd ed. Baltimore: Williams & Wilkins, 1997, with permission.)

**TABLE 22.1.** *Examples of enteric feeding tubes*

| | Retention mechanism | Sizes |
|---|---|---|
| Gastrostomy tubes | | |
| MIC gastrostomy tube | Balloon | 12–24 Fr |
| MIC KEY buttons | Balloon | 12–18 Fr |
| Carey-Alzate-Coons | Mushroom | 12 Fr |
| Ross Flexiflo | Balloon | 18 Fr |
| Gastrojejunostomy tubes | | |
| MIC GJ tube | Balloon | 14–24 Fr |
| Ross Flexiflo | Balloon | 18 Fr + 8 Fr |
| Cary-Alcer-Coons | Mushroom | 18 Fr + 8 Fr |
| MIC transgastric J tube | Balloon | 12 Fr |

- Gastrectomy and esophagojejunostomy
- Pyloric obstruction
- Severe gastric atony
- Recent gastric or esophageal surgery
- Spontaneous esophageal rupture treated conservatively
- Gastroesophageal reflux and aspiration

In addition, transpyloric enteric nutrition is an alternative to gastrostomy for patients who need decompression of the gastrointestinal tract and have contraindications for gastrostomy tube placement.

## Contraindications

The main contraindications for percutaneous transpyloric enteric procedures are inadequate access (due to organ interposition or poor definition of the target) and uncorrected coagulopathies.

## Technique

### Gastrojejunostomy

Gastrojejunostomy placement has been discussed previously (see Part A for a complete description of the techniques and materials used during this procedure). It is important to stress one of the main technical points that facilitate the placement of gastrojejunostomy tubes. The angulation of the puncture site must have an oblique orientation, pointing toward the pyloric region. This is clearly different from the puncture used for endoscopic or surgically placed gastrostomy tubes, which commonly have a more perpendicular orientation toward the gastric fundus.

Frequently, the tracts created surgically or by endoscopy point toward the gastric fundus, making conversion to gastrojejunostomy difficult. In some cases, conversion from a surgically or endoscopically placed gastrostomy to gastrojejunostomy is not possible, and repuncture with different angulation is needed. Placement of a gastrostomy using a surgical-endoscopic approach is comparably successful, but if a gastrojejunal feeding tube is to be placed, surgical-endoscopic placement lags far behind in terms of success.

Technical failure of converting a surgical or endoscopic gastrostomy to a gastrojejunostomy using fluoroscopic guidance has been reported as 18% to 22%, respectively. Conversely, almost all percutaneous gastrostomies performed under fluoroscopic guidance can be transformed to gastrojejunostomies without significant technical difficulty.

### Percutaneous Jejunostomy

In patients in whom the stomach has been surgically removed (gastrectomy, esophagojejunostomy) or in whom the gastric cavity is inaccessible because of tumor, organ interposition, or obstruction, postpyloric enteric feeding can be accomplished by direct access to the jejunum.

The technical approach to direct jejunostomy is similar to that of gastrostomy tube placement. A nasoenteric tube is advanced into the jejunum for insufflation with air and contrast material. Using rotational fluoroscopy, a loop of bowel adjacent to the abdominal wall is identified. Once the entry site is identified, the technique for catheter placement does not differ from that used for percutaneous gastrostomy. Because of the tendency of the bowel loop to move away during the introduction of catheters, the use of T-fasteners is mandatory to fix the bowel loop to the abdominal wall. Several modifications to this method have been proposed. Some authors have placed a large angioplasty balloon into the proximal jejunum close to the skin or a snare. Closeness to the skin is verified with rotational fluoroscopy, and the puncture is then made either into the inflated balloon or the open snare.

A modification to this approach calls for identification of a superficial small-bowel loop in the left upper quadrant with ultrasound. The punctures are then made under ultrasound guidance, and the anchors can actually be observed coming out of the needle with the ultrasound. Cook has come out with 0.018-in. anchors that are placed through a very small needle and are better for small-bowel puncture than the current 18-gauge needles. The procedure is then performed in the same manner as for gastrostomies.

### Translumbar Duodenostomy

A translumbar access to the duodenum for enteral feeding has also been described. This approach is useful for patients on peritoneal dialysis to avoid the risk of peritonitis from a surgical or percutaneous gastrostomy.

Because the second portion of the duodenum is in a retroperitoneal position, this access prevents the leakage of duodenal contents into the peritoneum. Initial puncture can be performed either under fluoroscopy (after adequate antegrade opacification of the duodenum) or under computed tomography guidance. The Seldinger technique is used for catheter placement.

## PERCUTANEOUS TRANSHEPATIC ENTERAL FEEDING

A percutaneous transhepatic catheter placement into the jejunum can be used for feeding purposes in patients with tumors of the head of the pancreas that are producing obstruction of the biliary tree and duodenum.

After the duodenum is reached through the transhepatic approach, two guidewires are introduced. One wire is advanced into the jejunum for feeding catheter introduction. The other wire is used to place a biliary drainage catheter.

## CONCLUSION

The different alternatives for enteric feeding provide nutritional support on a temporary or permanent basis. These procedures have in common a relatively low incidence of complications compared with those seen in more invasive surgical or endoscopic techniques.

# B. MANAGEMENT OF BENIGN ESOPHAGEAL STRICTURES

## Ho-Young Song

Since London and coworkers reported on the first report of the successful treatment of esophageal strictures with a Gruentzig-type balloon catheter in 1981, fluoroscopically guided balloon dilatation has been shown to be a safe, easy, and effective treatment for a variety of strictures in the esophagus. Theoretically, the esophageal rupture rate of balloon dilatation is lower than that of bougienage, not only because the bougienage exerts a shear stress force in comparison with the radial force in balloon dilatation, but also because in balloon dilatation under fluoroscopic control, the balloon is positioned across the stricture with the help of a guidewire carefully passed through the stenosis.

Fluoroscopic placement of self-expandable metallic stents is a relatively new method of treating benign esophageal strictures. In 12 patients with a benign esophageal stricture, the long-term results were reported to be discouraging: The rate of late complications caused by migration or formation of new strictures was as high as 100% (12 of 12 patients). However, on the basis of clinical observation in 12 patients (in whom three of the four stents that migrated within 2 months after placement caused recurrence of stricture, but none of the four stents that migrated after 2 months after stent placement showed stricture), it was hypothesized that benign esophageal strictures could be treated well if a covered stent was left in place for more than two months.

Part B deals with the technique of balloon dilatation and its results.

## INDICATIONS

Esophageal balloon dilatation is indicated for the following patients who had dysphagia to soft or solid food:

- Postoperative esophageal stricture
- Corrosive stricture
- Achalasia
- Esophageal web
- Peptic stricture
- Radiation stricture
- Congenital stenosis

## CONTRAINDICATIONS

There are no absolute contraindications, but patients with uncontrollable bleeding diathesis are considered to be relative contraindications.

## TECHNIQUE

The length, location, and severity of the stricture and any associated problems are evaluated before balloon dilatation by means of barium esophagography. Topical anesthesia of the pharynx with an aerosol spray can be supplemented by intramuscular or intravenous sedation with diazepam (Valium) if the patient is apprehensive. A small amount of water-soluble contrast medium is swallowed for opacification of the narrowed esophageal lumen. If necessary, the location of the stricture can be marked on the surface of the patient by putting clamps at the proximal and distal ends.

A 0.035-in. exchange guidewire is inserted, with or without the help of an angiographic catheter, through the mouth across the stricture into the distal esophagus or stomach. A deflated esophageal balloon catheter longer than the stricture (6 to 20 mm in diameter and 3 to 8 cm long) is passed over the guidewire to a position astride the stricture. The selection of balloon size is based on the diameter of the stricture and the caliber of the adjacent, unstrictured lumen. The deflated balloon is slowly inflated with a diluted water-soluble contrast medium until the "hourglass deformity" created by the stricture disappears from the balloon contour. Usually, a pressure of 4 to 6 atm is sufficient to dilate the stricture. The balloon is commonly left inflated for 2 to 5 minutes. The inflations are usually repeated two or three times. The balloon and the guidewire are then removed.

When a stricture is so severe, usually in longstanding corrosive stricture (Fig. 22.8) or in radiation stricture, that the patient complains of severe pain and the hourglass deformity does not disappear from the balloon contour, the dilatation is approached in a different fashion. Initially, dilatation to 4 to 6 mm is performed using balloon catheters. If dilatation with 4- to 6-mm balloons is easily accomplished and the patient tolerates the procedure well, the caliber of the balloon catheter is gradually increased on the same day. Further dilatation is discontinued if the stricture is unusually resistant to dilatation and the patient does not tolerate the procedure. In subsequent dilatations, the caliber of the esophagus is gradually increased until a suitable diameter is reached. The luminal diameter of the patient's esophagus is dilated until the patient can manage solid foods with little difficulty.

After completion of the procedure, a small amount of water-soluble contrast medium is swallowed by the patient to determine whether extravasation of the contrast media was identified or not. In case that extravasation is not identified, a barium esophagogram with a regular amount of barium is obtained. Patients are allowed a soft diet 2 hours after the procedure and encouraged to resume solid foods from the next day. If patients have no difficulty with solid foods, further dilatation is not performed. However, for patients who have dysphagia to solid food, the dilatation procedure is repeated on multiple different sessions 2 to 7 days apart, gradually increasing the amount of distention until a suitable diameter for patients to take solid foods is reached.

Most of the authors who have reported on esophageal balloon dilatation usually have targeted a luminal diameter of 20 mm for adults, 8 to 10 mm for infants, and 12 to 15 mm in older children with esophageal strictures. However, we believe that to dilate the esophagus over 20 mm is too much for some patients with corrosive stricture or radiation stricture. The reasons are as follows: (a) To dilate corrosive esophageal stricture up to 20 mm is very difficult and might cause rupture in more than half of patients; (b) some patients whose final diameter of the dilated esophagus is below 15 mm could intake most or all foods; and (c) every individual has a different optimal diameter for an esophageal stricture because it depends on the actual nature of the obstruction, its degree of elasticity, length, the power of the muscle pushing the bolus through, and the patient's threshold for tolerating discomfort in swallowing.

## RESULTS

The overall success rate and rupture rate have been reported to be 67.0% to 97.5% and 0% to 32% per patient, respectively. In our experience with patients with noncorrosive (n = 75) and corrosive (n = 28) esophageal stricture, dilatation of up to 17 mm was achieved in 71 of 75 patients (95%) with noncorrosive stricture. On the other hand, dilatation of up to 17 mm was possible in only six of 28 patients (21%) with corrosive stricture, which implies that dilatation of corrosive esophageal stricture is much more difficult than dilatation of noncorrosive esoph-

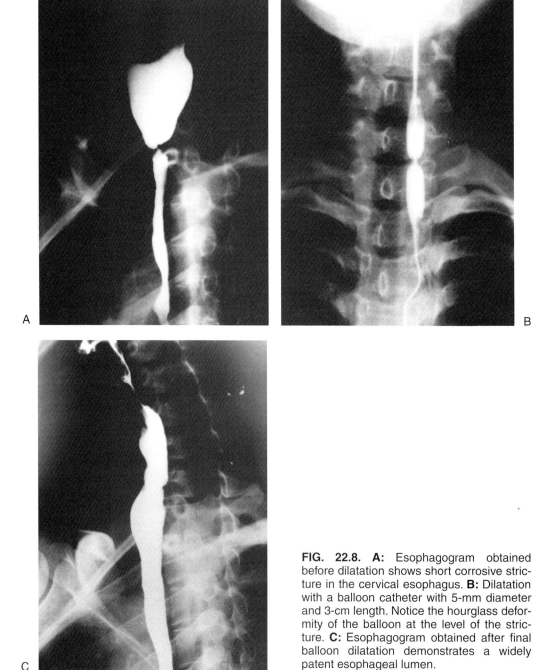

**FIG. 22.8. A:** Esophagogram obtained before dilatation shows short corrosive stricture in the cervical esophagus. **B:** Dilatation with a balloon catheter with 5-mm diameter and 3-cm length. Notice the hourglass deformity of the balloon at the level of the stricture. **C:** Esophagogram obtained after final balloon dilatation demonstrates a widely patent esophageal lumen.

ageal strictures. The recurrence rate in corrosive stricture is reported to be as high as 45%. In our opinion, the recurrence rate and the rupture rate in corrosive stricture are very high because of the dense scarring due to long duration, tendency of severe esophageal damage, and diffuse esophageal involvement. In patients with reflux esophagitis, dilatation is usually successful, but restenosis is common if the acid reflux was not corrected.

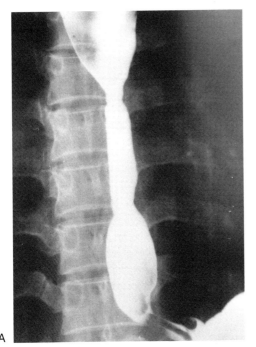
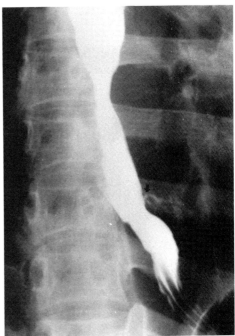
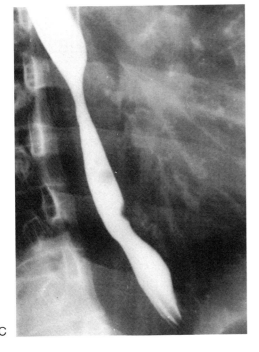

**FIG. 22.9. A:** Esophagogram obtained before dilatation shows moderate stricture in lower thoracic portion of the esophagus. **B:** Barium esophagogram demonstrates leakage of barium. **C:** Follow-up esophagogram obtained 14 days after esophageal rupture shows cessation of leakage.

In postoperative stenoses and esophageal web, the results of dilatation are generally excellent. In patients with achalasia, balloons of 20 mm diameter were successfully used with good results. In our experience, overdilatation of the stricture site with two 15- to 20-mm balloons side by side has been commonly necessary. We agree with the opinion that congenital esophageal stenosis is usually resistant to balloon dilatation and unsuccessfully treated.

## COMPLICATIONS

### Bleeding

Mild bleeding occurs in more than 50% of the patients as a result of trauma to the nasal mucosa or at the site of dilatation, which stops spontaneously. No one has reported about the sequelae.

### Esophageal Rupture

Esophageal rupture is the most serious complication. Overall rupture rates of esophageal balloon dilatation have been 0% to 9% in studies that have included more than 19 and fewer than 100 patients. Rupture rate of the esophagus, however, is relatively common in patients with corrosive stricture (Fig. 22.9) and congenital esophageal stricture. In 22 patients with corrosive stricture, esophageal rupture occurred in seven patients (32%). We prefer to use prophylactic antibiotics with ampicillin for 24 hours in patients with corrosive stricture because the esophageal perforation rate is very high.

The method of treatment for esophageal rupture remains controversial. Some authors emphasized that esophageal rupture should be treated surgically and promptly because they thought conservative or nonoperative treatment was attended by high morbidity and mortality. Esophageal rupture in achalasia may be more serious than in the other cases of esophageal stricture because of abundant retained food in the markedly dilated esophagus. However, others reported good results with medical treatment. Criteria for conservative treatment of esophageal rupture have been proposed: (a) minimal self-contained extravasation and adequate natural internal drainage into the esophagus, (b) no evidence of pleural contamination, (c) minimal symptoms, and (d) minimal evidence of clinical sepsis.

# C. MANAGEMENT OF MALIGNANT ESOPHAGEAL STRICTURES

## Ho-Young Song

Carcinomas of the esophagus and gastric cardia cause progressive dysphagia, and in the absence of treatment, starvation is a common cause of death. The best method of restoring the ability to ingest food in patients with dysphagia caused by nonresectable esophageal or gastric neoplasms remains controversial. Surgical bypass procedures using stomach, colon, or jejunum often have unsatisfactory results with considerable mortality. Radiotherapy is effective in 60% to 80% of the patients so treated, but abatement of symptoms is apparent only after 4 to 6 weeks. The patients need to be nourished for that period. Furthermore, radiotherapy is followed by dysphagia in more than 25% of cases, usually because of a fibrotic cicatricial narrowing. The use of neodymium: yttrium-aluminum-garnet (Nd:YAG) laser treatment is limited by its high cost, the requirement of frequent treatment sessions, and the frequency of tumor recurrence. In addition, submucosal or extrinsic compressing lesions are inaccessible to treatment with the Nd:YAG laser. Repeated balloon dilatations without intubation give only transient relief. Esophageal intubation is an attractively simple, rapid method for palliation of dysphagia caused by malignant neoplasms.

Fluoroscopic or endoscopic placement of a covered or uncovered expandable esophageal metallic stent is a new peroral intubation procedure that is increasingly being used for palliative treatment of esophagogastric strictures because the stent has been thought to overcome the considerable mortality and morbidity as well as the limited effectiveness in the relief of dysphagia associated with the conventional esophageal prostheses. The implantation of expandable metallic stents is better tolerated and safer than that of nonexpandable tubes because the diameters of the delivery systems are only 3 to 12 mm.

## INDICATIONS

Esophageal intubation is indicated for the following patients who have aphagia or dysphagia to soft food:

- Patients with nonresectable or inoperable esophagogastric neoplasms
- Patients with recurrent esophagogastric neoplasms after surgery and/or radiation therapy
- Patients with resectable esophageal malignancy who reject surgery
- Patients who absolutely need nourishment before surgery
- Patients with esophagorespiratory fistula due to a malignant tumor
- Patients with benign esophageal stricture that shows frequent recurrence after balloon dilatation

## CONTRAINDICATIONS

There are no absolute contraindications, but the followings are considered to be relative contraindications:

- Uncontrollable bleeding diathesis
- Severely ill patients with a very limited life expectancy
- Obstructive lesion of the small bowel due to peritoneal seeding

## TECHNIQUES

Although most patients may require hospitalization, the procedure can be performed on an outpatient basis. Local anesthesia of the hypopharynx can be supplemented by intravenous sedation with Valium if the patient is partially apprehensive. The site, severity, and length of the stricture are evaluated before balloon dilatation by means of barium esophagography.

Under fluoroscopic guidance, a small amount of barium (approximately 10 mL) is swallowed for opacification of the narrowed esophageal lumen. A 0.035-in. exchange guidewire is inserted, with or without the help of a straight, tapered angiographic catheter, through the mouth across the stricture into the distal esophagus or stomach. The location of the stenosis is marked on the patient's skin under fluoroscopic control if necessary. A deflated esophageal balloon catheter (6 or 8 cm long and 10 or 15 mm in diameter) is passed over the guidewire to a position astride the stricture. The deflated balloon is slowly inflated with a diluted water-soluble contrast medium until the hourglass deformity created by the stricture disappeared from the balloon contour. When a stricture is so severe that the patient complains of severe pain and the hourglass deformity does not disappear from the balloon contour, the inflation of the balloon should be discontinued. The balloon is then removed, and the guidewire is left in the esophagus.

With the patient in the left anterior oblique or supine position and in full extension of the neck, the delivery system with the stent inside, whose proximal part is lubricated with jelly, is passed over the guidewire into the esophagus and is advanced until the distal tip of the stent reaches beyond the stricture. The introducing sheath is slowly withdrawn in a continuous motion. This frees the stent and allows it to lie within the stricture and expand (Fig. 22.10). The delivery system and guidewire are removed. Esophagography can be performed immediately or 1 day after stent placement to verify the position and patency of the stent and to detect any esophageal perforation. Patients are allowed a liquid diet initially and then a soft and a solid diet. Patients are also advised to sit upright when eating, to take carbonated drinks with and after meals to clean food debris from the stents, to chew their food thoroughly, and to avoid large masses of food. Patients in whom the stent straddled the distal esophageal sphincter are advised to sleep in a semierect position to minimize the reflux and aspiration of gastric contents.

## RESULTS

Table 22.2 compares the results of different expandable esophageal stents. It is notable that all the technical success rates reported by the different investigators are 100%. Withdrawal of the half-olive–shaped catheter tip through the delivered Strecker stent (MEDITECH/Boston Scientific) is difficult in some patients with tight strictures because the tip tends to get caught in the mesh when the stent is not fully expanded. After stenting, 90% to 100% of the patients ingest soft or solids foods without dysphagia.

The incidence of esophagorespiratory fistula in patients with esophageal cancer ranges from 5% to 10%. Treatment should be instituted rapidly once the diagnosis is confirmed because most untreated

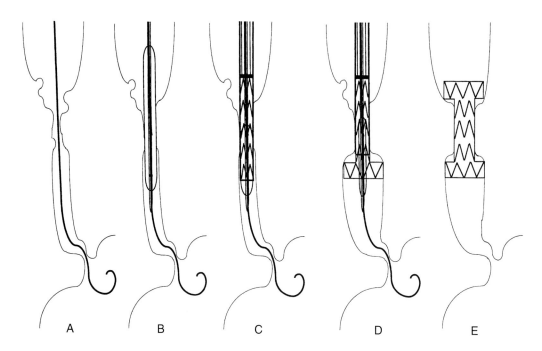

**FIG. 22.10.** Technical steps in the placement of an esophageal stent.

patients died within a month of development of the fistulas due to progressive aspiration, secondary pneumonia, lung abscess, and starvation. The results of covered expandable metallic stents for esophagorespiratory fistulae have been encouraging because they sealed off the fistula successfully (Fig. 22.11) as well as safely and easily. Uncovered expandable metallic stents, however, are not suitable for the treatment of esophagorespiratory fistula, and progressive tumor ingrowth through the openings between the wire filaments tends to cause progressive dysphagia. Schaer et al. reported tears of the silicone membrane after stent placement, which resulted in tumor ingrowth. Therefore, care-

**TABLE 22.2.** *Results of expandable esophageal stents*

| | Song | Gianturco-Rösh Z stent (Cook, Inc., Bloomington, IN) | Spiral Z | Strecker |
|---|---|---|---|---|
| No. of patients | 119 | 52 | 20 | 33 |
| Technical success (%) | 100 | 100 | 100 | 100 |
| Improvement of dysphagia (%) | 98 | 94 | ? | 100 |
| Esophageal perforation (%) | 0 | 2 | 5 | — |
| Bleeding (%) | 3 | 4 | 0 | 15 |
| Migration (%) | 10 | 10 | 30 | 0 |
| Blockage (%) | | | | |
|   Food impaction | 2 | 4 | 0 | ? |
|   Tumor overgrowth | 7 | 0 | 0 | 0 |
|   Tumor ingrowth | 0 | 2 | 0 | 12 |
| Mean survival (wks) | 16 | ? | ? | ? |

Data from 1993 Annual Meeting, Western Angiographic and Interventional Society and 5th International Radiology & New Vascular Imaging, 1994.

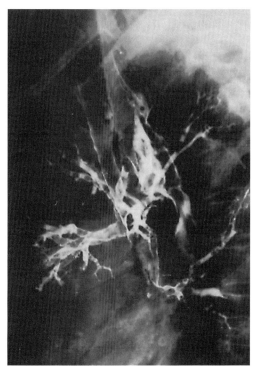 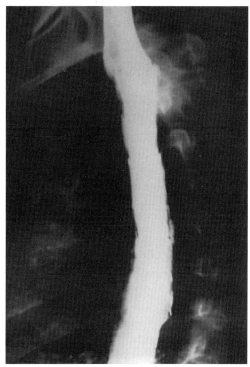

**FIG. 22.11. A:** Esophagogram shows a tracheoesophageal fistula. **B:** A Song esophageal endoprosthesis occludes the fistula and prevents aspiration of the contrast medium. (Reprinted from Castañeda-Zúñiga WR. *Interventional radiology*, 3rd ed. Baltimore: Williams & Wilkins, 1997, with permission.)

ful handling of the covering membrane is needed when using a covered metallic stent for esophagorespiratory fistula because the covering membrane could be damaged when the stent tube is being pushed into the introducing tube or when the introducing tube is being withdrawn.

Song et al. reported in 1994 that 104 of 119 patients died 2 to 80 weeks (mean of 16) after stent placement due to diffuse metastasis, cachexia, bleeding, or liver cirrhosis. Of the 104 patients, 27 (26%) lived more than 6 months, and six (6%) lived more than 1 year.

## COMPLICATIONS

### Esophageal Perforation

The perforation can occur during balloon dilatation of the stricture or during the process of stent placement by the formation of a false passage through or below the tumor. The rate of esophageal perforation associated with placement of an expandable stent has been reported to be 0% to 7%. Delayed esophageal perforation from pressure necrosis caused by the stent can occur after stent placement, and any penetration of the aorta or other mediastinal vessels leads to a massive and fatal hematemesis.

### Migration of the Stent

Migration is more common in covered stents than in uncovered stents. Migration is common in benign stricture as well as in soft and eccentric malignant strictures, especially when the stricture is in the esophagogastric junction.

### Blockage of the Stent

Blockage of the expandable stents can occur by tumor ingrowth, tumor overgrowth, and food impaction. Growth of the tumor may occlude the proximal or distal end of the stent. This can be easily managed by the placement of a second stent, overlapping the end of the first stent. The impacted food bolus can be displaced into the stomach with use of a balloon catheter.

### Reflux

Gastroesophageal reflux occured in 19% of the patients (nine of 47) who had a stent in the lower one-third of the esophagus bridging the gastroesophageal junction. The symptom can be relieved by taking antacids, sleeping with the head of the bed raised by approximately 30 degrees, and avoiding large meals before going to bed.

### Miscellaneous Complications

An expandable metallic stent placed in patients with cervical esophageal stricture can cause a foreign body sensation in the throat. Other complications such as pain, mucosal prolapse into the stent, and aspiration pneumonia in patients with stent placement in the esophagogastric junction can occur.

## D. INTERVENTIONAL TREATMENT OF ENTERIC STRICTURES

### Haraldur Bjarnason

Endoscopically guided dilatation of enteric strictures has its limitations. The endoscope can often not be passed beyond the stricture and does not provide a complete view of the stricture anatomy. Fluoroscopic technique provides a more functional view of the lesion and allows guidewires and catheters to be passed around bends and strictures more easily. Contrasts allow for dynamic evaluation of the bowel segment and the balloon dilatation can be monitored under fluoroscopy.

### ANTRAL AND PYLORIC STRICTURES

Strictures of the antrum and pylorus are usually caused by long-standing peptic or ingestion of corrosives such as acid or basic substances. If surgical repair is not possible, balloon dilatation may provide complete or partial resolution of symptoms.

Passing a catheter and guidewire through a narrowed antrum or pylorus poses considerably greater technical difficulties than are encountered in the esophagus. Although occasionally angiographic catheters will cannulate the pylorus, it is often impossible to gain enough control to allow these to pass through a stenotic pylorus. The endoscopist may be of invaluable aid in cases in which access is difficult. Although the endoscope itself is too large to traverse most of these strictures, the end-viewing capability may allow passage of a guidewire passed through the endoscope through the narrow area. This needs to be performed with fluoroscopy to secure safe passage of adequate amounts of guidewire past the stricture.

Selection of the proper balloon size poses some difficulties. When attempting to establish a fixed channel, some care should be exercised so that both gastric obstruction and rapid dumping of gastric contents into the duodenum are avoided (Fig. 22.12). One should attempt to dilate the pylorus to a diameter of 15 mm, but this is often not possible. In those cases smaller balloons such as 10-mm balloons can be used. If high-pressure balloons are left inflated for a long time (more than 15 minutes), ischemic necrosis may occur.

The results of antral and pyloric dilatation are impossible to assess from immediate postdilatation barium studies because considerable reactive edema accompanies the dilating. However, on long-term follow-up, 75% of the patients treated for gastric outlet obstruction have reported symptomatic relief and are able to ingest either a full liquid or a completely normal diet.

Griffin et al. reported in 1989 20 of 25 patients being asymptomatic at a mean follow-up of 9 months. Duodenal strictures can also be successfully dilated. Recurrent stricture after multiple attempts at dilatation can be treated by the placement of a self-expandable stent.

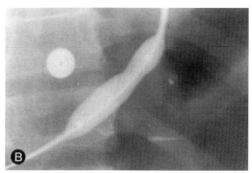
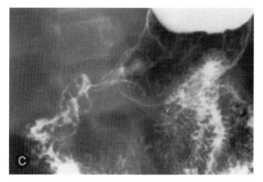

**FIG. 22.12. A:** An irregular stenosis of the gastric antrum in this child, who had accidentally ingested a corrosive substance. **B:** Initial dilatation was performed with an 8-mm balloon, shown inflated. **C:** After dilatation, there is marked improvement in the radiographic appearance of the antrum, with essentially normal gastric emptying. The same region was later dilated to a diameter of 12 mm. The patient was able to maintain a normal oral diet over 2 years of follow-up. (Reprinted from Castañeda-Zúñiga WR. *Interventional radiology*, 3rd ed. Baltimore: Williams & Wilkins, 1997, with permission.)

## ANASTOMOTIC STRICTURES

A variety of strictures may occur at sites of surgical anastomoses within the gastrointestinal tract. Many of these are ideally suited to balloon dilatation because they may be difficult or impossible to reach by conventional techniques of bougienage yet appear to respond extremely well to balloons. Strictures occurring in the esophagus at sites of esophagogastric, esophagojejunal, or esophagocolonic anastomoses are approached and dilated with the same techniques used for primary esophageal lesions (Fig. 22.13).

The anastomotic strictures commonly encountered outside the esophagus occur at sites of gastrojejunostomy. As with strictures of the pylorus, care should be exercised in dilating these strictures so as not to produce excessively rapid gastric emptying. This is particularly critical in patients who have undergone gastroplasty and gastrojejunostomy for morbid obesity because many surgeons believe that the rate of gastric emptying is one of the factors determining the ultimate weight loss. Passing a dilating balloon catheter through a narrow gastrojejunostomy may be difficult, and review of any prior contrast studies is invaluable before proceeding. Examination of the operative notes is necessary before locating and treating those strictures.

Usually a gently curved angiographic catheter is best suited for approaching the anastomosis. The catheter is brought to the anastomosis, and the stenotic area is gently probed with a guidewire. An angled torque responsive wire such as the Glidewire is very helpful for this purpose. Once the wire has passed the anastomosis, the catheter is advanced into the distal small bowel, and exchange is made for the appropriate-size balloon.

Generally, balloons of 12 to 15 mm in diameter are satisfactory. Many of these lesions are soft, and the balloon will inflate easily. It is often difficult to keep the balloon in place during the inflation because it will have tendency to migrate from the stricture during inflation. This is best handled by using a long balloon or by carefully centering the balloon over the stricture and being ready to pull on the balloon or advance as needed during the inflation.

Anastomotic as well as nonanastomotic strictures may also be dilated within the colon and at

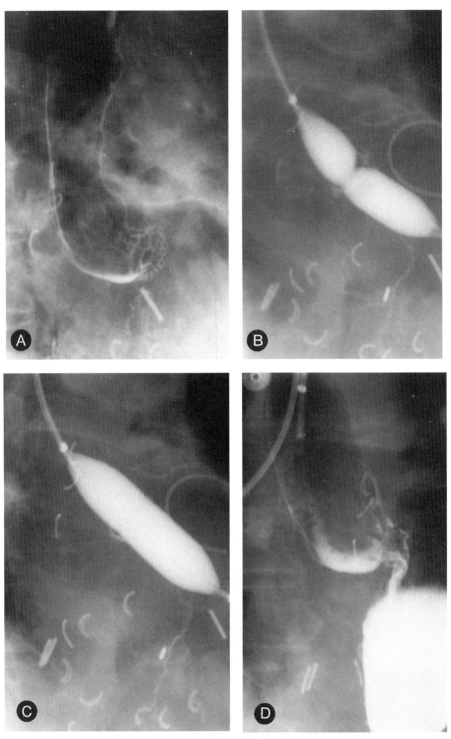

**FIG. 22.13. A:** Double-contrast barium study shows circumferential web at site of esophagogastric anastomosis. **B:** A 15-mm esophageal balloon has been passed through the lesion and is being inflated. **C:** As full inflation is reached, the margins of the balloon approximate those of the original anastomosis. **D:** After inflation, a repeat barium study shows free flow through the dilated lumen. (Reprinted from Castañeda-Zúñiga WR. *Interventional radiology,* 3rd ed. Baltimore: Williams & Wilkins, 1997, with permission.)

other sites within the gastrointestinal tract. Here as well, it may be very helpful to use a scope for lesions that are far away or where it is difficult to pass a wire across narrow area.

The appropriate balloon size must be judged on the basis of the lumen of normal adjacent bowel. Strictures low in the colon should be dilated to permit the passage of stool. Although a 20-mm balloon will suffice in some patients, patients with larger colons will need inflation of several balloons used side by side in order to achieve a functional diameter. Graham et al. reported dilatation of four rectosigmoidal strictures. In three of the patients, only minimal improvement was seen. Musher and Boyd reported successful management of an esophagocolonic anastomotic stricture with a proximal fistula. Stents are now commonly used in the colon (see Part E).

## CONCLUSION

Radiologically guided balloon dilatations of enteric strictures are uncommon procedures of considerable clinical usefulness. Many patients who can be quickly and safely treated by the methods described here have no other recourse than surgery to relieve their symptoms.

# E. COLONIC STENT PLACEMENT

Hector Ferral and Michael Wholey

Colonic obstruction is a frequent problem in patients with colorectal cancer. Approximately 30% of patients with colonic cancer will develop obstruction. Most of the obstructive neoplasms are located in the descending colon and rectosigmoid region. Acute large-bowel obstruction causes abdominal pain, discomfort, dehydration, electrolyte imbalance, and malaise in patients with colorectal cancer. The therapeutic alternatives include (a) emergency surgery for bowel decompression with the creation of a colostomy; (b) initial decompression with colostomy, followed by mass resection and colostomy closure; and (c) single-stage surgery with on-table bowel lavage.

The morbidity and mortality rates for emergency surgery are higher than those of elective surgery. The ideal management is to avoid an emergency surgery during the acute stage of obstruction and decompress the acutely obstructed colon.

## CONCEPT OF COLONIC STENTING

Doctors Antonio Mainar and Eloy Tejero described the use of metallic stents to treat acute colonic obstruction. The procedure consists of three basic steps:

1. The acute colorectal obstruction is treated with the transrectal placement of a self-expandable metallic stent. Stent placement may be performed either under direct fluoroscopic guidance or under colonoscopic and fluoroscopic guidance.

2. Once the acute obstruction is relieved, the general condition of the patient is improved with hydration and correction of electrolyte imbalance. The disease staging can be completed and colonic cleansing can be performed.

3. After adequate colonic preparation, elective surgery for mass resection and primary colonic anastomosis is performed. If the mass is considered to be nonresectable, the stent remains in place as a definitive palliative treatment.

## INDICATIONS AND CONTRAINDICATIONS

- One indication is intestinal obstruction caused by colonic neoplasms. Colonic stenting is ideally suited for tumors located in the descending colon and sigmoid. Tumors up to the distal transverse colon can be treated. Tumors located in the ascending colon are slightly more difficult to reach and stent placement is technically demanding. The presence and location of an obstructing mass must be confirmed by endoscopy or water-soluble contrast enema.
- Treatment of benign lesions in the colon must be approached with caution. In our experience, focal postsurgical strictures at colo-colonic

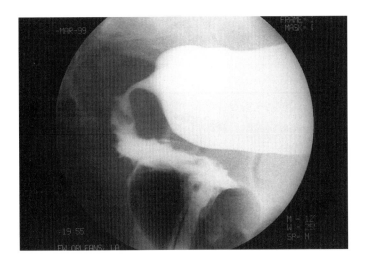

**FIG. 22.14.** Water-soluble contrast enema, lateral view, demonstrates a tight "apple core" stricture in the distal sigmoid colon.

anastomotic sites can be effectively treated with self-expandable metallic stents. Other benign lesions (ulcerative colitis, Crohn's disease, diverticulitis) are not well suited for this indication, mainly because there is poor response and a high risk of stent migration.
- Colonic stenting is not recommended for patients with extrinsic compression of the colon by pelvic or abdominal masses.
- The only absolute contraindication to colonic stenting is the clinical or radiologic evidence of bowel perforation.

## PATIENT PREPARATION

The patients are managed as if undergoing emergency surgery. If colonic stenting fails, the patient will need emergency surgery for bowel decompression.

- Chest x-ray
- Plain abdominal films
- Electrocardiogram
- Laboratory work-up (complete blood cell count, chemical analysis, coagulation profile, and tumor markers)
- Intravenous access (to be used for fluids, blood products, antibiotics, and sedation)
- Endoscopic procedure with biopsy to confirm the presence and location of the obstructing mass and determine its cellular component
- Insertion of nasogastric tube
- Informed consent

## TECHNIQUE

1. Once the diagnosis is confirmed and it is evident that the bowel obstruction is caused by the neoplastic process, the patient is considered a candidate for colonic stent placement.

2. The procedure is performed in the angiography or interventional radiology suite, under fluoroscopic guidance or combined colonoscopic and fluoroscopic guidance.

3. Intravenous sedation is used for patient comfort.

4. The patient may be placed supine or prone on the angiographic table. In our experience, placing the patient in a prone position has given better results.

5. A rectal examination is performed before inserting tubes or catheters per rectum.

6. A rectal tube is carefully advanced per rectum and an enema is performed using water-soluble contrast to identify the lesion (Fig. 22.14).

7. A 7-Fr multipurpose or vertebral curve catheter and 0.035-in. Glidewire combination is used to reach the obstruction and negotiate the occlusion.

8. The area of stenosis is crossed with the catheter-Glidewire combination. Once the catheter is advanced past the lesion, the guidewire is removed and a small amount of water-soluble contrast is injected to determine the length of the lesion and to rule out bowel perforation related to guidewire or catheter manipulation (Fig. 22.15).

9. Once it has been confirmed that no bowel

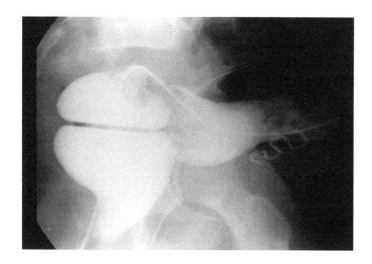

FIG. 22.15. Spot film radiograph during colonic stent placement. The guidewire and catheter combination has been successfully advanced across the stricture. Contrast injection shows no evidence of proximal colonic perforation.

perforation has been caused by catheter or guidewire manipulation, a 0.038-in. exchange length Amplatz Superstiff guidewire (MEDITECH/Boston Scientific) is advanced through the angiographic catheter.

10. A stent is then chosen for the lesion to be treated. We use large-diameter (20- to 22-mm), self-expandable Wallstents (MEDITECH/Boston Scientific). Stent length varies from 4.5 to 8.0 cm.

11. The location of the obstructing lesion is confirmed with fluoroscopy (Fig. 22.16) and the stent delivery system/catheter is advanced over the guidewire to its final position.

12. The stent is deployed in the usual fashion, according to the manufacturer's instructions.

13. After successful stent deployment, a repeat enema with water-soluble contrast is done to confirm stent patency (Fig. 22.17).

14. We do not recommend balloon dilatation of the stent, as balloon dilatation of the stent has been associated with an increased risk of bowel perforation.

15. The catheters are removed and the procedure is terminated.

## MANAGEMENT AFTER STENT PLACEMENT

- Plain abdominal radiographs are obtained daily to assess the progress of bowel decom-

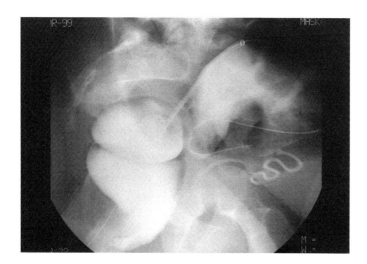

FIG. 22.16. Stent placement. Spot film demonstrates the self-expandable stent catheter, which has been placed across the colonic stricture.

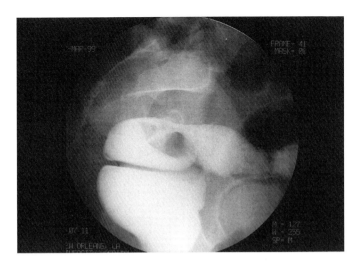

FIG. 22.17. Water-soluble contrast injection after successful stent placement demonstrates flow of contrast through the stent, which immediately after deployment, has an hourglass configuration. Stent dilatation is not necessary after stent deployment.

pression, assess the expansion of the stent, and rule out stent migration (Fig. 22.18).
- Disease staging can be performed after stent placement. Computed tomography scanning and even magnetic resonance image scanning can be performed immediately after Wallstent placement.
- Preparation of the colon for elective surgery can then be undertaken. The goal in these patients is to perform a single B stage surgery with tumor resection and creation of a primary colonic anastomosis.

## RESULTS

- Technical success rate is 91% to 92%. Most patients obtain complete bowel decompression within 48 hours after colonic stent placement. Stents achieve full expansion 48 to 72 hours after placement (Fig. 22.18B).
- Unsuccessful stent placement occurs in 7% to 8%.
- The inability to decompress the bowel after successful stent deployment occurs in 1% to 2%.
- Tumor resection and primary anastomosis are possible in approximately 70% of patients.
- The stent remains as a permanent palliative device in approximately 30% of patients.
- Surgical colostomy has been necessary in approximately 8% to 9% of patients in whom bowel decompression was not successful due to unsuccessful stent placement and of patients in whom, despite successful stent placement, inadequate bowel decompression was encountered.
- Complications include rectal bleeding, which is usually self limited (9%), stent migration (2%), and colonic perforation (1%).
- In our experience, we have found that stent migration and inadequate decompression of the colon tend to be more frequent when colonic stents are used in patients with benign strictures (Crohn's disease, diverticulitis) or with extrinsic compression of the colon by other pelvic malignancies (prostatic carcinoma or ovarian carcinoma).

## CONCLUSION

- The placement of metallic self-expandable stents to relieve acute bowel obstruction caused by colonic and sigmoid carcinomas offers several benefits to the patient.
- The acute obstructive problem is solved in a noninvasive fashion with excellent results.
- Colon cleansing and disease staging are possible after stent placement.
- Once the colon has been prepared for elective surgery, a single-stage procedure can be undertaken without the need for the creation of a colostomy.
- Complications of colonic stenting are minimal and of no major clinical consequence.

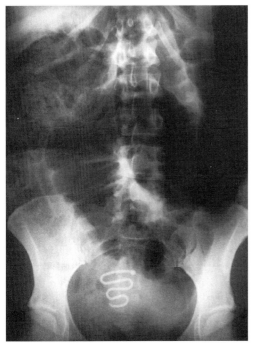 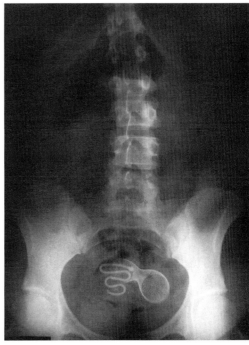

**FIG. 22.18. A:** Supine plain abdominal film before colonic stent placement demonstrates massive distention of bowel loops, consistent with bowel obstruction. No air is seen in the rectum. An intrauterine device is seen. **B:** Supine plain abdominal film 48 hours after stent placement demonstrates successful decompression of bowel loops. The colon appears now clean. The stent shows 70% to 80% expansion. Single-stage surgery with mass resection was performed 4 days after stent placement.

## F. PERCUTANEOUS CECOSTOMY FOR OGILVIE'S SYNDROME: LABORATORY OBSERVATIONS AND CLINICAL EXPERIENCE

### Haraldur Bjarnason

Sporadic case reports describing percutaneous cecostomy (PCC) suggest that the procedure may be beneficial for selected patients with cecal volvulus or colonic pseudoobstruction (Ogilvie's syndrome). Various access methods and catheters have been used in these early clinical cases. The aforementioned reports described a variety of routes and methods of access into the cecum.

### MATERIALS AND METHODS

Five patients—three men and two women—underwent PCC. Their ages ranged from 26 to 76 years. Broad-spectrum antibiotics were administered to all patients.

The indication for PCC was decompression of pseudoobstruction (Ogilvie's syndrome) in all five patients. The underlying problems were chronic obstructive pulmonary disease (three patients), oat cell carcinoma (one patient), and dementia and sepsis (one patient).

The sizes of the cecum in the patients undergoing PCC for decompression were 12, 15, 16, 18, and 19 cm. Colonoscopic decompression was unsuccessful in the four patients in whom it was attempted before PCC. In the patients in whom catheters were inserted, fluoroscopy was used for guidance in four patients and computed tomography was used in one patient. Catheters that were inserted included the Foley (8 Fr), Ingram (12 Fr), Foley (10 Fr), Wills-Oglesby (9 Fr), and Cope-type nephrostomy (9 Fr). All of the catheters had retention devices. Trocar insertion was used for two cases and modified

Seldinger technique was used for three patients. Eighteen-gauge, thin-walled needles were used for the Seldinger puncture. Once the retention catheters were in place, they were pulled snugly to the abdominal wall to help prevent leakage.

The following is a short description of placement of colostomies in cadavers and animals as presented in the third edition of the Castañeda book. This correlates with the procedure in patients.

To perform PCC in both the animals and cadavers, the colon was insufflated with air. The feasibility of attaching the cecum to the abdominal wall was tested by a fashioned Cope-type device. A suture was tied at the tip of a 6-Fr dilator. The suture was brought alongside the dilator for 2 cm, repunctured, and placed into the dilator. The proximal end of the suture was retracted from the hub of the dilator. After the dilator was placed into the cecum over a guidewire, the suture was pulled, tightened, and fixed by a stopcock, forming a loop at the distal part of the dilator. This permitted the dilator to move the cecum snugly up to the anterior abdominal wall for subsequent dilatation and catheter placement.

Catheters were inserted successfully in each patient. All patients benefited clinically from decompression of the colonic gas. In the five patients who underwent PCC for Ogilvie's syndrome, marked and rapid decrease in the cecal size occurred. None of these patients underwent operation. The catheters were removed uneventfully in four patients at 12, 14, 18, and 27 days. One patient died from respiratory arrest 8 days after PCC. The colon had been decompressed, and there were no catheter-related complications. Autopsy revealed that the transperitoneal tract was clean, and no infection was present. Another patient had transient back leakage along the catheter tract; the leakage ceased when the Foley balloon and catheter were positioned snugly against the abdominal wall and taped more securely.

## DISCUSSION

The initial clinical experience with PCC suggests that the technique is feasible and effective and seems to be safe. The procedure offers an important nonsurgical option for the treatment of Ogilvie's syndrome. Surgical therapy in pseudoobstruction of the colon has a high morbidity and mortality rate, in part because of the complicated accompanying medical problems in these typically elderly patients. In an analysis of 400 cases of Ogilvie's syndrome, the mortality rate in patients treated surgically was 26% if the bowel was viable, 44% with ischemic bowel, and 36% with perforated bowel. Other series reveal a 25% to 31% mortality rate with Ogilvie's syndrome, and it can be as high as 46% despite operative cecostomy if perforation has occurred. Speed in diagnosis, viability of the bowel, and underlying medical problems govern the prognosis.

Both transperitoneal and retroperitoneal approaches for PCC have been used in patients. There has been speculation that the retroperitoneal approach might be advantageous. The rationale for using a retroperitoneal route is that if fecal spillage from the cecum occurred, it would be contained in the retroperitoneum rather than soil the peritoneal cavity; thus, it has been reasoned, peritonitis might be avoided. However, in the majority of published clinical cases of PCC, an anterior approach has been used, and peritonitis has not been a problem. In addition, it is only speculation that bowel spillage into the peritoneum would be less toxic than spillage into the retroperitoneum. Stool in the retroperitoneum might cause a severe infectious fasciitis.

Aside from the advisability of a transperitoneal versus a retroperitoneal approach for the previously mentioned reasons, there is the morphologic feasibility of each. Both the anatomic studies in the cadavers and review of the intraperitoneal computed tomography contrast studies suggested that a retroperitoneal approach would be difficult, if not impossible, in most patients. The posterolateral extension of the peritoneum around the cecum, along with the overlying posterolateral iliac bone, make an anterior approach to the cecum far more feasible. Quinn et al. emphasized the anatomic limitations of computed tomography guidance for PCC; in their experience, an apparent retroperitoneal approach by computed tomography was in fact intraperitoneal at laparotomy. This failure of computed tomography guidance was probably due to the thin peritoneal reflection that surrounds the cecum; although not obvious on plain computed tomography, this anatomic juxtaposition was well displayed on the computed tomography intraperitoneal contrast studies when the potential peritoneal spaces were filled with fluid. Another study described difficulties in attempting a posterior approach into the cecum when the cecum is dilated, on a mesentery, and twisted anteromedially. The ascending colon was on a mesentery in 26% of patients in one series.

PCC for Ogilvie's syndrome may provide both immediate decompression and definitive therapy; in each of our five patients with Ogilvie's syndrome, the massively dilated colon was decompressed effectively. These patients were extremely ill from a variety of medical problems. PCC permits avoidance of laparotomy and general anesthesia in this high-risk group of patients.

Endoscopic decompression is another alternative. Success rates of 68% to 91% have been reported; recurrence occurs in up to 22%, and perforation itself may complicate the endoscopic approach. One report described 11 of 11 patients treated successfully by the endoscopic approach. The respective roles of endoscopic versus percutaneous therapy of Ogilvie's syndrome require further experience for evaluation. PCC was successful in all four patients in this series in whom attempted endoscopic decompression had been unsuccessful.

An optimal or preferred technique for PCC was not demonstrated from our study. Both Seldinger and trocar methods were feasible in the laboratory and in humans. All well-tapered catheters entered the dilated cecum easily, and relatively small-lumen catheters (7 to 9 Fr) per-

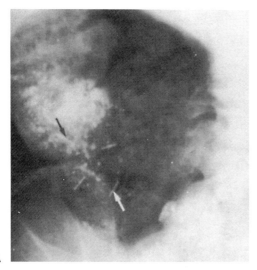

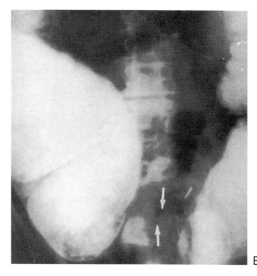

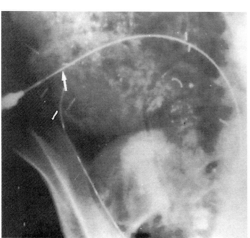

**FIG. 22.19.** Studies of a 71-year-old patient with metastases to the small and large bowels from metastatic cervical carcinoma and postoperative adhesions. **A:** Barium enema study demonstrates a large dilated cecum caused by obstruction in the sigmoid colon (*arrows*). **B:** Collimated spot radiograph of the dilated cecum after insertion of T-fasteners into the cecum (*arrows*). Note the square area delimited by the T-fasteners. **C:** Spot radiograph shows insertion of a 0.038-in. guidewire through the 18-gauge needle across the center (*arrow*) of the square area delimited by T-fasteners. (Reprinted from Castañeda-Zúñiga WR. *Interventional radiology,* 3rd ed. Baltimore: Williams & Wilkins, 1997, with permission.)

mitted adequate decompression of colonic gas. Whether methods to tack the stomach for percutaneous gastrostomy are applicable for colonic decompression may be tested in the future (Fig. 22.19). Although we used a modified dilator as a Cope loop to secure the cecum in the cadavers, it was unnecessary in patients.

In summary, PCC may be performed safely and effectively by a variety of instruments and methods, as is true for most interventional radiology procedures. Interventional radiologists should include PCC in their armamentarium of procedures for appropriate patients. PCC may be effective despite unsuccessful endoscopic attempts at colonic decompression. By avoiding general anesthesia and major surgery, PCC may be lifesaving in elderly or critically ill patients with pseudoobstruction.

## G. VOLVULUS OF THE SIGMOID COLON: TREATMENT BY TRANSRECTAL FLUOROSCOPIC CATHETERIZATION AND STENTING

### Haraldur Bjarnason

Volvulus of the sigmoid colon is a frequent cause of colonic obstruction in the adult, accounting for 1% to 9% of the cases in Western Europe and the United States and 11.6% to 54.6% in parts of Africa, Iran, and Russia. Twisting of the sigmoid mesocolon produces occlusion of the bowel lumen at two points, resulting in a closed-loop obstruction.

In 60% to 70% of cases, the diagnosis of sigmoid volvulus can be made from the analysis of a plain abdominal film. Barium enema is usually diagnostic.

Nonsurgical decompression, either by percutaneous needle deflation or by endoscopic cannulation when the colon is thought to be viable, is the first step in treatment and has a reported success rate from 71% to 100%.

### MATERIALS AND METHODS

1. Intravenous sedation is usually needed.
2. The patient is positioned prone and in anti-Trendelenburg position.
3. Previous bowel preparation is not required.
4. A barium enema confirms and localizes the distal colonic obstruction.
5. The obstruction is then crossed with an angiographic catheter and a Glidewire (Fig. 22.20).
6. Once the obstruction is crossed, water-soluble contrast medium is injected through the angiographic catheter to verify catheter position and absence of bowel perforation.
7. An Amplatz stiff guidewire is then inserted, and the catheter is removed.
8. A 14-Fr nasogastric catheter, with an end hole and sideholes on the distal 10 cm, is inserted over the guidewire and placed across the obstruction.
9. Once the stent is in good position, the external end of the tube is fixed to the medial side of one buttock.
10. The stent is connected to bag drainage.
11. This drainage system is left in place for 72 to 96 hours.

### RESULTS

Fluoroscopic intubation of the closed loop was undertaken in eight patients with sigmoid colon volvulus. The stents successfully decompressed the distended colon in all patients and were removed 72 to 90 hours after insertion. Symptomatic relief was almost immediate. Elective surgery was then planned for seven patients. All patients underwent sigmoidectomy with primary end-to-end anastomoses. No complications of the procedure occurred in these eight instances.

### DISCUSSION

Emergency surgery in the volvulus of the sigmoid colon has a high mortality (25% to 77%). Elective surgery after bowel preparation has

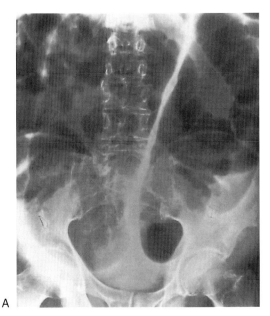
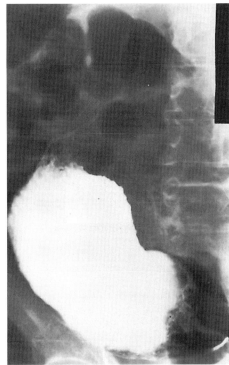
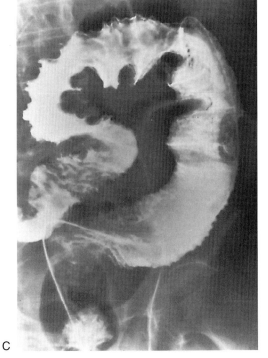

FIG. 22.20. A 72-year-old male with Parkinson's disease and an 8-day history of abdominal pain. **A:** Plain film of the abdomen demonstrates dilated closed loop of sigmoid with parietal edema. **B:** The barium enema verifies the volvulus. **C:** Immediately after stenting, note the presence of thumbprinting and parietal edema. (Reprinted from Castañeda-Zúñiga WR. *Interventional radiology*, 3rd ed. Baltimore: Williams & Wilkins, 1997, with permission.)

only a 3% to 10% mortality. Neither endoscopic decompression nor surgical procedures without resection of the redundant colon (i.e., simple detorsion, detorsion plus pexis, cecostomy) guarantee that the volvulus will not recur. The reason is that the affected segment has gross anatomic abnormality and chronic motility disturbance. Because of this, most authors advocate elective resection of the redundant sigmoid with end-to-end anastomosis. Typically, once the colon is decompressed, the sigmoid colon returns to its normal position. The stent position needs to be maintained for at least 3 days. Converting an emergency surgical procedure into an elective one with adequate bowel preparation is the main benefit of this approach. Possible complications of the procedure are bleeding and perforation.

# SUGGESTED READINGS

## A. Percutaneous Gastrostomy and Jejunostomy

Antinori CH, Andrew C, Villanueva DT, et al. A technique for converting a needle-catheter jejunostomy into a standard jejunostomy. *AJR Am J Roentgenol* 1992;164: 68–69.

Barkmeier JM, Treretola SO, Wiebke EA, et al. Percutaneous radiologic, surgical endoscopic, and percutaneous endoscopic gastrostomy/gastrojejunostomy: comparative study and cost analysis. *Cardiovasc Intervent Radiol* 1998;21:324–328.

Baskin WN. Advances in enteral nutrition techniques. *Am J Gastroenterol* 1992;87:1547–1553.

Chait PG, Weinberg J, Connolly BL, et al. Retrograde percutaneous gastrostomy and gastrojejunostomy in 505 children: a 4 1/2-year experience. *Radiology* 1996;201:691–695.

Coleman CC, Coons HG, Cope C, et al. Percutaneous enterostomy with the Cope suture anchor. *Radiology* 1990;174:889–891.

Edelman DS, Arroyo PS, Unger SW. Laparoscopic gastrostomy versus percutaneous endoscopic gastrostomy: a comparison. *Surg Endosc* 1994;8:47–49.

Gauderer MW. Gastrostomy techniques and devices. *Surg Clin North Am* 1992;72:1285–1298.

Halkier BK, Ho CS, Yee ACN. Percutaneous feeding gastrostomy with the Seldinger technique: review of 252 patients. *Radiology* 1989;171:359–362.

Hicks ME, Surratt RS, Picus D. Fluoroscopically guided percutaneous gastrostomy and gastroenterostomy: analysis of 158 consecutive cases. *AJR Am J Roentgenol* 1990;154:725–728.

Ho CS. Percutaneous gastrostomy for jejunal feeding. *Radiology* 1993;149:595–596.

Ho CS, Yeung EY. Percutaneous gastrostomy and transgastric jejunostomy. *AJR Am J Roentgenol* 1991;58:251–257.

Kanterman RY, Darcy MD. Complications of the fluoroscopically guided percutaneous gastrostomy. *Semin Intervent Radiol* 1996;13:317–327.

Koolpe HA, Dorfman D, Kramer M. Translumbar duodenostomy for enteral feeding. *AJR Am J Roentgenol* 1989; 153:299–300.

Lindberg CG, Ivancev K, Kan Z. Percutaneous gastrostomy. A clinical and experimental study. *Acta Radiol* 1991; 32:302–304.

Lu DSK, Mueller PR, Lee MJ, et al. Gastrostomy conversion to transgastric jejunostomy: technical problems, causes of failure and proposed solutions in 63 patients. *Radiology* 1993;187:679–383.

Marx MV, Williams DM, Perkins AJ, et al. Percutaneous feeding tube placement in pediatric patients: immediate and 30-day results. *J Vasc Intervent Radiol* 1996;7:107–115.

Maynar M, Reyes R, Pulido-Duque JM, et al. *Percutaneous outpatient gastrostomy*. Presented at the 17th Annual Scientific Meeting of the Society of Cardiovascular and Interventional Radiology. Washington, DC, April 5–9, 1992.

McNeely GF, Kickhous AJ, Colella JP, Hawkins IF. Percutaneous transhepatic choledochoenterostomy in a patient with a biliary obstruction. *Radiology* 1986;161:274–275.

Monturo CA. Enteral access device selection. *Nutr Clin Pract* 1990;5:207–213.

O'Keffe F, Carrasco CH, Charnsangavej C, et al. Percutaneous drainage and feeding gastrostomies in 100 patients. *Radiology* 1989;172:341–343.

Olson DL, Krubsack AJ, Stewart ET. Percutaneous enteral alimentation: gastrostomy versus gastrojejunostomy. *Radiology* 1993;187:105–108.

Ryan JM, Hahn PF, Boland GW, et al. Percutaneous gastrostomy with T-fastener gastropexy: results of 316 consecutive procedures. *Radiology* 1997;203:496–500.

Ryan JM, Hahn PF, Mueller PR. Performing radiologic gastrostomy or gastrojejunostomy in patients with malignant ascites. *AJR Am J Roentgenol* 1998;171:1003–1006.

Saini S, Mueller PR, Gaa J, et al. Percutaneous gastrostomy with gastropexy: experience in 125 patients. *AJR Am J Roentgenol* 1990;154:1003–1006.

Salim AS. Jejunostomy feeding for the conservative management of spontaneous rupture of the oesophagus. *Br J Clin Pract* 1991;45:37–40.

Sanchez RB, vanSonnenberg, E, D'Agostino HB, et al. CT guidance for percutaneous gastrostomy and gastroenterostomy. *Radiology* 1992;184:201–205.

Stuart SP, Tiley EH, Boland JP. Feeding gastrostomy: a critical review of its indications and mortality rate. *South Med J* 1993;86:169–172.

Tunca JC, Buchler DA, Mack EA, et al. The management of ovarian-cancer-caused bowel obstruction. *Gynecol Oncol* 1981;12:186–192.

vanSonnenberg E, Casola G, D'Agostino H. Radiologic percutaneous gastrostomy and gastroenterostomy. *Am J Gastroenterol* 1990;85:1561–1562.

vanSonnenberg E, Wittich GR, Brown LK, et al. Percutaneous gastrostomy and gastroenterostomy I. Techniques derived from laboratory evaluation. *AJR Am J Roentgenol* 1986;146:570–580.

vanSonnenberg E, Wittich GR, Cabrera OA, et al. Percutaneous gastrostomy and gastroenterostomy II: clinical experience. *AJR Am J Roentgenol* 1986;146:581–586.

Wasilijew BK, Ujiki GE, Beal JM. Feeding gastrostomy: complications and mortality. *Am J Surg* 1982;143:194–195.

Woolman B, D'Agostino HB. Percutaneous radiologic and endoscopic gastrostomy: a 3-year institutional analysis of procedure performance. *AJR Am J Roentgenol* 1997; 169:1551–1553.

## B. Management of Benign Esophageal Strictures

Ball WS, Strife JL, Rosenkrantz J, et al. Esophageal strictures in children: treatment by balloon dilation. *Radiology* 1984;150:263–264.

Chon DK, Song HY, Han YM, et al. Esophageal balloon dilatation: experiences in 100 patients. *J Korean Rad Soc* 1991;27:751–757.

Dawson SL, Mueller PR, Ferrucci JT, et al. Severe esophageal strictures: indications for balloon catheter dilatation. *Radiology* 1984;153:631–635.

de Lange EE, Shaffer HA. Anastomotic strictures of the upper gastrointestinal tract: results of balloon dilation. *Radiology* 1988;167:45–50.

de Lange EE, Shaffer HA, Daniel TM, Kron IL. Esophageal anastomotic leaks: preliminary results of treatment with balloon dilation. *Radiology* 1987;165:45–47.

Earlam R, Cunha-Melo JR. Malignant oesophageal strictures: a review of techniques for palliative intubation. *Br J Surg* 1982;69:61–68.

Goldthorn JF, Ball WS, Wilkinson LG, et al. Esophageal strictures in children: treatment by serial balloon catheter dilatation. *Radiology* 1984;153:655–658.

Hegedus V, Raaschou HO. Radiologically guided dilatation of stenotic gastroduodenal anastomosis. *Gastrointest Radiol* 1986;11:27–29.

Johnson A, Ingemann JL, Mauritzen K. Balloon-dilatation of esophageal strictures in children. *Pediatr Radiol* 1986;16:388–391.

Kim IO, Yeon KM, Kim WS, et al. Perforation complicating balloon dilation of esophageal strictures in infants and children. *Radiology* 1993;189:741–744.

LaBerge JM, Kerlan RK, Pogany AC, Ring EJ. Esophageal rupture: complication of balloon dilatation. *Radiology* 1985;157:56.

London RL, Trotman BW, Di Marino AJ, et al. Dilatation of severe esophageal strictures by an inflatable balloon catheter. *Gastroenterology* 1981;80:173–175.

Maroney TP, Ring EJ, Gordon RL, Pellegrini CA. Role of interventional radiology in the management of major esophageal leaks. *Radiology* 1989;170:1055–1057.

Maynar M, Guerra C, Reyes R, et al. Esophageal strictures: Balloon dilation. *Radiology* 1988;167:703–706.

McLean GK, Cooper GS, Hartz WH, et al. Radiologically guided balloon dilation of gastrointestinal strictures. Part I. Technique and factors influencing procedural success. *Radiology* 1987;165:35–40.

McLean GK, Cooper GS, Hartz WH, et al. Radiologically guided balloon dilation of gastrointestinal strictures. Part II. Results of long-term follow-up. *Radiology* 1987; 165:41–43.

Michel LUC, Grillo HC, Malt RA. Operative and nonoperative management of esophageal perforations. *Ann Surg* 1981;194:57–63.

Nashef SAM, Pagliero KM. Instrumental perforation of the esophagus in benign disease. *Ann Thorac Surg* 1987; 44:360–362.

Owman T, Lunderquist A. Balloon catheter dilatation of esophageal strictures: a preliminary report. *Gastrointest Radiol* 1982;7:301–305

Skinner DB, Little AG, Demesseter TR. Management of esophageal perforation. *Am J Surg* 1980;139:760–764.

Song HY, Han YM, Kim HN, et al. Corrosive esophageal strictures: safety and effectiveness of balloon dilation. *Radiology* 1992;184:373–378.

Song HY, Park SI, Do YS, et al. Expandable metallic stent placement in patients with benign esophageal strictures: results of long-term follow-up. *Radiology* 1997;203:131–136.

Starck E, Paolucci V, Herzer M, Crummy AB. Esophageal stenosis: treatment with balloon catheters. *Radiology* 1984;153:637–640.

Triggiani E, Belsey R. Oesophageal trauma: incidence, diagnosis and management. *Thorax* 1977;32:241.

Wichern WA. Perforation of the esophagus. *Am J Surg* 1970;119:534–537.

## C. Management of Malignant Esophageal Strictures

Acunas B, Rozanes I, Akpinar S, et al. Palliation of malignant esophageal strictures with self-expanding nitinol stents: drawbacks and complications. *Radiology* 1996; 199:648–652.

Angorn IB, Haffejee AA. Endoesophageal intubation for palliation in obstructing esophageal carcinoma. In: Manning TA, ed. *Interventional trends in general thoracic surgery*, Vol. 4. St. Louis: Mosby, Delarue and Eschapasse, 1988:410–419.

Caspers R, Welvaart K, Verkes R, et al. The effect of radiotherapy on dysphagia and survival in patients with esophageal cancer. *Radiother Oncol* 1988;12:15–23.

Cwikiel W, Stridbeck H, Tranberg KG, et al. Malignant esophageal strictures: treatment with a self-expanding nitinol stent. *Radiology* 1993;187:661–665.

Do YS, Song HY, Lee BH, et al. Esophagorespiratory fistula associated with esophageal cancer: treatment with a Gianturco stent tube. *Radiology* 1993;187:673–677.

Domschke W, Foerster EC, Matek W, Rodl W. Self-expanding mesh stent for esophageal cancer stenosis. *Endoscopy* 1990;22:134–136.

Earlam R, Cunha-Melo JR. Malignant esophageal strictures: a review of techniques for palliative intubation. *Br J Surg* 1982;69:61–68.

Fleischer D, Sivak M. Endoscopic Nd:YAG laser therapy as palliation for oesophagogastric carcinoma. *Gastroenterology* 1985;89:827–831.

Han YM, Song HY, Lee JM, et al. Esophagorespiratory fistulae due to esophageal carcinoma: palliation with a covered Gianturco stent. *Radiology* 1996;199:65–70.

Knyrim K, Wagner HJ, Bethge N, et al. A controlled trial of an expansile metal stent for palliation of esophageal obstruction due to inoperable cancer. *N Engl J Med* 1993;329:1302–1307.

Lolley DM, Ray JF, Ransdell HT, et al. Management of malignant esophagorespiratory fistula. *Ann Thorac Surg* 1978;25:516–520.

Miyayama S, Matsui O, Kadoya M, et al. Malignant esophageal stricture and fistula: palliative treatment with polyurethane-covered Gianturco stent. *J Vasc Intervent Radiol* 1995;6:243–248.

Neuhaus H. Therapeutic endoscopy in the esophagus. *Curr Opinion Gastroenterol* 1993;9:677–684.

Ogilvie AL, Dronfield MW, Percuson R, Atkinson M. Palliative intubation of oesophagogastric neoplasms at fiberoptic endoscopy. *Gut* 1982;23:1060–1067.

Saxon RR, Barton RE, Katon RM, et al. Treatment of malignant esophageal obstructions with covered metallic Z stents: long-term results in 52 patients. *J Vasc Intervent Radiol* 1995;6:747–754.

Schaer J, Katon RM, Ivancev K, et al. Treatment of malignant esophageal obstruction with silicone-coated metallic self-expanding stents. *Gastrointest Endosc* 1992;38:7–11.

Song HY, Choi KC, Cho BH, et al. Esophagogastric neoplasms: palliation with a modified Gianturco stent. *Radiology* 1991;180:349–354.

Song HY, Choi KC, Kwon HC, et al. Esophageal strictures: treatment with a new design of modified Gianturco stent. Work in progress. *Radiology* 1992;184:729–734.

Song HY, Do YS, Han YM, et al. Covered expandable esophageal metallic stent tubes: experiences in 119 patients. *Radiology* 1994;193:689–695.

Wagner HJ, Stinner B, Schwerk WB, et al. Nitinol prostheses for the treatment of inoperable malignant esophageal obstruction. *J Vasc Intervent Radiol* 1994;5:899–904.

Watkinson AF, Ellul J, Entwisle K, et al. Esophageal carcinoma: initial results of palliative treatment with covered self-expanding endoprostheses. *Radiology* 1995; 195:821–827.

## D. Interventional Treatment of Enteric Strictures

Ball WS Jr, Seigel RS, Goldthorn FJ, Kosloske AM. Colonic strictures in infants following intestinal ischemia. *Radiology* 1983;149:469–472.

Benjamin SB, Cattau EL, Glass RL. Balloon dilation of the pylorus: therapy for gastric outlet obstruction. *Gastrointest Endosc* 1982;28:253.

Cohen WN, Mason EE, Blommers TJ. Gastric bypass for morbid obesity. *Radiology* 1977;122:609–612.

Earlam R, Cunha-Melo JR. Benign oesophageal strictures: historical and technical aspects of dilatation. *Br J Surg* 1981;68:829–836.

Goautberg S, Afzelius L-E, Hambraeus G, et al. Balloon-catheter dilatation of strictures in the upper digestive tract. *Radiologe* 1982;22:479–483.

Graham DY, Tabibian N, Schwartz JT, Smith JL. Evaluation of the effectiveness of through-the-scope balloons as dilators of benign and malignant gastrointestinal strictures. *Gastrointest Endosc* 1987;33:432–435.

Griffen WO Jr, Bell RM. Surgical approaches to morbid obesity. *Contemp Surg* 1983;23:15–23.

Griffin SM, Chung SC, Leung JW, Li AK. Peptic pyloric stenosis treated by endoscopic balloon dilatation. *Br J Surg* 1989;76:1147–1148.

Hallisey MJ, Meranze SG. Interventional radiological approach to gastrointestinal strictures. *Semin Intervent Radiol* 1996;13:345–350.

McLean GK, Ring EJ, Freiman DB. Applications and techniques of gastrointestinal intubation. *Cardiovasc Intervent Radiol* 1982;5:108–116.

Merrell N, McCray RS. Balloon catheter dilation of a severe esophageal stricture. *Gastrointest Endosc* 1982;28:245–255.

Musher DR, Boyd A. Esophagocolonic stricture with proximal fistulae treated by balloon dilation. *Am J Gastroenterol* 1988;83:445–447.

Pinto IT. Malignant gastric and duodenal stenosis: Palliation by peroral implantation of a self-expanding metallic stent: *Cardiovasc Intervent Radiol* 1997;20:431–434.

Treem WR, Long WR, Friedman D, Watkins JB. Successful management of an acquired gastric outlet obstruction with endoscopy-guided balloon dilatation. *J Pediatr Gastroenterol Nutr* 1987;6:992–996.

## E. Colonic Stent Placement

Choo IW, Soo Do Y, Won Suh S, et al. Malignant colorectal obstruction: treatment with a flexible covered stent. *Radiology* 1998;206:415–422.

De Gregorio M, Mainar A, D'Agostino H, et al. Transanal metallic stents for malignant colonic obstruction: experience in 126 patients. *J Vasc Intervent Radiol* 1999; 10:219.

Lopera JE, Ferral H, Wholey MH, et al. Treatment of colonic obstructions with metallic stents: indications, technique and complications. *AJR Am J Roentgenol* 1997;169:1285–1290.

Maynar M, Ferral H, Wholey M, Castaneda-Zuniga WR. Treatment of malignant colonic obstruction with metallic stents: the Tejero-Mainar procedure. *J Vasc Intervent Radiol* 1997;8:139–141.

Wholey MH, Levine EA, Ferral H, Castaneda-Zuniga WR. Initial clinical experience with colonic stent placement. *Am J Surg* 1998;175:194–197.

## F. Percutaneous Cecostomy for Ogilvie's Syndrome: Laboratory Observations and Clinical Experience

Bode WE, Beart RW, Spencer RJ, et al. Colonoscopic decompression for acute pseudo-obstruction of the colon (Ogilvie's syndrome): report of 22 cases and review of the literature. *Am J Surg* 1984;147:243–245.

Casola G, Withers C, vanSonnenberg E, et al. Percutaneous cecostomy for decompression of the massively distended cecum. *Radiology* 1986;158:793–794.

Chait PG, Shandling B, Richards HM, Connolly BL. Fecal incontinence in children: treatment with percutaneous cecostomy tube placement—a prospective study. *Radiology* 1997;203:621–624.

Crass JE, Simmons RL, Mathis PF, Charles WM. Percutaneous decompression of the colon using CT guidance in Ogilvie syndrome. *AJR Am J Roentgenol* 1985;144:475–476.

Haaga JR, Bick JR, Zollinger RM Jr. CT-guided percutaneous catheter cecostomy. *Gastrointest Radiol* 1987;12:166–168.

Harig JM, Fumo DE, Loo FD, et al. Treatment of acute nontoxic megacolon during colonoscopy: tube placement versus simple decompression. *Gastrointest Endosc* 1988;34:23–27.

Johnson CD, Rice RP, Kelvin FM, et al. The radiologic evaluation of gross cecal distension: emphasis of cecal ileus. *AJR Am J Roentgenol* 1985;145:1211–1217.

Nanni G, Garbini A, Luchetti P, et al. Ogilvie's syndrome (acute colonic pseudo-obstruction): review of the literature and a report of four additional cases. *Dis Colon Rectum* 1982;25:157–166.

Nivatongs S, Vermeulen JD, Fang DT. Colonoscopic decompression of acute pseudo-obstruction of the colon. *Ann Surg* 1982;196:598–600.

Patel D, Ansari E, Berman MD. Percutaneous decompression of cecal volvulus. *AJR Am J Roentgenol* 1987;148:747–748.

Quinn SF, Jones EN, Maroney T. Percutaneous cecostomy in the management of cecal volvulus-report of a case. *J Intervent Radiol* 1987;2:137–139.

Rubenstein WA, Auh YH, Zirinsky K, et al. Posterior peritoneal recesses: assessment using CT. *Radiology* 1985;156:461–468.

Strodel WE, Nostrant TT, Eckhauser FE, Dent TL. Therapeutic and diagnostic colonoscopy in nonobstructive colonic dilatation. *Ann Surg* 1983;194:416–421.

Vanek VW, Al-Salti M. Acute pseudo-obstruction of the colon (Ogilvie's syndrome): an analysis of 400 cases. *Dis Colon Rectum* 1986;29:203–210.

vanSonnenberg E, Varney RR, Vasola G, et al. Percutaneous cecostomy for Ogilvie syndrome: laboratory observations and clinical experience. *Radiology* 1990;175:679–682.

Wojtalik RS, Lindenauer SM, Kahn SS. Perforation of the colon associated with adynamic ileus. *Am J Surg* 1973;125:601–606.

## G. Volvulus of the Sigmoid Colon: Treatment by Transrectal Fluoroscopic Catheterization and Stenting

Balthazar EJ. Congenital positional anomalies of the colon. In: Marshak RH, Lindner AE, Maklansky D, eds. *Radiology of the colon*. Philadelphia: WB Saunders, 1980:57–61.

Jones B. Cecosigmoid volvulus—a new entity? *Br J Radiol* 1978;51:466–469.

Mangiante EC, Croce MA, Fabian TC, et al. Sigmoid volvulus. A four decade experience. *Am Surg* 1989;55:41–44.

Prego X, Monton E, Prieto P, Costilla S. *Intubatcion perrectal fluoroscopica de los volvulos del colon sigmoide*. Presented at the annual meeting of the Sociedad Española Radiologia Medica, 1994.

Salim AS. Percutaneous deflation and colopexy for volvulus of the sigmoid colon: a new approach. *J R Coll Surg Edinb* 1990;35:356–359.

Salim AS. Management of acute volvulus of the sigmoid colon: a new approach by percutaneous deflation and colopexy. *World J Surg* 1991;15:68–72.

Scott Jones R, Schirner BD. Intestinal obstruction, pseudoobstruction and ileus. In: Sleisenger MH, Fordtran JS, eds. *Gastrointestinal disease*, 4th ed. Philadelphia: WB Saunders, 1989:373–377.

Theuer C, Cheadle WG. Volvulus of the colon. *Am Surg* 1991;57:145–150.

# 23
# Interventional Therapy of Infertility

Zhong Qian, Amy S. Thurmond, and Haraldur Bjarnason

## A. FALLOPIAN TUBE CATHETERIZATION FOR TREATMENT OF FEMALE INFERTILITY

*Infertility* is defined as the inability to conceive after 1 year of unprotected intercourse. The incidence of infertility in the United States is 15% to 20%, or one out of every five to six couples. The causes attributable to the female account for 30% to 40%. Fallopian tube disease is responsible for 30% to 40% of female infertility.

### TUBAL ANATOMY AND PATHOLOGY

Each fallopian tube is 8 to 15 cm long, located in the upper margin and between the two layers of the broad ligament. It runs laterally from the uterus to the uterine end of the ovary and forms the only anatomic channel from outside the body (via the vagina and uterus) to the peritoneal cavity. The tube is divided into four parts, which, counted from the uterus to the ovary, are (a) the interstitial or intramural part, (b) the isthmic part, (c) the ampullary part, and (d) the infundibulum or fimbrial portion (Fig. 23.1). Histologically, the tube consists of an innermost ciliated epithelium surrounded successively by a lamina propria, two longitudinal muscle layers with an intervening circular muscle layer, and a serosa. Blood supply comes from the uterine and ovarian arteries.

Disease processes that affect the fallopian tube are predominantly inflammatory and result in tubal occlusion, usually in the proximal interstitial portion or in the distal fimbriated end. Distal disorder and adhesions in the peritoneal cavity appear to be related to chlamydial or gonococcal infection or to endometriosis. Chronic obstruction at the fimbriated end could lead to dilatation, mucosal discharge, and destruction of the ciliated epithelium.

### TRANSCERVICAL FALLOPIAN TUBE CATHETERIZATION

Microsurgery has been the treatment of choice for tubal occlusion. However, surgical intervention is associated with high cost and morbidity. *In vitro* fertilization has been used to achieve conception as an alternative treatment. Prevalence of this new technique has been hampered by high cost, time-consuming procedure, and low (10% to 15%) pregnancy rates. Transcervical fallopian tube recanalization has proved to be effective in the management of tubal obstruction. With this minimally invasive technique, patients can now be treated on outpatient basis with a reasonably high success rate.

### PATIENT SELECTION

Women with unilateral or bilateral proximal tubal obstruction confirmed by hysterosalpingography or laparoscopy are candidates for the procedure. Severe uterine deformity or large uterine masses that would make catheterization impossible are a contraindication.

### PATIENT PREPARATION

The procedure is performed during the follicular phase of the patient's menstrual cycle, at least 3 days after bleeding has stopped (to prevent the flushing of blood into the fallopian tubes or peritoneum) and before the patient has ovulated (to ensure that patient is not pregnant). A dose of 100 mg of doxycycline is given orally twice a day for 5 days, starting 2 days before the procedure, or 200 mg orally just before the procedure followed

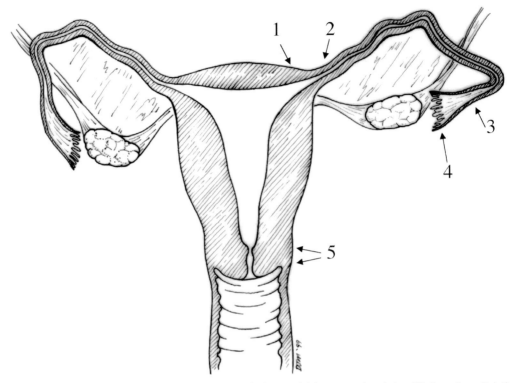

**FIG. 23.1.** Significant portions of the female genital tract. (*1*, intramural or interstitial portion; *2*, isthmic portion; *3*, ampullary portion; *4*, fimbria or the infundibulum; *5*, cervix.)

by 100 mg twice a day for 5 days. Aseptic technique is used. Small doses of midazolam (Versed) and fentanyl citrate may be given intravenously before and during the procedure as needed. Atropine, 0.5 mg intramuscularly, may be administrated before the procedure to prevent vasovagal symptoms.

## TECHNIQUES OF TUBAL CATHETERIZATION

A vacuum cup–type hysterosalpingography device [Thurmon-Rösch Hysterocath (Cook, Inc., Bloomington, IN)] is applied to the external cervix. It provides a sterile conduit through which catheters can be advanced and allows traction on the uterus without a tenaculum. As alternatives, modification of existing hysterosalpingography devices can also be used, but they have the disadvantage of requiring a cervical tenaculum.

After conventional hysterosalpingography is performed using water-soluble contrast medium, a coaxial catheter system is used to engage the tubal ostium and then to catheterize the proximal fallopian tube. A 9-Fr Teflon sheath and 5.5-Fr polyethylene catheter are advanced over a 0.035-in. J guidewire, with the 9-Fr sheath placed in the lower third of the uterus to stabilize the system and the 5.5-Fr catheter placed in the uterine cornua. For final positioning of the 5.5-Fr catheter in the tubal ostium, a 0.035-in. straight wire is helpful. The guidewire is removed and full-strength contrast agent injected to confirm positioning and to attempt visualization of the tube. If proximal tube obstruction persists, a 0.018-in. guidewire with a platinum tip and a 3-Fr Teflon catheter are advanced into the fallopian tube, and attempts are made to recanalize the obstruction with short, back-and-forth movements of the guidewire. If there is undue resistance to probing with the guidewire or if it is evident that there is an acute angulation in the fallopian tube, the small guidewire and catheter are exchanged for a softer, tapered system (Target Therapeutics, Santa Monica, CA). Both those systems are passed through the indwelling 5.5-Fr catheter in a coaxial manner. When the guidewire passes the obstruction, the wire is removed, and contrast agent is injected through the small catheter to visualize the distal anatomy. If injection indicates satisfactory recanalization, the small catheter is withdrawn, and contrast agent is injected through the 5.5-Fr catheter still in place in the tubal ostium to better visualize the entire tube, particularly the site of recanalization.

If the contralateral site is also blocked, the procedure is repeated by reinserting the 0.035-in. J guide and directing the 5.5-Fr catheter to the contralateral cornua. At the end of the procedure, the 5.5-Fr catheter is pulled back into the uterus for the postprocedural hysterosalpingography. The average radiation dose to the ovaries during the procedure has been reported to be equivalent to the estimated ovarian doses for a barium enema or excretory urogram.

## POSTPROCEDURAL CARE

Patients usually have mild vaginal bleeding for 2 to 3 days and occasionally mild cramping and bloating after the procedure. Most of them return to normal daily life, including intercourse, the next day. The patients should consult their infertility specialist for consideration of treatment of any other infertility problems, such as ovulatory dysfunction or sperm disorders.

## OUTCOME

Reported success rates for catheterization of the fallopian tubes and visualization of distal tubal anatomy range from 71% to 89% with the above-mentioned technique. Tubal perforation occured in 2% of patients, usually those with a history of proximal tubal surgery or severe salpingitis isthmica nodosa. After the procedure, one-third of patients had normal appearing tubes; another one-third had normal tubes but peritubal adhesions; and of the remaining cases, approximately 8% each had a slight dilatation at the site of the recanalization, frank salpingitis isthmica nodosa, or a concomitant distal occlusion. Fifty-eight percent of intrauterine pregnancies were reported in patients with bilateral proximal obstruction after the procedure; a 33% intrauterine rate and a 6% ectopic pregnancy rate were seen in a more clinical heterogeneous group of patients with unilateral or bilateral proximal tubal obstruction.

Transcervical fallopian tubal catheterization results in improved diagnosis of tubal disease and has been proved to be a successful treatment for infertility resulting from tubal obstruction. Pregnancy and tubal reocclusion rates with this technique are comparable to those with microsurgery. The low patient morbidity and cost make the minimally invasive technique more attractive. For these reasons, it should be the first intervention in patients in whom proximal fallopian tubal obstruction has been confirmed. This approach has been supported by the American Fertility Society.

## B. TRANSCATHETER EMBOLIZATION OF THE INTERNAL SPERMATIC VEIN FOR TREATMENT OF VARICOCELE

*Varicocele* is defined as abnormal dilatation and tortuosity of the veins of the pampiniform plexus, which is responsible for 40% of male infertility. An abnormal diameter on ultrasound is 3 mm or larger. Pathologically, it is characterized by retrograde blood flow into the internal spermatic vein down to the pampiniform plexus. Classic varicoceles result from absent or inadequate valves, although valvular abnormalities can also be present without being clinically symptomatic, a condition known as *subclinical varicocele*. Several etiologic factors have been suggested as playing a role in varicocele development, including valvular incompetence, elevated hydrostatic pressure in the left renal vein and internal spermatic veins, and mechanical pressure from the superior mesenteric artery as it crosses the left renal vein (nutcracker syndrome). The absent or incompetent valves are, however, the most important factor contributing to the pathogenesis of varicocele. Varicocele-related damage to the testis has been thought to be associated with testicular hyperthermia caused by decreased blood flow through the testis.

Varicocele is found in 10% to 15% of the normal population. In subfertile males the incidence is approximately 40%. Eight to 9% of patients have bilateral varicocele, and only a small proportion has isolated right varicocele. The left side is much more prevalent as the sole varicocele.

## DIAGNOSTIC VENOGRAPHY

Three approaches to spermatic vein catheterization have been proposed: femoral, jugular, and basilic. The standard femoral approach is adequate for routine venography, although it is difficult to advance a catheter deep into either the right or the left spermatic vein if embolization is to be attempted. It is, however, adequate for the injection of sclerosing agents or detachable balloon. A jugular approach facilitates deep catheterization of both spermatic veins; thus, this technique has been suggested whenever embolization is planned.

## ANATOMY

The left spermatic vein usually originates from the inferior surface of the left renal, 2 to 3 cm from the junction of the renal vein and the inferior vena cava (IVC). In some cases, the spermatic vein has a common origin with lumbar veins that arise from the inferior surface of the renal vein, just to the left of the lumbar vertebral bodies (Fig. 23.2). In rare cases, the left spermatic vein arises from intrarenal venous branches or from elsewhere in the IVC. The right spermatic vein arises at the inferior border of the right renal vein from the IVC. Anatomic variations in the spermatic system are common. Anastomoses of the internal spermatic veins with other retroperitoneal veins, such as the capsular, periureteral, and internal iliac veins, are frequently seen. Such collateral anastomoses make the precise placement of occluding device imperative.

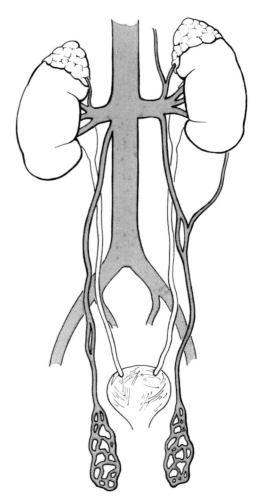

**FIG. 23.2.** Anatomic variations of the spermatic veins. The right spermatic vein most commonly comes off the inferior vena cava just below or at the renal vein level. The anatomy is, however, highly variable. The left spermatic vein has a more consistent origin from the left renal vein, as shown. There is a plethora of collateral interconnection on both sides with the retroperitoneum and renal capsular veins.

## DIAGNOSIS

If a competent valve is found and there are no clinical or Doppler findings indicative of a varicocele the vein is considered normal, and procedure on that side is terminated. If the valve is competent but the physical or Doppler examination demonstrates clear-cut evidence of varico-

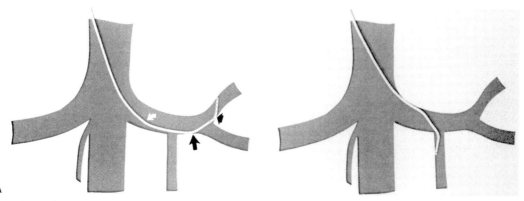

**FIG. 23.3. A:** Left catheter is shaped so that it will easily enter the left renal vein. Primary (*arrow*) and secondary curves (*arrowhead*) are unaltered. The tertiary curve (*curved arrow*) is reversed. **B:** When rotated 180 degrees, the long tertiary curve holds the tip firmly against the inferior wall of the vein, which makes entering the spermatic vein easy. (Reproduced from Hunter DW, Castañeda-Zúñiga WR, Coleman CC, et al. Spermatic vein embolization with hot contrast medium or detachable balloons. *Semin Intervent Radiol* 1984;1;163–169.)

cele, the vein is presumed to be abnormal based on the presence of another orifice or through the existence of retroperitoneal or pelvic collateral vessels. When the presence of the incompetent connections to the spermatic vein is confirmed, the spermatic vein anatomy should be fully delineated to guide catheter placement for embolization or sclerotherapy.

### Femoral Approach

After a femoral access is established, a gently curved 7-Fr modified Headhunter 1 (Cook, Inc.) mor cobra catheter is advanced over a guidewire into the IVC. Left renal venography is performed during a Valsalva maneuver to identify reflux into the left spermatic vein. If free reflux is observed, confirming incompetence of the vein, gentle selective spermatic venography is performed to evaluate valve competence further distal in the vein. The right spermatic vein is difficult to catheterize from the femoral approach. If the right spermatic vein enters the right renal vein, the Sidewinder catheter (Cook, Inc.) can be used to probe the inferior surface of the right renal vein.

### Jugular Approach

A transjugular approach is recommended to facilitate catheterization and embolization of the spermatic veins. Ultrasound-guided access is recommended, and the use of either a 6- or 7-Fr introducer sheath is helpful. Catheterization on the left side can usually be accomplished with a 5- to 6-Fr modified Headhunter catheter with the lordotic tertiary curve (Fig. 23.3). A reshaped Headhunter with a 70- to 90-degree reverse curve or a modified cobra catheter is recommended for the right spermatic vein (Fig. 23.4).

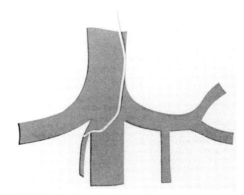

**FIG. 23.4.** Catheterization of the right spermatic vein. The right-sided catheter tip points inferiorly. No other changes are made to convert a left catheter into a right catheter. (Reproduced from Hunter DW, Castañeda-Zúñiga WR, Coleman CC, et al. Spermatic vein embolization with hot contrast medium or detachable balloons. *Semin Intervent Radiol* 1984;1;163–169.)

### Basilic Vein Approach

A higher technical success rate has been achieved with the basilic approach than with the femoral approach. The right basilic vein is usually chosen as an access to the venous system. There are no reported serious complications related to this technique.

## INDICATIONS FOR INTERVENTIONAL THERAPIES

The main indications for intervention are the following: (a) large varicocele interfering with a patient's daily life, (b) symptomatic varicocele, and (c) clinical or subclinical varicocele associated with infertility.

## INTERVENTIONAL THERAPIES

### Transcatheter Embolization with Occluding Devices

Embolization of the spermatic veins can be carried out at the time of venography. The method of occlusion includes stainless steel coils and compressed polyvinyl alcohol sponge (Ivalon). If coils are used, their diameter should be slightly larger than that of the spermatic vein. The position of the coil should be at least 6 cm distal to the junction of the spermatic vein with the renal vein. The introduction of a spider in a proximal portion of the vein reduces the possibility of coil or Ivalon plug migration. A spider combined with an Ivalon plug can be used for a one-step embolization. Because duplication and collateralization of the spermatic vein system are common, embolic devices are usually placed at the level of the inguinal canal and also as near as possible to the level of the renal vein.

### Embolization with Tissue Adhesive

The tissue adhesive is a rapidly polymerizing agent that hardens immediately on contact with an ionic solution such as blood. The most commonly used agent is *n*-butyl cyanoacrylate (NBCA). NBCA is usually injected via a 3-Fr catheter introduced coaxially through a 5- to 6-Fr catheter. The tip of the 3-Fr catheter must be advanced to the exact site selected for embolization. This site should be distal to the lowermost collateral and preferably superior to any bifurcation. Patients are examined on a remote-controlled tilting table in a 2- to 10-degree anti-Trendelenburg position to stop the circulation in the internal spermatic vein completely. This position excludes accidental reflux into the renal vein and pampiniform plexus.

A tuberculin syringe loaded with 1 mL of 10% glucose is connected to the three-way stopcock of the embolization catheter, which is filled with 5% glucose. The glucose solution slows down the polymerization of the tissue adhesive. With a second syringe, 0.2 to 0.3 mL of Lipiodol (iodized oil used as a contrast medium) is mixed with 0.4 to 0.6 mL of NBCA just before the embolization. The amount of NBCA should never exceed more than one ampule (0.6 mL). Mixing it with the contrast medium not only helps to make the mixture opaque but also slows down the polymerization process. The second syringe loaded with the adhesive is then connected to the stopcock. After the adhesive has been injected into the target vessel, the catheter is flushed with 0.5 mL of 10% glucose by using the first syringe. The coaxial embolization catheter should be immediately pulled back 2 to 3 cm, and the remaining 0.5 mL of glucose is injected to remove all NBCA. Glucose is used as a moistening and rinsing solution. Physiologic saline is absolutely contraindicated because it will result in premature polymerization of the NBCA. Never try to perform more than one embolization through the same inner catheter. As soon as the tissue adhesive comes in contact with blood, it polymerizes to form a permanent occlusion. The inner catheter should be completely withdrawn from the outer one immediately after embolization. Inadequate catheter flushing with a nonionic solution or too rapidly flushing will cause unwanted consequences, such as polymerization of the glue in the catheter and renal or lung embolization resulting from reflux.

### Sclerotherapy

The most popular sclerosing agent is Varicocid 5%, the salt of a fatty acid from cod liver oil, which has been successfully used in Europe but is not available in the United States. Other agents used for sclerotherapy include 80% iothalamate sodium, 3% ethoxylsclerol, and 3% sodium tetradecyl sulfate (Sotradecol). The use of Varicocid mixed with 1 mL of air in one syringe has been recommended because some foam can be formed by shaking the syringe. The foamy air is believed to slow down the reflux of sclerosant to the pampiniform plexus and produces a better thrombotic stimulus by displacement of blood, therefore allowing contact between the subsequent undiluted sclerosant and the vessel wall.

Approximately 3 mL of a sclerosing agent is introduced into the distal third of the spermatic vein, left in place for 5 to 15 minutes, and then aspirated. The patient is instructed to perform a Valsalva maneuver during injection of the sclerosant to facilitate its passage down the spermatic vein. Venography is repeated 15 to 30 minutes after the procedure, and if necessary, sclerosing is repeated. Patients should be kept under observation in a supine position for 2 hours after the procedure. The advantages of sclerotherapy include ease of delivery, minimal patient discomfort, and a higher success rate. The disadvantages are possible reflux of agent into the renal vein and testicular thrombophlebitis if the agent fills the pampiniform plexus.

### Detachable Balloon

Selective occlusion of the spermatic vein can also be achieved with detachable balloons. Depending on the size of the spermatic vein, 4- to 6-mm detachable balloons are passed through a catheter placed in the spermatic vein. The balloons are inflated with iohexol 140, which has lower osmolality, and a test injection is made through the introducing catheter to determine if the balloons are optimally placed. The advantages of detachable balloons are the same as with other mechanical occluding agents, primarily selective occlusion of the spermatic vein related to collaterals. The disadvantages include their high cost and the possible migration of a balloon to the pulmonary circulation.

### Thermal Vessel Occlusion

Hot contrast medium has been used for embolization of the spermatic veins. With this technique, patients are usually sedated with intravenous injection of Versed (dose range of 2 to 10 mg given in 1-mg increments up to an average dose of 4 mg per treated vein) and fentanyl citrate (dose range, 50 to 450 µg per treated vein). Locally, intravenous injection of 3 to 4 mL of 4% lidocaine can be supplemented with sedation to achieve greater anesthesia. A steam-reshaped, 5- to 6-Fr Headhunter catheter with one to two sideholes in the distal 1 cm is used for catheterization of the left renal and spermatic veins. Before the sclerotherapy, room-temperature contrast medium should be used to demonstrate no reflux into the pampiniform plexus while the vein is compressed firmly against the superior pubic ramus with a specially developed radiolucent compressor. Reflux into the pampiniform plexus may cause scrotal edema and possible testicular atrophy. To achieve maximal therapeutic effects, blood in the treated vein is aspirated with the catheter for 5 to 10 seconds. During sclerotherapy, the tip of the catheter is positioned at the level of the middle to upper part of the sacroiliac joint. Then, with a standard 12-mL syringe, 8 to 12 mL of boiling, high-osmolar contrast medium is vigorously injected into the emptied vein. No contrast medium should escape below the compressor. At least three such injections are needed, and the total dose should be around 24 to 30 mL. The total amount of hot contrast medium injected depends on several factors, including the size of the vein, number of intercommunicating veins, and the point at which stasis occurred in the vein after the release of inguinal compression. Two minutes are permitted to elapse between injections, allowing time for the patient to recover from the discomfort of the procedure and become sedated again. The advantages of hot contrast medium over other agents include ease of introduction, precise visualization, ability to obliterate parallel communicating veins and collaterals, lack of long-term toxicity, wide availability, and low cost.

The major disadvantages are severe pain during injection, which is blunted by good conscious sedation, and possible scrotal discomfort or swelling. This complication can be avoided by adequate compression and careful fluoroscopic monitoring.

# SUGGESTED READINGS

### A. Fallopian Tube Catheterization for Treatment of Female Infertility

Confino E, Friberg J, Gleicher N. Preliminary experience with transcervical balloon tuboplasty. *Am J Obstet Gynecol* 1988;159:370–375.

Pittaway De, Winfield AC, Maxson W, et al. Prevention of acute pelvic inflammatory disease after hysterosalpingography: efficacy of doxycycline prophylaxis. *Am J Obstet Gynecol* 1983;147:623.

Serafini P, Batzofin J. Diagnosis of female infertility, a comprehensive approach. *J Reprod Med* 1989;34:29–40.

Society for Assisted Reproductive Technology of the American Fertility Society. Assisted reproductive technology in the United States and Canada: 1991 results from the Society of Assisted Reproductive Technology generated from the American Fertility Society Registry. *Fertil Steril* 1993;59:956–962.

Sulak PJ, Letterie GS, Coddington CC, et al. Histology of proximal tubal occlusion. *Fertil Steril* 1987;48:437–440.

Thurmond AS, Rosch J. Nonsurgical fallopian tube recanalization for treatment of infertility. *Radiology* 1990;174:371–374.

Winfield AC, Wentz AC. *Diagnostic imaging of infertility.* Baltimore: Williams & Wilkins, 1987:105–125.

Yussman MA, Vermesh M. Guideline for tubal disease. *Am Fertil Soc* 1993;1:1–8.

### B. Transcatheter Embolization of the Internal Spermatic Vein for Treatment of Varicocele

Ahlberg NE, Bartley O, Chidekel N. Right and left gonadal veins: an anatomic and statistical study. *Acta Radiol [Diagn] (Stockh)* 1966;4:593.

Coolsaet B, The varicocele syndrome: venography determining the optimal level for surgical management. *J Urol* 1980;124:833.

Dubin L, Amelar RD. Varicocelectomy: 968 cases in a twelve year study. *Urology* 1977;10:446–449.

Gonzalez R, Narayan P, Castañeda-Zúñiga WR, Amplatz. Transvenous embolization of the internal spermatic veins for the treatment of varicocele scrotic. *Urol Clin North Am* 1982;9:177–184.

Kim SH, Park JH, Han MC, Paick JS. Embolization of the internal spermatic vein in varicocele: significance of venous pressure. *Cardiovasc Intervent Radiol* 1992;15:102–107.

Kuroiwa T, Hasuo K, Yasumori K, et al. Transcatheter embolization of testicular vein for varicocele testis. *Acta Radiol* 1991;32:311–314.

Pollak JS, Egglin TK, Rosenblatt MM, et al. Clinical results of transvenous systemic embolotherapy with a neuroradiologic detachable balloon. *Radiology* 1994;191:482.

Sigmund S, Bahren W, Gall H. Idiopathic varicocele: feasibility of percutaneous sclerotherapy. *Radiology* 1987;164:161–168.

Takihara A, Sakatoku J, Cockett ATK. The pathophysiology of varicocele in male infertility. *Fertil Steril* 1991; 55:861–868.

Walsh PC, White RI. Balloon occlusion of the internal spermatic vein for the treatment of varicocele. *JAMA* 1981; 246:1701–1702.

White RI, Kaufman SL, Barth KH, et al. Occlusion of varicoceles with detachable balloon. *Radiology* 1994; 191: 482.

# 24

# Lacrimal Duct Intervention

## Ho-Young Song

Epiphora, the imperfect drainage of tears through the lacrimal passages so that they fall over the lid margin onto the cheek, is a common ophthalmic problem, comprising 3% of clinic visits. Although there are many causes of lacrimal outflow obstruction, most cases are due to idiopathic inflammation and scarring of the nasolacrimal system. It is always a source of annoyance to the patient and is frequently accompanied by pain and discomfort in addition to the embarrassment of constant tearing. Although probing is effective for treatment of congenital stenosis of the lacrimal system, it is usually ineffective after the age of 2 years. Silicone intubation is disagreeable because the tubes must be left in place for long periods, which may cause complications such as recurrent dacryocystitis, granuloma formation, and canaliculi erosion. External dacryocystorhinostomy—that is, edge-to-edge anastomosis of the lacrimal sac mucosa to nasal mucosa over the margins of a hole made through the lacrimal bone—has remained the accepted method of correction, yielding success rates of 89% to 95% in primary repairs. Although they are uncommon, complications of dacryocystorhinostomy include complete or partial closure of the fistulous tracts, hypertropic facial scar, and regrowth of mucous membrane over the nasolacrimal opening.

Two kinds of interventional procedures—balloon dacryocystoplasty and placement of an expandable metallic stent or plastic stent—have been advocated for the treatment of epiphora to overcome some of these problems of dacryocystorhinostomy.

## BALLOON DACRYOCYSTOPLASTY AND LACRIMAL STENT PLACEMENT

### Indications

Patients with epiphora due to obstruction of the lacrimal system are candidates for the procedure.

### Contraindications

Acute inflammation is the only contraindication.

### Techniques

#### Balloon Dacryocystoplasty

The site and severity of the obstruction are evaluated before balloon dilatation by means of dacryocystography (Fig. 24.1A). For local anesthesia and decongestion of the nasal mucosa, a nasal pack is placed in the inferior nasal meatus using two to three cotton pledgets moistened with equal parts of cocaine hydrochloride (10%) and epinephrine (1 to 100,000) for 3 to 5 minutes. After removal of the cotton pledgets, the upper two-thirds of the face is cleansed with boric acid solution. Topical anesthesia of the eyes is accomplished with 0.5% proparacaine, and an infratrochlear nerve block over the medial canthal and lacrimal sac areas is accomplished with 2% lidocaine.

The superior punctum is dilated with use of a punctal dilator when necessary. A 0.018-in. ball-tipped guidewire (Cook, Queensland, Australia) is introduced through the superior punctum into the canaliculus (Fig. 24.2A through D). As the ampulla portions of the lacrimal systems are vertically oriented for approximately 2 mm, the

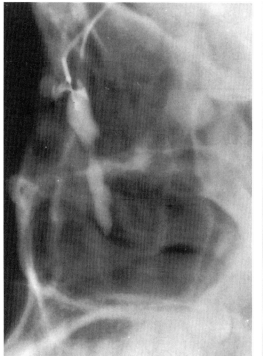 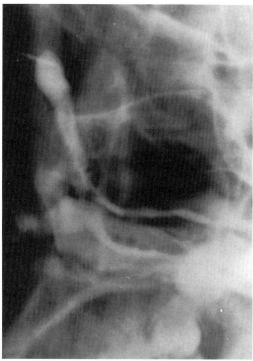

**FIG. 24.1. A:** Lateral predilatation dacryocystogram shows complete occlusion at the lower level of the nasolacrimal duct. **B:** Lateral postdilatation dacryocystogram demonstrates resolution of the obstruction. (Reprinted from Castañeda-Zúñiga WR. *Interventional radiology*, 3rd ed. Baltimore: Williams & Wilkins, 1997, with permission.)

guidewire is first introduced vertically into the punctum and then rotated horizontally in the same plane 90 degrees, conforming to the bend in the first portion of the canaliculus. With lateral tension on the lid to prevent kinking of the canaliculus, the guidewire is then advanced until it hits bony firmness, which means that it has reached the nasal wall of the lacrimal sac. The guidewire is then slightly withdrawn and rotated upward 90 degrees vertically to point the tip of the guidewire caudad and then angulated to point 15 to 20 degrees posteriorly. Under fluoroscopic guidance, the guidewire is advanced gently across the obstruction into the inferior meatus of the nasal cavity. At the obstruction, the guidewire meets some resistance.

A hook (Cook) under fluoroscopic guidance is entered into the nasal cavity inferiorly and laterally toward the inferior meatus to grasp the guidewire. When the tip of the hook touches the ball-tipped guidewire, a hook on the tip of the guidewire is felt and soft metallic contact can also be heard. After grasping the guidewire with the hook or hemostat, the guidewire is pulled out of the external naris and then the hook is removed from the guidewire by cutting the ball with wire cutting scissors (Storz, St. Louis). A deflated angioplasty balloon catheter of 2 to 4 mm in diameter (Balt, Rue Croix Vigneron, Montmorency, France), 2- or 3-mm balloon catheter for canalicular obstruction, or 3- or 4-mm balloon catheter for the lacrimal sac or nasolacrimal duct obstruction is passed retrograde over the guidewire, through the inferior meatus until it straddles the obstruction under fluoroscopic control.

In balloon dilatation of the canalicular obstruction, a 3-Fr tapered dilator (Cook) is passed retrograde over the guidewire through the superior punctum and then removed before passing the balloon catheter.

Dilatation is performed by inflating the balloon with water-soluble contrast media for a period of 30 seconds to 2 minutes with use of an appropriate syringe and a one-way stopcock. The inflations can be repeated twice. The balloon catheter is removed inferiorly and the guidewire superiorly.

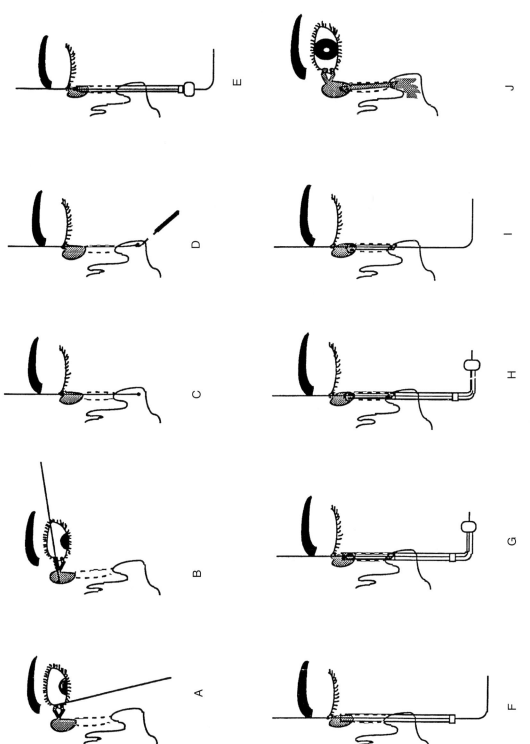

**FIG. 24.2.** Technical steps in the placement of a polyurethane nasolacrimal duct stent. (Reprinted from Castañeda-Zúñiga WR. *Interventional radiology*, 3rd ed. Baltimore: Williams & Wilkins, 1997, with permission.)

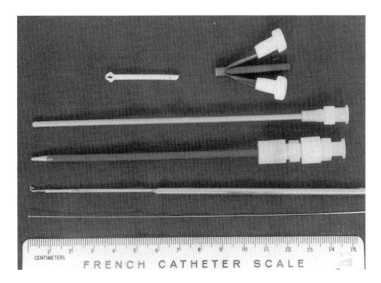

FIG. 24.3. (*Top to bottom*) A polyurethane nasolacrimal duct stent (*left*) and a stent loader (*right*), a pusher catheter, a 6-Fr sheath containing a dilator, a hook, and a ball-tipped guidewire. (Reprinted from Castañeda-Zúñiga WR. *Interventional radiology*, 3rd ed. Baltimore: Williams & Wilkins, 1997, with permission.)

Dacryocystography is performed immediately, 7 days, and 6 months after the procedure to assess the patency of the lacrimal system (Fig. 24.1B). Patients are given prophylactic ampicillin orally for 24 hours before the procedure. Ampicillin is continued for 7 days after the procedure, along with topical antibiotic-steroid eyedrops.

### Stent Placement

The techniques for providing local anesthesia of the eyes and nasal mucosa and for introducing the guidewire through the punctum into the inferior meatus of the nasal cavity, as well as the trapping of the ball-tipped guidewire out of the nostril, have been described in the discussion of balloon dacryocystoplasty.

The polyurethane stent [Song Naso-Lacrimal Duct Stent (Cook)] is 35 mm long and has a radioopaque proximal tip of 0.5 mm length. The proximal end of the stent just distal to the radioopaque portion has a mushroom tip (5 mm in diameter and 5 mm long). The distal end of the stent is beveled and has two small round holes (1 mm in diameter) just proximal to the beveled portion to improve tear drainage. The polyurethane stent is implanted with use of a stent set [Song Naso-Lacrimal Duct Stent Set (Cook)]. The set consists of a 6-Fr polyurethane stent, a 6-Fr introducer set (a dilator, a sheath, a stent loader, and a pusher catheter), a ball-tipped guidewire, a hook, and a dacryocystography needle (Fig. 24.3).

Under fluoroscopic guidance, a 6-Fr sheath with a dilator from the nasolacrimal duct stent set is passed retrograde over the guidewire and advanced across the lesion until the proximal tip of the dilator is lying in the dilated sac (Figs. 24.2 and 24.4). To place the tip of the sheath more properly in the lacrimal sac, the sheath is advanced approximately 3 mm into the lacrimal sac while the dilator is withdrawn from the sheath. After the dilator is removed from the sheath, a stent is introduced over the guidewire into the sheath by help of a stent loader and advanced until the tip of the stent is located at the sheath tip, by using a pusher catheter. The pusher catheter is held in place while the sheath is withdrawn. This frees the stent, allowing the mushroom tip of the nylon or polyurethane stents to expand and to lie within the dilated lacrimal sac and the body portion of the stent to lie within the nasolacrimal duct slightly protruding into the inferior meatus of the nasal cavity. After the stent is released from the sheath, the sheath containing a pusher catheter is pulled out inferiorly and the guidewire superiorly.

In placement of the canalicular stents, a 3-Fr tapered dilator (Cook) is passed retrograde over the guidewire through the superior punctum to dilate the canalicular obstruction. After the dilator is removed from the guidewire, the 6-Fr sheath with a dilator from the nasolacrimal duct stent set is passed retrograde over the guidewire and advanced across the lesion until the proxi-

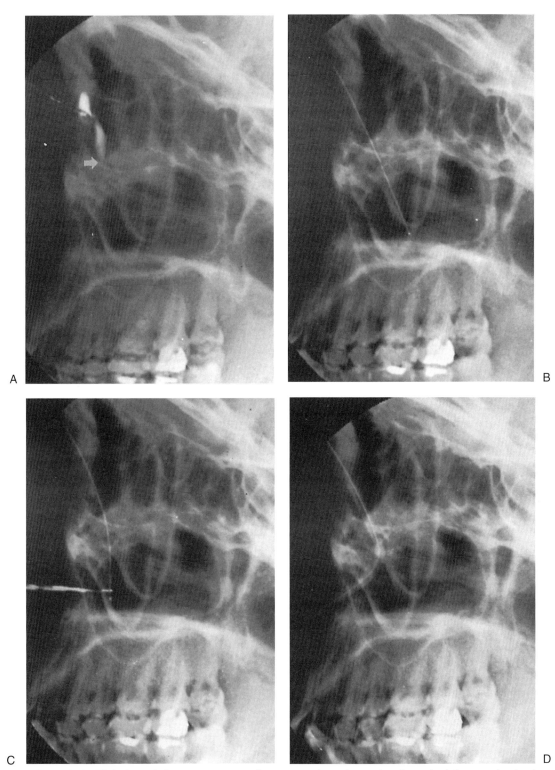

**FIG. 24.4. A:** Right lateral dacryocystogram shows complete obstruction (*arrow*) of the nasolacrimal system at the junction between the lacrimal sac and nasolacrimal duct. **B:** Antegrade placement of a ball-tipped guidewire through the obstruction into the inferior meatus of the nasal cavity. **C:** Grasping the guidewire with a hook. **D:** Retrograde placement of a 6-Fr sheath containing a dilator. (*Continued.*)

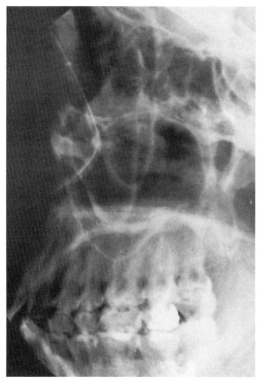 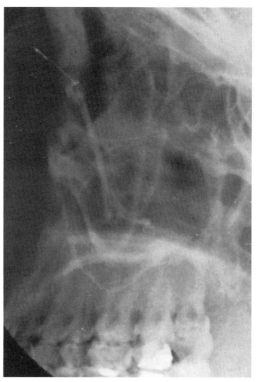

**FIG. 24.4.** *Continued.* **E:** Progressive deployment of the stent from the sheath permits spontaneous expansion of the mushroom tip. **F:** Postprocedure dacryocystogram demonstrates a good passage of contrast medium. (Reprinted from Castañeda-Zúñiga WR. *Interventional radiology*, 3rd ed. Baltimore: Williams & Wilkins, 1997, with permission.)

mal tip of the dilator is lying in the dilated sac. After the dilator is removed from the sheath, a canalicular stent (Cook) (Fig. 24.5) is introduced over the guidewire into the sheath by help of a stent loader. The stent is advanced with a pusher catheter until its tip of the stent is located in the superior canaliculus, 2 to 3 mm from the superior punctum. The next step to finish the canalicular stent placement is the same as the one done in the placement of the nasolacrimal duct stent.

Clinical examinations and dacryocystography is performed immediately, 1 week, and 6 months after the procedure to verify the position and patency of the stent (Fig. 24.6). Patients are given prophylactic ampicillin orally for 24 hours before the procedure. Ampicillin is continued for 7 days after the procedure, along with topical antibiotic-steroid eyedrops.

The radiologist who undertakes nasolacrimal stent placement should have a precise knowledge of the anatomy of the lacrimal drainage system and nose. In our experience, four important things must be borne in mind in placing the stent successfully. First, the ball tip of a guidewire should be properly entered in the lacrimal sac. If the guidewire rotates vertically before its tip enters the sac, it may damage the canaliculus to make a false passage. Second, in some patients with tight obstruction of the nasolacrimal system, it may be difficult to negotiate the tight obstruction with a ball guidewire since the guidewire is relatively flexible to recanalize the obstruction. For this patient, we advise using a Bowman's probe (Storz) or a stylet of 0.035-in. hole punch (Cook), which is stiff enough to open the obstruction. Third, occasionally an introducer set may be difficult to introduce retrograde through the tight obstruction. For this, we advise the use of a balloon catheter to dilate the tight obstruction before placement of the introducer set. Fourth, the hook should be directed to the

FIG. 24.5. A polyurethane canalicular stent. Scale indicates centimeters.

undersurface of the inferior turbinate for the operator to meet, to feel and to hear the sound of the hook on the tip of the guidewire. After successful grasping, the hook should be pulled upward out of the external nare.

### Results

#### *Balloon Dacryocystoplasty*

The technical success rate of 89% to 95% has been reported. The technical failure rate is higher in patients with traumatic obstruction than in those with idiopathic obstruction. No major complications have been reported. The majority of patients express mild pain during the inflation of the balloon and complain of slightly blood tinged nasal discharge for 1 to 72 hours after the procedure. After balloon dilatation, 57% to 86% show clinical improvement in epiphora. In incomplete obstruction of the lacrimal system, the technical success rate and the initial improvement rate of epiphora after balloon dacryocystoplasty are higher than those in complete obstruction of the lacrimal system. As for the obstruction site of the lacrimal system, the results are best in the lacrimal duct and worst in the canaliculus.

#### *Stent Placement*

##### Nasolacrimal Duct Stent

Stent placement technically failed in 13 of the 283 systems (5%). Technical failure rate was 29% (10 of 34 systems) for traumatic causes and 1% (three of 249 systems) for nontraumatic causes. The results of the procedure were usually immediately evident; tearing and discharge stopped. At 7 days after stent placement, 235 of 270 systems (87%) with successful placement demonstrated complete resolution of epiphora, 27 (10%) partial resolution, and the remaining eight (3%) no resolution.

The majority of the patients had slightly blood-stained nasal discharge for 1 to 48 hours after the procedure, which stopped spontaneously. Epistaxis occurred in three patients and was controlled by a nasal pack. Three patients had extravasation of the contrast media from the canaliculus during follow-up dacryocystography. They had a bluish skin discoloration at the site of extravasation, which disappeared spontaneously 5 to 9 days later. Migration of the stent occurred in two patients. The migrated stents were removed and a second stent with a wider mushroom tip was placed in each patient.

During the follow-up of 52 to 134 (mean of 95) weeks, 81 of the 262 improved systems (31%) showed recurrence of epiphora 3 to 83 (mean of 16) weeks after stent placement. The causes of recurrence were complete obstruction of the stent in 77 systems and complete obstruction of the common canaliculus in the other four systems. The recurrence rate in the lacrimal sac obstruction (64%; 28 of 44 improved systems) was much higher than that in the obstruction at the junction (26%; 48 of 185 improved systems) between the lacrimal sac and the nasolacrimal duct or in the nasolacrimal duct obstruction (15%; five of 33 improved systems). Twenty-five of the 77 obstructed stents were recanalized by forceful irrigation with saline through the supe-

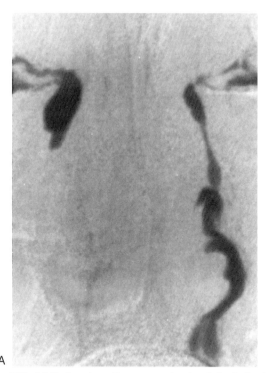 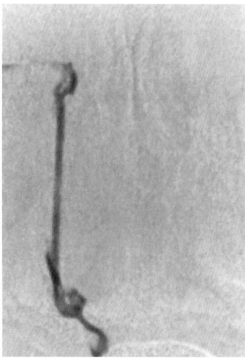

**FIG. 24.6.** A polyurethane nasolacrimal duct stent. **A:** Anteroposterior subtraction dacryocystogram obtained before stent placement demonstrates normal left lacrimal system and complete obstruction of the right lacrimal system at the junction between the lacrimal sac and nasolacrimal duct. **B:** Right dacryocystogram obtained 2 months after placement of a nasolacrimal duct stent shows patent nasolacrimal system. (Reprinted from Castañeda-Zúñiga WR. *Interventional radiology,* 3rd ed. Baltimore: Williams & Wilkins, 1997, with permission.)

rior punctum. The other 52 obstructed stents were removed with use of a hemostat under headlight. In the four systems with recurrence by common canalicular obstruction, the stents were also removed. Therefore, a total of 56 stents were removed. There was a mucoid material in 26 of the 56 removed stents, and granulation tissue in the remaining 30 ones.

Forty-one of the 56 lacrimal systems became patent after removal of the occluded stents. However, during the follow-up of 10 to 81 (mean of 42) weeks after removal of the stent, 24 of the 41 lacrimal systems became obstructed. In the 15 systems obstructed immediately and in the 24 systems obstructed during the follow-up after removal of the occluded stent, 31 underwent placement of a second stent, three underwent dacryocystorhinostomy or conjunctival dacryocystorhinostomy, and the other five who had had grade 1 or 2 epiphora with mucopurulent discharge before stent placement had no further treatment because their low-grade epiphora with improved mucopurulent discharge was not troublesome to them.

*Canalicular Stent*

Stent placement technically failed in one of the 31 eyes. Punctal slitting occurred in one eye at the time of retrograde dilatation of the obstruction with use of a tapered 4-Fr dilator. In this eye, there has been no evidence of closure of the punctum during follow-up. Most patients expressed a foreign body sensation when they blinked or moved the eyes for 1 to 10 days after stent placement, which disappeared spontaneously.

At 7 days after stent placement, all eyes with technical success showed improvement of epiphora with patency of the stent on dacryocysto-

Song HY, Lee CH, Park SS, et al. Lacrimal canaliculus obstruction: nonsurgical treatment with a newly designed polyurethane stent. *Radiology* 1996;199:280–282.

Song HY, Lee CH, Park SS, et al. Lacrimal canaliculus obstruction: safety and effectiveness of balloon dilation. *J Vasc Intervent Radiol* 1996;7:929–934.

Thornton SP. Nasolacrimal duct reconstruction with the nasolacrimal duct prosthesis: an alternative to standard dacryocystorhinostomy. *Ann Ophthalmol* 1977;9:1575–1582.

Traquair HM. Chronic dacryocystitis: its causation and treatment. *Arch Ophthalmol* 1941;26:165–180.

Welham RAN, Henderson P. Failed dacryocystorhinostomy. *Trans Am Acad Ophthalmol Otolaryngol* 1974;78:824–828.

# 25

# Interventional Procedures in the Thorax

Haraldur Bjarnason

## A. BIOPSIES OF THORACIC STRUCTURES

Percutaneous transthoracic needle biopsy was initially reported in 1883 for the diagnosis of infection and in 1886 for the diagnosis of malignancy. It is now a common diagnostic procedure to determine the etiology of focal abnormality in the lung, pleura, mediastinum, hilum, or chest wall. Lesions amenable to percutaneous biopsy may be soft tissue or fluid. Diffuse pulmonary disease and diffuse pleural thickening have also proven amenable to percutaneous biopsy.

### PREOPERATIVE PREPARATION OF THE PATIENT

- Review medical chart and radiographic film record.
- Look up clotting factors (international normalized ratio of more than 1.3 and platelet count of more than 100,000).
- Check pulmonary function tests if available.
- Obtain informed consent.
- Have a pathologist present to verify the sample.

### CONTRAINDICATIONS

There are no absolute contraindications to percutaneous chest biopsy. Relative contraindications include the following:

- Abnormal coagulation (international normalized ratio of more than 1.3)
- Pulmonary hypertension
- Inability to cooperate
- Positive pressure ventilation

Platelet function is decreased in uremia, liver disease, and hematologic neoplasms and during use of acetylsalicylic acid.

Location of a lesion near a large pulmonary vein increases the risk of systemic air embolism (cerebral).

Computed tomography (CT) examination of the chest before biopsy is helpful.

### IMAGING THE LESION FOR BIOPSY

Percutaneous biopsy may be performed with the guidance of fluoroscopy, CT/CT fluoroscopy, or ultrasound. Use of magnetic resonance imaging is still experimental.

#### Fluoroscopy

Fluoroscopy permits rapid identification of the lesion and needle placement. The biopsy procedure may be faster with fluoroscopic guidance than with CT guidance. It is the preferable method of guidance in patients who have difficulty cooperating with the procedure because of their dyspnea, immaturity, or psychiatric disease.

If a lesion is not clearly seen with fluoroscopy, CT should be used.

C-arm or U-arm equipment is desirable to permit anteroposterior (AP), lateral, and oblique views. Cranial or caudal angulation may also assist in identifying and localizing such a lesion. Biplane fluoroscopy may be helpful.

Further confirmation by examination in the orthogonal view is desirable before sampling. This permits the interventionist to move the needle slightly or vibrate it gently under fluoroscopy. Motion of a nodule in concert with the needle confirms that it is the lesion impaled by the needle.

For all except very large lesions, every effort should be made to perform fluoroscopically guided biopsies in the angiography suite.

## Computed Tomography

CT provides excellent anatomic details of thoracic structures. The depth of a pulmonary lesion can be measured at the console (Fig. 25.1). Angulation of the gantry may help in planning access to a lesion. The use of a 20-degree angulation places the beam perpendicular to the anterior intercostal spaces and facilitates negotiation of the thoracic cage. CT fluoroscopy has added a new dimension to the procedure allowing for "life" observation of the needle and "life" observation of the needle passage into the lesion.

By changing the gantry angulation and the patient's position on the table, new access paths may come apparent and extrapulmonary paths to the mediastinum may appear. Iatrogenic pneumothorax or pleural fluid may act as an extrapulmonary path to mediastinal lesion. Injection of saline in the soft tissues close to the recesses pushing the lung away can be helpful in creating extrapulmonary paths (Fig. 25.2). Extrapleural routes to the prevascular, paratracheal, and pretracheal spaces exist from the suprasternal notch and parasternal locations (Fig. 25.3). The extrapleural paraspinous route permits biopsy of posterior mediastinal masses, including subcarinal lymph nodes, without pulmonary injury.

However, CT has limitations. These include the volume-averaging effect that results from voxel depth and the relatively slow progression of the procedure due to delay by image reconstruction. This is overcome with CT fluoroscopy, which now has become commonly available and greatly improves the technique of CT guided procedures.

## Ultrasound

Like fluoroscopy, ultrasound provides real-time imaging. Air intervening between the transducer and the lesion prevents its use in the majority of pulmonary lesions. When a pulmonary process abuts the pleura or has arisen within the pleura, the chest wall, or a portion of the mediastinum contiguous with the chest wall, ultrasound may be the method of choice for localizing the lesion and guiding the biopsy needle. This may be especially true for apical pulmonary and extrapulmonary lesions for which CT and fluoroscopy cannot demonstrate an easy access route. Because ultrasound is portable, biopsy can be performed at the bedside.

## BIOPSY METHODS

After the lesion has been identified and the patient placed in the preferred position, the skin site is prepped and draped in sterile fashion. Lidocaine 1% (Xylocaine 1%) is used for local anesthesia. It is imperative to anesthetize well the vicinities of the periosteum and the parietal pleura, as well as the soft tissues in the planned needle tract. One should wait for a few minutes to allow the anesthesia to work. The parietal pleura should not be traversed with the anesthesia needle.

A skin incision is then made at the skin puncture site.

There are two methods for acquiring tissue at biopsy:

1. Aspiration, which produces a cellular slurry for cytologic examination
2. Core biopsy, which produces a piece of tissue with histologic integrity

### Fine Needle Aspiration

Fine needle aspiration (FNA) may be used for sampling solid or fluid lesions. It causes less pulmonary and pleural trauma than does cutting-needle biopsy. It produces a highly diagnostic sample that may be processed for cytologic analysis or by cell-block technique. FNA can be performed using Chiba (Becton Dickinson, Rutherford, NJ), E-Z-EM (E-Z-EM, Westbury, NY), Greene (Cook, Inc., Bloomington, IN), or Turner (Cook, Inc.) needles ranging in gauge from 21 to 23.

Chiba needles have a 25-degree bevel. By turning the needle to orient the bevel, the needle

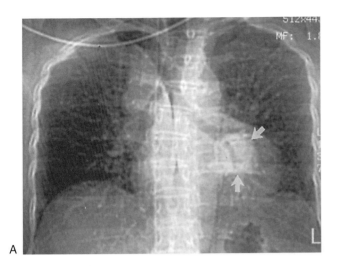
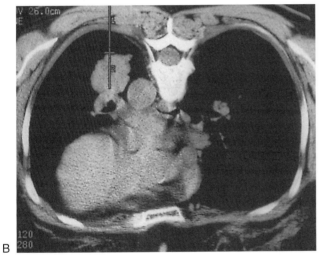
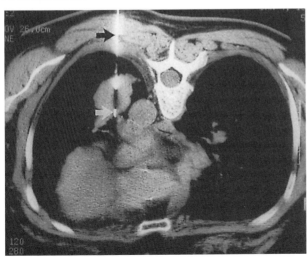

**FIG. 25.1.** Coaxial, computed tomography–guided core needle biopsy. **A:** Scanogram of the chest demonstrates a large mass in the left lower lobe (*arrows*). **B:** The depth to the lesion has been measured (*1*) and the diameter of the lesion has been measured (*2*). **C:** Scan obtained during coaxial needle biopsy shows the 19-gauge guide needle traversing the chest wall and lung (*black arrow*) and the 20-gauge automated cutting biopsy needle (*white arrow*) within the mass. (Reprinted from Castañeda-Zúñiga WR. *Interventional radiology*, 3rd ed. Baltimore: Williams & Wilkins, 1997, with permission.)

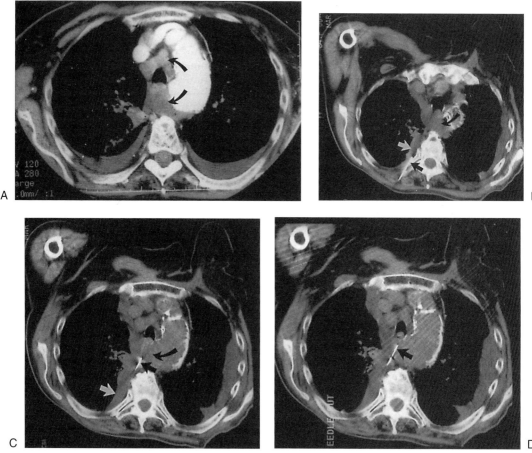

**FIG. 25.2.** Use of an extrapulmonary approach for a biopsy of a mediastinal mass. **A:** Contrast-enhanced computed tomography scan in a 73-year-old woman demonstrates pretracheal and retroesophageal lymph node enlargement (*curved arrows*) and bilateral small pleural effusions. **B,C:** Computed tomography scan with the patient in the left lateral decubitus position shows the right pleural effusions (*white arrow*) creating an extrapulmonary access to the posterior mediastinal nodes (*black curved arrow*). The guiding needle (*straight black arrow*) is seen coursing through the pleural fluid with its tip at the edge of the lesion (*curved arrow*). (Scans rotated for comparison with **A**.) **D:** Scan obtained after coaxial needle placement confirms the biopsy needle tip (*arrow*) within the posterior nodes. (Image rotated for comparison with **A**.) (Reprinted from Castañeda-Zúñiga WR. *Interventional radiology*, 3rd ed. Baltimore: Williams & Wilkins, 1997, with permission.)

can be made to follow a mildly oblique path away from the angled surface.

It is desirable to have a cytopathologist or cytotechnologist attend the procedure if aspirated samples are planned to determine that a diagnostic sample has been obtained before the procedure is terminated.

Coaxial technique is sometimes used. A larger needle (19 gauge or greater) is first placed. Then the stylet of that needle is removed and replaced by the aspiration needle. Greene needles 10 cm in length are most commonly used through the outer needle in coaxial aspirations.

The outer coaxial needle can be positioned in two manners:

1. The tip of the outer needle is placed peripheral to the inner margin of the chest wall. Only the aspiration needle will puncture the pleura.

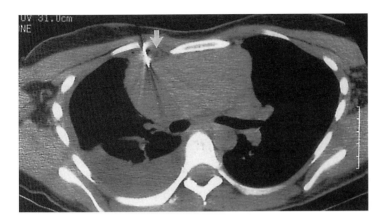

FIG. 25.3. Computed tomography–guided biopsy of anterior mediastinal mass. Computed tomography scan during biopsy of large anterior mediastinal mass. Note needle placement lateral to the right internal mammary vessels (*arrow*) and medial to medial left pleural reflection. (Reprinted from Castañeda-Zúñiga WR. *Interventional radiology*, 3rd ed. Baltimore: Williams & Wilkins, 1997, with permission.)

2. The outer needle may be placed across the parietal and visceral pleura with its tip in the pulmonary lesion or at its margin. This method produces a hole in the visceral pleura equal to the gauge of the large outer needle, but it ensures that access to the lesion will be maintained during consecutive aspirations. If a lesion is located deep in the lung, this access to the lesion by the transpleural outer needle can be particularly valuable because of the difficulty that may occur when directing the slender inner needles over a long distance.

Visceral pleural invagination along the biopsy needle is easily detected by CT but may be overlooked when fluoroscopic guidance is used. In the latter circumstance it may therefore be an unsuspected cause of a nondiagnostic biopsy.

Either of two biopsy maneuvers may be used to withdraw cellular aspirates into the Chiba needle. Both techniques may be used in the same biopsy procedure:

1. The patient is permitted several easy breaths and then again instructed to suspend respiration while the stylet of the outer needle is withdrawn and replaced by the aspiration needle, whose tip is positioned at or slightly beyond the tip of the outer lesion. With the patient again suspending respiration, the stylet is removed and the needle passed in and out of the lesion several times in a gentle fashion. The aspiration needle is removed, still without its inner stylet, and the stylet of the outer needle is replaced, after which the patient is permitted to breathe. This method produces cellular material that, although of a small quantity, has a high diagnostic yield because of the minimal sanguinous staining that occurs with this rather atraumatic technique. Extrusion of the material from the needle onto a microscope slide may require passing the stylet through the needle's lumen.

2. With the alternate maneuver, negative pressure is applied with a Luer-Lok–tipped syringe. With the tip of the aspiration needle in the lesion, the plunger of the syringe is withdrawn and the needle moved around in the lesion or passed in and out of it. Pressure is released before the needle is withdrawn from the chest. This method is more traumatic and causes greater staining of the sample by blood. It produces a larger quantity of material. The presence of blood does not interfere with diagnostic abilities. Material from liquefied lesions can be sent for histology. To sample a solid lesion for microbiologic analysis, add several milliliters of nonbacteriostatic normal saline to the syringe and vigorously aspirate the lesion with negative pressure until sanguinous staining of the saline occurs; the result is a specimen adequate for microbiologic analysis or culture.

### Core Biopsy

Material obtained from core biopsy preserves the histologic integrity of the tissue. The structural relationships of cellular elements and the presence and distribution of noncellular matrix may

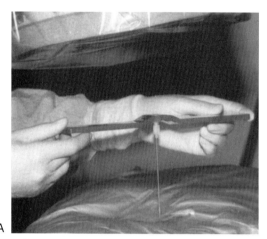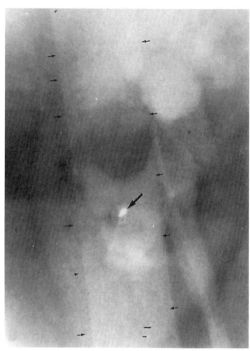

**FIG. 25.4.** "Down-the-barrel" puncture of collecting system. **A:** Anteroposterior plane position of needle held by plastic handle. **B:** Spot film on anteroposterior projection during advancement of needle under continuous fluoroscopy. Needle appears as a dot (*large arrow*) advancing toward infundibulum of lower-pole calix. Plastic handle is faintly seen (*small arrows*). Note ureteral catheter with tip in upper-pole calix. (Reprinted from Castañeda-Zúñiga WR. *Interventional radiology*, 3rd ed. Baltimore: Williams & Wilkins, 1997, with permission.)

provide useful pathologic information and consequently improve the certainty of a diagnosis.

Examples of core biopsies are Tru-Cut needles (Baxter Healthcare, McGaw Park, IL), Sure-Cut needles (Meadox Surgimed, Oakland, NJ), and Franseen needles (Cook, Inc.). These are available in a variety of sizes, with 21 through 19 gauge.

These needles come either with or without a guiding needle. A guiding needle allows for multiple passes through the lesion but only once passing the visceral pleura. The pathologist who is at the site can tell if the acquired sample is adequate, and if not, repeated biopsies can be performed through the guiding needle.

If fluoroscopic guidance is used for the biopsy, a T-bar that is mounted on the needle hub can be used (Fig. 25.4). This consists of a plastic bar that is attached to the hub of the needle. The fluoroscope is then positioned such that the needle is seen end on and projected on the lesion. The operator can hold the bar and keep the hands out of the radiation beam at the same time as he or she can advance the needle in toward the lesion. With intermittent oblique rotations of the fluoroscopic tube the depth of the needle can be judged.

## MARKER NEEDLE FOR THORACOSCOPY

In some patients, removal of intrapulmonary nodules under thoracoscopic guidance may be desirable. Intrapulmonary lesions often do not change the overlying visceral pleura and are not palpable at thoracoscopy. It is therefore often desirable to place a visible marker on the pleural surface to localize the lesion.

When placing a localization wire, it is preferable to pierce the lesion entirely. In some circumstances, however, it may be necessary to place the wire next to the lesion.

Injection of methylene blue at the level of the pleura has been used to indicate an underlying pulmonary lesion either alone or in supplement to wire marker localization. The dye tends to diffuse, however, making the localization imprecise.

## PROCESSING OF BIOPSY SAMPLES

For percutaneous biopsy by FNA or even core biopsies, it is very important to have the pathologist at the scene for verification of adequate material and to ensure proper handling of the specimen.

Samples for microbiology can be obtained by inoculating several milliliters of normal, nonbacteriostatic saline in a syringe via FNA. A sanguineous blush in the saline usually indicates that it has been sufficiently inoculated.

Such material may be displayed on slides and stained for microbiologic analysis. It may also be used for culture.

## DIAGNOSTIC YIELD OF PERCUTANEOUS BIOPSY

FNA for the diagnosis of malignancy was found by Conces et al. in 1987 to have positive predictive value of 98.6% and a negative predictive value of 96.7% in 222 needle aspirations with the cytopathologist present for verification. This included all cell types of bronchogenic carcinoma, Hodgkin's and non-Hodgkin's lymphoma, germ cell tumors, melanoma, carcinoid, and mesothelioma. Definitely benign lesions included granuloma, infarction, hamartoma, fibrosis, and a wide range of infections.

Core biopsy increases the diagnosis of lymphoma over that obtained with FNA (62% versus 12%), as shown by Goralnik et al., and is an important procedure in patients where the cytopathologist is not able to make a definitive diagnosis on FNA.

A high level of accuracy is also achieved in percutaneous biopsy of mediastinal lesions. Sensitivity is as high as 91%, specificity 98%, positive predictive value 83%, and negative predictive value 99%. In 28 patients found to have thymoma, diagnosis at percutaneous aspiration had a sensitivity of 71%, specificity of 94%, positive predictive value of 77%, and negative predictive value of 92%. Tumors that are potentially thymic or lymphomatoid in nature are more easily diagnosed with core tissue samples. However, the mediastinal location of these lesions can make the placement and activation of cutting needles a daunting prospect, and many will choose to first attempt diagnosis with aspiration techniques.

The size of the lesion affects the accuracy of diagnosis. In one study, FNA biopsy was positive in 96% of lesions of more than 1.5 cm as compared with 74% in those more than 1.5 cm in diameter.

## ROUTINE POSTOPERATIVE CARE OF THE BIOPSY PATIENT

- The patients are observed for 1 to 4 hours.
- The patients are on bedrest with the biopsy site dependent if tolerated.
- Vital signs are monitored and routine postoperative care provided.
- Inpatients can be observed on regular patients' ward.
- Oxygen saturation is monitored.
- Vital signs observed every 20 minutes for the follow-up period.
- Immediate and follow-up chest radiographs are obtained at 1 to 2 and 4 hours after the procedure.
- An immediate postprocedure upright chest radiograph allows for comparison of pneumothorax size with later films.
- Increase in the size of pneumothorax would indicate persistent pleural leak.
- Increasing pneumothorax on consequent films or pneumothorax that is symptomatic and occupies more than 30% of the chest volume should be considered for drainage.
- Asymptomatic, stable pneumothorax requires no intervention.

## COMPLICATIONS OF THORACIC BIOPSIES

### Pneumothorax

The most common complication from percutaneous pulmonary biopsy is pneumothorax. Air may be introduced into the pleural space.

- Along the needle tract
- From the lung because the visceral pleura fails to close after the needle removal

Pneumothorax may be diagnosed at the time of biopsy. Once air is present in the pleural space, negative intrathoracic pressure is lost, and further perforation of the visceral pleura may be impossible. At this point it may be necessary to discontinue biopsy attempts even if a diagnosis has not yet been achieved, unless the pneumothorax is evacuated by a percutaneously placed catheter.

Reports of the rate of pneumothorax in patients undergoing percutaneous needle biopsy of pulmonary lesions have varied from 14% to 57%. Factors that influence the rate of pneumothoraces include the following:

- The method used to recognize the presence of pleural air
- The biopsy technique
- The duration of the procedure (influenced in turn by the imaging method used for guidance)
- Patient compliance
- Concomitant diseases such as emphysema
- Depth of lesion (length of needle pass)
- Size of lesion

The rate of pneumothoraces requiring evacuation has been reported as 4% to 13%.

Two clinical studies have demonstrated that positioning the patient with the biopsy site dependent for the recovery period resulted in a decrease in pneumothoraces requiring percutaneous drainage. Of those patients in whom the biopsy site was positioned dependently, 17.9% had detectable pneumothorax, with 0.4% requiring a drainage tube versus 34% and 10% of patients without positional restriction. Only 1.9% of the patients subject to positional restriction developed a delayed pneumothorax compared with 9.1% of those without restrictions.

Sealing the pleural puncture site is advocated by some interventionists. Collagen and coagulated autologous blood have been used. In a series of 50 patients, the rate of pneumothorax was reduced from 28% in the control group to 8% in the treated group. However, the rate of pneumothorax requiring evacuation was unchanged with treatment (8% of patients in each group required evacuation). No ill effects consequent to the collagen plug were observed in the treated patients.

Postbiopsy pneumothorax can be treated on an outpatient basis as described by Gurley et al.

## Pleural Pain

Patients may experience pleural pain after biopsy and is most common in patients with pneumothorax, hemothorax, or pleural hematoma. It can be managed with mild doses of an analgesic such as codeine.

## Bleeding and Hemoptysis

Mild hemoptysis is not uncommon in patients who have undergone biopsy. Westcott reported mild hemoptysis (less than 60 mL of blood) in 7% of 100 patients undergoing hilar or mediastinal biopsies with no instance of major hemoptysis. He also identified mediastinal widening in one patient (1%).

If abrupt hemoptysis occurs during or after the procedure, the patient should be rolled into a decubitus position with the affected side dependent to prevent introduction of blood or clots into the nonbiopsied lung. The episode of bleeding, although dramatic, is usually self-limited. An antitussive such as codeine is useful in these patients.

Bleeding consequent to pulmonary biopsy may be massive, uncontrollable, and fatal. Normal clotting ability does not exclude hemorrhage (Fig. 25.5). Hemopericardium with tamponade has been reported and is a rare complication of thoracic biopsy.

The incidence of hemoptysis may be increased in the following patients:

- In patients who undergo vigorous aspiration or cutting-needle biopsy
- In patients whose pulmonary lesion is located in the perihilar region
- In patients whose lesion is in a nondependent location relative to the draining airways

## Rare Complications of Percutaneous Thoracic Biopsy

During lung biopsy, air may be introduced into the systemic circulation, causing severe air emboli

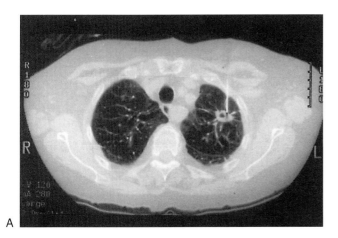

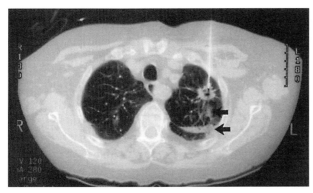

**FIG. 25.5.** Parenchyma bleeding complicating transthoracic needle biopsy. **A:** Computed tomography scan during biopsy of a left upper-lobe cavitary nodule shows the needle tip in the lateral portion of the mass. Cultures yielded *Mycobacterium tuberculosis*. **B:** Computed tomography scan after two needle passes demonstrated air space density in the dependent portion of the lobe (*arrows*) representing pulmonary hemorrhage. (Reprinted from Castañeda-Zúñiga WR. *Interventional radiology,* 3rd ed. Baltimore: Williams & Wilkins, 1997, with permission.)

even with fatal consequences. This results from introduction of air into the pulmonary venous.

The risk of air embolism is increased in the following patients:

- In patients on positive pressure ventilation
- In patients who cough while the biopsy needle is within the lung
- In patients whose lung is abnormal and noncompliant due to underlying disease

The treatment consists of the following:

- Administration of 100% oxygen to promote nitrogen exchange
- Keeping the patient in the supine, prone, or Trendelenburg position, which will avoid further embolization of air retained in the cardiac chambers or large vessels

Seeding of the needle tract with tumor has been reported, although it does not appear to affect the 5-year survival rates in patients with bronchogenic carcinoma.

Acute hemopericardium due to aortopericardial fistula after biopsy of a lesion adjacent to or within the mediastinum has been reported, as has right ventricular perforation.

# B. PERCUTANEOUS DRAINAGE OF PLEURAL COLLECTIONS AND PNEUMOTHORAX

## Timothy V. Myers

Nonsurgical drainage of pleural collections is a widespread and well-tolerated procedure. The most common settings are treatment of pneumothorax and diagnosis and/or treatment of pleu-

**TABLE 25.1.** *Transudate versus exudate—chemical and cytologic determinants*

| Component | Transudate | Exudate |
|---|---|---|
| Protein | <3.0 g/dL | >3.0 g/dL |
| Pleural fluid/serum protein ratio | <0.5 | >0.5 |
| Lactate dehydrogenase | <250 mg/dL | >250 mg/dL |
| Pleural fluid/serum lactate dehydrogenase ratio | <0.6 | >0.6 |
| Glucose | Approximately the same as blood | <40 mg/dL |
| pH[a] | >7.3 | <7.1 |
| Red blood cell count | <10,000 µL[b] | >100,000 µL[c] |
| White blood cell count | <1,000 µL | >1,000 µL[d] |

[a]pH values between 7.1 and 7.3 should be considered indeterminate, and repeat thoracentesis should be considered *when indicated.*
[b]If traumatic, tap can be excluded.
[c]Consider pulmonary embolus, malignancy, trauma.
[d]Should be >50% polymorphonuclear white blood cells for pyogenic/acute inflammatory process.

ral fluid collections. Most commonly, clinicians perform drainage at the bedside without the aid of imaging guidance. Imaging guidance, initially reserved for the more difficult cases, is now being used to assist with an ever-increasing variety of these procedures. In this part, discussions regarding the different types of pleural collections, imaging modalities available, drainage procedures, and patient management are presented.

## DEFINITIONS

Under normal circumstances there are 5 to 15 mL of fluid present within the pleural space. This fluid is produced by the parietal surface of the pleura as a transudate and reabsorbed by the visceral surface. Under the influence of a variety of disease states, this balance can be upset and fluid can accumulate in the pleural space. These abnormal collections are most usually described by their chemical and cytologic characteristics (Table 25.1). In general, effusions that contain less than 3 g per dL of protein are considered to be transudative. Transudates occur as a result of processes that do not involve the pleural surfaces directly (Table 25.2). Effusions that contain more than 3 g per dL of protein are considered to be exudative. Exudates are secondary to processes that involve the pleural surfaces or lymphatics (Table 25.3). Other terms used include *parapneumonic* (when associated with a parenchymal infection), *empyema* (when gross purulence is present), and *malignant* (when malignant cells are obtained from the fluid).

Although defining a fluid collection using these terms is helpful clinically in evaluating for an etiology, it is not necessarily helpful in determining which fluid collections will need to be drained and which can be treated conservatively. Criteria suggested for drainage of effusions include evidence of purulence with or without a positive Gram's stain or culture, a low glucose level, a low pH, a high lactate dehydrogenase level, the presence of loculations, and the presence of grossly bloody effusions, and in some cases of symptomatic malignant effusions. Inadequate or delayed drainage of effusions with these risk factors can result in a significant increase in morbidity and mortality.

**TABLE 25.2.** *Common causes of transudates*

1. Due to an increase in hydrostatic pressure
   - Congestive heart failure
   - Constrictive pericarditis
   - Pancreatitis[a]
   - Meig's syndrome
   - Postcardiotomy syndrome
   - Mesothelioma[b]
   - Collagen vascular diseases[b]
2. Decreased colloid-oncotic pressure
   - Cirrhosis/liver failure with ascites (usually right>left when effusion is present)
   - Nephrotic syndrome
   - Hypothyroidism
3. Chylothorax[b]

[a]Check amylase, effusion usually left more often than right.
[b]Fluid analysis can be transudative or exudative.

**TABLE 25.3.** *Common causes of exudates/empyema*

1. Infectious
   - Pyogenic: aerobic and anaerobic
   - Mycobacterial
   - Fungal
   - Viral
   - Mycoplasmal
   - Parasitic (amoeba, echinococcus)
2. Carcinoma
   - Bronchogenic
   - Mesothelioma
   - Lymphoma (may be chylous)
   - Metastatic
3. Pulmonary embolism
4. Trauma
5. Miscellaneous
   - Rupture of the esophagus
   - Aortic rupture
   - Subphrenic abscess
   - Chylous
   - Pseudochylous: cholesterol crystals (rare, seen with tuberculosis and rheumatoid associated lung disease)
   - Collagen vascular diseases
   - Ascites from any cause

Thoracoscopy, limited or formal open thoracotomy, and/or placement of large-bore chest tubes may be needed to treat more complicated effusions. Other adjunctive measures sometimes needed include placement of a second or more tubes in cases of loculations or separate fluid collections, the use of lytic therapy with urokinase or streptokinase, and the use of sclerotherapy.

Pneumothorax results when air enters the pleural space. Pneumothoraces are either loculated or free within the pleural space. Pneumothoraces can be further divided into three categories depending on their etiology: traumatic, iatrogenic, and spontaneous. A traumatic pneumothorax occurs as a result of penetrating or blunt chest trauma. Iatrogenic pneumothoraces are seen as a complication of thoracentesis, central venous catheterization, or transthoracic lung biopsy; after bronchoscopic procedures or biopsies; and after some intraabdominal surgical procedures, including laparoscopy. Spontaneous pneumothoraces are most often related to the rupture of intrapleural blebs, underlying cystic or interstitial lung disease. Other causes of spontaneous pneumothorax include cocaine and "crack" cocaine use and during chemotherapy for some tumors.

## RADIOLOGIC EVALUATION

Patients with known or suspected pulmonary disease will most commonly come to the attention of the radiologist after plain film studies of the chest. To perform a thorough evaluation of the chest, posteroanterior (PA), lateral, and decubitus views should be obtained. Typically PA and lateral studies will require as much as 100 to 200 mL or more of fluid to be present before detection. Lateral decubitus views can demonstrate as little as 15 mL of fluid. The combination of these studies will help to evaluate the size and amount of free fluid and the size and distribution of possible loculations or separate fluid collections. Pleural fluid collections, however, can assume a variety of appearances on plain film studies and a thorough knowledge of these variations is needed for accurate diagnosis. Pneumothoraces will require the use of expiratory AP or PA films and occasionally lateral and/or decubitus views for a complete evaluation.

Ultrasound is frequently used as an adjunct in evaluating the presence and amount of pleural fluid present. Ultrasound is excellent in locating free and loculated fluid collections, determining their size and extent, and searching for septations. It is relatively inexpensive and easy to use, both in the radiology department and at the bedside. Direct ultrasound-guided access into collections is the safest, most effective use for ultrasound in treatment of fluid collections. Some physicians prefer that a marker be placed on the chest overlying the area of the most prominent pocket/collection of fluid before drainage. In this second instance, careful attention is needed to note the depth of the collection and the position of the patient. The patient must be maintained in the position in which he or she was scanned for best accuracy, particularly in the presence of free fluid or if fluid collections are near the diaphragm or other vital structures.

CT is usually reserved for more complicated fluid collections or when there is a combination of pleural and parenchymal disease. CT can be of help in differentiating transudates from exudates and empyema versus parenchymal abscess. CT is the best method for obtaining a general evaluation of the lungs and anatomic relationships and defining the extent of the pleural fluid collection and loculations. CT can also be used as a tool for guiding drainage, particularly in the presence of loculations and when loculations/collections are adjacent to mediastinal structures or the diaphragm.

Fluoroscopy is used infrequently for primary evaluation or treatment of pleural collections. It can be helpful, however, when repositioning previously placed tubes or when placing a tube for drainage of a pneumothorax.

## TECHNIQUE

### Imaging Guidance

The choice of imaging guidance for drainage of pleural collections should be determined by the amount, location, and type of fluid/air to be drained. It should be kept in mind that in most instances bedside placement of drainage tubes without imaging guidance will be done for drainage of large and/or simple effusions and pneumothoraces. Imaging guidance is used more commonly in the special circumstances of complicated pneumothorax, empyema, and malignant, hemorrhagic, and postoperative effusions, particularly when multiple loculations are present. Therefore, CT and ultrasound are used most commonly. When ultrasound is used, fluoroscopy can be of help with control and verification of tube placement.

### Drainage Catheters

Traditionally, most complicated pleural collections have been treated with large-bore (28- to 32-Fr) chest tubes or with open drainage via thoracotomy or more recently with thoracoscopy. Most collections, however, will respond to the much smaller (8- to 14-Fr) catheters, usually placed with imaging guidance. Catheters used for drainage should have a single lumen and sideholes large enough to adequately drain the material that is present. Locking and nonlocking pigtail catheters can be obtained from a wide variety of sources in the sizes recommended above.

### Procedure for Fluid Collection

The insertion site should be chosen with a clear anatomic window to the fluid collection. The depth of the effusion from the skin surface and the distance to the center of the fluid collection should be noted. The site should be chosen so the course of the catheter will be over the top of the rib, which will prevent damage to the neurovascular bundle that is present along the undersurface of the rib.

A diagnostic thoracentesis should be performed as the initial step for drainage of fluid collections. It can be done as part of the procedure if a modified Seldinger (needle and guidewire) technique is to be used or before insertion of a trocar-and-catheter combination. With the patient positioned for the drainage and the collection located, the skin over a wide area is then sterilely prepared and draped. At the insertion site the skin, subcutaneous tissue, and chest wall to the pleural surface should be anesthetized with local anesthetic. A needle of appropriate size (usually 18 or 19 gauge) is advanced with constant suction until fluid is returned or the needle has passed to the center of the collection as noted on the preliminary study. If fluid is not returned, a reevaluation of the placement should be done. If the needle is clearly within the expected site and no fluid can be returned, the material is unlikely to respond to closed drainage and an open procedure should be considered. If fluid is returned, it should be evaluated with the criteria for drainage discussed previously. Nearly all collections with these attributes should be drained. Uncomplicated fluid collections, usually transudates, can be treated effectively with aspiration without the placement of a drainage catheter. If catheter drainage is to be performed, only a small amount of fluid should be removed for evaluation, as complete drainage through the needle will make placement of a drainage catheter difficult.

After evaluating the fluid and determining that drainage is needed, the method of drainage should be chosen. If formal drainage with a catheter is believed to be indicated, either the modified Seldinger or trocar techniques are used. If a modified Seldinger approach is to be used, a floppy-tipped guidewire [usually a 0.035-in. Bentson (Cook, Inc.)] is inserted through the needle. The tract is dilated in 2-Fr increments up to the diameter of the drainage catheter that will be placed. Overdilatation can lead to pneumothorax, air leak around the tube, or both. The catheter is then inserted over the guidewire into a dependent position and the guidewire is then removed. If a locking catheter has been placed, the pigtail should now be locked.

If a trocar approach is used, the catheter and trocar will be advanced as a unit to the level of the collection. The catheter will then be separated from the trocar. With the trocar held in position, the catheter is advanced into the fluid.

With the catheter in position the fluid collection should now be manually aspirated. If fluid was not sent to the laboratory previously, it should be done now. When the fluid has been completely removed the collection should be reevaluated to ensure there are no undrained pockets or loculations remaining. If loculations are present, additional tubes can be placed. Also, a decision will be made as to whether the catheter will be placed to suction or to dependent drainage. Nearly all collections will require suction with a suction device, such as a Pleur-Evac (Deknatel, Fall River, MA) to prevent pneumothorax. Some chronic empyema loculations, however, can be treated with the catheter to dependent drainage and left in place for sometimes several months while the infection is treated.

### Procedure for Air Collection or Pneumothorax

Pneumothoraces can be roughly divided into two varieties. The typical and most commonly seen are free and associated with varying degrees of lung collapse. The second type is loculated. In most patients with asymptomatic pneumothoraces there will be gradual reexpansion of the lung without the need for catheter placement, and these patients can be safely observed. Indications for drainage of pneumothoraces include progressive enlargement during observation, dyspnea, and severe chest pain. Although some typical pneumothoraces can be treated with a simple aspiration, most are treated by placement of a small-bore drainage catheter. The trocar technique is most commonly used, although the modified Seldinger technique described above can also be used. If a loculated pneumothorax is present, it can be drained in the same way as loculated fluid collections described above using CT or fluoroscopic guidance. In these cases suction will commonly be needed to remove the air and reexpand the lung in the area.

Treatment of nonloculated pneumothoraces can be accomplished through the use of one of several commercially available small-gauge catheters. These include the Cook pneumothorax catheter (Cook, Inc.), the Sacks catheter (Electro-Catheter, Rahway, NJ), the Arrow pneumothorax catheter (Arrow International, Reading, PA), and the Tru-Close Thoracic Vent (UreSil, Skokie, IL). As any of these catheters provide good drainage of pneumothoraces, a choice of one of these units will depend on availability and the individual clinician's familiarity with each of the available products.

To drain a typical pneumothorax, an anterior approach through the second intercostal space in the midclavicular line is used (Fig. 25.6). With

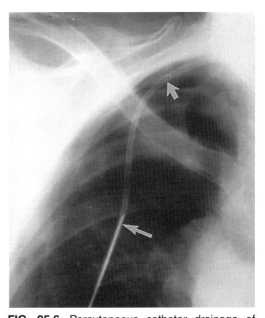

**FIG. 25.6.** Percutaneous catheter drainage of pneumothorax. Spot radiograph obtained during fluoroscopically guided drainage of pneumothorax as a complication of lung biopsy. The catheter (*short arrow*) has been advanced over the inner cannula (*long arrow*) and curves over the apex of the right lung. (Reprinted from Castañeda-Zúñiga WR. *Interventional radiology*, 3rd ed. Baltimore: Williams & Wilkins, 1997, with permission.)

the patient positioned supine, the skin over a wide area is then sterilely prepared and draped. At the insertion site the skin, subcutaneous tissue, and chest wall to the pleural surface should be anesthetized with local anesthetic. A small skin incision is then made over the second rib and the catheter/trocar combination is advanced through the chest wall over the superior edge of the second rib and into the pleural space. The intrapleural position is confirmed when air can be aspirated. The catheter/trocar combination is then advanced approximately 1 cm further. With the trocar held in position, the catheter is advanced toward the lung apex. After the catheter is sutured into position, petroleum gauze is wrapped around the catheter at the insertion site and an occlusive dressing is applied. The catheter is then attached via an adapter to a flutter (Heimlich) valve or to a suction device with water seal such as a Pleur-Evac.

## COMPLICATIONS

If care is taken, complications secondary to imaging guided pleural drainages are uncommon or rare. Intrapleural or extrapleural bleeding due to intercostal vessel injury, damage to the lung, or mediastinal and subdiaphragmatic structures can also occur.

## MANAGEMENT

### Fluid Collections

The patient is assessed daily to evaluate the therapeutic response to drainage. The amount of drainage is monitored. Daily chest radiographs are obtained to assess the size of any residual collections and determine the need for additional manipulations or alternative therapies. If needed to maintain catheter patency, small amounts of saline can be flushed through the catheter as needed or on a scheduled basis.

The duration of catheter drainage is determined by the daily output. The time required for complete drainage ranges from 1 to 45 days or longer; however, most collections will need only 5 to 10 days of treatment. The catheter can be removed when drainage has diminished to less than 10 mL daily, the patient's fever and white blood cell count have diminished, and the pleural collection has resolved. A CT scan is helpful when obtained before catheter removal, as this can better assess the adequacy of drainage and detect any residual collections.

If, after assessment, it is determined that drainage is inadequate or if the catheter is continuously occluded by the fluid being drained, there are several options. First, some catheters may require repositioning to improve drainage. Second, new drainage tubes may need to be placed if it is clear that loculated areas are not being drained. Additionally, manipulation using a diagnostic catheter can be done to break up loculations. Exchanging to a catheter with a larger diameter can be helpful for drainage of thick pus or bloody or formed material. Administration of fibrinolytics may also aid in drainage of loculated collections. For this, more recent reports have centered on the use of urokinase, although streptokinase has also been used (Table 25.4 and Fig. 25.7).

**TABLE 25.4.** *Use of intrapleural urokinase*

| Units administered | Time left in pleural space |
|---|---|
| 80,000–100,000 units in 100 mL of sterile water or saline[a] | 30 min to 12 h |
| 1,000 units/mL in 80–150 mL aliquots; repeat after drainage is complete[b] | 1–2 h |
| 100,000–250,000 units/mL instilled 1–4 installations performed per day until drainage is complete[c] | 1–4 h |

[a]Diethelm L, Klein JS, Xu H. Interventional procedures in the thorax. In: Castañeda-Zúñiga WR, et al., eds. *Interventional radiology*, 3rd ed. Baltimore: Williams & Wilkins, 1997:1785–1816.
[b]Moulton JS, Moore PT, Mencini RA. Treatment of loculated pleural effusions with transcatheter intracavitary urokinase. *AJR Am J Roentgenol* 1989;153:941–945.
[c]Moulton JS, Benkert RE, Weisiger KH, Chambers JA. Treatment of complicated pleural fluid collections with image-guided drainage and intracavitary urokinase. *Chest* 1995;108:1252–1259.

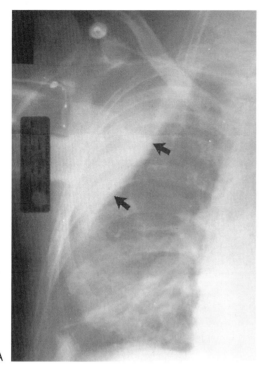 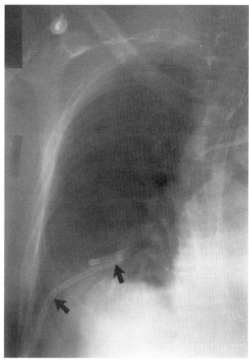

**FIG. 25.7.** Intrapleural fibrinolysis for loculated empyema. **A:** A collection of fluid is present in the right pleural space, with a loculated component present laterally. **B:** After percutaneous placement of a drainage catheter under ultrasound guidance and three daily instillations of urokinase, the fluid has resolved. (Reprinted from Castañeda-Zúñiga WR. *Interventional radiology,* 3rd ed. Baltimore: Williams & Wilkins, 1997, with permission.)

### Air Collections

Most patients with indwelling pneumothorax catheters will be managed as inpatients, although this is not required in some cases. Patients who have complete lung reexpansion and resolution of symptoms after evacuation can be considered candidates for management as outpatients with a Heimlich valve or thoracic vent. Most of these pneumothoraces will be resolved and ready for tube removal within 2 to 3 days; however, to evaluate treatment, these patients should return to the hospital in 24 to 48 hours for a follow-up chest radiograph. If there is no residual pneumothorax the tube should be clamped for 1 hour. If, after clamping of the tube, there is no residual pneumothorax, the tube can be safely removed. After catheter removal another chest radiograph should be obtained to evaluate for any residual or newly accumulated pneumothorax.

All other patients, including those who are debilitated, have severe obstructive lung disease, have incomplete reexpansion, have persistent air leak, or have other complicating factors, are best treated as inpatients. These patients should also be treated with suction and not a Heimlich valve. Daily chest radiographs are obtained to assess the size of any residual air collections. The amount of residual pneumothorax is monitored.

When the lung has completely reexpanded, the tube should be clamped for 1 hour and another chest radiograph should be obtained to exclude air leaks. If there is no evidence of air leak, the catheter can then be removed. If reexpansion does not occur, further clinical evaluation should be done and CT considered. Failure to reexpand the lung can be due to kinking of the catheter, occlusion of the catheter, the catheter "backing out" or being pulled from the pleural space, or air leaks. Patients who do not have complete reexpansion, continued air leaks, bronchopleural fistula, significant loculations with or without fluid collections, or other complicating factors may require thoracoscopic, limited, or open surgical procedures.

## RESULTS

Successful treatment of pleural collections is defined by the completeness of evacuation and resolution of the patient's pulmonary symptoms. Success rates for radiologically guided drainage of fluid collections are typically reported in the 72% to 88% range, which compares favorably with success rates being reported in the surgical literature. The success rate of image-guided drainage of pneumothorax, as reported in the literature, is from 87% to 93%.

## CONCLUSION

The role of interventional radiology in patient care continues to expand. The treatment and diagnosis of thoracic diseases are no exception. Percutaneous radiologic interventional procedures are now performed in place of many surgical and bronchoscopic procedures. It should be remembered, however, that not all patients can or should be treated in this manner, and the indications for bronchoscopy, thoracoscopy, and surgery should be as familiar to the radiologist as they are to the surgeon and pulmonologist. It is also important to note that interventional radiologic treatment of patients does not end with the procedure. All patients treated in the radiology department should be followed by radiologists as well as the appropriate clinicians until the treatment is completed. Also, the interventional radiologist should be encouraged to institute cooperation and consultation with our clinical colleagues to ensure good management and follow up of the patient.

## SUGGESTED READINGS

Alfageme I, Munoz F, Pena N, Unibria S. Empyema of the thorax in adults. *Chest* 1993;103:839–843.

Ashbaugh DG. Empyema thoracis. Factors influencing morbidity and mortality. *Chest* 1991;99:1162–1165.

Austin JH, Cohen MB. Value of having a cytopathologist present during percutaneous fine-needle aspiration biopsy of lung: report of 55 cancer patients and metaanalysis of the literature. *AJR Am J Roentgenol* 1993;160:175–177.

Bressler EL, Kirkham JA. Mediastinal masses: alternative approaches to CT-guided needle biopsy. *Radiology* 1994;191:391–396.

Casola G, vanSonnenberg E, Keightley A, et al. Pneumothorax: radiologic treatment with small catheters. *Radiology* 1988;166:89–91.

Cohen ML, Finch IJ. Transcatheter intrapleural urokinase for loculated pleural effusion. *Chest* 1994;105:1874–1876.

Colquhoun SD, Rosenthal DL, Morton DL. Role of percutaneous fine-needle aspiration biopsy in suspected intrathoracic malignancy. *Ann Thorac Surg* 1991;51:390–393.

Conces DJ, Schwenk R, Doering PR. Thoracic needle biopsy: improved results utilizing the team approach. *Chest* 1987;91:813–922.

Conces DJ, Tarver RD, Gray WC, Pearcy EA. Treatment of pneumothoraces utilizing small-caliber chest tubes. *Chest* 1988;94:55–57.

De Sanctis JT, Mueller PR. Thoracic interventions: the role of ultrasound guidance for diagnostic and therapeutic procedures. *Semin Intervent Radiol* 1997;14:429–448.

Engeler CE, Hunter DW, Castañeda-Zúñiga W, et al. Pneumothorax after lung biopsy: prevention with transpleural placement of compressed collagen foam plugs. *Radiology* 1992;184:787–789.

Ferguson AD, Prescott RJ, Selkon JB, et al. The clinical course and management of thoracic empyema. *QJM* 1996;89:285–289.

Gardner D, vanSonnenberg E, D'Agostino HB. CT-guided transthoracic needle biopsy. *Cardiovasc Intervent Radiol* 1991;14:17–23.

Gazelle GS, Haaga JR. Biopsy needle characteristics. *Cardiovasc Intervent Radiol* 1991;14:13–16.

Goff BA, Mueller PR, Muntz HG, Rice LW. Small chest-tube drainage followed by bleomycin sclerosis for malignant pleural effusions. *Obstet Gynecol* 1993;81:993–996.

Goralnik CH, O'Connell DM, El Yousef SB. CT-guided cutting-needle biopsies of selected chest lesions. *AJR Am J Roentgenol* 1988;151:903–907.

Grinnan NP, Lucena FM, Romero JV. Yield of percutaneous needle lung aspiration in lung abscess. *Chest* 1990;97:69–74.

Gurley MB, Richli WR, Waugh KA. Outpatient management of pneumothorax after fine-needle aspiration: economic advantages for the hospital and patient. *Radiology* 1998;209:718.

Heffner JE, Brown LK, Barbieri C, DeLeo JM. Pleural fluid chemical analysis in parapneumonic effusions. A meta-analysis. *Am J Respir Crit Care Med* 1995;151: 1700–1708.

Henke CA, Leatherman JW. Intrapleurally administered streptokinase in the treatment of acute loculated nonpurulent parapneumonic effusions. *Am Rev Respir Dis* 1992;145:680–684.

Himelman RB, Callen PW. The prognostic value of loculations in parapneumonic pleural effusions. *Chest* 1986; 90:852–856.

Hopper KD, Abendroth CS, Sturtz KW, et al. CT percutaneous biopsy guns: comparison of end-cut and side-notch devices in cadaveric specimens. *AJR Am J Roentgenol* 1995;164:195–199.

Jacobsen FL, Shaffer K, Mentzer S, et al. CT-guided tattoo for lesion localization before thoracoscopy. *Radiology* 1992;185(P):293.

Jereb M, Us-Krasove M. Transthoracic needle biopsy of mediastinal and hilar lesions. *Cancer* 1977;40:1354–1357.

Katada K, Kato R, Anno H, et al. Guidance with real-time CT fluoroscopy: early clinical experience. *Radiology* 1996;200:851.

Kazerooni EA, Lim FT, Mikhail A, Martinez FJ. Risk of pneumothorax in CT-guided transthoracic needle aspiration biopsy of the lung. *Radiology* 1996;198:371.

Komiya T, Kusunoki Y, Kobayashi M, et al. Transcutaneous needle biopsy of the lung. *Acta Radiol* 1997;38:821.

Li H, Boiselle PM, Shepard JO, et al. Diagnostic accuracy and safety of CT-guided percutaneous needle aspiration biopsy of the lung: comparison of small and large pulmonary nodules. *AJR Am J Roentgenol* 1996; 167:105.

Light RW. Parapneumonic effusions and empyema. *Clin Chest Med* 1985;6:55–62.

McCarroll KA, Roszler MH. Lung disorders due to drug abuse. *J Thorac Imaging* 1991;6:30–35.

Mercier C, Page A, Verdant A, et al. Outpatient management of intercostal tube drainage in spontaneous pneumothorax. *Ann Thorac Surg* 1976;22:163–165.

Merriam NM, Cronan IJ, Dorfman GS, et al. Radiologically guided percutaneous catheter drainage of pleural fluid collections. *AJR Am J Roentgenol* 1988;151:1113–1116.

Meyer CA, White CS, Wu J et al. Real-time fluoroscopy: usefulness in thoracic drainage. *AJR Am J Roentgenol* 1998;171:1097.

Milner LB, Ryan K, Gullo J. Fatal intrathoracic hemorrhage after percutaneous aspiration lung biopsy. *AJR Am J Roentgenol* 1979;132:280–281.

Molina PL, Solomon SL, Glazer HS. A one-piece unit for treatment of pneumothorax complicating needle biopsy: evaluation in 10 patients. *AJR Am J Roentgenol* 1990;155:31–33.

Moore E, LeBlanc J, Montesi S. Effects of patient positioning after needle aspiration lung biopsy. *Radiology* 1991; 181:385–387.

Moore EH. Technical aspects of needle aspiration lung biopsy: personal perspective. *Radiology* 1998; 208:303.

Moore EH, Shepard JO, McLoud TC. Positional precautions in needle aspiration lung biopsy. *Radiology* 1990; 175:733–735.

Moulton JS, Benkert RE, Weisiger KH, Chambers JA. Treatment of complicated pleural fluid collections with image-guided drainage and intracavitary urokinase. *Chest* 1995;108:1252–1259.

Moulton JS, Moore PT, Mencini RA. Treatment of loculated pleural effusions with transcatheter intracavitary urokinase. *AJR Am J Roentgenol* 1989;153: 941–945.

O'Brien J, Cohen M, Solit R, et al. Thoracoscopic drainage and decortication as definitive treatment for empyema thoracis following penetrating chest injury. *J Trauma* 1994;36:536–539; discussion 539–540.

Omenaas O, Moerkve O, Thomassen L. Cerebral air embolism after transthoracic aspiration with 0.6-mm (23-gauge) needle. *Eur Respir J* 1989;2:908–910.

O'Moore PV, Mueller PR, Simeone JF, et al. Sonographic guidance in diagnostic and therapeutic interventions in the pleural space. *AJR Am J Roentgenol* 1987;149:1–5.

Parker LA, Charnock GC, Delany DJ. Small bore catheter drainage and sclerotherapy for malignant pleural effusions. *Cancer* 1989;64:1218–1221.

Perlmutt LM, Braun SD, Newman GE. Timing of chest film follow-up after transthoracic needle aspiration. *AJR Am J Roentgenol* 1986;146:1049–1050.

Peters J, Kubitscliek KR. Clinical evaluation of a percutaneous pneumothorax catheter. *Chest* 1984;86: 714–717.

Poe RH, Marin MG, Israel RH, Kallay MC. Utility of pleural fluid analysis in predicting tube thoracostomy/decortication in parapneumonic effusions. *Chest* 1991; 100:963–967.

Pollak JS, Passik CS. Intrapleural urokinase in the treatment of loculated pleural effusions. Intrapleural urokinase in the treatment of loculated pleural effusions. *Chest* 1994;105:868–873.

Pothula V, Krellenstein DJ. Early aggressive surgical management of parapneumonic empyemas. *Chest* 1994; 105:832–836.

Robinson LA, Moulton AL, Fleming WH, et al. Intrapleural fibrinolytic treatment of multiloculated thoracic empyemas. *Ann Thorac Surg* 1994;57:803–813; discussion 813–814.

Salazar AM, Westcott JL. The role of transthoracic needle biopsy for the diagnosis and staging of lung cancer. *Clin Chest Med* 1993;14:99–110.

Seyfer AE, Walsh DS, Graeber GM. Chest wall implantation of lung cancer after thin-needle aspiration biopsy. *Ann Thorac Surg* 1989;48:284–286.

Shepard JO, Mathiesen DJ, Muse VV, et al. Needle localization of peripheral lung nodules for video-assisted thoracoscopic surgery. *Chest* 1994;105: 1559–1563.

Silverman SG, Mueller PR, Saini S. Thoracic empyema: management with image-guided catheter drainage. *Radiology* 1988;169:5–9.

Stein ME, Haim N, Drumea K, et al. Spontaneous pneumothorax complicating chemotherapy for metastatic seminoma. A case report and a review of the literature. *Cancer* 1995;75:2710–2713.

Swerdlow DR, Thaete FL. CT-guided core biopsy of difficult lesions: a modified approach. *AJR Am J Roentgenol* 1994;163:195–196.

Swischuk JL, Castaneda F, Patel JC, et al. Percutaneous transthoracic needle biopsy of the lung: review. *J Vasc Intervent Radiol* 1998;9:347.

vanSonnenberg E, Casola G, Ho M. Difficult thoracic lesions: CT-guided biopsy experience in 150 cases. *Radiology* 1988;167:457–461.

vanSonnenberg E, Wittich GR, Goodacre BW, et al. Percutaneous drainage of thoracic collections. *J Thorac Imaging* 1998;13:74.

Westcott JL. Direct percutaneous needle aspiration of localized pulmonary lesions: results in 422 patients. *Radiology* 1980;137:31–35.

Yu CJ, Yang PC, Chang DB, Luh KT. Diagnostic and therapeutic use of chest sonography: value in critically ill patients. *AJR Am J Roentgenol* 1992;159:695–701.

# 26
# Percutaneous Biopsy of Abdominal Masses

Joseph L. Higgins, Jr., and Janis Gissel Letourneau

Percutaneous biopsy of abdominal masses is a commonly performed radiologic procedure. Indications include initial diagnosis of a solid lesion, staging of known disease, and follow-up in cancer surveillance. Diagnostic aspiration of fluid collections is also commonly performed. This chapter provides a brief overview of percutaneous intraabdominal biopsy procedures, including methods of imaging guidance, patient preparation, biopsy techniques, specimen evaluation, and postprocedure management.

## LOCALIZATION

Accurate localization of the target lesion is critical to the performance of a successful radiologically guided biopsy. The modality of choice for radiologic guidance is, in general, the modality that best visualizes the target lesion.

Ultrasound (US) and computed tomography (CT) are the main modalities for localization and guidance in the abdomen. Fluoroscopic target localization in the abdomen plays a small role and is now limited to intraluminal biopsy in sites such as the biliary tree and major veins, the liver via the hepatic veins, and the urinary collecting system. Magnetic resonance imaging (MRI) localization for percutaneous biopsy of abdominal masses is uncommon, as most intraabdominal lesions can be localized by US or CT.

US is an important mode of guidance for percutaneous biopsy of abdominal lesions. It is fast, generally more readily available than CT, provides real-time monitoring of needle position, and easily allows angulated needle paths to avoid intervening structures.

The simplest method of US guidance is to plan a needle path perpendicular to the skin surface that is also either parallel or perpendicular to the tabletop. This can be accomplished in most instances by placing the US probe perpendicular to the skin along the planned needle path; the patient is then positioned so this path is perpendicular or parallel to the tabletop. Once the needle path has been determined, the next step is to determine the site from which the needle will be observed as it is passed into the target. The positioning of the US probe relative to the needle and lesion can be a limiting step; if a satisfactory observation position is not available, it may require replanning of the needle path so that an observation position is available. At the US frequencies commonly used in percutaneous biopsies (3.5 MHz to 7.0 MHz), the needle acts as a specular reflector; thus, the angle of reflected sound waves from the needle equals the angle of incident sound waves. Therefore, the optimal position for observing the needle is one perpendicular to the needle path. This provides for the greatest needle visibility. Needle visibility is also a function of transducer frequency; higher frequency transducers improve needle visibility. Thus, the highest frequency transducer that will allow adequate visualization of the target lesion should be used during US-guided percutaneous biopsy. Needle visualization is also improved by keeping the entire length of the needle in the scanned plane. This also allows for quicker perception of the needle trajectory. Localization of the needle is often improved by small in-and-out movements of the needle.

Needle guides are available for most US transducer configurations; however, the freehand technique allows greater freedom in planning needle paths and also allows the needle free movement when the patient is breathing.

CT localization is commonly used for percutaneous biopsy of both intraperitoneal and retroperitoneal abdominal masses. CT often provides visualization of lesions that are not clearly defined by any other imaging modality. There are technical limitations in the use of CT localization for biopsy. The main limitation is that the needle tip position cannot be monitored continuously. Also, rarely, it is not possible to verify the exact position of the needle tip, specifically in those situations in which the combination of lesion position, patient size, needle length, and gantry aperture diameter do not allow repositioning of the patient within the CT scanner for repeat studies.

Careful evaluation of the diagnostic examinations and planning of potential needle paths and patient positioning before the patient's arrival will speed up the performance of the procedure. The patient is then placed in the predetermined position that is thought to optimize target lesion localization. For CT-guided biopsy, a transaxial plane for biopsy is selected from diagnostic images and a grid or marker is placed on the patient's skin at that level. A repeat scan with the marker in place allows for precise definition of the puncture site, the depth of the target lesion, and the degree of angulation of the needle tract in that transaxial plane (Fig. 26.1). Once the needle is inserted using an aseptic technique, the tip position is verified by repeat scanning. Alternatively, an angled needle course may be required to avoid intervening structures. This may be accomplished with tilting of the CT scanner gantry. With this technique, the target is localized directly and the needle is inserted in the same plane as the plane of gantry tilt.

## PATIENT PREPARATION

Patient cooperation is a necessity for the procedure. If it cannot be ensured in apprehensive or pediatric patients, premedication may need to be given. Although most patients are candidates for percutaneous biopsy, the procedure may be contraindicated in some patients because of extreme agitation or uncontrollable movement.

A screening history for possible coagulation disorders and recent use of aspirin or other medication that decreases platelet function should precede all procedures. Laboratory testing should be obtained routinely before the performance of percutaneous biopsy and should include a complete blood cell count with platelet quantification (more than 70,000), prothrombin time (less than 15 seconds), and partial thromboplastin time (less than 1.5 times control) determinations. An attempt to correct abnormal bleeding parameters should be made before the initiation of the biopsy procedure.

In patients undergoing US lesion localization, fasting reduces the presence of intestinal bowel gas, which can obscure the target lesion. CT biopsy may require administration of oral contrast depending on the lesion location. The necessity of oral contrast can be determined by evaluation of the diagnostic examination.

Informed consent should be obtained in all patients before the administration of any sedative medications, if they are to be used. Prophylactic antibiotics may be indicated in some patients before the aspiration or biopsy of suspected infected lesions. They are not necessary in all patients, and the decision to use antibiotics should be made in conjunction with the clinicians caring for the patient.

Once a puncture site is determined, the skin is marked and cleaned with an antiseptic preparation. The puncture site is draped, and liberal local anesthesia (lidocaine 1%) is given. The value of patient cooperation is most apparent during needle placement, which should be done in suspended or quiet breathing depending on lesion location.

## BIOPSY TECHNIQUE

Two basic types of biopsy needles are available: aspiration and cutting needles. In general, fine needles—22-gauge—should be used when there is a risk of entering a vital structure, such as a major blood vessel or bowel loop. Passing the needle through the spleen should be avoided because of the risk of laceration and hemorrhage. Larger needles may be selected for percutaneous biopsy when no vital structures intervene along the needle course. Biopsy guns are also available for obtaining specimens for histologic evaluation. Biopsy guns are preferred when needle path and lesion allow their safe use.

The creation of a small skin nick at the puncture site with a scalpel blade decreases resistance

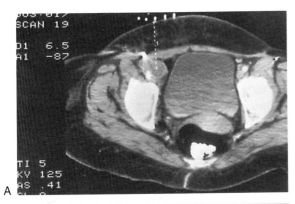
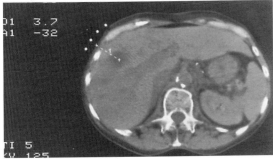
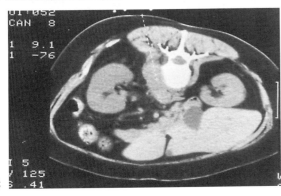

**FIG. 26.1.** Electronic determination of biopsy needle course by computed tomography. **A:** Biopsy tract for these iliac lymph nodes enlarged by metastases of squamous cell carcinoma of the vulva was nearly perfectly perpendicular. Angle of the needle tract (*A1*) was 87 degrees from the horizontal plane; distance of the nodes from the skin surface (*D1*) was 6.5 cm. **B:** A more horizontal needle tract was planned for this hepatic metastasis from squamous cell carcinoma of the cervix. Needle was angled 32 degrees (*A1*) from the horizontal plane; the center of the lesion was 3.7 cm (*D1*) from the skin. **C:** Posterior needle course was determined in another patient with metastases from squamous cell carcinoma. Depth measurement to the middle of the lesion was 9.1 cm (*D1*) at an angle of 76 degrees (*A1*). (Reprinted from Castañeda-Zúñiga WR. *Interventional radiology*, 3rd ed. Baltimore: Williams & Wilkins, 1997, with permission.)

and deviation of the needle path during passage through the skin and subcutaneous tissues. If needle advancement is to take place without continuous visualization, monitoring of the needle angulation by an observer is helpful. With US guidance, the needle tip position can be monitored continuously during placement. As continuous visualization is not possible with CT or MRI guidance, the needle should be advanced to the predetermined depth in a single, smooth movement. Patient breathing is then resumed during repeat scanning to define the needle tip position. Ideal positioning of the needle tip for percutaneous biopsy will depend in part on the radiographic appearance of the lesion. Biopsy should be performed from the periphery of a solid lesion and from the center of a cystic or inflammatory lesion.

The technique for biopsy with an aspiration-type needle, such as the Chiba (Becton Dickinson, Rutherford, NJ) or Turner (Cook, Inc., Bloomington, IN) (Fig. 26.2), is simple. Aspiration and nonaspiration methods have been used. The basic steps of both methods are the same: The needle tip

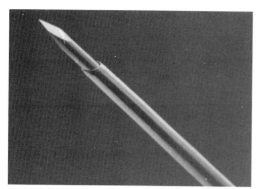

**FIG. 26.2.** The Turner needle features a diamond-tipped stylet and a beveled biopsy needle. (Reprinted from Castañeda-Zúñiga WR. *Interventional radiology,* 3rd ed. Baltimore: Williams & Wilkins, 1997, with permission.)

is advanced to the margin of the lesion with the stylet in place. The stylet is removed once the needle is in position. The specimen is acquired by moving the needle back and forth within the periphery of the lesion while at the same time it is rotated. The methods differ, as their name implies, in whether or not suction is placed on the needle while acquiring the specimen.

Suction can be applied, when performing the aspiration method, by attaching a 10- to 12-mL syringe to the needle and withdrawing the plunger while acquiring the specimen. Connecting the syringe to the needle via polyethylene connecting tubing allows one to control the needle with one hand while maintaining suction with the other.

There are two basic varieties of cutting-type needles. Some cutting needles have a specially designed tip that facilitates the coring of tissue fragments; these include needles such as the Greene (Cook), Sure-Cut (Meadox Surgimed, Oakland, NJ), and Franseen (MEDITECH/Boston Scientific, Natick, MA) needles (Fig. 26.3). Other cutting needles are designed with a biopsy window in the distal portion; these include the fine-gauge Westcott and the larger Lee and Tru-Cut (Baxter Healthcare, McGaw Park, IL) needles.

Biopsy with the former variety of cutting needle is accomplished with a rotary coring motion

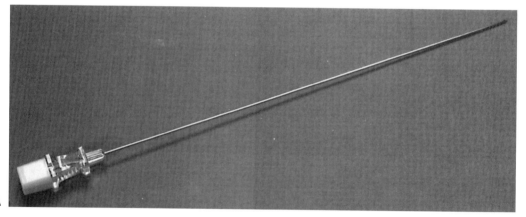

**FIG. 26.3. A:** Franseen needle (Crown biopsy needle) with stylet in place. **B:** Close-up view of needle tip with stylet removed. (Courtesy of MEDITECH/Boston Scientific, Natick, MA.) (Reprinted from Castañeda-Zúñiga WR. *Interventional radiology,* 3rd ed. Baltimore: Williams & Wilkins, 1997, with permission.)

as it is advanced and retracted slightly within the lesion. For all needles except the Sure-Cut variety, the aspiration or nonaspiration method can be used. With biopsy using the Sure-Cut needle, the stylet is partially retracted and suction applied by pulling back on the syringe. A plastic stop holds the syringe barrel in place. Biopsy is accomplished as the needle is rotated and passed back and forth. With this system, suction is usually released before the needle is withdrawn.

Both the Lee and the Tru-Cut needles have a cutting window on the inner needle. With the Tru-Cut needle (Fig. 26.4), the biopsy needle and outer cannula are advanced to the margin of the lesion. Holding the position of the inner needle fixed, the outer cannula is retracted to expose the biopsy window. The needle is then passed into the lesion and the outer cannula is quickly advanced over the window to cut the specimen. The entire assembly is then withdrawn in a locked position.

Modifications of this general technique have been made, primarily in an attempt to optimize lesion localization and to minimize the risks of the biopsy procedure. These include the tandem, coaxial, and modified coaxial techniques. The tandem technique uses a fine needle to guide sub-

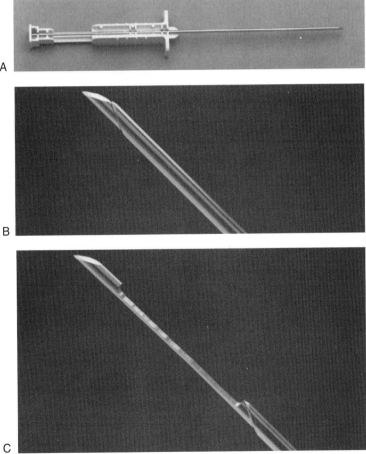

FIG. 26.4. A: Tru-Cut needle in closed position for needle advancement. T-shaped portion of handle apparatus moves back to expose biopsy window. B: Specimen window covered by outer cannula. C: Close-up view of biopsy window exposed by retracting outer cannula. (Reprinted from Castañeda-Zúñiga WR. *Interventional radiology,* 3rd ed. Baltimore: Williams & Wilkins, 1997, with permission.)

sequent passes by the biopsy needle. Once the tip of the guiding fine needle is documented to lie within the target lesion, multiple passes with the biopsy needle can be made along its side. The biopsy needle can be of fine or larger gauge.

The coaxial technique for percutaneous biopsy also facilitates accurate sampling by the biopsy needle. This involves the placement of a thin-walled 18- or 19-gauge needle near the margin of the lesion with biopsy accomplished by passing a longer needle through the lumen of the larger needle. Multiple passes of the fine needle can be made for repeated sampling. Gentle pressure applied to the hub of the larger needle in the horizontal or sagittal plane as the fine needle is advanced will permit sampling of different regions of the target lesion. An aspiration sample can also be obtained through the outer guiding needle as it is withdrawn. The coaxial technique is useful with all guidance modalities. It allows repeat scanning to confirm needle location before placement of the biopsy gun into position.

A modified coaxial technique allows for initial localization with a fine needle rather than an 18- or 19-gauge needle. A fine needle is advanced to the margin of the lesion. The hub is then removed by cutting or unscrewing, allowing the placement of a larger needle over the fine needle. When the tip of the larger needle is fixed in place, the fine localizing needle is withdrawn. The biopsy is obtained using multiple passes of another fine needle through the larger needle in a coaxial fashion.

## LABORATORY EXAMINATION OF THE BIOPSY MATERIAL

The final step of percutaneous biopsy entails the processing of the material obtained. This must be done carefully to maximize the success of the procedure. Laboratory examination of the biopsy material must be determined in part by the requesting physicians. Chemical analyses and bacteriologic studies are of value when fluid is aspirated. Protein, bilirubin, amylase, and lactate dehydrogenase values may be useful in determining the etiology of a fluid collection. Cell counts in the fluid specimen may also be obtained. A vast array of microbiologic studies can be ordered on aspirated fluid. These include special stains, such as Gram's and potassium hydroxide preparations, and aerobic and anaerobic, fungal, mycobacterial, and viral cultures. If anaerobic culture is desired, the specimen must be transported to the laboratory in an airless, capped syringe or in special transport medium.

The presence of the cytologist during the procedure ensures adequate treatment of the specimen. However, this is not always possible, and familiarity with a variety of means of handling the biopsy material is then necessary. The aspirated material is expelled from the syringe onto the frosted glass slides. The aspirated material is spread using the edge of a second glass slide, taking care to distribute the cellular material obtained. The slides are then quickly fixed in 95% ethanol for Papanicolaou staining. This particular method of rapid fixation preserves nuclear detail. Ethanol fixation can be continued for 24 hours to allow for transport of the material if necessary. Alternatively, air drying of the slides can be done. This is followed by Wright-Giemsa or May-Grünwald-Giemsa staining to preserve the tissue pattern and to permit staining of the tissue stroma.

Residual material within the needle and syringe should be expelled using sterile nonbacteriostatic saline or a solution of 50% Ringer's and 50% ethanol. This material can then be processed as a cell block.

When cutting or large needles are used, core biopsy fragments of tissue are obtained for histologic examination. These tissue fragments should be fixed immediately in 10% formalin.

## POSTPROCEDURE MANAGEMENT

Both inpatients and outpatients are candidates for percutaneous biopsy procedures. Careful monitoring of patients in both categories is required for 2 to 4 hours after biopsy, depending on the biopsy site and needle size. Vital signs should be obtained at 15-minute intervals for the first hour, followed by 30-minute intervals the second hour and each hour thereafter in patients who are stable. Consequently, outpatients should be held in a facility within the hospital that permits frequent monitoring of vital signs. An outpatient should then be released in the care of a responsible person. Postprocedure chest radiograph should be obtained to evaluate for pneumothorax if there is the possibility the pleura was transgressed.

## COMPLICATIONS

The complication rate for guided percutaneous biopsy in the abdomen is low, ranging from 1% to 3%.

Complications of directed hepatic biopsy include hematoma, pneumothorax, and possibly pancreatitis. Hemorrhagic complications with fine needles, even of a cutting variety, are uncommon and were reported in only two of a series of 139 patients. Additionally, in one study of liver biopsies, no increase in complications was reported with the use of 18-gauge needles compared with 22-gauge needles.

With pancreatic biopsy, complications include pancreatitis, peritonitis or sepsis, and needle tract and cutaneous seeding. There have been six reported deaths as a result of percutaneous biopsy of the pancreas; five resulted from pancreatitis and one from sepsis. A series of 269 pancreatic biopsies performed under CT and US guidance reported no deaths and an overall major complication rate of approximately 1%; major complications included pancreatitis and pancreatic duct laceration. Mesenteric hematomas have also been reported.

Percutaneous fine needle biopsy of renal masses is complicated by transient hematuria, hemorrhage, or extravasation of either urine or cyst contents in approximately 10% of patients. Serious complications are reported in less than 1% of patients and include hemorrhage necessitating transfusion, persistent hematuria, arteriovenous communications, and pneumothorax. These problems are associated primarily with the use of larger biopsy needles. Complications of adrenal biopsy occur at a rate on the order of 10% and include hemorrhage, infrequently requiring transfusion, and pneumothorax.

Thus, although many different types of complications have been reported with percutaneous biopsy in the abdomen, these complications are infrequent. Serious complications with percutaneous biopsy procedures are very rare. The mortality rate associated with the procedure, particularly with the use of fine needles, is also low. Careful attention to patient selection, preprocedural laboratory evaluation, and lesion localization and needle placement should minimize the occurrence of these complications. Tract embolization techniques may minimize complications in potentially high-risk situations.

## CONCLUSION

Performance of radiologic biopsy of abdominal masses is a common procedure performed on nearly a daily basis in large hospital settings. The reasons for this include its high rate of diagnostic accuracy and the low rate of complications. Percutaneous biopsy is often performed to evaluate a potentially malignant lesion. However, it need not be confined to this clinical setting and can be used to diagnose inflammatory and hemorrhagic processes as well. Accurate lesion localization is the cornerstone of a successful procedure and can be performed with fluoroscopic, US, CT, and, more recently, MRI guidance. Advances in catheter and instrument technology have provided the radiologist with new access routes and approaches to guided biopsy and with new means of minimizing complications in high-risk settings. Nonetheless, successful performance of these procedures requires close cooperation between the patient, radiologist, pathologist, and ordering clinician.

## SUGGESTED READINGS

Bankoff MS, Belkin BA. Technical note: percutaneous fine needle aspiration biopsy using a modified coaxial technique. *Cardiovasc Intervent Radiol* 1989; 12:43–44.

Barbaric ZL, MacIntosh PK. Periureteral thin-needle aspiration biopsy. *Urol Radiol* 1981;2:181–185.

Bernardino ME, Walther MM, Phillips VM. CT-guided adrenal biopsy: accuracy, safety and indications. *AJR Am J Roentgenol* 1985;144:67–69.

Bisceglia M, Matalon TAS, Silver B. The pump maneuver: an atraumatic adjunct to enhance US needle tip localization. *Radiology* 1990;176:867–868.

Brandt KR, Charboneau JW, Stephens, DH, et al. CT- and US-guided biopsy of the pancreas. *Radiology* 1993;187:99–104.

Charboneau JW, Reading CC, Welch TJ. CT and sonographically guided needle biopsy: current techniques and new innovations. *AJR Am J Roentgenol* 1990;154:1–10.

Corr P, Beningfield SJ, Davey N. Transjugular liver biopsy: a review of 200 biopsies. *Clin Radiol* 1992;45:238–239.

Cozens NJA, Murchison JT, Allan PL, Winney RJ. Conventional 15G needle techniques for renal biopsy compared with ultrasound-guided spring-loaded 18G needle biopsy. *Br J Radiol* 1992;65:594–597.

Cropper LD, Gold RE. Simplified brush biopsy of the bile ducts. *Radiology* 1984;148:307–308.

Diaz-Buxo JA, Donadio JV Jr. Complications of percutaneous renal biopsy: an analysis of 1000 consecutive biopsies. *Clin Nephrol* 1975;4:223–227.

Duckweiler G, Lufkin RB, Hanafee WN. MR-directed needle biopsies. *Radiol Clin North Am* 1989;27:255–263.

Evans WK, Ho CS, McLoughlin MJ, Tao LC. Fatal necrotizing pancreatitis following fine-needle aspiration biopsy of the pancreas. *Radiology* 1981;141:61–62.

Ferrucci JT Jr, Wittenberg J. CT biopsy of abdominal tumors: aids for lesion localization. *Radiology* 1978;129:739–744.

Ferrucci JT Jr, Wittenberg J, Mueller PB. Diagnosis of abdominal malignancy by radiologic fine-needle aspiration biopsy. *AJR Am J Roentgenol* 1980;134:323–330.

Hopper KD, Abendroth CS, Sturtz KW, et al. Fine-needle aspiration biopsy for cytopathologic analysis: utility of syringe handles, automated guns, and the nonsuction method. *Radiology* 1992;185:819–824.

Jackson JE, Adam A. Percutaneous transcaval tumor biopsy using a 'road-map' technique. *Clin Radiol* 1991;44:195–196.

Jacobsen GK. Aspiration biopsy cytology. In: Holm HH, Kristensen JK, eds. *Ultrasonically guided puncture technique*. Philadelphia: WB Saunders, 1980.

Kinney TB, Lee MJ, Filomena CA, et al. Fine-needle biopsy: prospective comparison of aspiration versus nonaspiratin techniques in the abdomen. *Radiology* 1993;186:549–552.

Lieberman RP, Hafaz GR, Crummy AB. Histology from aspiration biopsy: Turner needle experience. *AJR Am J Roentgenol* 1982;138:561–564.

Lundquist A. Fine-needle aspiration biopsy of the liver. *Acta Med Scand* 1971;520:1–28.

Martino CR, Haaga JR. Percutaneous biopsy of the liver. *Semin Intervent Radiol* 1985;2:245–253.

McGahan JP. Percutaneous biopsy and drainage procedures in the abdomen using a modified coaxial technique. *Radiology* 1984;153:257–258.

Moulton JS, Moore PT. Coaxial percutaneous biopsy technique with automated biopsy device: value in improving accuracy and negative predictive value. *Radiology* 1993;186:515–522.

Mueller PR, Miketic LM, Simeone JF, et al. Severe acute pancreatitis after percutaneous biopsy of the pancreas. *AJR Am J Roentgenol* 1988;151:493–494.

Pagani JJ. Biopsy of focal hepatic lesions. *Radiology* 1983;147:673–675.

Rashleigh-Belcher HJC, Russell RCG, Lees WR. Cutaneous seeding of pancreatic carcinoma by fine-needle aspiration biopsy. *Br J Radiol* 1986;59:182–183.

Silverman SG, Mueller PR, Pfister RC. Hemostatic evaluation before abdominal interventions: an overview and proposal. *AJR Am J Roentgenol* 1990;154:233–238.

Silverman SG, Mueller PR, Pinkney LP, et al. Predictive value of image-guided adrenal biopsy: analysis of results of 101 biopsies. *Radiology* 1993;187:715–718.

Smith EH. Complications of percutaneous abdominal fine-needle biopsy. *Radiology* 1991;178:253–258.

Soost HJ. Requirements in gaining and treating biopsy material. In: Anacker H, Gullotta U, Rupp N, eds. *Percutaneous biopsy and therapeutic vascular occlusion*. New York: Thieme-Stratton, 1980.

vanSonnenberg E, Lin AS, Deutsch AL, Mattrey RF. Percutaneous biopsy of difficult mediastinal, hilar, and pulmonary lesions by computed tomographic guidance and a modified coaxial technique. *Radiology* 1983;148:300–302.

Welch TJ, Sheedy PF II, Johnson CD, et al. CT-guided biopsy: prospective analysis of 1,000 procedures. *Radiology* 1989;171:493–496.

Welch TJ, Sheedy PF, Stephens DH, et al. Percutaneous adrenal biopsy: review of a 10-year experience. *Radiology* 1994;193:341–344.

Wittenberg J, Mueller PR, Ferrucci JT Jr. Percutaneous core biopsy of abdominal tumors using 22-gauge needle: further observations. *AJR Am J Roentgenol* 1982;139:75–80.

Yankaskas BC, Staab EV, Craven MB, et al. Delayed complications from fine-needle biopsies of solid masses of the abdomen. *Invest Radiol* 1986;21:325–328.

Yueh N, Halvorsen RA, Letourneau JG, Crass JR. Gantry tilt technique for CT-guided biopsy and drainage. *J Comput Assist Tomogr* 1989;13:182–184.

# 27

# Drainage of Abdominal Abscesses and Fluid Collections

Gregory B. Snyder, George C. Scott, and Haraldur Bjarnason

In spite of the use of potent systemic antibiotic therapy, undrained abdominal abscesses carry an 80% mortality rate. Early diagnosis of abdominal fluid collections is important for successful treatment. Image-guided percutaneous drainage of abdominal fluid collections have become the modality of choice for both diagnostic and therapeutic purposes because of the low morbidity and mortality and the high success rates associated with these procedures. Differential diagnosis of the fluid collections includes hematoma, biloma, pancreatic pseudocyst, abscess, lymphocele, seroma, urinoma, and loculated ascites.

## IMAGING MODALITIES

Multiple modalities can be used to detect fluid collections. Ultrasound (US), because it is safe, quick, and relatively inexpensive, is often used for initial screening. US, however, has significant technical limitations (e.g., obesity, bowel gas) and computed tomography (CT) remains the diagnostic method of choice for a thorough evaluation of the abdominal cavity. Magnetic resonance imaging (MRI) and radionuclide imaging can also be used.

### Ultrasound Guidance

US guidance is a powerful technique allowing for direct real-time visualization of the needle. It is multiplanar as opposed to CT, in which one is limited to an axial plan, even though CT fluoroscopy has changed that significantly. Alternatively, US may be used to identify the appropriate location for a blind puncture of large, superficial collections. The typical findings are a sonolucent mass with irregular margins and few internal echoes. Depending on the internal composition, it may also appear as a solid or complex mass (Fig. 27.1). A gas-containing abscess may appear as an echogenic mass with or without acoustic shadowing.

Limitations include abscesses located behind gas-filled bowel loops, under open wounds, or under surgical dressings. A displaced or deformed urinary bladder can be mistaken for an abscess, and care must be taken to repeat scanning either after the patient voids or after the bladder has been distended.

### Computed Tomography and Computed Tomography Fluoroscopy

CT provides excellent anatomic detail, including information about location and involvement of adjacent structures (Fig. 27.2). CT can clearly demonstrate multifocal fluid collections, loculations, and extraluminal air. CT fluoroscopy allows for near real-time imaging in the axial plane, which is proving to be very useful for biopsies of smaller foci or drainages with limited approach windows.

Limitations include difficulty in differentiating from normal structures due to incomplete encapsulation, mottled soft tissue density, or poor oral contrast opacification of adjacent bowel loops.

### Conventional Radiography and Fluoroscopy

Conventional methods (radiographs, upper gastrointestinal examinations, barium enemas, or intravenous urograms) have been used to detect intraabdominal fluid collections. These present as soft tissue masses that may contain mottled

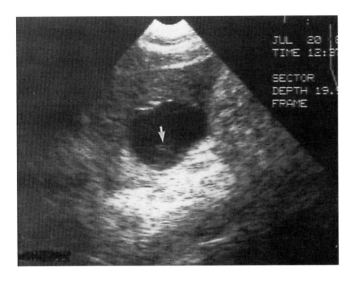

**FIG. 27.1.** Longitudinal sonogram shows subhepatic fluid collection after cholecystectomy; note internal echoes within cystic mass (*arrow*). (Reprinted from Castañeda-Zúñiga WR. *Interventional radiology*, 3rd ed. Baltimore: Williams & Wilkins, 1997, with permission.)

air or air fluid levels. Contrast can be used to look for fistulas and sinus tracts or for anastomotic leaks in postoperative patients.

Fluoroscopy is generally used in conjunction with CT or US for fluid collection and drainage catheter management (i.e., tube exchange/repositioning, fistulogram/sinogram, cavity evaluation).

### Magnetic Resonance Imaging

Although not routinely used at this time, rapid advances may make MRI the modality of choice in the future.

### INDICATIONS FOR DRAINAGE

Indications for drainage include any fluid collection that cannot be reached or treated by simple incision and drainage. This can be divided into two categories:

1. Diagnostic sampling, requiring aspiration of a small quantity of fluid sufficient for laboratory analysis
2. Therapeutic management, removal of as much fluid as practical for therapeutic or palliative purposes, often requiring placement of a drainage catheter for days or weeks

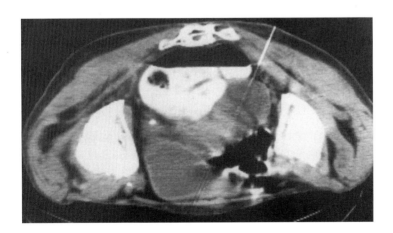

**FIG. 27.2.** Computed tomography scan with patient in prone decubitus position; pelvic abscess has been entered with 22-gauge needle for aspiration. (Reprinted from Castañeda-Zúñiga WR. *Interventional radiology*, 3rd ed. Baltimore: Williams & Wilkins, 1997, with permission.)

## CONTRAINDICATIONS FOR DRAINAGE

- Lack of safe access route
- Uncorrectable coagulopathy
- Intrahepatic echinococcal cysts

## PREPROCEDURAL EVALUATION

- Obtain informed consent.
- Review history and previous radiologic studies.
- Confirm acceptable laboratory profiles. The following parameters are followed:
  - International normalized ratio (INR) greater than 1.5 can be corrected with administration of fresh frozen plasma and vitamin K if time allows.
  - Partial thromboplastin time (PTT) of less than 45 seconds. Hold heparin for 2 hours before the procedure. Protamine sulfate can be given for heparin reversal.
  - Platelet count of more than 50,000 per μL. Platelets should be given immediately before or during the procedure.
- Establish intravenous accesses for hydration, sedation, and possible emergency intervention.
- Consider preprocedural antibiotic coverage, specifically if abscess is suspected.
- Patients should have nothing by mouth for 6 hours before the procedure to reduce risk of aspiration.

## PROCEDURE

1. Position patient for optimal access. Transvaginal and transrectal routes may be used for pelvic collections.
2. Establish vital signs.
3. Adhere to sterile technique as by your institution standards.
4. Administer sedation and local anesthesia. One percent lidocaine may be buffered with sodium bicarbonate at a 1 to 9 ratio (i.e., 0.5 mL sodium bicarbonate per 4.5 mL lidocaine) will reduce pain on injection.
5. Aspirate can be obtained through the needle for immediate Gram's stain, culture, sensitivity, and cytology.
6. A guidewire is then advanced through the needle into the collection.
7. The collection can be emptied there or a drain left in place.

## CATHETER TYPES FOR ABSCESS DRAINAGE

### Nonsump Catheters and Sump Drains with Double Lumen

Nonsump catheters have a pigtail configuration, which limits risk of wall cavity perforation and inhibits premature dislodgment. This catheter is effective for purulent and serous fluid drainage.

Sump catheters have double lumens, which allow for continuous aspiration and irrigation. This is effective for managing purulent collections or for the administration of antibiotic lavage therapy. In some cases, the smaller lumen is left open to air, which will help inhibit catheter adhesion to the abscess wall. This was the initial intent for the sump catheter.

### Trocar Catheter

The trocar catheter has a sharp stylet within that sticks out of the catheter end and allows it to be placed directly into the lumen of the abscess without a wire. Trocar catheters are best used for large abscesses that are superficial and contact the abdominal wall.

## PUNCTURE TECHNIQUES

### Ultrasound-Guided Puncture

US offers real-time imaging and direct visualization of the needle during the puncture. Two methods are used: freehand and guided-assisted puncture. The freehand method calls for one hand to hold and control the US probe while the other hand aligns the needle within the plane of the transducer beam. Specific guides have been developed that fit on to the transducer where the needle is then fit. This will steer the needle into the path of the US field. The US image has a line projected onto the screen, which draws the estimated path of the needle so that the probe can be set in the correct position before the needle is advanced. Alternatively, if the fluid collection is large, the US can be used to identify the appropriate location for puncture and a blind puncture can then be performed.

## Computed Tomography and Computed Tomography Fluoroscopic–Guided Puncture

CT offers the advantage of anatomic detail and clear visualization of the collection and surrounding viscera. With the recent advent of CT fluoroscopy, near real-time imaging is now possible during the procedure. The Seldinger method may be used after the needle has entered the abscess cavity, or direct trocar method may be used. In most cases the entire procedure is performed on the CT scanner, including the placement of the drainage catheter. In more difficult cases, the patient is transferred to the fluoroscopy suite as soon as access has been gained with the CT scanner and a wire passed into the collection. Guidewire manipulation is easier using fluoroscopy. After the catheter placement, contrast medium is injected to verify that the sideholes are within the cavity and are draining properly. Sideholes within the tract can cause leakage of the purulent material into the peritoneal cavity or to the skin.

## Magnetic Resonance Imaging

Technically, the drainage procedure itself is identical to the CT-guided drainage. Advantages include the lack of ionizing radiation and possibility of real-time multi-planar visualization. Specific tools (nonferromagnetic) and devices are currently being created as this technology rapidly advances.

# DIAGNOSTIC ASPIRATION FOR CULTURE AND CYTOLOGY

Fine needle aspiration of a fluid collection enables rapid characterization of the nature of the fluid. Gram's staining and culture allow for specific identification of the organism and antibiotic selection. Typically, 22- or 23-gauge needles are used. The skin is scrubbed with antiseptic and infiltrated with 1% lidocaine. A percutaneous puncture is performed with the needle inserted to the appropriate depth. Chemical analysis and cell counts help in differentiating fluid collections. Specimen material is sent for Gram's staining and aerobic/anaerobic cultures, fungal/mycobacterial cultures, or other cultures based on specific concerns.

# PERCUTANEOUS DRAINAGE BY CATHETER

When an abscess cavity is detected, several items must be addressed before the preferred treatment can be determined:

- Abscess size: One may attempt to treat small abscesses (less than 5 cm) with aspiration alone.
- Abscess extent: Large, loculated abscesses may be best treated surgically, which allows for evacuation of all locules and removal of necrotic debris.
- Abscess location: Is a safe route possible?
- Fistulization: Every effort should be made to diagnose enteric fistula to the abscess cavity. Enteric fistulae require prolonged catheter drainage, effective diversion of intestinal contents, and specialized catheter manipulation.
- Fungal infection: Abscesses caused by or superinfected with *Candida albicans* are commonly unresponsive to percutaneous drainage and antibiotic therapy, particularly in the immunosuppressed patient. Surgical debridement may be indicated.
- Operative candidate—Surgery may not be possible due to the patient's critical condition. Therefore, percutaneous drainage may be planned as a temporizing measure, if the drainage is expected to help, even though operation may be ultimately necessary for definitive treatment.

# ASPIRATION FOR CULTURE AND TREATMENT

A small abscess may be treated by complete needle aspiration of the infected material. Complete aspiration may require an 18- to 20-gauge needle or temporary placement of a small drainage catheter.

Saline irrigation of the abscess facilitates the evacuation of viscous material. This is continued until the return is clear. One gram of ampicillin per 1 g cephaloridine in 10 mL of sterile water can be instilled before needle removal. One study showed that with aspiration, 60% of abscesses were cured whereas a comparison cohort treated with drainage catheter had a 100% success. The hospital stay was no different and the average time taken for total resolution was the same for both cohorts. The initial abscess size was similar between the two cohorts.

Patients who have had an abscess aspirated should be reexamined by CT or US within 48 to 72 hours to evaluate for fluid recurrence.

## DRAINAGE METHODS AND TECHNIQUES

### General Considerations

The shortest route to the cavity that avoids organs or vital structures is selected. The access route should permit gravity drainage of the most dependent portion of the abscess cavity.

The cavity should not be emptied before catheter placement because collapsing of the cavity makes guidewire and catheter manipulation more difficult and allows the displaced viscera to shift around, possibly obliterating the safe access route.

Several specific techniques can be used for catheter drainage.

### Cope Method

A 22-gauge needle is placed into the abscess and diagnostic aspiration is performed. Then, a 0.018-in. guidewire is threaded through the needle. The needle is then replaced with a special 5-Fr Teflon dilator with sidehole and curve, which is advanced over the guidewire. This dilator has a tapered tip with a sidehole along the inside of the curve, through which a 0.035-in. guidewire exits. Subsequent dilators and larger catheters can be introduced over this guidewire. This method is fast and relatively atraumatic and only requires one puncture for placement of the 0.035-in. wire.

### Tandem Technique

A 22-gauge needle is advanced into the abscess. Then a larger 18-gauge needle is inserted next to it with the same angle and depth. Through the larger needle, a guidewire is passed and coiled in the cavity. Dilators are passed over the guidewire and ultimately the drainage catheter is placed.

### Single 18- or 19-Gauge Needle Technique

In superficial collections, a direct puncture with an 18- or 19-gauge needle can be performed with imaging guidance. The 0.035-in. wire is advanced and standard Seldinger technique used to place the catheter.

### Coaxial System

Access is gained with the 22-gauge needle and the 0.018 guidewire passed into the cavity. The triple-axial dilator, which is composed of an inner metal stiffener, a 3-Fr catheter on the stiffener, and a 6-Fr dilator externally, is then inserted over the wire. When the stiffener and 3-Fr catheter are removed, a 0.035-in. wire can then be passed through the dilator alongside the 0.018-in. wire and a drainage catheter can then be placed. This allows for keeping the 0.018-in. wire in for safety during the placement. That wire is then removed when the drain is safely in.

### Trocar Technique

A trocar catheter (combination of catheter, cannula, and trocar) may be inserted directly into a superficial abscess cavity in one pass. When the assembly has entered the abscess cavity, the trocar is removed; then the cannula is removed while the catheter is advanced into the cavity.

## ABSCESS CAVITY MANAGEMENT

After placement of a drainage catheter, the abscess cavity is drained, aspirated empty, and then flushed out with saline until the aspirate is clear. Contrast may be instilled to determine size of cavity and to access for fistula. One should be careful not to distend the cavity with the flushing and install lesser amount of fluid than initially drained. This is to minimize the risk of bacteremia. In patients who are septic, one should not manipulate more than absolutely needed and flushing may not be necessary. The catheter should be sutured to the skin to prevent accidental removal. Most or all drainage catheters have a self-retaining mechanism such as a balloon or a locking pigtail configuration. Even those catheters should be sutured in place or secured to the skin with one of many skin adhesive systems available. The catheter is connected to gravity drainage and outputs need to be monitored. Continuous suction should not be used because the walls of the abscess will be sucked against the catheter, preventing the fluid to flow toward the catheter. Intermittent wall suction can be used. We have had good experience with Jackson-Pratt Reservoir (Allegiance Healthcare, McGaw Park, IL), which is a silicon container that, by expansion force, causes negative pressure. When the container has filled, the contents are expelled and vacuum reestablished.

### Size and Exchange of the Drainage Catheter

Larger catheters provide better drainage than smaller ones. Seven- to 13-Fr catheters are most commonly used and have the best flow characteristics; however, larger catheters are available if needed. Smaller catheters that do not drain adequately can easily be exchanged for larger ones over a wire.

Occlusion of the catheter by debris can usually be resolved or prevented by regular irrigation with saline or, if occluded, by clearing the catheter with a guidewire. In some cases, injection of dilute proteolytic solutions (5% to 10% acetylcysteine) or thrombolytic agent such as streptokinase can be helpful. If these techniques fail, the catheter must be replaced.

Suction may be used to assist with the drainage. Suction may cause the tissue to collapse around the catheter and cause obstruction of the sideholes, leading to failure of drainage. Having a vent within the tube or using a sump catheter can prevent this problem.

### Irrigation of the Abscess Cavity after Catheter Insertion

The practice of regular irrigation of drainage catheters is debatable. It is likely not necessary as long as the purulent material is draining. Small volume irrigations are useful in keeping the catheters patent. Proteolytic agents (acetylcysteine) and thrombolytic agents have all been used. Several papers have shown that installment of thrombolytic agents into abscess cavities accelerates resolution of the abscess as compared with those not treated with installment.

### Catheter Removal

The clinical signs of infection are crucial in determining when the catheter should be removed. Most patients become afebrile 24 to 48 hours after drainage of an abscess. Follow-up CT or US examinations are used to ascertain that the abscess is getting smaller and that the tube remains in good position. A sinogram is indicated to look for a fistula in patients with persistent fever and leukocytosis. Patients with fistulas require long-term catheter drainage because of persistent flow through the fistula. If the abscess cavity is unchanged and the purulent material is not draining adequately, a larger catheter or a second catheter is indicated.

The guidelines for catheter removal include the following:

- Drainage of less than 25 mL per day
- Change in the draining material from purulent to clear
- Resolution of abscess cavity on follow-up imaging
- Clinical improvement of the patient's symptoms

## FAILURE OF PERCUTANEOUS DRAINAGE

The following are the most common causes for failed drainage:

- Too large abscess
- Presence of necrotic material
- Multiloculated abscess
- Poorly defined parenchymal abscesses
- Tumor necrosis simulating abscess
- Gastrointestinal tract fistula

- Diffuse microabscesses and phlegmonous collections
- Superinfection with *C. albicans* or other opportunistic agent

## COMPLICATIONS OF PERCUTANEOUS DRAINAGE

The overall complication rate of abscess fine needle aspiration is less than 10%. The main complications are sepsis, tube migration/dislodgment, bowel perforation, hemorrhage, and peritonitis. To minimize complications, avoid transversing loops of bowel, minimize wire/catheter manipulation, and irrigate cavities gently to minimize septicemia.

## SPECIFIC ANATOMIC LOCATIONS AND DRAINAGE OF ABSCESSES

### Pyogenic Hepatic Abscesses

Approximately one-half of the hepatic abscesses are related to hepatic/biliary surgery or biliary inflammatory/obstructive disease. Many of these patients have comorbid conditions such as diabetes, alcoholism, immunosuppression, malignancy, and cirrhosis. Pyogenic microorganisms reach the liver along several different routes (biliary duct stasis, hematogenous spread, portal vein phlebitis, or direct extension). Ten percent of liver abscesses are secondary to traumatic injury of the liver (bacterial growth in subcapsular hematoma) and approximately 5% of abscesses have no obvious cause and are termed *cryptogenic abscesses*, which may be the result of bacterial seeding via a silent intraabdominal source.

### Hepatic Amebic Abscess

The incidence of liver abscess from intestinal amebiasis is 1% to 5%. This incidence is low in the United States and Western countries and highest in Southeast Asia. Only one-half of the patients have symptoms referable to intestinal amebiasis.

Abscesses are typically single and located in the posterosuperior or anteroinferior aspect of the right lobe. The principal morbidity is associated with abscess rupture, which occurs in 2% of cases. Rupture may occur into the peritoneal cavity, pleural cavity, lung/bronchial tree, and pericardium.

Debate exists in the literature with regard to the need for drainage.

Amebic abscesses can be diagnosed by serology. The aspirate from an amebic abscess has an "anchovy sauce" appearance. Ameba can be recovered in two-thirds of the cases, either by biopsy of the wall or from the cellular material obtained by brushing the walls.

### Splenic Abscesses

The traditional treatment of macroscopic splenic abscesses has consisted of antibiotic therapy and splenectomy. Some authors have shown that fine needle abscess aspiration aids in recovery. There are reports of placement of drainage catheters into splenic abscesses in cases in which other options are not feasible.

### Subphrenic Abscesses

Nearly all subphrenic abscesses occur after operation with bowel contamination of the peritoneal cavity. These abscesses can usually be drained by a subcostal extraserous approach; however, an intercostal transpleural approach can be performed.

Subphrenic abscesses commonly drain readily and may resolve completely in a matter of days. Larger cavities may require drainage for periods of as long as 2 weeks. If there is a persistent leak with continuing drainage after 2 weeks, it usually must be controlled surgically.

### Enteric Fistulae

Enteric fistulae carry a mortality rate of as high as 25%. Selective catheterization of all sinus tracts should be performed. A drain through the bowel perforation that diverts the bowel content away from the fistula, creating a controlled fistula, may help allow the tracts to close down around the catheter until the drainage catheter can be changed intermittently to a smaller one and finally removed. Fistulas will not close if there is a persistent infection, distal obstruction, or underlying tumor. Several authors have demonstrated spontaneous closure rates ranging from 57% to 88% after percutaneous abscess drainage. Longer duration of drainage is required for abscesses with enteric communication than for abscesses without fistulae. Generally, abscesses with a fistula can be managed with a single drainage catheter if the sideholes can be placed at the site of the fistulous communication within the abscess cavity. Several catheters may be required for adequate control of fistulous output in difficult cases (e.g., high-output fistula, multiple fistulae), and direct catheterization of the fistulous tract(s) may be necessary. Treatment may last for several months and necessitates concomitant parenteral nutrition.

### Crohn's Disease

Abscesses form spontaneously in 10% to 28% of patients with Crohn's disease as a result of transmural inflammatory disease and fistulae. Postsurgical abscesses occur in 16% to 20% of patients secondary to anastomotic leaks. Conventional therapy for spontaneous and postoperative abscesses in Crohn's disease is surgical; however, successful percutaneous treatment has been described.

### Diverticular Abscesses

Diverticulitis can present with only microperforations to full-fledged peritonitis. Surgical treatment includes abscess drainage with diverting colostomy, resection of affected segment, and ultimately bowel reanastomosis.

### Pelvic Abscesses

CT and US are useful for draining pelvic abscesses. Transabdominal approach is limited because of interposing bowel, vessels, bladder, uterus, or large amounts of abdominal or pelvic soft tissue. Additional access techniques have been developed such as transgluteal (transsciatic), paracoccygeal-infragluteal, transvaginal, transperineal, and transrectal approaches (Fig. 27.3).

Transrectal pelvic abscess drainage has become an acceptable alternative for gaining access to deep pelvic fluid collections and is preferred over transvaginal access by some authors. Direct fluoroscopy, endosonography, or CT can be used to guide the transrectal approach. Regardless of the image guidance used, the transrectal approach appears to be a safe, well-tolerated, and effective procedure.

### Acute Pancreatic Fluid Collections

CT is the modality best-suited for the diagnoses of acute pancreatic fluid collections. These fluid collections usually resolve spontaneously in 50% of cases, and subsequent percutaneous drainage is not necessary unless they persist or become infected.

### Pancreatic Pseudocysts

Pseudocysts are well-circumscribed, round or oval, encapsulated, low-density collections with a thin, nonenhancing rim or a thick wall demonstrating intravenous contrast enhancement on CT imaging. They can be classified as infected and noninfected on the basis of fine needle puncture and Gram's staining. Noninfected pseudocysts less than 5 cm in diameter often resolve spontaneously and should be followed. Indications for drainage of noninfected pseudocysts are the following:

- Diameter greater than 5 cm
- Increasing diameter over time
- Severe pain
- Gastrointestinal/biliary obstruction due to compression

Percutaneous needle evacuation has a 5% 1-year recurrence rate, which may be lower with direct catheter drainage. Transgastric technique and internalization with stent placement into the stomach have also been described.

### Pancreatic Abscesses

Pancreatic abscesses are associated with a mortality rate of 70% to 80%. The traditional treatment has been surgical, but percutaneous drainage has been shown to be successful in 50% to 79% of cases. Pancreatic enzymes generate large amounts of necrotic debris and catheter drainage is often incomplete. Several microorganisms are commonly isolated from pancreatic abscesses: *Escherichia coli*, *Klebsiella pneumoniae*, *Proteus mirabilis*, and *Pseudomonas*,

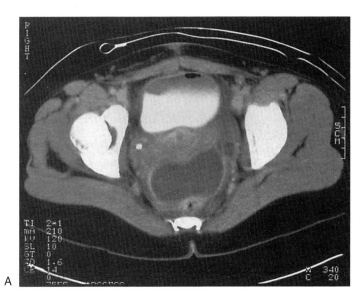
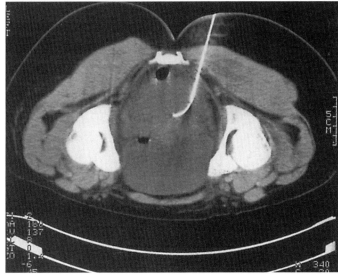

**FIG. 27.3.** Presacral abscess. **A:** Supine axial computed tomography demonstrates rim-enhancing low-density, presacral abscess collection. **B:** Axial computed tomography with patient in prone position displays transgluteal placement of catheter within presacral abscess. (Reprinted from Castañeda-Zúñiga WR. *Interventional radiology*, 3rd ed. Baltimore: Williams & Wilkins, 1997, with permission.)

sometimes in mixed culture. The amylase concentration in the fluid is usually elevated (20,000 to 40,000 IU per mL).

### Iliopsoas Abscesses

Located posterior to the retroperitoneum, an abscess in this compartment is of varied etiologies, including previous surgery, penetrating trauma, hematogenous/contiguous spread from spinal osteomyelitis, renal infection, tuberculosis, or inflammatory bowel disease. Treatment of such abscesses by percutaneous drainage has shown promising results and should be attempted.

## RENAL AND PERIRENAL FLUID COLLECTIONS, ABSCESSES, AND PYONEPHROSIS

### Renal Abscess

A renal abscess is a well-encapsulated lesion that has not extended beyond the confines of the renal parenchyma. It is usually due to pyelonephritis or multifocal bacterial nephritis. In most cases, gram-negative bacteria are the offending agents.

### *Diagnosis*

Renal abscesses present early with fever and flank pain associated with guarding. This progresses to a chronic state in which malaise, vague back pain, and weight loss can occur. Nonspecific leukocytosis and bacteriuria may occur.

US findings range from a homogeneous mass with few internal echoes to a highly echogenic mass. The walls of the abscesses may be thick and have irregular margins.

On CT, the abscess appears as a low-density, space-occupying lesion. The presence of gas bubbles is the hallmark of an abscess. After administration of contrast medium, a well-defined, enhancing wall is commonly seen.

### *Treatment*

Conventional treatment of renal abscess is surgical incision with drainage. Nonfunctioning abscessed kidneys are removed. Percutaneous transretroperitoneal drainage of the abscess can reduce the significant surgical morbidity and mortality of the procedure.

### Perinephric Abscesses

Perinephric abscess is an infrequent condition that appears as an extension of infection from pyogenic conditions in the kidney (pyelonephritis, infected cyst, or abscess). Obstructive uropathy is frequently associated. The collection of pus is confined to the perinephric space (not penetrating the renal capsule), but it can spread from this location in several locations.

### *Diagnosis*

The most common clinical presentation includes fever and weight loss with low back pain, flank pain, or pain/mass in the groin.

CT allows for a detailed evaluation of the extent of the abscess. Gas bubbles within a perinephric fluid collection are easily identified. Extension into the underlying psoas muscle or adjacent organs can also be assessed.

### *Treatment*

The classic treatment of perinephric abscess has been high doses of parenteral antibiotics and open surgical incision/drainage with or without nephrectomy. The mortality rate associated with surgical drainage/nephrectomy varies from 6% to 30%. Percutaneous drainage provides an attractive alternative.

Percutaneous access allows irrigation of the cavity with antibiotics. Frequent cultures of the fluid should be obtained so that antibiotic coverage can be altered to best treat the changing character of the bacterial flora. In abscesses secondary to urolithiasis, primary percutaneous drainage may be followed by a nephrostolithotomy. Frequently, drainage of purulent material from within the confines of Gerota's fascia will relieve the pressure on the renal tissue and allow the kidney to recover.

## ABDOMINAL ABSCESSES

### Indications

The operative drainage of abdominal abscesses has been associated with a mortality rate of 20% to 40%. As a result of this significant complication rate, the development of image-guided percutaneous diagnosis and management of intraabdominal fluid collections has permitted a dramatic reduction in patient mortality while maintaining a high success rate. Primary goals include the following:

- Diagnostic aspiration—to determine the nature and origin of unspecified fluid collections (e.g., hematoma, abscess, seroma, lymphocele, malignancy).
- Therapeutic drainage—to provide relief of systemic inflammatory symptoms, mass effect, or pain.

### Contraindications

- Lack of a safe and medically acceptable access and approach to the area of interest will

typically forego percutaneous drainage as an option. Each procedure must be individualized to the concerns of the patient and all efforts to prospectively and thoroughly review any pre-procedure imaging studies should be made.
- Uncorrectable coagulopathy profiles also will typically negate percutaneous interventions.
- The potential leakage of suspected echinococcal liver cyst contents into the peritoneal cavity can lead to anaphylactic shock reaction. Thus, such a procedure should be avoided.

**Preprocedural Evaluation**

1. Review of all prior imaging studies ensures the appropriateness of image-guided percutaneous intervention and determines the most suitable image-guidance modality.
2. Confirm acceptable laboratory profiles:
   a. Prothrombin time within 3 seconds of control (international normalized ratio of less than 1.5).
   b. PTT of less than 45 seconds.
   c. Platelet level greater than 50,000 per µL.
   
   Elevated prothrombin time can be corrected by administering vitamin K (intravenously or subcutaneously) or using fresh frozen plasma (10 to 20 mg per kg). An elevated PTT can usually be corrected by discontinuation of heparin for at least 3 to 4 hours or by the administration of protamine sulfate (slow intravenous administration at 2 mg per minute). Also, platelet levels below 50,000 per µL may require platelet transfusion.
3. Obtain informed or emergency consent.
4. Establish intravenous access for any potential procedure related sedation or as a precautionary measure in the event of an emergency.
5. Discuss the option of providing preprocedural antibiotic coverage (either broad-based or specific coverage for a known or suspected organism).
6. Have the patient on nothing by mouth status at least 3 to 4 hours before the procedure.

**Image-Guidance Modalities**

1. Fluoroscopy is most commonly used in conjunction with CT or US for fluid collection and drainage catheter management (i.e., tube exchange/repositioning, fistulogram/sinogram, cavity evaluation).
2. US guidance.
   a. Indirect needle guidance is a two-part procedure in which US is used initially to localize a fluid collection and determine an adequate approach. Once this information is known, a skin mark is usually made and a needle/catheter unit is directed to the desired location without direct observation by the US unit. This method is primarily used in large, superficial fluid collections.
   b. Freehand/needle guidance technique allows for continuous imaging guidance of the needle as it is advanced toward the fluid collection.
3. CT guidance is, overall, the most commonly used modality and generally preferred for small, deep fluid collections or collections surrounded by bowel structures.

**Catheter Types and Drainage Methods**

- Nonsump catheters (6 to 14 Fr) have a standard pigtail configuration that limits the perforation of the cavity wall and inhibits premature dislodgment. This catheter is effective for both purulent and serous fluid contents.
- Sump drains with a double lumen (12 to 18 Fr) allow for the continuous circulation of air within the drain and minimize the adherence of the catheter sideholes to the cavity walls and contents.
- With the direct trocar method, the catheter assembly (catheter, stiffening cannula, inner trocar) can be inserted as a single unit into the fluid collection without the aid of a guidewire. The trocar is removed and an aspirate is obtained to confirm the return of contents. Next the stiffening cannula is gradually removed while advancing the catheter into the fluid collection. This

method is generally used for large superficial fluid collections.

- With the Seldinger method, an 18-gauge needle is advanced into the fluid collection and an aspirate is obtained to verify needle location. A 0.035- to 0.038-in. guidewire is passed through the needle and coiled in the cavity. Dilators are passed over the guidewire up to a size 1 or 2 Fr larger than the drainage catheter. Afterward, the catheter is advanced over the guidewire and positioned within the cavity.
- With needle aspiration of small fluid collections or microabscesses (any cavity not large enough for formation of pigtail loop), an 18- to 22-gauge needle can be directed into the fluid collection with subsequent complete aspiration of cavity contents.
- For the coaxial system using a 22-gauge needle and a triaxial dilator, more than one vendor offers an access set, which includes a 15-cm-long 22-gauge needle, a 20-cm-long 5-Fr triaxial dilator, and a 0.018-in. guidewire. Access is gained with the 22-gauge needle and the guidewire is passed into the cavity. A fluid sample can first be drawn from the needle. The three-axial dilator, which is composed of an inner metal stiffener that accepts a 0.018-in. wire, a 3-Fr catheter on the metal stiffener, and a 6-Fr dilator on the outside, is then passed into the cavity over the wire. The metal stiffener and the 3-Fr catheter are then removed, keeping the wire in place. A 0.035-in. wire can then be passed through the 6-Fr dilator beside the 0.018-in. wire and a drainage catheter can then be placed.

### General Considerations and Procedure

1. In general, one should choose the safest and shortest possible route to the fluid collection.

2. Typical image-guided approaches include transabdominal (transperitoneal), transgluteal (transsciatic), paracoccygeal-infragluteal, transvaginal, transperineal, or transrectal routes.

3. The skin and entry site are prepped and cleaned using sterile technique. A local 1% to 2% lidocaine anesthetic solution is given for local pain control (intravenous sedation may also be necessary, as discussed previously).

4. The needle, catheter, or both are subsequently directed into the fluid collection with the aid of image guidance.

5. Aspirates can be obtained for immediate Gram's stain, culture, and sensitivity and cytology or entire contents can be removed as a therapeutic maneuver.

### Abscess Cavity Management

- On initial complete evacuation of the cavity, sterile saline irrigation should be performed until clear aspirates are received. Daily irrigations are debatable. However, if contents are particularly viscous, fibrinolytic agents can be used to reduce viscosity by breaking down fibrin connections in the fluid and improving flow transit time.
- Sinogram can be performed initially and repeated throughout the patient's stay to assess for cavity resolution and evaluate for enteric communication or fistulous tracts. Water-soluble contrast (an amount not to exceed the initial aspirate volume, thus avoiding risk of sepsis and bacteremia) is introduced via the existing tube with image documentation (CT or fluoroscopy) of contrast distribution. Patients with an intestinal fistula require larger periods of drainage secondary to persistent flow through the fistula.
- Depending on the results of the initial catheter drainage, small or nonsump catheters may require exchanging over guidewires for a larger and more effective drainage catheter.
- Catheters should be checked for patency at least daily and flushed with 3 to 5 mL saline every 8 to 24 hours.
- Considerations for catheter removal include the following:
  - A decrease in the amount of drainage per day
  - A change in the appearance of contents from purulent to clear
  - A decrease in the size of the abscess cavity
  - An improvement in patient symptoms

**Special Situations**

- Amebic abscess: Based on an increased morbidity and mortality associated with the rupture of cavity contents into the peritoneal, pericardial, and pleural cavities, medical management is considered the best treatment option.
- Subphrenic abscess: Although generally avoided, an intercostal transpleural approach can be used for access. Such a route should be attempted only when a catheter can be directed through a region of fibrosed pleural lining.
- Enteric fistula: This fistula will not close in the setting of an underlying tumor, distal obstruction, or persistent infection. As such, catheter management may require placement of the catheter sideholes directly at the site of fistulous communication or necessitate several catheter placements for adequate control of fistulous output. Also, given the high rate of reoccurrence of fistula in patients with Crohn's disease, catheter drainage is considered only a temporizing measure prior to surgical intervention.
- Noninfected pancreatic pseudocyst: This pseudocyst spontaneously resolves in approximately 50% of cases and often does not require percutaneous/surgical management. Although controversial, proposed indications for percutaneous drainage of noninfected pseudocyst can include the following:
  - Greater than 5 cm diameter in size
  - Increasing diameter
  - Severe pain
  - Gastrointestinal or biliary obstruction
- Pyonephrosis: In the past, the treatment of choice was surgical nephrectomy. However, presently, a more conservative approach with percutaneous nephrostomy should be considered initially, unless end-stage pyonephrosis in a nonfunctioning kidney exists.
- Pelvic abscesses: These abscesses are common and there is a vide variety of access options. The anterior approach is often not possible because of bowel and bladder in front. Several unique approaches have been described:
  - Transrectal access, usually performed with a combination of US and fluoroscopic guidance, has been found to be a well-tolerated and effective way of draining deep pelvic fluid collections.
  - Transvaginal drainage is also performed with combined US and fluoroscopic guidance. This method has been found to be effective and well tolerated as well. The vaginal wall is tough and it can be difficult to pass the drainage catheter over the wire. Stiff wires and predilatation with dilators can be helpful.

## SUGGESTED READINGS

Alazraki NP. Radionuclide imaging in the evaluation of infections and inflammatory disease. *Radiol Clin North Am* 1993;31:791–792.

Alexander AA, Eschelman DJ, Nazarian LN, Bonn J. Transrectal sonographically guided drainage of deep pelvic abscesses. *AJR Am J Roentgenol* 1994;162:1227.

Ariel IM, Karzrian KK. *Diagnosis and treatment of abdominal abscess*. Baltimore: Williams & Wilkins, 1971.

Baek SY, Lee M, Cho KS, et al. Therapeutic percutaneous aspiration of hepatic abscesses: effectiveness in 25 patients. *AJR Am J Roentgenol* 1993;160:799.

Balfe DM. *CT of the retroperitoneum*. Syllabus: ARRS, Categorical course syllabus (body CT) 1994:103–105.

Balthazar EJ, Freeny PC, vanSonnenberg E. Imaging and intervention in acute pancreatitis. *Radiology* 1994; 193:297.

Barbaric ZL, Wood BP. Emergency percutaneous nephropyelostomy: experience with 34 patients and review of the literature. *AJR Am J Roentgenol* 1977;128:453.

Boland GW, Mueller PR. Update on abscess drainage. *Semin Intervent Radiol* 1996;13:27.

Butch RJ, Mueller PR, Ferrucci JT, et al. Drainage of pelvic abscesses through the greater sciatic foramen. *Radiology* 1986;158:487.

Carroll G. Nontuberculous infections of the urinary tract. In: Campbell MF, ed. *Urology*. Philadelphia: WB Saunders, 1963:36.

Chung T, Hoffer FA, Lund DP. Transrectal drainage of deep pelvic abscesses in children using a combined transrectal sonographic and fluoroscopic guidance. *Pediatr Radiol* 1996;26:874.

Cwikiel W. Percutaneous drainage of abscess in psoas compartment and epidural space—case report and review of the literature. *Acta Radiol* 1991;32:159.

D'Costa H, Bloor C, Maskell GF, et al. Technical report: perirectal abscess drainage—simple modification in technique using a vascular sheath. *Clin Radiol* 1997;52:469.

Decosse JJ. Subphrenic abscess. *Surg Gynecol Obstet* 1974;138:841.

Do H, Lambiase RE, Deyoe L, et al. Percutaneous drainage of hepatic abscesses: comparison of results in abscesses with and without intrahepatic biliary communication. *AJR Am J Roentgenol* 1991;157:1209.

Eisenberg PJ, Lee MJ, Boland GW, Mueller PR. Percutaneous drainage of a subphrenic abscess with gastric fistula. *AJR Am J Roentgenol* 1994;162:1233.

Fabiszewski NL, Sumkin JH, Johns CM. Contemporary radiologic percutaneous abscess drainage in the pelvis. *Clin Obstet Gynecol* 1993;36:452.

Freeny PC, Lewis GP, Traverso LW, Ryan JA. Infected pancreatic fluid collections: percutaneous catheter drainage. *Radiology* 1988;167:435.

Gazelle GS, Haaga JR, Stellato TA, Gauderer MWL. Pelvic abscesses: CT-guided transrectal drainage. *Radiology* 1991;181:49.

Gerzof SG. Percutaneous catheter drainage of abdominal abscesses: 5-year experience. *N Engl J Med* 1981; 305:653.

Gobien RP. Computed tomographic guidance of percutaneous needle aspiration and drainage of abdominal abscess. *J Comput Assist Tomogr* 1982;6:127.

Hadas-Halpren I, Hiller N, Dolberg M. Percutaneous drainage of splenic abscesses: an effective and safe procedure. *Br J Radiol* 1992;65:968.

Hovsepian DM. Transrectal and transvaginal abscess drainage. *J Vasc Intervent Radiol* 1997;8:501.

Hoyt AC, D'Agostino HB, Carrillo AJ, et al. Drainage efficiency of double-lumen sump catheters and single-lumen catheters: in vitro comparison. *J Vasc Intervent Radiol* 1997;8:267.

Jeffrey RB Jr. *Enteric abscesses: imaging and intervention.* Syllabus: a categorical course in diagnostic radiology (interventional radiology) 1991:76.

Kang M, Gupta S, Gulati M, et al. Ilio-psoas abscess in the pediatric population: treatment by US-guided percutaneous drainage. *Pediatr Radiol* 1998;28:478.

Kastan DJ, Nelson KM, Shetty PC, et al. Combined transrectal sonographic and fluoroscopic guidance for deep pelvic abscess drainage. *Ultrasound Med* 1996;15:235.

LaBerge JM, Kerlan RK Jr, Gordon RL, Ring EJ. Nonoperative treatment of enteric fistulas: results in 53 patients. *J Vasc Intervent Radiol* 1992;3:353.

Lahorra JM, Haaga JR, Stellato T, et al. Safety of intracavitary urokinase with percutaneous abscess drainage. *AJR Am J Roentgenol* 1993;160:171.

Lambiase RE, Deyoe L, Cronan JJ, Dorfman GS. Percutaneous drainage of 335 consecutive abscesses: results of primary drainage with 1-year follow-up. *Radiology* 1992;184:167.

Lambiase RE, Cronan JJ, Dorfman GS, et al. Percutaneous drainage of abscesses in patients with Crohn's disease. *AJR Am J Roentgenol* 1988;150:1043.

Lambiase RE. Percutaneous abscess and fluid drainage: a critical review. *Cardiovasc Intervent Radiol* 1991;14:143.

Lang EK. Renal, perirenal and pararenal abscesses: percutaneous drainage. *Radiology* 1990;174:109.

Lang EK, Price ET. Redefinitions of indications for percutaneous nephrostomy. *Radiology* 1983;147:419.

Lee MJ, Rattner DW, Legemate DA, et al. Acute complicated pancreatitis: redefining the role of interventional radiology. *Radiology* 1992;183:171.

Longo JM, Bilbao JI, deVilla VH, et al. CT-guided paracoccygeal drainage of pelvic abscesses. *J Comput Assist Tomogr* 1993;17:909.

Martin EC, Fankuchen EI, Neff RA. Percutaneous drainage of abscess: a report of 100 patients. *Clin Radiol* 1983; 28:97.

McLean GK, Mackie JA, Frieman DB, Ring EJ. Enterocutaneous fistulae: interventional radiologic management. *AJR Am J Roentgenol* 1982;138:615.

Michalson AE, Brown BP, Warnock NG, Simonson TM. Presacral abscesses: percutaneous transperineal drainage with use of bone landmarks and fluoroscopic guidance. *Radiology* 1994;190:574.

Mueller PR, vanSonnenberg E, Ferrucci JT. Percutaneous drainage of 250 abdominal abscesses and fluid collections: current procedural concepts. *Radiology* 1984; 151:343.

Neff CC, vanSonnenberg E, Casola G, et al. Diverticular abscesses: percutaneous drainage. *Radiology* 1987;163:15.

Nosher JL, Winchman HK, Needell GS. Transvaginal pelvic abscess drainage with US guidance. *Radiology* 1987; 165:872.

Park JK, Kraus FC, Haaga JR. Fluid flow during percutaneous drainage procedures: an in vitro study of the effects of fluid viscosity, catheter size, and adjunctive urokinase. *AJR Am J Roentgenol* 1993;160:168.

Pereira JK, Chait PG, Miller SF. Deep pelvic abscesses in children: transrectal drainage under radiologic guidance. *Radiology* 1996;198:393.

Pombo F, Martin-Egana R, Cela A, et al. Percutaneous catheter drainage of tuberculous psoas abscesses. *Acta Radiol* 1993;34:366.

Pruett TL, Rotstein OD, Crass J, et al. Percutaneous aspiration and drainage for suspected abdominal infection. *Surgery* 1984;96:731.

Rajak CL, Gupta S, Jain S, et al. Percutaneous treatment of liver abscesses: needle aspiration versus catheter drainage. *AJR Am J Roentgenol* 1998;170:1035.

Ralls PW, Barnes PF, Johnson MB, et al. Medical treatment of hepatic amebic abscess: rare need for percutaneous drainage. *Radiology* 1987;165:805.

Ryan JM, Murphy BL, Boland GW, et al. Use of the transgluteal route for percutaneous abscess drainage in acute diverticulitis to facilitate delayed surgical repair. *AJR Am J Roentgenol* 1998;170:1189.

Schuster MR, Crummy AB, Wojtowycz MM, McDermott JC. Abdominal abscesses associated with enteric fistulas: percutaneous management. *J Vasc Intervent Radiol* 1992;3:359.

Schwerk WB, Gorg C, Gorg K, Restrepo I. Ultrasound-guided percutaneous drainage of pyogenic splenic abscesses. *J Clin Ultrasound* 1994;22:161.

Shah H, Harris VJ. Saline injection into the perirectal space to assist transgluteal drainage of deep pelvic abscesses. *J Vasc Intervent Radiol* 1997;8:119.

Sperling DC, Needleman L, Eschelman DJ, et al. Deep pelvic abscesses: transperineal US-guided drainage. *Radiology* 1998;208:111.

Van Allen RJ, Katz MD, Johnson MB, et al. Uncomplicated amebic liver abscess: prospective evaluation of percutaneous therapeutic aspiration. *Radiology* 1992; 183:827.

vanSonnenberg E. Percutaneous drainage of abscesses and fluid collections: techniques, results, and applications. *Radiology* 1982;142:1.

vanSonnenberg E, D'Agostino HB, Casola G, et al. Percutaneous abscess drainage: current concepts. *Radiology* 1991;181:617.

vanSonnenberg E, D'Agostino HB, Casola G, et al. US-guided transvaginal drainage of pelvic abscesses and fluid collections. *Radiology* 1991;181:53.

# Appendix 1
# Angiography: Infusion Rates

### Haraldur Bjarnason

Diagnostic angiography is usually performed with direct injection of contrast into the vascular bed to be studied. If it is the arterial anatomy that is of most interest, as it is when one is looking for stenosis or other vascular abnormalities, one wants to replace at least one-half the vessel volume and ideally reflux just into the ostium of the vessel to ensure that the entire vessel is seen, including the ostium. When a 4- to 5-Fr catheter with a 0.035- to 0.038-in. lumen is used, a hard hand injection using a 10-mL syringe gives an injection rate of 5 to 7 mL per second, which can be used as a reference before deciding how fast the injection needs to be. A test injection is made, and depending on how that fills the vessel, one can adjust the power injector. A 2-second injection is usually used for diagnostic angiograms. This will give a good demonstration of the vessel. Occasionally, longer injections are needed. Bleeding, especially in the gastrointestinal tract, requires longer injections to allow enough contrast to escape into the bowel to be detected on the x-rays.

Full-strength contrast was used when regular films were used, but now, with digital subtraction techniques (digital subtraction angiography), half-strength contrast is usually sufficient. Table A1.1 provides suggested injection rates and volumes.

**TABLE A1.1.** *Suggested injection rates and volumes*

| Vessel | Injected volume/injection rate | Contrast dilution |
|---|---|---|
| Aortic arch | 70–100 mL at 35–50 mL/sec | 50% with DSA |
| Abdominal aorta | 40 mL at 20 mL/sec | 50% with DSA |
| Pelvic arteries | 30 mL at 15 mL/sec | 50% with DSA |
| Lower extremity | Depends on technique used | Variable |
| Celiac axis | 12–16 mL at 6–8 mL/sec | 50% with DSA |
| Superior mesenteric artery | 20–40 mL at 7–8 mL/sec | 50% with DSA |
| Inferior mesenteric artery | 6–8 mL at 3–4 mL/sec | 50% with DSA |
| Renal arteries | Usually hand injection 12–16 mL at 6–8 mL/sec | 50% with DSA |
| Pulmonary arteries | 40–50 mL at 20–25 mL/sec | Full strength even with DSA |
| Inferior vena cava | 40 mL at 20 mL/sec | 50% with DSA |

DSA, digital subtraction angiography.

# Appendix 2
# Normal Laboratory Values

## Darren K. Postoak

Values vary by laboratory. Check with your local laboratory for normal ranges.

| Blood chemistry | Normal range |
| --- | --- |
| Sodium | 138–148 mEq/L |
| Potassium | 3.8–5.2 mEq/L |
| Chloride | 96–107 mEq/L |
| Carbon dioxide | 24–32 mEq/L |
| Urea nitrogen | 7–25 mg/dL |
| Creatinine | 0.8–1.6 mg/dL |
| Glucose | 60–120 mg/dL |
| Ammonia | 3–37 mmol/L |
| Calcium | 8.6–10.5 mg/dL |
| Phosphorus | 2.6–4.7 mg/dL |
| Magnesium | 1.5–2.6 mg/dL |
| Total protein | 6.0–7.9 gm/dL |
| Albumin | 3.4–4.7 gm/dL |
| Total bilirubin | 0.1–1.2 mg/dL |
| Direct bilirubin | 0.0–0.3 mg/dL |
| Aspartate aminotransferase | 10–40 IU/L |
| Alanine aminotransferase | 2–46 IU/L |
| Alkaline phosphatase | 41–116 IU/L |
| Lipase | 0–60 IU/L |
| Amylase | 25–100 U/L |
| Total cholesterol | 120–200 mg/dL |
| Triglycerides | 50–200 mg/dL |

| Hematologic studies | Normal range |
| --- | --- |
| White blood cell count | 4,500–11,000/μL |
| Hematocrit | 39–47% |
| Platelets | 150,000–400,000/μL |

| Coagulation studies | Normal range |
|---|---|
| Prothrombin time | 10.2–13.5 sec |
| Partial thromboplastin time | 22.9–33.1 sec |
| Ivy bleeding time, 5-mm wound | <9 min |
| Fibrinogen | 175–350 mg% |
| Fibrin split products | 10 µg/mL |

| Arterial blood gases | Normal range |
|---|---|
| pH | 7.38–7.44 |
| $PCO_2$ | 35–45 mm Hg |
| $PO_2$ | 80–100 mm Hg |
| $HCO_3$ | 21–28 mEq/L |
| $O_2$ saturation | 93–99% |

## SUGGESTED READINGS

Fauci AS, et al., eds. *Harrison's principles of internal medicine*, 14th ed. New York: McGraw-Hill, 1998.

Tietz NW, ed. *Clinical guide to laboratory tests*, 3rd ed. Philadelphia: WB Saunders, 1995.

# Subject Index

Note: Page numbers followed by *f* indicate figures; page numbers followed by *t* indicate tables.

## A

Abciximab, in interventional radiology, 8
Abdominal abscess, drainage of
  abscess cavity management in, 389–390, 396
  aspiration in, for culture and treatment, 388–389
  catheter types for, 387, 395–396
  considerations in, 396
  contraindications to, 387, 394–395
  diagnostic aspiration in, for culture and cytology, 388
  goals of, 394
  imaging modalities in, 385, 395
  indications for, 386, 394
  methods of, 389, 395–396
  percutaneous
    by catheter, 388
    complications of, 391
    failure of, 390–391
  preprocedural evaluation in, 387, 395
  procedure for, 387, 396
  puncture techniques in, 387–388
Abdominal angina, pathophysiology of, 89
Abdominal aorta
  angiography of, injection rates and volume for, 401t
  percutaneous transluminal angioplasty of, 105–107, 105, 106f
Abdominal mass
  localization of, 377–378, 379f
  percutaneous biopsy of, 377–384
    complications of, 383
    materials obtained by, laboratory examination of, 382
    patient preparation for, 378
    postprocedure management, 382–383
    technique for, 378–382, 380f, 381f

Abscess(es). *See also specific type*
  amebic, 397
  Crohn's disease and, 392
  hepatic
    amebic, 391
    pyogenic, 391
  iliopsoas, 393
  pancreatic, 392–393
  pelvic, 392, 393f, 397
  perinephric, 394
  renal, 393–394
  subphrenic, 391, 397
Absolute ethanol, for embolotherapy, in children, 256
Access
  central venous. *See* Central venous access
  devices for, 19–23
  hemodialysis. *See* Hemodialysis access
  sites for, 19
  techniques for, 19–23
    Seldinger technique. *See* Seldinger technique, for access
Acetylcysteine, for irrigation of drainage catheters, 390
Activated protein C resistance, hypercoagulable state and, 192
Adriamycin, in transcatheter arterial chemoembolization, 48, 48t
Airway(s), assessment of, in intravenous sedation, 2, 3t
Alanine aminotransferase, normal range of, 403
Albumin, normal range of, 403
Alcohol
  for embolotherapy, 28t
  polyvinyl, for embolotherapy, 28t, 30–31, 30f

Alkaline phosphatase, normal range of, 403
Amebic abscess, 397
Amebic hepatic abscess, 391
Amikacin, prophylactic, 5
Ammonia, normal range of, 403
Ampicillin
  abdominal abscess drainage and, 389
  balloon dacryocystoplasty and, 350
  fine needle percutaneous transhepatic cholangiography and, 284
  lacrimal stent placement and, 352
  prophylactic, 5
Amplatz Extra Stiff guidewire, 24
Amplatz gooseneck snare, for intravascular foreign body retrieval, 237, 238f
Amplatz Super Stiff guidewire, 24
Amplatz thrombectomy device, 203–205, 204f
  for occluded hemodialysis access, 124–125, 125f, 126f
Amylase, normal range of, 403
Anastomotic bleeding, postsurgical, 59
Anastomotic stricture, interventional treatment of, 323–325, 324f
Aneurysm(s)
  aorto-iliac stent-graft for. See Stent-graft, aorto-iliac, for aneurysm
  atherosclerosis and, 77–78
  false, after angiography, management of, 25
  hemodialysis access and, 119
Angina, abdominal, pathophysiology of, 89
Angiography
  after transcatheter arterial chemoembolization, 49
  aorto-iliac stent-graft and, 157
  bedrest after, 24
  carbon dioxide digital subtraction. See Carbon dioxide digital subtraction angiography
  in children, 251
  complications of, 25–26
  digital, radiation management during, 15
  digital subtraction
    carbon dioxide. See Carbon dioxide digital subtraction angiography
    exposure during, 15
  gastrointestinal bleeding and, 55–56
  hemobilia and, 60
  hemodialysis access and, 120–123
  infusion rates in, 401, 401t
  pulmonary, pulmonary embolism and, 213

vascular malformations and, 39
AngioJet rheolytic thrombectomy device, 205, 205f
Angioplasty. See also Percutaneous transluminal angioplasty
  for occluded hemodialysis access, 124
    complications of, 127
    metallic stents after, 124
  transluminal, intravascular ultrasound and, 183
Angio-Seal Device, 25
Ankle-brachial index, for peripheral vascular disease, 167
Antibiotic(s)
  aorto-iliac stent-graft and, 159
  in interventional radiology, 4–6
    reasons for, 4
  prophylactic, in interventional radiology. See Antibiotic prophylaxis
Antibiotic prophylaxis
  for biliary drainage, 5
  defined, 4
  for genitourinary system, 5
  indications for, 4–5
  for transjugular intrahepatic portosystemic shunt, 5–6
  for tunneled port, 5
Antibiotic therapy, defined, 4
Anticoagulation
  contraindications to, inferior vena cava filter and, 213
  in interventional radiology, 6–8. See also specific drug
    antiplatelet drugs, 7–8
    heparin, 6–7
  oral, in interventional radiology, 7
Anticoagulation agents, 190–191. See also specific agent
Antiphospholipid syndrome, hypercoagulable state and, 193
Antithrombin III
  coagulation cascade and, 187–188, 190f
  deficiency of, hypercoagulable state and, 193
Antral stricture, interventional treatment of, 323, 323f
Aorta
  abdominal
    angiography of, injection rates and volume for, 401t

percutaneous transluminal angioplasty of, 105–107, 105f, 106f
intravascular stents in, 149, 150f, 151
intravascular ultrasound evaluation of, 182–183
Aortic arch, angiography of, injection rates and volume for, 401t
Aortofemoral bypass graft, anastomoses of, stenoses at, percutaneous transluminal angioplasty for, 113
Aortoiliac bypass graft, anastomoses of, stenoses at, percutaneous transluminal angioplasty for, 113
Aorto-iliac stent-graft, for aneurysm. *See* Stent-graft, aorto-iliac, for aneurysm
Aortoplasty, technique for, 105–107
APACHE II score, 69
Apron lead, 14
Arrhythmia(s), central venous access and, 232
Arrow-Trerotola percutaneous thrombectomy device, 206, 208f
Arrow-Trerotola percutaneous thrombolytic device, for occluded hemodialysis access, 125
Arterial malformations, embolotherapy of, 39
Arterial Sealing Device, 25
Arterial thrombosis, acute
atherosclerosis and, 77
of lower extremity, thrombolysis for. *See* Thrombolysis
Arterial wall, normal, intravascular ultrasound evaluation of, 180, 181f
Arteriogram, percutaneous transluminal angioplasty and, 101–102
Arteriovenous fistula
after angiography, management of, 25
color Doppler imaging and, 168
embolotherapy for, 42, 43f
pulmonary, embolotherapy of, 38–39
surgically created, hemodialysis access and, 118f, 119
Arteriovenous malformation
in children, 256, 258
of colon, 63–64
embolotherapy for, 40–42, 41f
Arteritis, Takayasu's, 84, 84f
in children, 254
Artery(ies). *See also specific artery*
puncture of, Seldinger technique for, 20, 20f, 21f

Ascites
gastrostomy and, 310
jejunostomy and, 310
Aspartate aminotransferase, normal range of, 403
Aspiration, fine needle. *See* Fine needle aspiration
Aspiration thrombectomy, 203
Aspirin
iliac artery intravascular stent placement and, 148
in interventional radiology, 7
intravascular stents and, 143
percutaneous transluminal angioplasty and
of abdominal aorta and iliac artery, 107
of aortic, iliac, and peripheral arteries, 104
of carotid artery, 85, 87
of celiac artery, 89
of renal artery, 98
of subclavian artery, 83
of superior mesenteric artery, 89
of vertebral artery, 88
renal artery intravascular stent placement and, 145
Trac-Wright catheter and, 135
Atherectomy
directional, intravascular ultrasound in, 183
percutaneous, 129–135
Auth rotational atherectomy device for, 129–130, 130f
defined, 129
Simpson catheter for, 130–132, 131f
transluminal extraction catheter for, 132–133, 132f
Trac-Wright catheter for, 133–135, 134f
Atherosclerosis
manifestations of, 77
renal, percutaneous transluminal angioplasty for, results of, 99
Atherosclerotic plaques, 77–78
Atropine, transcervical fallopian tube catheterization and, 340
Auth rotational atherectomy device, for percutaneous atherectomy, 129–130, 130f
Autologous blood clot, for embolotherapy, 28–29
Axillary artery, percutaneous transluminal angioplasty of, 83–84

## B

Balloon(s)
detachable

Balloon(s) (*continued*)
   for embolotherapy, 34, 35f
      in children, 256
      for spermatic vein occlusion, 345
   for embolotherapy, 28t
   in percutaneous transluminal angioplasty of aortic, iliac, and peripheral arteries, 103
   size of, in percutaneous transluminal angioplasty, 79–80
Balloon angioplasty, percutaneous transluminal. *See* Percutaneous transluminal angioplasty
Balloon catheter
   compliance of, dilating force and, 82
   crossability of, 80
   dilating force of, 81, 81f
   for intravascular foreign body retrieval, 242
   new developments in, 82
   pushability of, 80
   trackability of, 80
Balloon dacryocystoplasty, for epiphora
   contraindications to, 347
   indications for, 347
   results of, 353
   technique for, 347–348, 348f, 349f, 350
Balloon dilatation
   of antral and pyloric strictures, 322, 323f
   esophageal. *See* Esophageal balloon dilatation
   of mesocaval shunt, in children, 252
   percutaneous, for biliary strictures, 297–298, 299f
Basket(s), for intravascular foreign body retrieval, 238, 240, 241f
Bedrest, after angiography, 24
Bentson guidewire, 24
Benzodiazepine(s), in intravenous sedation, 4
Berman catheter, 23
Bile ducts
   demonstration of, 288–289, 290f
   nondilated, fine needle percutaneous transhepatic cholangiography in, 286
   strictures of, interventional techniques in, 297–298, 299f
Biliary drainage
   antibiotic prophylaxis for, 5
   percutaneous transhepatic. *See* Percutaneous transhepatic biliary drainage
Biliary tract
   disease of, fine needle percutaneous transhepatic cholangiography in, 286
   interventional techniques in, 283–304. *See also* Hepatobiliary system, interventional techniques in
Biliary tree, metallic stents in, 296
Bilirubin
   direct, normal range of, 403
   total, normal range of, 403
Biopsy
   of abdominal mass. *See* Abdominal mass, percutaneous biopsy of
   core, for thoracic biopsy, 363–364, 364f
   hepatic, complications of, 383
   liver, transjugular, in children, 253
   pancreatic, complications of, 383
   renal, complications of, 383
   thoracic, 359–367. *See also* Thoracic biopsy
Bird's Nest filter, 220–222, 221f
Bleeding. *See also* Hemorrhage
   after angiography, management of, 25
   after thoracic biopsy, 366, 367f
   after thrombolysis, 201
   after transcatheter arterial chemoembolization, management of, 50
   anastomotic, postsurgical, 59
   esophageal balloon dilatation and, 318
   gastrointestinal. *See* Gastrointestinal bleeding
   mechanical thrombectomy and, 208, 210
Blood clot, autologous, for embolotherapy, 28–29
Blood vessel, damage to, mechanical thrombectomy and, 210
Blue digit syndrome, 107
Blue toe syndrome, warfarin and, 7
Bone, embolotherapy of, preoperative, 37–38
Boren-McKinney retriever, for intravascular foreign body retrieval, 237–238, 239f
Brachial artery, access of, 19
Brescia-Cimino fistula, 117, 118f
   failing, venous fistulogram and, 120–121, 122f, 123
Bridge graft, 117, 118f
Bronchial artery, embolotherapy of, 35–37
   anatomic considerations in, 35–37, 36f
   complications of, 37
   techniques for, 37
Budd-Chiari syndrome
   in children, percutaneous transluminal angioplasty for, 252–253

venous stent placement for, 153–154
Bullet(s), intravascular, removal of, 243, 245f
N-Butyl cyanoacrylate, for embolization of spermatic vein, 344

## C

Calcium, normal range of, 403
Canalicular stent, for epiphora, results of, 354–355, 355f-356f
Capillary malformations, embolotherapy of, 39
Carbon dioxide
　delivery of, 177–178, 177f
　as imaging agent, 175. *See also* Carbon dioxide digital subtraction angiography
　normal range of, 403
　properties of, 175
Carbon dioxide digital subtraction angiography, 175–178
　complications of, 176–177
　contraindications to, 176
　history of, 175
　indications for, 175–176
　precautions in, 176–177
　procedure for, 178
Carcinoma. *See specific type*
Carotid artery
　internal, fibromuscular dysplasia of, 84, 85f
　percutaneous transluminal angioplasty of, 84–87, 84f–87f
　puncture of, central venous access and, 231
Catheter(s)
　for abdominal abscess drainage, 387, 395–396
　　management of, 390, 396
　angiography, described, 23
　balloon. *See* Balloon catheter
　Berman, 23
　Cathlink, 225, 226f
　central venous
　　fragmentation of, 232, 232f
　　totally implanted, 230–231
　　transhepatic, percutaneous, placement of, 230, 230f
　　translumbar, percutaneous, placement of, 229, 229f
　complications of, in endovascular treatment of gastrointestinal bleeding, 64
　dual-lumen hemodialysis, 127–128
　for embolotherapy, 34–35
　Groshong, 225
　Hickman, 225, 226f
　for intravascular foreign body retrieval, 242
　MTI Thrombolytic Brush, for occluded hemodialysis access, 125
　nonsump, for abdominal abscess drainage, 387, 395
　Oasis, 206, 207f
　P.A.S. port, 225, 226f
　for percutaneous cholecystostomy, care and removal of, 302
　percutaneous drainage of abdominal abscess by, 388
　for percutaneous drainage of pleural fluid collections and pneumothorax, 370
　for percutaneous transhepatic biliary drainage, 287
　peripherally inserted central, 225, 227f, 230
　pigtail, for intravascular foreign body retrieval, 242
　placement of, Seldinger technique for, 21–23, 22f
　Port-a-Cath, dislodgement of removal of, 243, 225, 226f, 246f, 247
　Simpson, for percutaneous atherectomy, 130–132, 131f
　sump, with double lumen, for abdominal abscess drainage, 387, 399
　Trac-Wright, for percutaneous atherectomy, 133–135,134f
　transluminal extraction, for percutaneous atherectomy, 132–133, 132f
　Trocar, for abdominal abscess drainage, 387, 389, 395–396
Catheter arterial chemoembolization, 49
Catheterization
　hepatic vein, portal venous pressure and, 67
　for percutaneous transhepatic biliary drainage. *See* Percutaneous transhepatic biliary drainage, catheter placement in
　retrograde, of ureter, without endoscopic assistance, 272–273, 277f
　spermatic vein. *See* Spermatic vein catheterization, approaches to
　transrectal, stenting and, for volvulus of sigmoid colon, 332–334, 333f
Cathlink catheter, 225, 226f
Cecostomy, percutaneous, for Ogilvie's syndrome. *See* Ogilvie's syndrome, percutaneous cecostomy for

Cecum, ulceration of, in immunocompromised patients, 64
Cefazolin
　percutaneous gastrostomy and jejunostomy and, 305–306
　prophylactic, 5, 6
Cefoxitin, prophylactic, 5
Ceftizoxime, for transjugular intrahepatic portosystemic shunt, 69
Ceftriaxone, prophylactic, 5
Celiac artery, percutaneous transluminal angioplasty of, 89, 90f, 91, 91f, 92
Celiac axis, angiography of, injection rates and volume for, 401t
Center for Devices and Radiological Health, within Food and Drug Administration, 12
　stent-grafts and, 157
Central vein occlusion, hemodialysis-related, venous stent placement for, 154–155
Central venous access
　categories of, 225
　catheter(s) for, 225, 226f, 227f
　　peripherally inserted, 225, 227f, 230
　　totally implanted, 230–231
　catheter fragmentation and, 232, 232f
　catheter misplacement and, 232
　catheter removal and, 234
　catheter rescue in, 233–234, 234f
　complications of, 231–233
　internal jugular vein for, 228
　long-term, in children, 260
　patient preparation for, 225–226
　percutaneous transhepatic, 230, 230f
　percutaneous translumbar, 229, 229f
　risks associated with, 226
　subclavian vein for, 227–228
　subcutaneous tunnel in, 228–229
　uses for, 225
Central venous catheter
　totally implanted, 230–231
　transhepatic, percutaneous, placement of, 230, 230f
Cephaloridine, abdominal abscess drainage and, 389
Cephalosporin, prophylactic, 5
Check-Flo sheath, for intravascular foreign body retrieval, 237, 239f

Chemoembolization, transcatheter arterial. *See* Transcatheter arterial chemoembolization
Chiba needle, for abdominal biopsy, 379
Child-Pugh score, 69, 69t
Children
　angiography in, 251
　　complications of, 252
　　indications for, 252
　arteriovenous malformation in, 256, 258
　Budd-Chiari syndrome in, percutaneous transluminal angioplasty for, 252–253
　central venous access in, long-term, 260
　embolotherapy in, 256, 257f
　fibrinolytic therapy in, 254, 256
　gastrointestinal bleeding in, 258
　gastrostomy in, 309
　hemangiomas in, embolotherapy of, 39–40
　hemoptysis in, 259
　hemorrhage in, 258–259
　intraluminal foreign body retrieval in, 259, 260
　liver transplantation in, hepatic artery stenosis after, 254
　mesocaval shunt in, balloon dilation of, 252
　metallic stent in, 254, 255f
　percutaneous nephrostomy in, 268
　percutaneous transluminal angioplasty in
　　renal artery, 251
　　renal transplantation stenosis, 251–252
　portal hypertension in, 253–254
　Takayasu's arteritis in, 254
　transjugular intrahepatic portosystemic shunt in, 253–254
　transjugular liver biopsy in, 253
　trauma in, 258–259
Chloride, normal range of, 403
Cholangiography, transhepatic, percutaneous, fine needle. *See* Fine needle percutaneous transhepatic cholangiography
Cholangiopancreatography, endoscopic retrograde, vs. fine needle percutaneous transhepatic cholangiography, 286
Cholecystitis
　after transcatheter arterial chemoembolization, management of, 49–50
　percutaneous cholecystostomy for, 302

Cholecystostomy, percutaneous, 300–303, 300f, 301f
　catheter care and removal in, 302
　for cholecystitis, 292–302
　complications of, 302–303
　transhepatic technique of, 300–301
　transperitoneal technique of, 301–302
Cholesterol, total, normal range of, 403
Cholesterol embolization, after angiography, management of, 26
Cinefluoradiography, hemodialysis access and, 117, 119
Cinefluorography, radiation management during, 15
Cisplatin, in transcatheter arterial chemoembolization, 48, 48t
Clinical outcome, defined, 99
Coagulation cascade, 187–189, 188f–190f
　inhibitors of, 189
Coagulopathy, interventional radiology and, 8–9
Coaxial system, for abdominal abscess drainage, 389, 396
Cocaine hydrochloride, balloon dacryocystoplasty and, 347
Coil(s)
　for embolotherapy, 28t
　misplaced, retrieval of, 242, 242f
　stainless steel, for embolotherapy, 32, 33f, 34
　　in children, 256
Colon
　arteriovenous malformation of, 63–64
　sigmoid, volvulus of, transrectal fluoroscopic catheterization and stenting for, 332–334, 333f
Colonic obstruction, stent placement for. *See* Colonic stent, placement of
Colonic stent, placement of, 325–328
　concept of, 325
　contraindications to, 325–326
　indications for, 325–326
　management after, 327–328, 328f, 329f
　patient preparation for, 326
　results of, 328
　technique for, 326–327, 326f–329f
Color Doppler imaging
　for deep venous thrombosis, of upper extremity, 171–173, 172f
　hemodialysis access and, 120, 121f

　for peripheral vascular disease, 167–168, 168f, 169f
　of portal veins, 173
　of renal arteries, 174
　of transjugular intrahepatic portosystemic shunt, 173–174, 174f
　transjugular intrahepatic portosystemic shunt and, 74
Common femoral artery
　access of, 19
　　goal of, 21
　percutaneous transluminal angioplasty of, 111
Common iliac artery
　bifurcation of, percutaneous transluminal angioplasty of, 109–110, 109f–111f
　stenoses of, percutaneous transluminal angioplasty of, 107–109, 108f
Compliance, of balloon catheter, dilating force and, 82
Computed tomography (CT)
　abdominal abscess drainage and, 385, 386f, 388, 395
　abdominal mass and, 377, 379f
　aorto-iliac stent-graft and, 157
　in percutaneous drainage of pleural fluid collections and pneumothorax, 369–370
　pulmonary embolism and, 213
　in thoracic biopsy, 360, 361f, 362f, 363f
　urolithiasis and, 278
　vascular malformations and, 39
Computed tomography fluoroscopy, abdominal abscess drainage and, 385, 386f, 388
Congestive heart failure, hemodialysis access and, 119
Conscious sedation, 2–4. *See also* Sedation, intravenous
Contamination, of carbon dioxide, 176
Contrast media
　in central venous access, 227
　decreased renal function and, after angiography, management of, 26
　hot, for embolotherapy, 31
　risks associated with, 9–10
　subintimal injection of, during angiography, management of, 26
Contrast venography, deep vein thrombosis and, 213

Cook retrieval forceps, for intravascular foreign body retrieval, 237–238, 240f
Cope method, for abdominal abscess drainage, 389
Core biopsy, for thoracic biopsy, 363–364, 364f
Coulombs per kilogram unit, 11, 12t
Covered metal stent, for transjugular intrahepatic portosystemic shunt, 163–165
    indications for, 164
    outcome of, 164–165
    techniques for, 164
Cragg EndoPro System I, 161
Creatinine, normal range of, 403
Crohn's disease, abscesses and, 392
Crossability, of balloon catheter, 80
Cure, defined, 99
Cyanoacryl, for embolotherapy, 28t
Cystic fibrosis, hemoptysis due to, 259
Cystic hygroma, embolotherapy for, 44

## D

Dacryocystoplasty, balloon. *See* Balloon dacryocystoplasty, for epiphora
Dacryocystorhinostomy, 347
Deep femoral artery
    percutaneous transluminal angioplasty of, 112–113
    stenoses of, percutaneous transluminal angioplasty for, 111–112, 112f
Deep venous thrombosis
    incidence of, 213
    inferior vena cava filter for, 213. *See also* Inferior vena cava filter
    lower extremity, noninvasive evaluation of, 170–171, 170f, 171f
        Doppler ultrasound, 170–171, 170f, 171f
    noninvasive evaluation of, 170–173
        Doppler ultrasound, 170
    risk factors for, 170
    upper extremity
        causes of, 171
        complications of, 171
        noninvasive evaluation of, 171–173
            color Doppler imaging, 171–173, 172f
            gray-scale imaging, 171
            pulsed Doppler imaging, 171–173, 172f
            ultrasound, 171
        thrombolysis for, 200
Detachable balloons
    for embolotherapy, 34, 35f
        in children, 256
    for spermatic vein occlusion, 345
Deterministic effects, of radiation, 12
Diabetes mellitus, interventional radiology and, 9
Diazepam
    esophageal balloon dilatation and, 315
    Trac-Wright catheter and, 135
Dicumarol, 191
Digital angiography, radiation management during, 15
Digital subtraction angiography
    carbon dioxide. *See* Carbon dioxide digital subtraction angiography
    exposure during, 15
Digital subtraction angiography venogram, hemodialysis access and, 117, 119
Dilatation
    balloon. *See* Balloon dilatation
    and metallic stent placement, for ureteral stricture, 275, 277
Dilating force, of balloon catheter, 81, 81f
    compliance and, 82
Diphenhydramine, Trac-Wright catheter and, 135
Dipyridamole
    in interventional radiology, 7–8
    percutaneous transluminal angioplasty and
        of aortic, iliac, and peripheral arteries, 104
        of renal artery, 98
Directional atherectomy, intravascular ultrasound in, 183
Diverticular abscess, 392
Diverticulitis, 62–63, 63f
Diverticulum, Meckel's, 64
Doppler ultrasound
    for deep venous thrombosis, 170
        of lower extremity, 170–171, 170f, 171f
    percutaneous transluminal angioplasty and, 101
    before percutaneous transluminal angioplasty of vertebral artery, 88
    of renal transplant, 174–175
    transjugular intrahepatic portosystemic shunt and, 69
    vascular malformations and, 39
Dose, radiation, 11, 12t
    patient exposure to, management of, 15–16

staff exposure to, monitoring of, 12–13
Dotter retrieval set, for intravascular foreign body retrieval, 238, 240, 241f
Double-J stent, for nephrostomy, 272, 273f–276f
  exchange of, fluoroscopy-guided, 273–275, 278f
Doxorubicin hydrochloride, in transcatheter arterial chemoembolization, 48, 48t
Doxycycline, transcervical fallopian tube catheterization and, 339–340
Dual-lumen hemodialysis catheters, 127–128
Duodenal peptic ulcer disease, 58
Duodenostomy, translumbar, technique for, 313–314
Duplex ultrasound, hemodialysis access and, 120, 121f

E

Effort thrombosis, 200
Electronic array devices, for intravascular ultrasound, 180, 180f
Embolism, pulmonary. *See* Pulmonary embolism
Embolization
  after angiography, management of, 26
  central venous access and, 232, 232f
  defined, 27
  distal, after thrombolysis, 201
  mechanical thrombectomy and, 210
  of spermatic vein
    with occluding devices, 344
    with tissue adhesive, 344
Embolotherapy. *See also specific indication*
  agents for, 28–34
    absorbable materials, 28–30
      autologous blood clot, 28–29
      Gelfoam, 29–30, 29f
      oxycel, 30
      thrombin, 29
    classification of, 28
    nonabsorbable materials, 30–31
    nonparticulate, 32–34
    sclerosing, 31–32
  approach to, 35
  of bone, preoperative, 37–38
  bronchial artery, 35–37
  catheter selection in, 34–35
  in children, 256, 257f

  for gastrointestinal bleeding, 57–58
  for hemobilia, 60–61
  historical aspects of, 27, 28t
  of kidneys, preoperative, 38
  patient care in, 27–28
  postoperative management in, 28
  preoperative, 37–38
  preoperative management in, 27–28
  of pseudoaneurysm, 38
  of pulmonary arteriovenous fistula, 38–39
  technical aspects of, 34–35
  of unintended location, 65
  of vasculogenic malformations, 39
Empyema
  causes of, 368, 369t
  defined, 368
Encephalopathy(ies), hepatic, after transjugular intrahepatic portosystemic shunt, 75
Endoleak, aorto-iliac stent-graft and, 159, 159t
Endoprosthesis(es). *See also* Stent(s)
  Wallgraft, 161
Endoscopic retrograde cholangiopancreatography, vs. fine needle percutaneous transhepatic cholangiography, 286
Endoskeleton, 161
Endovascular prosthesis, Hemobahn, 161, 163f
Energy-assisted thrombectomy, 203
Enteral nutrition
  feeding tubes for
    obstruction of, 311
    types of, 309, 310f
  gastrostomy for, 305–314. *See also* Gastrostomy, percutaneous
  indications for, 305
  jejunostomy for, 305–314. *See also* Jejunostomy, percutaneous
  percutaneous transhepatic, 314
  small-bowel, 311–314
  transpyloric, 311–314
Enteric fistula, 391–392, 397
Enteric strictures, interventional treatment of, 322–325. *See also specific type*
Epinephrine, balloon dacryocystoplasty and, 347
Epiphora, 347–357
  balloon dacryocystoplasty for. *See* Balloon dacryocystoplasty, for epiphora
  causes of, 347

Epiphora (*continued*)
  defined, 347
  lacrimal stent placement for. *See* Lacrimal stent, placement of, for epiphora
Epirubicin hydrochloride, in transcatheter arterial chemoembolization, 48, 48t
Esophageal balloon dilatation
  complications of, 318
  contraindications to, 314
  indications for, 314
  results of, 315
  technique for, 314–315, 316f
Esophageal metallic stent
  blockage of, 321–322
  complications of, 321–322
  contraindications to, 319
  indications for, 318–319
  migration of, 321
  results of, 319–321, 320t
  techniques for, 319, 320f
Esophageal perforation, esophageal metallic stent placement and, 319
Esophageal rupture, esophageal balloon dilatation and, 318, 317f
Esophageal stricture
  benign, management of, 314–318. *See also* Esophageal balloon dilatation
  malignant, management of, 318–322. *See also* Esophageal metallic stent
Esophageal tears, Mallory-Weiss, 59
Esophageal varices, bleeding, 59
Esophagorespiratory fistula, 319, 321f
Ethanol
  absolute, for embolotherapy, in children, 256
  for embolotherapy, 31
  in embolotherapy, 27
  hepatocellular carcinoma and, 50–51
Ethoxysclerol, in sclerotherapy for spermatic vein, 345
Exposure, radiation units and, defined, 11, 12t
External iliac artery, stenoses of, percutaneous transluminal angioplasty of, 107–109, 108f
Extrinsic coagulation cascade, 187, 188f
Exudate
  causes of, 368, 369t
  defined, 368
  vs. transudate, chemical and cytologic determinants of, 368, 368t

**F**
Factor V Leiden, hypercoagulable state and, 192
Factor XI, deficit of, 187
Fallopian tube
  anatomy of, 339, 340f
  pathology of, 339
  transcervical catheterization of, for infertility. *See* Transcervical fallopian tube catheterization, for infertility
False aneurysm, after angiography, management of, 25
Female infertility, treatment of, 339–341. *See also* Transcervical fallopian tube catheterization, for infertility
Femoral artery
  common
    access of, 19
      goal of, 21
    percutaneous transluminal angioplasty of, 111
  deep
    percutaneous transluminal angioplasty of, 112–113
    stenoses of, percutaneous transluminal angioplasty for, 111–112, 112f
  puncture of, technique for, 103
  superficial, stenoses of, percutaneous transluminal angioplasty for, 111–112, 112f
Femoral balloon catheter system, 94–95, 94f
Femoropopliteal artery, intravascular stents in, 151–152
Fentanyl
  embolotherapy and, 27–28
  in intravenous sedation, 4
  mechanical thrombectomy and, 209
  percutaneous transhepatic biliary drainage and, 288
  thermal vessel occlusion of spermatic vein and, 345
  transcervical fallopian tube catheterization and, 340
  transjugular intrahepatic portosystemic shunt and, 69
Fertility interventions, 339–346. *See also* Spermatic vein catheterization, approaches to; Transcervical fallopian tube catheterization, for infertility
Fever, after transcatheter arterial chemoembolization, management of, 49–50

Fibrin
  dissolution of, 191, 191f
  formation of, 188
Fibrin split products, normal range of, 404
Fibrinogen
  conversion to fibrin, 188
  normal range of, 404
Fibrinolytic therapy. *See also* Thrombolysis
  anticoagulation agents, 190–191
    coumarin, 191, 191t
    heparin, 190–191
  in children, 254, 256
  coagulation cascade and, 187–189, 188f–190f
    inhibitors of, 189
  complications of, 201
  hypercoagulable state, 192–193
  laboratory tests in, 200–201
  thrombolysis. *See* Thrombolysis
  thrombolytic agents, 191–192, 191f
    pro-urokinase, 192
    streptokinase, 191
    tissue plasminogen activator, 192
    urokinase, 191–192
Fibromuscular dysplasia
  of internal carotid artery, 84, 85f
  renal, percutaneous transluminal angioplasty for, results of, 99, 100f
Film badge, 13
Filter(s), inferior vena cava. *See* Inferior vena cava filter
Fine needle aspiration
  of abdominal abscess, for culture and cytology, 388
  for abdominal abscess drainage, 396
  for thoracic biopsy, 360, 362–363
Fine needle percutaneous transhepatic cholangiography
  complications of, 286
  contraindications to, 283
  indications for, 283
  patient preparation for, 283–284
  in patients with suspected biliary disease and nondilated bile ducts, 286
  procedure for, 284–285, 284f–286f
  vs. endoscopic retrograde cholangiopancreatography, 286
Fistula(ae). *See also specific type*
  arteriovenous. *See* Arteriovenous fistula

  Brescia-Cimino, 117, 118f
    failing, venous fistulogram and, 120–121, 122f, 123
  enteric, 391–392, 397
  esophagorespiratory, 319, 321f
Fistulography, hemodialysis access and, 120–121, 122f, 123
Fluoroscopy
  abdominal abscess drainage and, 385–386, 395
  abdominal mass and, 377
  in central venous access, 227
  computed tomography, abdominal abscess drainage and, 385, 386f, 388
  in exchange of double-J stent, for nephrostomy, 273–275, 278f
  in percutaneous drainage of pleural fluid collections and pneumothorax, 370
  for percutaneous nephrostomy puncture, 267
  percutaneous transhepatic biliary drainage and, 288
  pulsed, 13
  in thoracic biopsy, 359–360
Food and Drug Administration
  Center for Devices and Radiological Health within, 12
  stent-grafts and, 157
Forceps, for intravascular foreign body retrieval, 237–238, 239f, 240f
Foreign bodies
  intraluminal, retrieval of, in children, 259, 260
  intravascular, retrieval of, 237–249
    clinical applications of, 242–243, 247–248
    failure of, reasons for, 248
    tools for, 237–242, 238f–241f
Franseen needle, for abdominal biopsy, 380, 380f
Fresh-frozen plasma
  abdominal abscess drainage and, 387, 395
  in coumarin reversal, 191
  in interventional radiology, 8

# G
Gall bladder
  anatomic relationships of, 299–300
  interventional techniques in, 299–303

Gall bladder (*continued*)
    percutaneous cholecystostomy, 300–303, 300f, 301f
    ischemia of, after transcatheter arterial chemoembolization, management of, 49–50
Gastric ulcer, 58–59
Gastritis, 58, 58f
Gastroesophageal reflux, esophageal metallic stent placement and, 322
Gastrointestinal bleeding
  in children, 258
  diagnosis of, 55–56
    angiographic, 55–56
    nuclear medicine, 55
  embolotherapy for, 57–58
  endovascular treatment of, 56
    complications of, 64–65
  lower, causes of, 61–64
    arteriovenous malformation, 63–64
    cecal ulceration, in immunocompromised patients, 64
    diverticular disease, 62–63, 63f
    inflammatory bowel disease, 64
    Meckel's diverticulum, 64
    tumors, 64
  pharmacotherapy for, 56
  upper, causes of, 58–61
    bleeding esophageal varices, 59
    duodenal peptic ulcer disease, 58
    gastric ulcer, 58–59
    gastritis, 58, 58f
    hemobilia, 60–61
    Mallory-Weiss esophageal tears, 59
    pancreatic disease, 61, 62f
    postsurgical anastomotic bleeding, 59
    vascular tumors, 59–60
Gastrointestinal tract, interventional techniques in, 305–337. *See also specific disorder or technique*
Gastrojejunostomy, technique of, 313
Gastrostomy, percutaneous
  complications of, 311
  contraindications to, 305
  indications for, 305
  outcome of, 310–311
  postprocedure care, 310
  preprocedure considerations, 305
  procedure for, 306–310, 306f–312f, 312t

Gastrostomy button, 309, 311f, 312f
Gelfoam
  for embolotherapy, 29–30, 29f
    in children, 256
  for gastrointestinal bleeding, 57
Gelfoam particles, for embolotherapy, 28t
Gentamicin
  fine needle percutaneous transhepatic cholangiography and, 284
  prophylactic, 5
Gianturco stent, for transjugular intrahepatic portosystemic shunt, 164
Glucagon, percutaneous gastrostomy and jejunostomy and, 306
Glucose
  for embolization of spermatic vein, 344
  normal range of, 403
Glue, tissue, for embolotherapy, 32
Graft(s)
  bridge, 117, 118f
  bypass, occluded, thrombolysis for, 195–196, 195f–196f
  polytetrafluoroethylene, 117
  stent-graft and. *See* Stent-graft
Gray unit, 11, 12t
Gray-scale imaging, for deep venous thrombosis, of upper extremity, 171
Greene needle, for abdominal biopsy, 380
Greenfield filter
  misplaced, management of, 243, 244f
  titanium, 219–220, 220f
  12-French stainless steel, 220
  24-French stainless steel, 219, 220f
Groshong catheter, 225
Guided coaxial balloon catheter system, 95–96, 95f, 96f
Guidewire(s)
  Amplatz Extra Stiff, 24
  Amplatz Super Stiff, 24
  Bentson, 24
  described, 23
  Glidewire, 24
  placement of, Seldinger technique for, 20–21
  Wholey, 24

**H**
$HCO_3$, normal range of, 404
Hemangioma(s)
  embolotherapy of, 39

intramuscular, embolotherapy for, 44
pediatric, embolotherapy of, 39–40
Hematocrit, normal range of, 403
Hematoma(s)
　after angiography, management of, 25
　after renal artery percutaneous transluminal
　　angioplasty, 99
　mechanical thrombectomy and, 209–210
Hemobahn endovascular prosthesis, 161, 163f
Hemobilia, 60–61
Hemodialysis, prevalence of, 117
Hemodialysis access, 117–128
　artificial, 119–120
　complications of, 119–120
　considerations in, 117
　desperation intervention in, 128
　dual-lumen, 127–128
　evaluation of, 120–123
　interventional radiologic salvage of, 123–126
　　angioplasty, 124
　　complications of, 126–127
　　mechanical thrombectomy, 124–125, 125f,
　　　126f
　　percutaneous balloon-assisted aspiration
　　　thrombectomy, 125–126, 127f
　　thrombolysis, 123–124
　long-term, types of, 117, 118f
　occluded, mechanical thrombectomy for, 209
　preoperative evaluation in, 117, 119
　venous stenoses of, venous stent placement
　　for, 154–155
Hemolysis, mechanical thrombectomy and,
　207–208, 210
Hemoptysis
　after thoracic biopsy, 366
　in children, 259
Hemorrhage. *See also* Bleeding
　in children, 258–259
　intraplaque, atherosclerosis and, 77
　percutaneous nephrostomy and, 270
Hemostasis, puncture, 24–25
Hemostatic cascade, 187–189, 188f–190f
　inhibitors of, 189
Heparin, 190–191, 190f
　abdominal abscess drainage and, 387, 395
　acute lower extremity arterial thrombosis and,
　　194
　aorto-iliac stent-graft and, 159
　deep vein thrombosis and, 213

　iliac artery intravascular stent placement in, 148
　in interventional radiology, 6–7
　intravascular stents and, 143
　low-molecular-weight, in interventional radi-
　　ology, 7
　mechanical thrombectomy and, 209
　for occluded hemodialysis access, 124
　percutaneous transluminal angioplasty and
　　of abdominal aorta and iliac artery, 106
　　of aortic, iliac, and peripheral arteries, 104
　　of carotid artery, 87
　　of celiac artery, 91
　　of renal artery, 94, 98
　　of subclavian artery, 83
　　of superior mesenteric artery, 91
　　of vertebral artery, 88
　renal artery intravascular stent placement and,
　　145, 146
　Trac-Wright catheter and, 135
　upper extremity deep vein thrombosis and,
　　200
　venous stent placement and, 152
Hepatic abscess
　amebic, 391
　pyogenic, 391
Hepatic artery stenosis, after liver transplanta-
　tion, in children, 254
Hepatic biopsy, complications of, 383
Hepatic encephalopathy, after transjugular
　intrahepatic portosystemic shunt, 75
Hepatic failure, after transcatheter arterial
　chemoembolization, management of,
　49–50
Hepatic vein catheterization, portal venous
　pressure and, 67
Hepatobiliary system, interventional techniques
　in, 283–299. *See also specific tech-
　nique*
　biliary metal stents, 296–297
　for biliary strictures, 297–298, 299f
　fine needle percutaneous transhepatic cholan-
　　giography, 283–286
　for malignant obstructive jaundice, 296
　percutaneous biliary drainage procedures,
　　286–297
Hepatocellular carcinoma
　mortality of, 47
　percutaneous treatment of, 47–53. *See also
　　specific treatment*

Hickman catheter, 225, 226f
Homocystinuria, hypercoagulable state and, 193
Hoop stress, of balloon catheter, 81, 81f
Hydraulic thrombectomy, 203
Hydrolyser, 205–206, 206f
Hygroma(s), cystic, embolotherapy for, 44
Hypercoagulable state, 192–193
Hypertension
    interventional radiology and, 9
    portal. *See* Portal hypertension
    prevalence of, 93
    renovascular, defined, 93
    venous, hemodialysis access and, 119
Hypogastric artery, percutaneous transluminal angioplasty of, 109–110, 109f–111f

## I

Iliac artery
    common
        bifurcation of, percutaneous transluminal angioplasty of, 109–110, 109f–111f
        stenoses of, percutaneous transluminal angioplasty for, 107–109, 108f
    external, stenoses of, percutaneous transluminal angioplasty for, 107–109, 108f
    intravascular stents in, 147–149
    occluded, recanalization of, 111
    percutaneous transluminal angioplasty of, 105–107, 105f, 106f
Iliofemoral venous thrombolysis, 198–200, 199f
Iliopsoas abscess, 393
Imaging, in percutaneous drainage of pleural fluid collections and pneumothorax, 370
Immunosuppression, cecal ulceration and, 64
Improvement, defined, 99
In vitro fertilization, 339
Infection
    central venous access and, 232–233
    hemodialysis access and, 119
    percutaneous nephrostomy and, 270
Inferior mesenteric artery, angiography of, injection rates and volume for, 401t
Inferior vena cava
    angiography of, injection rates and volume for, 401t
    occlusion of, after inferior vena cava filter placement, 218, 218f
    oversized, inferior vena cava filter placement and, 216
    penetration of, after inferior vena cava filter placement, 219
Inferior vena cava filter
    Bird's Nest, 220–222, 221f
    complications of
        immediate, 216–217, 216f
        long-term, 217–219
    components of, fracture of, 219
    contraindications to, 214
    guidewire entrapment and, 219
    history of, 213
    migration of, 218–219
    misplaced, management of, 243, 244f
    oversized inferior vena cava and, 216
    placement of
        access route selection in, 214–215
        in superior vena cava, 216
        suprarenal, 216
        technique for, 214–216
    postplacement management, 217
    preprocedural venacavography and, 215–216, 215f
    retrievable, 223
    Simon Nitinol, 222, 222f
    suprarenal, placement of, 246
    temporary, 223, 223f
    titanium Greenfield, 219–220, 220f
    12-French stainless steel Greenfield, 220
    24-French stainless steel Greenfield, 219, 220f
    Vena Tech-LGM, 222–223, 223f
Inferior vena cava syndrome, venous stent placement for, 152–153, 153f
Infertility
    defined, 339
    female, treatment of, 339–341. *See also* Transcervical fallopian tube catheterization, for infertility
    male, treatment of, 341–346. *See also* Spermatic vein catheterization, approaches to
Inflammatory bowel disease, 64
Informed consent
    for embolotherapy, 28
    in interventional radiology, 1
    in intravenous sedation, 3
Infrapopliteal percutaneous transluminal angioplasty, 113
Infusion rates, for angiography, 401, 401t

Innominate artery, percutaneous transluminal angioplasty of, 84–87, 84f–87f
Internal jugular vein, for central venous access, 228
Interventional radiology
  patient preparation for, 1
  postprocedure care, 10
Intraarterial I-131 iodized oil radiotherapy, for hepatocellular carcinoma, 47–48, 52
Intramuscular hemangioma, embolotherapy for, 44
Intramuscular venous malformations, embolotherapy for, 44
Intravascular foreign body, retrieval of. *See* Foreign bodies, intravascular, retrieval of
Intravascular stent
  balloon-expandable, 137–141
    Palmaz stent, 137–140, 138f, 138t, 139f
    Strecker stent, 138t, 140–141
  characteristics of, 138t
  intravascular ultrasound and, 183
  self-expanding, 141–143
    Wallstent, 138t, 141–142, 141f
  in supraaortic vessels, 143–152. *See also specific vessel*
    aorta, 149, 150f, 151
    contraindications to, 143
    femoropopliteal arteries, 151–152
    iliac arteries, 147–149
    indications for, 143
    postprocedure care, 143, 145
    renal artery, 145–147, 147f
    results of, 145
    technique for, 143, 144f
  venous. *See* Venous stents
Intravascular ultrasound
  clinical applications of, 182–183
  as diagnostic adjunct, 182–183
  in directional atherectomy, 183
  electronic array devices for, 180, 180f
  and intravascular stents, 183
  mechanical devices for, 179, 179f
  technical considerations in, 179
  as therapeutic adjunct, 183
  transluminal angioplasty and, 183
  of vascular anatomy, 180–181, 181f, 182f
Intravenous pyelogram, urolithiasis and, 278

Intrinsic coagulation cascade, 187, 188f
Intubation, silicone, for epiphora, 347
Iothalamate sodium, for sclerotherapy for spermatic vein, 345
Irrigation, of drainage catheters, 390
Ischemia, gall bladder, after transcatheter arterial chemoembolization, management of, 49–50
Ivalon
  for arteriovenous malformation, in children, 258
  for embolotherapy, in children, 256, 257f
Ivy bleeding time, 5-mm wound, normal range of, 404

## J

Jackson-Pratt Reservoir, 390
Jaundice, malignant, obstructive, biliary drainage techniques for, 296
Jejunostomy, percutaneous
  complications of, 311
  contraindications to, 305
  indications for, 305
  outcome of, 310–311
  postprocedure care, 310
  preprocedure considerations, 305
  technique of, 313
Joint Commission on Accreditation of Healthcare Organizations standards for anesthesia care: 1995, 2, 2t
Jugular vein
  access of, in transjugular intrahepatic portosystemic shunt, 71–72, 72f
  internal, for central venous access, 228

## K

Kasabach-Merritt syndrome, 40
Kidney(s)
  anatomy of, 263, 264f, 265f, 266f
  embolotherapy of, preoperative, 38
Kinking, of central venous catheters, 232
"Kissing balloon" technique, in renal artery percutaneous transluminal angioplasty, 97

## L

Labetalol, in interventional radiology, 9
Laboratory tests
  in fibrinolytic therapy, 200–201

Laboratory tests (*continued*)
  before interventional radiology, 1
  during thrombolysis, 200–201
Laboratory values, normal ranges of, 403–404
Lacing, for acute lower extremity arterial thrombosis, 194
Lacrimal duct, interventional techniques in, 347–357. *See also* Epiphora
Lacrimal stent, placement of, for epiphora
  canalicular stent, results of, 354–355, 355f–356f
  contraindications to, 347
  indications for, 347
  nasolacrimal duct stent, results of, 353–354
  technique for, 350–353, 350f–354f
Lead aprons, 14
Lee needle, for abdominal biopsy, 380
Lidocaine
  abdominal abscess drainage and, 387
  balloon dacryocystoplasty and, 347
  thermal vessel occlusion of spermatic vein and, 345
  thoracic biopsy and, 360
  thrombolysis and, 197
Lipase, normal range of, 403
Lipiodol, for embolization of spermatic vein, 344
Liver
  coagulant factors produced by, vitamin K–dependent, 191t
  transjugular biopsy of, in children, 253
  vascular anatomy of, transjugular intrahepatic portosystemic shunt and, 70–72
Liver transplantation, hepatic artery stenosis after, in children, 254
Liver tumor. *See* Hepatocellular carcinoma
Loop snare, in intravascular foreign body retrieval, 237, 238f, 239f
Lower extremity, angiography of, injection rates and volume for, 401t
Lymphatic malformations, embolotherapy for, 44

## M

Magnesium, normal range of, 403
Magnetic resonance angiography (MRA), for peripheral vascular disease, 168, 170
Magnetic resonance imaging (MRI)
  abdominal abscess drainage and, 386, 388
  abdominal mass and, 377
  aorto-iliac stent-graft and, 157
  inferior vena cava filter and, 217
  vascular malformations and, 39
Male infertility, treatment of, 341–346. *See also* Spermatic vein catheterization, approaches to
Malignant obstructive jaundice, biliary drainage techniques for, 296
Mallory-Weiss esophageal tear, 59
Mechanical devices, for intravascular ultrasound, 179, 179f
Mechanical thrombectomy. *See* Thrombectomy, mechanical
Meckel's diverticulum, 64
Meperidine, for embolotherapy, 28
Mesenteric artery
  inferior, angiography of, injection rates and volume for, 401t
  superior
    angiography of, injection rates and volume for, 401t
    percutaneous transluminal angioplasty of, 89, 90f, 91, 91f, 92
    thrombolysis of, 197
Mesocaval shunt, balloon dilation of, in children, 252
Metallic stent
  after angioplasty, for occluded hemodialysis access, 124
  in biliary system, 296–297
  in children, 254, 255f
  esophageal. *See* Esophageal metallic stent
  placement of, dilatation of, for ureteral stricture, 275, 277
Metformin, in interventional radiology, 10
Midazolam
  for embolotherapy, 28
  mechanical thrombectomy and, 209
  for percutaneous transhepatic biliary drainage, 288
  thermal vessel occlusion of spermatic vein and, 345
  transcervical fallopian tube catheterization and, 340
  for transjugular intrahepatic portosystemic shunt, 69
Migration, of inferior vena cava filter, 218–219
Mitomycin, in transcatheter arterial chemoembolization, 48, 48t
MTI Thrombolytic Brush catheter, for occluded hemodialysis access, 125

## N

Nasolacrimal duct stent, for epiphora, results of, 353–354
National Council on Radiation Protection and Measurements, 12, 12t
Necrosis, skin, warfarin and, 7
Needle(s)
  for abdominal abscess drainage, 389
  for abdominal mass biopsy, 377, 378
  marker, for thoracoscopy, in thoracic biopsy, 364–365
  for percutaneous transhepatic biliary drainage, 287
Needle aspiration, fine. *See* Fine needle aspiration
Nephrostomy
  double-J stent for, 272, 273f–276f
    exchange of, fluoroscopy-guided, 273–275, 278f
  percutaneous
    in children, 268
    complications of, 270
    emergency bedside, 268
    puncture for, 265–267
      contraindications to, 266
      indications for, 265–266
      patient evaluation in, 266
      preprocedural preparation, 266
      technique for, 266–267
    for stone removal, 277–280. *See also* Urolithiasis, percutaneous nephrostomy access for
    in transplant kidney, 268–270, 269f
Nephrostomy tube, percutaneous
  dislodged, replacement of, 270–271
  occluded, removal of, 271–272
  placement of, 267–268, 268f
  Whitaker test in, 270, 271f
Nifedipine
  in interventional radiology, 9
  mechanical thrombectomy and, 209
  percutaneous transluminal angioplasty and, of aortic, iliac, and peripheral arteries, 104
Nitroglycerin
  carbon dioxide delivery and, 177
  carbon dioxide digital subtraction angiography and, 178
  in interventional radiology, 9

percutaneous transluminal angioplasty and
  of aortic, iliac, and peripheral arteries, 103, 104
  infrapopliteal, 113
  of subclavian artery, 83
  renal arterial branch spasm and, 98
  renal artery intravascular stent placement and, 145
  thrombolysis and, 197
Noninvasive evaluation
  of abdominal structures, 173–175
  of deep venous thrombosis. *See* Deep venous thrombosis, noninvasive evaluation of
  of peripheral vascular disease. *See* Peripheral vascular disease, noninvasive evaluation of
Nonrecirculation thrombectomy, 203
Nonsump catheter, for abdominal abscess drainage, 387, 395
Nothing by mouth policy, 3, 3t
Nuclear medicine, gastrointestinal bleeding and, 55
Nutrition, enteral. *See* Enteral nutrition

## O

$O_2$ saturation, normal range of, 404
Oasis catheter, 206, 207f
Occlusion, stent, after transjugular intrahepatic portosystemic shunt, 75
Ogilvie's syndrome, percutaneous cecostomy for, 329–332, 331f
  materials for, 329–330
  methods for, 329–330
Osler-Weber-Rendu syndrome, 38
Oxycel, for embolotherapy, 30

## P

Pacemaker, wires of, migration of, retrieval of, 247–248
Pain, pleural, after thoracic biopsy, 366
Palmaz stent, 137–140, 138f, 138t, 139f
Pancreas, bleeding of, 61, 62f
Pancreatic abscess, 392–393
Pancreatic biopsy, complications of, 383
Pancreatic fluid collections, acute, 392
Pancreatic pseudocyst, 392
  noninfected, 397

Papaverine
    percutaneous transluminal angioplasty and, of aortic, iliac, and peripheral arteries, 103
    thrombolysis and, 197
Parapneumonic, defined, 368
Partial thromboplastin time, normal range of, 404
P.A.S. port catheter, 225, 226f
$PCO_2$, normal range of, 404
Pelvic abscess, 392, 393f, 397
Pelvic arteries, angiography of, injection rates and volume for, 401t
Peptic ulcer disease, duodenal, 58
Perclose, 25
Percutaneous atherectomy, 129–135. *See also* Atherectomy, percutaneous
Percutaneous balloon-assisted aspiration thrombectomy, for occluded hemodialysis access, 125–126, 127f
Percutaneous biopsy, of abdominal mass. *See* Abdominal mass, percutaneous biopsy of
Percutaneous ethanol injection, in hepatocellular carcinoma, 47, 50–51, 51f, 52f
    goal of, 47
Percutaneous Hemostasis Device, 25
Percutaneous mechanical thrombectomy. *See* Thrombectomy, mechanical
Percutaneous transhepatic biliary drainage
    approaches to, 291–292
    biliary duct demonstration in, 288–289, 290f
    candidates for, 287
    catheter placement in
        complications of, 295
        dislodged, 295
        drainage dysfunction and, 295
        for internal-external drainage, 294–295
        patient follow-up and, 296
        technique for, 292–296, 293f, 294f
    contraindications to, 288
    indications for, 287–288
    instrumentation for, 287
    monitoring in, 288
    patient evaluation in, 288
    patient preparation for, 288
    puncture techniques in, 289, 291
Percutaneous transhepatic central venous catheter, placement of, 230, 230f

Percutaneous transhepatic cholangiography, fine needle. *See* Fine needle percutaneous transhepatic cholangiography
Percutaneous translumbar central venous catheter, placement of, 229, 229f
Percutaneous transluminal angioplasty, 77–116
    aortic, iliac, and peripheral arteries, 101–114. *See also specific artery*
        advantages of, 101
        antiplatelet therapy, 104–105
        arteriogram in, 101–102
        complications of, 114
        contraindications to, 101–102
        indications for, 101–102
        intraprocedural care, 103–104
        lesions suitable for, 102
        patient selection in, 101–102
        postangioplasty care, 104
        preangioplasty care, 102–103
        results of, 114
    balloon catheter technology in, 80–81, 81f
    for Budd-Chiari syndrome, in children, 252–253
    of celiac artery, 89, 90f, 91, 91f, 92
    infrapopliteal, 113
    mechanism of, 78–80
    morphologic changes after, 78, 79f
    principles of, 77–78
    of renal arteries. *See* Renal arteries, percutaneous transluminal angioplasty of
    of renal transplantation stenosis, in children, 251–252
    size of balloon in, 79–80
    of superior mesenteric artery, 89, 90f, 91, 91f, 92
    of supraaortic vessels, 82–89
        axillary artery, 83–84
        carotid artery, 84–87, 84f–87f
        innominate artery, 84–87, 84f–87f
        preliminary investigations before, 82–83
        subclavian artery, 83–84
    of vertebral artery, 87–89
Perinephric abscess, 394
Peripheral arteries, embolotherapy of, 38
Peripheral occlusive disease, stent-graft for. *See* Stent-graft, for peripheral occlusive disease

Peripheral vascular disease, noninvasive evaluation of
  ankle-brachial index, 167
  color Doppler imaging, 167–168, 168f, 169f
  magnetic resonance angiography, 168, 170
  segmental limb pressure, 167
  toe pressure measurement, 167
  transcutaneous oxygen pressure, 167
Peripherally inserted central catheter, 225, 227f, 230
pH, normal range of, 404
Phosphorus, normal range of, 403
Pigtail catheter, for intravascular foreign body retrieval, 242
Piperacillin, prophylactic, 5
Plaque(s)
  atherosclerotic, 77–78
  formation of, history and theories of, 77–78
  morphology of, intravascular ultrasound evaluation of, 180–181
Platelet(s), normal range of, 403
Pleural fluid collections. *See also* Exudate; Transudate
  malignant, defined, 368
  normal amounts of, 368
  percutaneous drainage of, 367–374
    complications of, 372
    management of, 372–374
      fluid collections, 372
      urokinase for, 372, 372t, 373f
    procedure for, 370–372
    radiologic evaluation in, 369–370
    results of, 374
Pleural pain, after thoracic biopsy, 366
Pneumothorax
  after thoracic biopsy, 365–366
  categories of, 369
  central venous access and, 231
  defined, 369
  percutaneous drainage of, 367–374
    complications of, 372
    management of, air collections, 373–374
    procedure for, 371–372, 371f
    radiologic evaluation in, 369–370
    results of, 374
$PO_2$, normal range of, 404
Polytetrafluoroethylene graft, 117
  failing, 123

for transjugular intrahepatic portosystemic shunt, 164
Polyvinyl alcohol, for embolotherapy, 28t, 30–31, 30f
Polyvinyl alcohol sponge, for embolotherapy, in children, 256, 257f
Port-a-Cath catheter, 225, 226f
  dislodgement of, removal of, 243, 246f, 247
Portal hypertension, 67–75
  in children, 253–254
  noninvasive evaluation of, 173
  pathophysiology of, 67, 68f
  portal venous pressure in, evaluation of, 67
  transjugular intrahepatic portosystemic shunt for, 67–75. *See also* Transjugular intrahepatic portosystemic shunt
  treatment of, 67
Portal vein
  identification of, in transjugular intrahepatic portosystemic shunt, 71, 71f
  portal-hepatic vein pressure gradient measurement in, 72–74, 73f
Portal vein thrombosis, in children, 253
Portal venous pressure, evaluation of, 67
Portosystemic shunt, intrahepatic, transjugular. *See* Transjugular intrahepatic portosystemic shunt
Postembolic syndrome, 65
Postembolization syndrome, after transcatheter arterial chemoembolization, management of, 49–50
Postimplantation syndrome, after stent-graft, 163
Postthrombotic syndrome, 198
Potassium, normal range of, 403
Pressure, measurement of, in percutaneous transluminal angioplasty of aortic, iliac, and peripheral arteries, 103
Procedural failure, defined, 99
Profile, of balloon catheter, 80
Proparacaine, balloon dacryocystoplasty and, 347
Prostate, biopsy of, ultrasound-guided, 280–281, 281f
Prosthesis(es), endovascular, Hemobahn, 161, 163f
Protamine sulfate
  abdominal abscess drainage and, 387, 395
  for heparin reversal, 6–7, 190
  percutaneous transluminal angioplasty and, of aortic, iliac, and peripheral arteries, 104

Protein(s), total, normal range of, 403
Protein C
  coagulation cascade and, 187
  deficiency of, hypercoagulable state and, 192
Protein C resistance, activated, hypercoagulable state and, 192
Protein S
  coagulation cascade and, 187
  deficiency of, hypercoagulable state and, 192
Prothrombin time, normal range of, 404
Pro-urokinase, 192
Pseudoaneurysm(s)
  color Doppler imaging and, 168
  embolotherapy of, 38
  hemodialysis access and, 119
  mechanical thrombectomy and, 209–210
Pseudocyst(s), pancreatic, 392
  noninfected, 397
Pulmonary angiography, pulmonary embolism and, 213
Pulmonary arteries, angiography of, injection rates and volume for, 401t
Pulmonary arteriovenous fistulae, embolotherapy of, 38–39
Pulmonary embolism
  incidence of, 213
  inferior vena cava filter for, 213. *See also* Inferior vena cava filter
  mortality of, 213
  recurrent, after inferior vena cava filter placement, 218
Pulsed Doppler imaging, for deep venous thrombosis, of upper extremity, 171–173, 172f
Pulsed fluoroscopy, 13
Pulsed-spray pharmacomechanical thrombolysis, for acute lower extremity arterial thrombosis, 194
Puncture, of femoral artery, technique for, 103
Puncture hemostasis, 24–25
PURPOSE trial, of pro-urokinase, 192
Pushability, of balloon catheter, 80
Pyelogram, intravenous, urolithiasis and, 278
Pyloric stricture, interventional treatment of, 322, 323f
Pyogenic hepatic abscess, 391
Pyonephrosis, 397

# R

Radiation
  biological effects of, 11–12
  maximum permissible dose equivalents of, 12t
  scatter, reduction of, 13–14, 14f
Radiation control
  basic protection, 13–15
    distance, 13–14, 14f
    shielding, 14–15
    time, 13
  management of patient radiation dose in, 15–16
  radiation protection regulations and, 12, 12t
  radiation units in, 11, 12t
  staff radiation dose monitoring in, 12–13
Radiation units, 11, 12t
Radiography
  abdominal abscess drainage and, 385–386
  after colonic stent placement, 327, 328f, 329f
  in percutaneous drainage of pleural fluid collections and pneumothorax, 369–370
Recanalization, of occluded iliac arteries, 111
Recirculation thrombectomy, 203
Red blood cells, technetium Tc 99m-labeled, gastrointestinal bleeding and, 55
Rem unit. *See* Roentgen unit.
Renal abscess, 393–394
Renal arteries
  angiography of, injection rates and volume for, 401t
  branch of, spasm of, after renal artery percutaneous transluminal angioplasty, 98–99
  intravascular stents in, 145–147, 147f
  percutaneous transluminal angioplasty of, 93–101
    in children, 251
    complications of, 98–99
    indications for, 93
    periprocedural management in, 98
    results of, 99–101
    techniques for, 93–97
      axillary approach, 96, 96f
      femoral approach, 94–95, 94f
      guided coaxial balloon catheter system, 95–96, 95f, 96f
      "kissing balloon," 97
      sidewinder approach, 97, 97f

stenosis of
  noninvasive evaluation of, 174
  percutaneous transluminal angioplasty of, 93
  thrombolysis of, 197–198
  thrombosis of, after renal artery percutaneous transluminal angioplasty, 98
Renal function, decreased, contrast-induced, after angiography, management of, 26
Renal insufficiency
  renal percutaneous transluminal angioplasty for, results of, 100–101
  transient, after renal artery percutaneous transluminal angioplasty, 98
Renal mass, biopsy of, complications of, 383
Renal transplant
  noninvasive evaluation of, 174–175
  percutaneous nephrostomy in, 268–270, 269f
  stenoses in, renal percutaneous transluminal angioplasty for, results of, 99–100
  stenoses of, percutaneous transluminal angioplasty of, in children, 251–252
Renovascular hypertension, defined, 93
Retrograde catheterization, of ureter, without endoscopic assistance, 272–273, 277f
Roentgen unit, 11, 12t

## S

Saphenous vein bypass grafts
  anastomoses of, stenoses at, percutaneous transluminal angioplasty for, 113
  stenoses in, renal percutaneous transluminal angioplasty for, results of, 100
Scatter radiation, reduction of, 13–14, 14f
Sclerotherapy, for spermatic vein, 344–345
Sedation, intravenous, 2–4. *See also specific drug*
  defined, 2
  for embolotherapy, 27–28
  equipment for, 3–4
  guidelines for, 2, 2t
  informed consent in, 3
  mechanical thrombectomy and, 209
  monitoring in, 3
  multiple agents in, 4
  patient evaluation before, 2, 3t
  patient preparation for, 3, 3t
  recovery from, 4
Segmental limb pressure, for peripheral vascular disease, 167

Seldinger technique
  for abdominal abscess drainage, 396
  for access, 19–23
    arterial puncture in, 20, 20f, 21f
    catheter placement in, 21–23, 22f
    guidewire placement in, 20–21
Sepsis, percutaneous nephrostomy and, 270
Septicemia, after transjugular intrahepatic portosystemic shunt, 75
Serpins, 190
Shielding, in radiation control, 14–15
Shunt(s)
  mesocaval, balloon dilation of, in children, 252
  portosystemic, intrahepatic, transjugular. *See* Transjugular intrahepatic portosystemic shunt
Sidewinder technique, for renal artery percutaneous transluminal angioplasty, 97, 97f
Sievert unit, 11, 12t
Silicone intubation, for epiphora, 347
Silk sutures, for embolotherapy, 28t
Simon Nitinol filter, 222, 222f
Simpson catheter, for percutaneous atherectomy, 130–132, 131f
Sinogram, abdominal abscess drainage and, 396
Skin, necrosis of, warfarin and, 7
Small-bowel enteral nutrition, 311–313
Snare(s)
  Amplatz gooseneck, for intravascular foreign body retrieval, 237, 238f
  loop, for intravascular foreign body retrieval, 237, 238f, 239f
Sodium
  abdominal abscess drainage and, 387
  normal range of, 403
Sodium tetradecyl sulfate
  in embolotherapy, 27, 31–32
    in children, 256
  in sclerotherapy for spermatic vein, 345
Sotradecol, for embolotherapy, 28t, 31–32
  in children, 256
Special energy-assisted thrombectomy, 203
Spermatic vein
  anatomy of, 342, 342f
  embolization of
    with occluding devices, 344
    with tissue adhesive, 344
  occlusion of, detachable balloon for, 345

Spermatic vein (*continued*)
   sclerotherapy for, 344–345
   thermal vessel occlusion of, 345–346
Spermatic vein catheterization, approaches to
   basilic vein, 344
   femoral, 343
   jugular, 343, 343f
Splenic abscess, 391
Stainless steel coil, for embolotherapy, 32, 33f, 34
   in children, 256
*Staphylococcus aureus*, hemodialysis access and, 119
*Staphylococcus epidermidis*, hemodialysis access and, 119
Steal syndrome, hemodialysis access and, 119
Stenosis(es)
   hepatic artery, after liver transplantation, in children, 254
   iliac artery, percutaneous transluminal angioplasty for, 107–109, 108f
   renal artery
     noninvasive evaluation of, 174
     percutaneous transluminal angioplasty of, 93
   in renal transplant
     percutaneous transluminal angioplasty of
       in children, 251–252
     renal percutaneous transluminal angioplasty for, results of, 99–100
   in saphenous bypass grafts, renal percutaneous transluminal angioplasty for, results of, 100
   stent, after transjugular intrahepatic portosystemic shunt, 75
   venous, hemodialysis-related, venous stent placement for, 154–155
Stent(s). *See also specific type*
   covered metal. *See* Covered metal stent, for transjugular intrahepatic portosystemic shunt
   double-J, for nephrostomy, 272, 273f–276f
     exchange of, fluoroscopy-guided, 273–275, 278f
   Gianturco, for transjugular intrahepatic portosystemic shunt, 164
   intravascular, 137–156. *See also* Intravascular stent
   lacrimal. *See* Lacrimal stent, placement of, for epiphora
   metallic. *See* Metallic stent

   nasolacrimal duct, for epiphora, results of, 353–354
   for occluded hemodialysis access, complications of, 127
   Palmaz, 137–140, 138f, 138t, 139f
   stenosis of, after transjugular intrahepatic portosystemic shunt, 75
   Strecker, 138t, 140–141
   transrectal catheterization and, for volvulus of sigmoid colon, 332–334, 333f
Stent-graft
   aorto-iliac, for aneurysm, 157–160
     complications of, 159, 159t
     indications for, 157
     outcome of, 160
     techniques for, 157–159, 158f
   for peripheral occlusive disease, 160–163
     complications of, 162–163
     indications for, 160–161
     materials for, 161, 162f, 163f
     outcomes of, 161–162
     techniques for, 161, 162f, 163f
   for transjugular intrahepatic portosystemic shunt, 163–165. *See* Covered metal stent, for transjugular intrahepatic portosystemic shunt
Steroid(s), for Kasabach-Merritt syndrome, 40
Stochastic effects, of radiation, 11–12
Strecker stent, 138t, 140–141
Streptokinase, 191
   abdominal abscess drainage and, 396
   for occluded central venous catheter, 233–234
Stricture(s)
   bile duct, interventional techniques in, 297–298, 299f
   enteric, interventional treatment of, 322–325. *See also specific type*
   esophageal. *See* Esophageal stricture
   of ureter, dilatation and metallic stent placement for, 275, 277
Subclavian artery
   percutaneous transluminal angioplasty of, 83–84
   puncture of, central venous access and, 231
Subclavian vein
   for central venous access, 227–228
   puncture of, central venous access and, 231

Subclinical varicocele, 341
Subcutaneous tunnel, in central venous access, 228–229
Subintimal contrast injection, during angiography, management of, 26
Subintimal dissection, after renal artery percutaneous transluminal angioplasty, 98
Subintimal passage, of guidewire or catheter, in angiography, management of, 26
Subphrenic abscess, 391, 397
Sump catheter, with double lumen, for abdominal abscess drainage, 387, 395
Superficial femoral artery, stenoses of, percutaneous transluminal angioplasty for, 111–112, 112f
Superior mesenteric artery
  angiography of, injection rates and volume for, 401t
  percutaneous transluminal angioplasty of, 89, 90f, 91, 91f, 92
Superior vena cava, inferior vena cava filter placement in, 216
Superior vena cava syndrome, venous stent placement for, 152–153, 153f
Supraaortic vessels
  intravascular stents in, 143–152. *See also specific vessel*
  percutaneous transluminal angioplasty of. *See Percutaneous transluminal angioplasty, of supraaortic vessels. See also specific vessel*
Suprarenal inferior vena cava filter, placement of, 216
Sure-Cut needle, for abdominal biopsy, 380
Surgical gelatin sponge, for embolotherapy, in children, 256
Suture(s), silk, for embolotherapy, 28t

**T**

Takayasu's arteritis, 84, 84f
  in children, 254
Tandem technique, for abdominal abscess drainage, 389
Technetium Tc 99m sulfur colloid, gastrointestinal bleeding and, 55
Technetium Tc 99m-labeled red blood cells, gastrointestinal bleeding and, 55
Technical success, defined, 99
Tenth value layers, 14

Thermal vessel occlusion, of spermatic vein, 345–346
Thoracentesis, diagnostic, for percutaneous drainage of pleural fluid collections, 370
Thoracic biopsy
  complications of, 365–367
  contraindications to, 359
  diagnostic yield of, 365
  imaging in, 359–360
  methods of, 360–364
    core biopsy, 363–364, 364f
    fine needle aspiration, 360, 362–363
  patient preparation for, 359
  postprocedure care, 365
  samples obtained during, processing of, 365
  thoracoscopy in, marker needle for, 364–365
Thoracic outlet thrombosis, 200
Thoracoscopy
  marker needle for, in thoracic biopsy, 364–365
  for pleural effusion, 369
Three-layered image, in intravascular ultrasound of arterial wall, 180, 181f
Thrombectomy
  mechanical, 203–211
    aspiration, 203
    blood loss and, 208
    clinical applications of, 209
    complications of, 209–210
    device designs for, 203–206
      Amplatz thrombectomy device, 203–205, 204f
      AngioJet, 205, 205f
      Arrow-Trerotola percutaneous thrombectomy device, 206, 208f
      classification of, 203
      Hydrolyser, 205–206, 206f
      Oasis catheter, 206, 207f
    endpoint of procedure, 208–209
    energy-assisted, 203
    hemolytic effect of, 207–208
    hydraulic, 203
    medications during, 209
    monitoring during, 209
    nonrecirculation, 203
    for occluded hemodialysis access, 124–125, 125f, 126f
    recirculation, 203

Thrombectomy (*continued*)
    technical considerations in, 206–208
    percutaneous balloon-assisted aspiration, for occluded hemodialysis access, 125–126, 127f
    special energy-assisted, 203
Thrombin, for embolotherapy, 29
Thrombolysis. *See also* Fibrinolytic therapy
    for acute lower extremity arterial thrombosis, 193–198
        lacing technique for, 194
        pulsed-spray pharmacomechanical technique for, 194
    development of, 193
    iliofemoral venous, 198–200, 199f
    laboratory tests during, 200–201
    mesenteric artery, 197
    of occluded bypass grafts, 195–196, 195f–196f
    for occluded hemodialysis access, 123–124
        complications of, 126–127
    renal artery, 197–198
    upper extremity, 197
    upper extremity deep vein thrombosis, 200
    vs. surgery, for lower extremity occlusive disease, 196–197
Thrombolysis and Peripheral Arterial Surgery Trial, 194
Thrombolytic agents, 191–192, 191f. *See also specific agent*
    for irrigation of drainage catheters, 390
Thrombosis
    acute arterial, atherosclerosis and, 77
    after angiography, management of, 26
    arterial, acute, of lower extremity, thrombolysis for, 193–198. *See also* Thrombolysis, for acute lower extremity arterial thrombosis
    central venous access and, 233
    effort, 200
    hemodialysis access and, 119
        early warning signs of, 120
    insertion vein, after inferior vena cava filter placement, 217
    pericatheter, after thrombolysis, 201
    portal vein, in children, 253
    recurrent, mechanical thrombectomy and, 210
    of renal artery, after renal artery percutaneous transluminal angioplasty, 98

    thoracic outlet, 200
    venous, deep. *See* Deep venous thrombosis
Thrombus(i)
    characteristics of, 181, 182f
    formation of, after renal artery intravascular stent placement, 146
Ticlopidine
    in interventional radiology, 8
    intravascular stents and, 143
    percutaneous transluminal angioplasty and, of aortic, iliac, and peripheral arteries, 105
    renal artery intravascular stent placement and, 146
Tissue adhesive, for embolization of spermatic vein, 344
Tissue glue, for embolotherapy, 32
Tissue plasminogen activator, 192
    acute lower extremity arterial thrombosis and, 194
    in children, 254
    iliofemoral venous thrombolysis and, 198, 200
    mechanical thrombectomy and, 209
    upper extremity deep vein thrombosis and, 200
Toe pressure measurement, for peripheral vascular disease, 167
Tolazoline
    percutaneous transluminal angioplasty and, of aortic, iliac, and peripheral arteries, 103
    thrombolysis and, 197
Totally implanted central venous catheter, 230–231
Trackability, of balloon catheter, 80
Trac-Wright catheter, for percutaneous atherectomy, 133–135, 134f
Transcatheter arterial chemoembolization
    for hepatocellular carcinoma, 47
        anticancer drug and embolic agent in, preparation of, 48, 48t
        complications of, prevention and management of, 49–50
        follow-up for, 49
        indications for, 47
        postprocedure care, 49
        preparation for, 48
    segmental, for hepatocellular carcinoma, technique for, 48–49, 49f

Transcervical fallopian tube catheterization, for infertility, 339–341
  outcome of, 341
  patient preparation for, 339–340
  patient selection for, 339
  postprocedure care, 341
  techniques of, 340–341
Transcutaneous oxygen pressure, for peripheral vascular disease, 167
Transhepatic biliary drainage, percutaneous, 286–297. *See* Percutaneous transhepatic biliary drainage
Transhepatic central venous catheter, percutaneous, placement of, 230, 230f
Transhepatic cholangiography, percutaneous, fine needle. *See* Fine needle percutaneous transhepatic cholangiography
Transjugular intrahepatic portosystemic shunt, 67–75
  antibiotic prophylaxis for, 5–6
  in children, 253–254
  complications of, 74–75
  contraindications to, 68–69
  covered metal stent for. *See* Covered metal stent, for transjugular intrahepatic portosystemic shunt
  follow-up in, 74
  indications for, 68
  noninvasive evaluation of, 173–174, 174f
  portal-hepatic vein pressure gradient measurement in, 72–74, 73f
  results of, 74
  stenosis of, 74
  technique for, 69–72
    identification of portal vein in, 71, 71f
    jugular vein access in, 71–72, 72f
    materials for, 70, 70f
    patient preparation for, 69, 69t
    vascular anatomy of liver and, 70–72
  tract dilation in, 72–74, 73f
Transjugular liver biopsy, in children, 253
Translumbar central venous catheter, percutaneous, placement of, 229, 229f
Transluminal extraction catheter, for percutaneous atherectomy, 132–133, 132f
Transpyloric enteral nutrition, 311–313
Transrectal fluoroscopic catheterization, stenting and, for volvulus of sigmoid colon, 332–334, 333f

Transudate
  causes of, 368, 368t
  defined, 368
  vs. exudate, chemical and cytologic determinants of, 368, 368t
"Trapping," of carbon dioxide, 176
Trauma, in children, 258–259
Triglyceride(s), normal range of, 403
Trocar catheter, for abdominal abscess drainage, 387, 389, 395–396
Tru-Cut needle, for abdominal biopsy, 380, 381f
Tumor(s). *See specific type*
Tunnel, subcutaneous, in central venous access, 228–229
Tunneled port, antibiotic prophylaxis for, 5
Turner needle, for abdominal biopsy, 379, 380f

**U**
Ulcer(s)
  duodenal peptic, 58
  gastric, 58–59
Ultrasound
  abdominal abscess drainage and, 385, 386f, 387–388, 395
  abdominal mass and, 377
  in central venous access, 227–228, 230
  for deep venous thrombosis, of upper extremity, 171
  Doppler. *See* Doppler ultrasound
  Duplex, hemodialysis access and, 120, 120f
  intravascular. *See* Intravascular ultrasound
  in percutaneous drainage of pleural fluid collections and pneumothorax, 369
  percutaneous nephrostomy puncture and, 266–267
  percutaneous transhepatic biliary drainage and, 288, 289
  prostate biopsy and, 280–281, 281f
  in thoracic biopsy, 360
  transjugular intrahepatic portosystemic shunt and, 74
  vascular malformations and, 39
  venous
    deep vein thrombosis and, 213
    inferior vena cava filter placement and, 214
Unistep, Wallstent and, 141
Urea nitrogen, normal range of, 403

Ureter
  retrograde catheterization of, without endoscopic assistance, 272–273, 277f
  stricture of, dilatation and metallic stent placement for, 275, 277
Urokinase, 191–192
  abdominal abscess drainage and, 396
  acute lower extremity arterial thrombosis and, 193, 194
  in children, 254
  iliofemoral venous thrombolysis and, 198, 200
  intrapleural, 372, 372t, 373f
  for irrigation of drainage catheters, 390
  mechanical thrombectomy and, 209
  for occluded central venous catheter, 233–234
  for occluded hemodialysis access, 123–124
  upper extremity deep vein thrombosis and, 200
  venous stent placement and, 152
Urolithiasis, percutaneous nephrostomy access for, 277–280
  complications of, 280
  contraindications to, 277–278
  operative management of, 278–279
  patient evaluation in, 278
  preoperative management of, 278–279
  radiologic studies in, 278
  technique for, 279–280, 279f, 280f
  tract dilatation in, 279–280, 279f, 280f
Uroradiology, percutaneous techniques for, 263–270. See also Nephrostomy; Nephrostomy tube, percutaneous

V

Vancomycin, prophylactic, 5, 6
Varice(s), esophageal, bleeding, 59
Varicocele(s)
  defined, 341
  diagnosis of, 342–343
  spermatic vein catheterization for, 341–346. See also Spermatic vein catheterization, approaches to
  subclinical, 341
  treatment of, indications for, 344
Varicocid, for sclerotherapy for spermatic vein, 344–345

Vascular access. See Access
Vascular embolotherapy, 27–45. See also Embolotherapy
Vascular malformations, types of, embolotherapy of, 39–40
Vascular tumors, 59–60
Vasculogenic malformations, types of, embolotherapy of, 39–40
Vasopressin
  complications of, in endovascular treatment of gastrointestinal bleeding, 64
  for gastric ulcer, 58–59
  for gastrointestinal bleeding, 56
  in children, 258
VasoSeal, 25
Vein(s). See also specific type
Vena Tech-LGM filter, 222–223, 223f
Venacavography
  after inferior vena cava filter placement, 217
  before inferior vena cava filter placement, 215–216, 215f
Venogram, digital subtraction imaging, hemodialysis access and, 117, 119
Venography
  contrast, deep vein thrombosis and, 213
  diagnostic, spermatic vein catheterization and, 342
  hemodialysis access and, 117, 119
  vascular malformations and, 39
  venous malformation and, 43
Venous catheter, central, totally implanted, 230–231
Venous hypertension, hemodialysis access and, 119
Venous malformations
  embolotherapy for, 42–44
  intramuscular, embolotherapy for, 44
Venous stenoses, hemodialysis-related, venous stent placement for, 154–155
Venous stents
  indications for, 152–155
    Budd-Chiari syndrome, 153–154
    hemodialysis-related venous stenoses, 154–155
    malignant caval obstruction, 152–153, 153f
  technique for, 152

Venous thrombosis, deep. *See* Deep venous thrombosis
Venous ultrasound
  deep vein thrombosis and, 213
  inferior vena cava filter placement and, 214
Ventilation-perfusion lung scan, pulmonary embolism and, 213
Verapamil
  percutaneous transluminal angioplasty and, infrapopliteal, 113
  renal arterial branch spasm and, 98
  renal artery intravascular stent placement and, 145
Vertebral arteries, percutaneous transluminal angioplasty of, 87–89
Visceral arteries, embolotherapy of, 38
Vitamin K
  abdominal abscess drainage and, 387, 395
  coagulant factors dependent on, produced by liver, 191t
  in coumarin reversal, 191
  in interventional radiology, 8

Volvulus, of sigmoid colon, transrectal fluoroscopic catheterization and stenting for, 332–334, 333f

## W

Wallgraft endoprosthesis, 161
Wallstent, 138t, 141–142, 141f
  misplaced, retrieval of, 247, 247f
  for transjugular intrahepatic portosystemic shunt, 164
Warfarin, 191
  central venous access and, 233
  deep vein thrombosis and, 213
  in interventional radiology, 7
  renal artery intravascular stent placement and, 146
Westcott needle, for abdominal biopsy, 380
Whitaker test, 270, 271f
White blood cell count, normal range of, 403
Wholey guidewire, 24

## X

X-ray beam size, in radiation control, 151